AF575425

Architecting EDI for SAP S/4HANA®

Marek Piaseczny, Agasthuri Doss

Architecting EDI for SAP S/4HANA®

Editor Rachel Gibson
Acquisitions Editor Hareem Shafi
Copyeditor Melinda Rankin
Cover Design Graham Geary
Photo Credit iStockphoto: 536510443/© zodebala
Layout Design Vera Brauner
Production Eric Wyche
Typesetting III-Satz, Germany
Printed and bound in Canada, on paper from sustainable sources

ISBN 978-1-4932-2458-6
© 2025 by Rheinwerk Publishing, Inc., Boston (MA)
1st edition 2025

Library of Congress Cataloging-in-Publication Data
Names: Doss, Agasthuri, author. | Piaseczny, Marek, author.
Title: Architecting EDI for SAP S/4HANA / by Agasthuri Doss and Marek Piaseczny.
Description: 1st edition. | Bonn ; Boston : Rheinwerk Publishing, 2024. | Includes index.
Identifiers: LCCN 2024034022 | ISBN 9781493224586 (hardcover) | ISBN 9781493224593 (ebook)
Subjects: LCSH: Electronic data interchange--Computer programs. | SAP HANA (Electronic resource)
Classification: LCC HF5548.33 .D57 2024 | DDC 658/.0557--dc23/eng/20240820
LC record available at https://lccn.loc.gov/2024034022

Contents at a Glance

Contents

Preface

We have written *Architecting EDI for SAP S/4HANA* to address the growing need for a structured, in-depth resource on integrating electronic data interchange (EDI) with SAP S/4HANA. As businesses increasingly adopt SAP S/4HANA for its advanced capabilities, there is a critical demand for clear guidance on how to effectively design and implement EDI solutions within this complex environment. This book aims to bridge the knowledge gap by providing a comprehensive framework for EDI architecture, including best practices, practical examples, and strategies for overcoming common integration challenges. By offering a detailed and practical approach, the book seeks to support IT professionals, architects, developers, project managers, and system administrators in optimizing EDI processes, ensuring successful deployments, and leveraging SAP S/4HANA to its fullest potential. In addition, it aids in digital transformation by exploring integration with cloud solutions and enhancing data-driven decision-making while ensuring compliance and security.

Objective of This Book

If you look for books related to SAP and IT system integration, then you will be able to find at least a few. Many of them are worth recommending but may not be what you expect. Many of these publications will work great as an introduction and even as a technical manual for a specific SAP integration product, but hardly any will present an end-to-end solution for a specific integration use case. This book is different: it still has the theoretical part, but the core part of the book is about demonstrating the implementation of a B2B EDI scenario. Of course, the form and volume of the book forced us to use certain simplifications and omit certain aspects, but no aspect of the end-to-end integration has been skipped completely. If you follow the configuration chapters, you will be able to recreate the whole scenario within your landscape. That should be considered a bonus beyond the fact that by doing this, you will learn about multiple aspects of B2B EDI integration as well as different technical components and their configuration.

The second unique aspect of this book is that it involves a few SAP components that are pretty new. Using them all for an EDI solution as presented in this book is probably not something you could encounter in an actual SAP project before now.

This book aims to serve as a comprehensive guide for designing and implementing EDI solutions within the SAP S/4HANA environment. It seeks to clarify how EDI integrates with SAP S/4HANA by explaining key concepts, protocols, and technologies. It provides detailed strategies for architecting EDI solutions, shares best practices for configuration and deployment, and addresses real-world challenges through practical examples and case studies. It also offers insights into optimizing EDI processes for performance and scalability while facilitating effective knowledge transfer for IT professionals, developers, project managers, and system administrators involved in EDI and SAP S/4HANA projects.

Target Audience

The book covers so many aspects that it is hard to specify one single group that can benefit from reading it, but we'll start with consultants involved directly in the SAP integration, in any role: architects, developers, analysts, and so on. For this audience, we recommend reading the whole book and not skipping any chapters. We would even encourage you to rebuild the scenario in some kind of sandbox environment and then extend it by implementing your own improvements and requirements. But integration consultants are not the only group that can benefit: any other nonintegration consultant or even project manager should also benefit from reading some of the chapters. You don't necessarily need to go through all the configuration details, but having an overall view of SAP's integration capabilities would be a great experience.

As a whole, this book is relevant to IT professionals and consultants responsible for implementing and managing EDI solutions, SAP users and administrators involved in maintaining and configuring ERP systems, and business analysts seeking to optimize workflows. It also serves supply chain and logistics professionals using SAP to coordinate with partners, developers, and technical architects focused on the technical aspects of EDI integration, project managers overseeing EDI-related projects, and training and support personnel guiding users. ERP professionals involved in enhancing SAP systems with EDI capabilities also will find this book valuable.

Structure of This Book

We start the book with a high-level introduction to SAP S/4HANA and SAP integration technologies, with a short overview of the EDI concept and related standards. After this introduction, we go to the most extensive part of the book: step-by-step instructions for implementing EDI in an SAP S/4HANA on-premise landscape. What makes this book special is the fact that the content covers not only the integration-related aspects of the SAP S/4HANA system but also an overall, end-to-end solution for EDI processing. This means that in this book you can find a comprehensive guideline to the SAP Integration Suite services that act as our middleware, but also, most importantly, we present how all these different pieces may work together to create a reliable and future-proof solution for EDI processing.

The first part of the book is structured as follows:

- **Chapter 1** provides a high-level overview of SAP S/4HANA, its deployment options, and its implementation options. You will learn what distinguishes SAP S/4HANA from SAP ERP as a next-generation ERP system.
- In **Chapter 2**, you will learn about the integration technologies available within SAP S/4HANA and the integration services that are a part of SAP Business Technology Platform (SAP BTP). Services such as those of SAP Integration Suite are recommended for integration in an SAP S/4HANA landscape.
- **Chapter 3** provides an introduction to EDI. You will learn about EDI generally, which will help you to follow along with the rest of the book. The chapter will explore standards for messages and communication, with an emphasis on UN/EDIFACT and AS2, which will play a role in the book's running example scenario.
- **Chapter 4** introduces the recommended technical solution for EDI implementation in an SAP S/4HANA landscape. You will learn about the sample business use case, understand the function performed by each unique component and technology of the technical solution, and see how the solution works as a whole. Over the course of the book, you will learn to implement this solution in step-by-step detail.

The next part of the book continues from the technical solution presented in Chapter 4. By this point, you should have a good foundation in the theory behind EDI integration. The next step is to practice, using step-by-step instructions and detailed explanations for enabling EDI in the SAP S/4HANA landscape. This part of the book will go through all the required configuration steps up to the execution of the sample business scenario:

- Before configuration can begin, it is important to collect and verify some information. **Chapter 5** explains the prerequisites that must be checked before beginning an EDI implementation. We will review the main assumptions and items that must be taken care of at the start of the project. This chapter will also describe the fictitious data that is being used for the book's sample EDI scenario.
- In **Chapter 6**, you will learn to perform initial checks and actions on the different technical components involved in an EDI implementation. You will also configure the connection between SAP S/4HANA and SAP BTP.
- At this stage, all necessary components have been activated and connected to each other and the EDI scenario implementation can begin. **Chapter 7** describes the required configuration within SAP S/4HANA itself, beginning with defining and customizing IDocs. Then you will learn to configure EDI communication with these IDocs. Finally, the chapter explains how to configure the new output control.
- **Chapter 8** introduces SAP Application Interface Framework, a component that has been available long before SAP S/4HANA but only recently has been given its due attention and become an integral part of SAP S/4HANA. You will learn to use SAP Application Interface Framework in EDI implementations to harmonize developments, enable central monitoring, and enable error handling.

- SAP Integration Suite is the integration platform selected for the sample SAP S/4HANA landscape. In addition to being the SAP-recommended solution, it is a highly useful integration platform as a service (iPaas). In **Chapter 9**, you will learn to perform each step of the configuration process, from message and mapping definition in the Integration Advisor to partner management in Trading Partner Management, and ending with Cloud Integration, where the end-to-end iFlow is enabled.
- Then it's time to check the results of the configurations performed thus far. Processing the first messages in an EDI implementation is an essential step, which is why a full chapter is dedicated to the topic. In **Chapter 10**, you will learn to simulate the business scenario and then check the EDI processing step by step.
- In real-life scenarios, you will have more than one EDI partner and more than a few basic transactions. Testing must also be performed on multiple different occasions. **Chapter 11** will provide a recommended approach for such situations using automated tests.
- This book focuses on an IDoc-based solution for EDI implementation, but SAP S/4HANA also allows for other options. In **Chapter 12**, we'll teach you how to use OData APIs for B2B EDI integration in certain situations.
- In **Chapter 13**, we quickly review additional aspects and best practices related to B2B EDI integration pattern.

Acknowledgments

I would like to start by expressing my gratitude to SAP PRESS and Int4 for giving me the opportunity to write a book. Although I had never considered authoring a book before, I couldn't miss such an opportunity, especially on a topic with which I am comfortable. I began my SAP career in an EDI support team, making this subject particularly relevant to my experience.

A big thank you to Agasthuri: writing a book is not always a pleasant and enjoyable task, but having somebody in the same boat made things easier, and mutual support was crucial to complete this publication.

Last but not least—actually, most importantly: I would like to express my profound gratitude to my entire family. This begins with my parents, who raised me and instilled in me a curiosity about the world. I am also deeply grateful to the most important person in my life, my wife, Magdalena, who supports me in every situation and with whom I share the daily hardships and joys of life. I must also mention my children, Piotr and Julia, who I thank for their constant inspiration and love.

—Marek Piaseczny

Writing a book has long been a dream of mine, and many individuals have played a crucial role in making this journey possible.

First and foremost, I want to express my deepest gratitude to my coauthor, Marek Piaseczny. Your patience, support, and guidance have been invaluable. And to my publisher, Rheinwerk Publishing/SAP PRESS, thank you for giving me the chance to write this book and for your unwavering support and guidance. I have come to deeply appreciate the complexity of the publishing process, and I admire the dedication and expertise you bring to your work.

My heartfelt thanks go to my parents, Baladandapani and VijayaKumari, for the sacrifices you made and the hard work you put in throughout my childhood. Your love, support, and guidance have shaped me into the person I am today. To my wife, Kathryn, I am profoundly grateful for your tireless proofreading and constant support. To my son, Christopher, thank you for the joy you bring to our lives and for always making us laugh.

Finally, to the reader, thank you for your interest in this topic. I hope this book serves as a valuable resource for your career, and I wish you every success in your endeavors.

—Agasthuri Doss

1

Chapter 1
Introduction

This chapter provides a high-level overview of SAP S/4HANA, its architecture, and its characteristics, including its place as modern technology, shown with its user experience, business processes, and intelligence and analytics. We will delve into the deployment options and look at the implementation options, including greenfield, brownfield, and hybrid approaches.

SAP S/4HANA is crucial for electronic data interchange (EDI) due to its real-time data processing capabilities, which enable quick handling of orders, invoices, and shipments, enhancing efficiency and responsiveness. It offers robust integration with various EDI standards, facilitating seamless connections with trading partners. Its advanced analytics features provide insights into EDI transactions, aiding organizations in performance monitoring and data-driven decision-making. The modern user interface improves user experience, while automation reduces manual errors and accelerates transaction cycles. SAP S/4HANA also supports cloud and hybrid deployments for flexibility and ensures compliance with security regulations, making it essential for businesses that rely on EDI.

This chapter presents an overview of SAP S/4HANA, highlighting its key features, deployment methods, and implementation approaches when compared to its predecessor, SAP ERP. It begins by introducing SAP S/4HANA as a next-generation ERP system designed to leverage modern technology and enhance user experience. The chapter explores its architecture, focusing on its use of contemporary technologies, improved user interfaces, optimized business processes, and advanced capabilities in intelligence and analytics.

Furthermore, it discusses various deployment options available for SAP S/4HANA, including cloud, on-premise, and hybrid models, allowing organizations flexibility in choosing the right setup for their needs. This chapter also delves into different implementation strategies such as greenfield, brownfield, and hybrid approaches, each offering distinct advantages depending on the organization's existing infrastructure and business requirements.

Overall, this introductory chapter serves as a foundational guide for readers to understand the essence of SAP S/4HANA and its potential as a transformative ERP solution in the modern business landscape.

1.1 Introduction to SAP S/4HANA

SAP S/4HANA represents SAP's next-generation ERP suite, built on the advanced in-memory database SAP HANA. The integrated applications unify all aspects of a business into an intelligent suite on a fully digital platform, replacing legacy, process-driven systems. Today, SAP boasts over 200 million cloud users, offers more than 100 solutions encompassing all business functions, and maintains the largest cloud portfolio among providers.

SAP S/4HANA is an enterprise resource planning system designed by SAP to help businesses manage their business processes. It is designed to be cloud-enabled and mobile-friendly. With its modern advanced technologies, SAP S/4HANA can help businesses drive innovation and improve customer satisfaction. SAP S/4HANA's integrated analytics capabilities can provide real-time insights into all aspects of operations.

This is the latest generation of SAP's ERP solutions and was created with a focus on simplifying the user experience. SAP HANA (with HANA derived from *high-performance analytic appliance*) is a multimodel database that stores data in its memory with a column-oriented, in-memory database design. This allows a business to run advanced analytics in a single system. By storing data in column-based tables in the main memory and bringing online analytical processing (OLAP) and online transactional processing (OLTP) together, SAP HANA is significantly faster, with near-zero latency, and it can query data in an instant versus other traditional database management systems.

OLAP versus OLTP

There are two acronyms that often create confusion, OLAP and OLTP:

- *OLAP* is a computing method that enables users to easily and selectively extract and query data to analyze it.
- *OLTP* is a type of data processing that consists of executing a few transactions occurring concurrently, like order entry and banking.

SAP S/4HANA with EDI enhances data exchange capabilities, improves process efficiency, and supports business growth by leveraging real-time data and automation of the structured transmission of data between computer systems, following agreed-upon message standards. It replaces traditional paper-based methods with electronic equivalents, enabling businesses to exchange documents such as purchase orders, invoices, shipping notices, and more, directly between their ERP systems or with external partners. Integration of SAP S/4HANA and EDI has several advantages, like data exchange efficiency enabled by SAP S/4HANA's real-time processing capabilities. These complement EDI by allowing faster transaction processing and response times, improving overall efficiency in data exchange, and facilitating automation of business processes by reducing the need for manual intervention, while EDI automates data entry and document processing, reducing errors and processing time. SAP S/4HANA's real-time analytics capabilities also enhance EDI transactions by providing insights into transaction volumes, processing times, and other key metrics. This enables businesses to optimize operational efficiency.

Table 1.1 details SAP S/4HANA and its predecessor, SAP ERP. These are both enterprise resource planning systems developed by SAP, but they differ in several key aspects. The major difference is that SAP ERP is the traditional client-server (disk-based) system, which SAP intends to eventually phase out, while SAP S/4HANA is its next-generation counterpart.

Key Aspects	SAP S/4HANA	SAP ERP
Database platform	Runs exclusively on the SAP HANA database, which is an in-memory database-management system. This allows for faster data processing and real-time analytics.	Primarily runs on traditional relational database management systems (RDBMS) like Oracle, Microsoft SQL Server, IBM DB2, and others.
Data model	Utilizes a columnar data storage model optimized for SAP HANA's capabilities, enhancing performance for analytical and transactional processing.	Uses a traditional row-based data storage model.
User interface and experience	Offers a modern and intuitive SAP Fiori user interface, designed for enhanced usability and accessibility across devices.	Typically runs a traditional SAP GUI presentation for users.
Functionality and features	Builds upon SAP ERP functionalities, but also integrates advanced features such as machine learning, Internet of Things (IoT) capabilities, and real-time analytics.	Includes comprehensive ERP functionalities covering financials, logistics, human resources, and so on.
Architecture	Has a simplified two-tier architecture due to its integration with SAP HANA, reducing data redundancy and improving performance.	Follows a three-tier client-server architecture.
Performance	In general, provides faster processing speeds and quicker response times due to its in-memory computing capabilities.	Performance can vary depending on the underlying database and hardware setup.

Table 1.1 Key Differences between SAP S/4HANA and SAP ERP

Key Aspects	SAP S/4HANA	SAP ERP
Deployment options	SAP S/4HANA offers on-premise and cloud deployment options, with SAP S/4HANA increasingly geared toward cloud-first deployments and supporting hybrid scenarios more seamlessly.	SAP ERP primarily offers on-premises deployment options.
Roadmap and support	SAP is focusing its development efforts on SAP S/4HANA, SAP's evolution toward a more streamlined, real-time, and intelligent ERP solution leveraging in-memory computing and modern UI/UX paradigms.	SAP intends to eventually phase out SAP ERP.

Table 1.1 Key Differences between SAP S/4HANA and SAP ERP (Cont.)

SAP S/4HANA comes with built-in intelligent technologies, including artificial intelligence (AI), machine learning, and advanced analytics capabilities that allow any company to run an integrated and intelligent digital business in real time. SAP S/4HANA has many innovations over SAP ERP, which is an on-premise ERP system. Key changes include the development of a brand-new code base, utilization of SAP HANA's in-memory computing architecture, providing the ability to access real-time business statistics and analytics, and moving beyond the transactional processes toward gathering data. It edges toward a more innovative approach to provide dynamic decision guidance for business users in real-time, depending on data sourced from various platforms.

SAP S/4HANA Cloud provides SAP S/4HANA functionality by placing a customer's instance of the solution on a partitioned cloud server alongside other customers, and private cloud deployment provides SAP S/4HANA functionality without the need for the customer to maintain the technical backend. It is great for companies running SAP ERP who already have some cloud experience and infrastructure; with this, customers can host the solution in their own cloud without having to share space with other tenants. A hybrid SAP S/4HANA instance provides a combination of both on-premise and cloud deployments. This may be done for a number of reasons, including testing the feasibility of both instances or hosting an on-premise deployment at headquarters while utilizing a cloud deployment in a satellite office.

SAP S/4HANA integrating with EDI offers several significant benefits to organizations, like improved accuracy and compliance, ensuring that data is transmitted in a structured, standardized format and thus reducing errors associated with manual data entry. SAP S/4HANA's integration allows businesses to enforce compliance with industry standards and regulations more effectively, and SAP S/4HANA allows for customization to adapt EDI processes according to specific business requirements and industry standards, thus enhancing flexibility in business operations.

Figure 1.1 represents the SAP HANA transformation to SAP S/4HANA, which shows the changes made throughout the years. Prior to SAP HANA, SAP software was run on third-party databases such as SQL and others. SAP HANA is SAP's own in-memory platform, and it enables faster analysis and access to data.

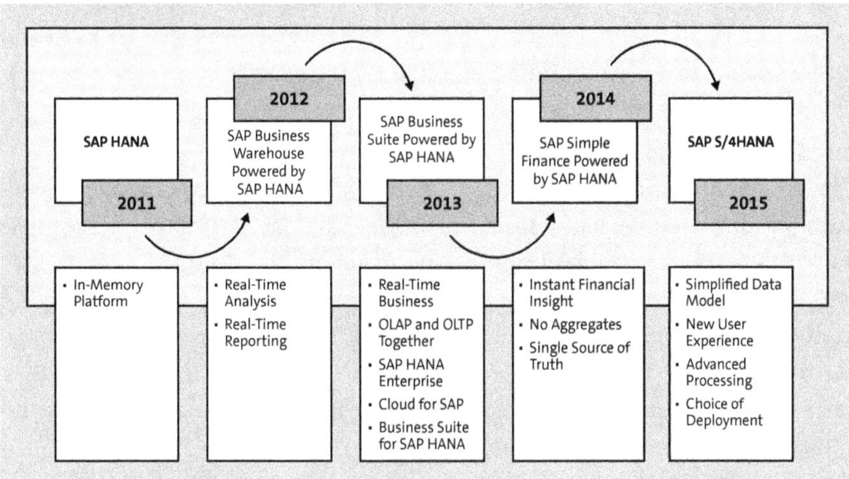

Figure 1.1 Brief History of SAP S/4HANA

SAP S/4HANA

SAP S/4HANA stands for SAP Business Suite generation 4 for the high-performance analytic appliance.

SAP S/4HANA represents a significant advancement in ERP technology, offering businesses the tools to streamline operations, enhance decision-making with real-time insights, and adapt to evolving market demands. Whether deployed on premise, in the cloud, or through a hybrid model, SAP S/4HANA aims to drive efficiency, innovation, and growth for organizations across various industries. This is a high-level overview; we'll discuss this architecture in more detail in Section 1.2.

Some of the key features and benefits of SAP S/4HANA include the following:

- **In-memory computing**
  SAP S/4HANA leverages SAP HANA's in-memory computing capabilities, enabling real-time data processing and analytics. This accelerates business processes and provides instant insights for decision-making.
- **Simplified data model**
  SAP S/4HANA introduces a simplified data model that reduces data redundancy and improves data processing efficiency. This allows for faster transaction processing and analytics. SAP S/4HANA is agile and innovative, and businesses can innovate

faster and respond quickly to market changes with flexible deployment options and advanced analytics capabilities.

- **Modern user experience**
  SAP Fiori, the user interface for SAP S/4HANA, offers a responsive and intuitive user experience across devices. It simplifies navigation and enhances productivity. SAP Fiori is a design system with enhanced user productivity. It provides a modern and intuitive interface, improving user adoption and productivity across the organization. It enables business to create apps with a consumer-grade user experience.
- **Advanced analytics and reporting**
  With embedded analytics and reporting capabilities, SAP S/4HANA enables businesses to perform complex analyses on real-time data, facilitating better strategic decisions. SAP S/4HANA enables improved performance by allowing a user to plan, execute, and generate reports and analytics based on live data, as well as triggering period closing and providing improved predictions. It also enables organizations to provide better service for customer-centric applications. With real-time insights, businesses can make informed decisions quickly with real-time data analytics and reporting capabilities.
- **Cloud-first approach**
  SAP S/4HANA supports both on-premise and cloud deployment options, allowing businesses to choose the deployment model that best suits their needs. Cloud deployment offers scalability, flexibility, and reduced IT infrastructure costs.
- **Industry-specific solutions**
  The system provides industry-specific functionalities and solutions tailored to meet the unique requirements of various sectors, such as manufacturing, retail, healthcare, and more.
- **Compliance and security**
  Compliance and security are a benefit as SAP S/4HANA incorporates robust security measures and compliance standards to protect data integrity and ensure regulatory compliance.
- **Integration capabilities**
  SAP S/4HANA integrates seamlessly with other SAP and third-party applications, facilitating end-to-end business processes and enabling a unified view of operations. Integrating SAP S/4HANA with EDI combines the advanced capabilities of the SAP system with the streamlined data exchange processes of EDI. A few of the key features and benefits are that SAP S/4HANA integrates various business functions into a single platform, including finance, sales, procurement, manufacturing, and more. This integration ensures seamless data flow across departments and facilitates EDI integration across business processes in a unified platform. SAP S/4HANA offers robust integration capabilities with external systems and trading partners. It supports various communication protocols (e.g., AS2, FTP, SFTP) and EDI document standards, facilitating seamless data exchange. Considering the importance of EDI,

SAP S/4HANA provides flexibility to customize EDI processes according to specific business needs and industry requirements. This includes configuring mappings, transformations, and validation rules for EDI messages.

1.2 Architecture and Characteristics

In this section, we'll discuss an architecture overview, modern technology, user experience, business processes, and intelligence and analytics.

1.2.1 Architecture Overview

The core components of the SAP S/4HANA architecture are the SAP Fiori UX, applications and extensions, and the SAP HANA in-memory database. As the presentation layer, the SAP Fiori UX makes a significant change to SAP's approach to user experience by offering a clean, intuitive, and personalized user experience, with improved security policies and efficiency. In the application layer, the advanced analytics and real-time data processing ability can handle complex computations and analytics in real time. The SAP HANA in-memory computing engine, the powerhouse at the heart of SAP S/4HANA, revolutionizes data processing in the enterprise world. As an in-memory computing engine, it allows for fast processing of large volumes of data.

Figure 1.2 shows the core components of the SAP S/4HANA architecture, which are designed to deliver real-time data processing, enhanced user experiences, and support for emerging technologies such as AI and IoT. The integration layer enables smooth interactions with other systems and underpins efforts in digital transformation.

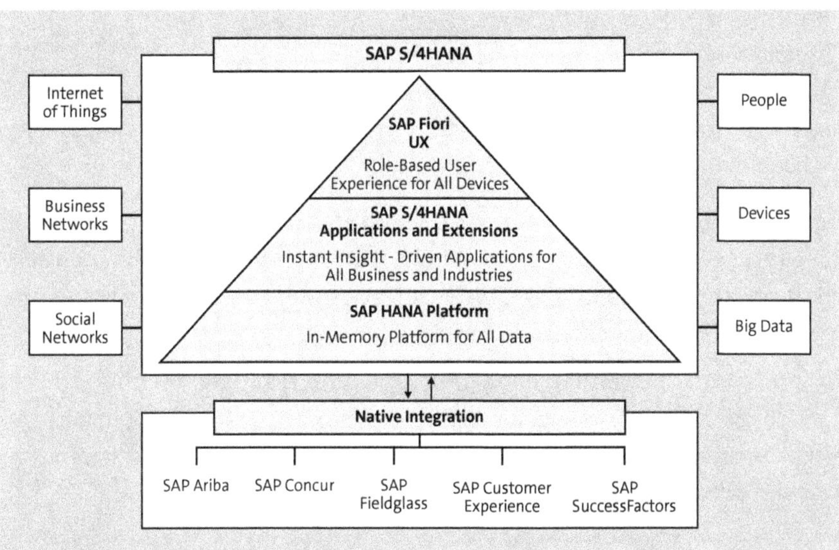

Figure 1.2 SAP S/4HANA High-Level Architecture

The SAP HANA system is composed of three components: a host, an instance, and a system. A *host* is the operating environment where you will find the resources and services (CPU, memory, network, and operating system) that the SAP HANA database requires. An *instance* is the smallest operational unit of a host, whereas a *system* has one or more instances.

The SAP S/4HANA architecture is simple, business-driven, agile, and scalable to help businesses run more efficiently and effectively with a high-performance, in-memory database that can handle large volumes of data quickly and efficiently. SAP S/4HANA utilizes SAP HANA as its underlying database, which is designed for in-memory computing. This architecture enables real-time data processing, analytics, and transactions by storing and processing data in RAM rather than in traditional disk-based storage. SAP HANA is an in-memory, column-oriented, relational database-management system. It was initially released in 2010 and its primary function, as the software that runs a database server, is to store and retrieve data from applications as requested. The following are its key characteristics:

- **Simplified data model**
  SAP S/4HANA introduces the Universal Journal, which integrates financial and managerial accounting data into a single table. This simplification reduces data redundancy, improves data processing speed, and provides a unified view of financial data across different modules.
- **Modular structure**
  SAP S/4HANA has functional modules that are organized into modular components such as finance (SAP S/4HANA Finance), supply chain management, manufacturing, sales and distribution, and so on. Each of these components encapsulates specific business processes and functionalities.
- **Cloud-first architecture**
  SAP S/4HANA supports both on-premise and cloud deployment options, with a focus on cloud-first architecture. This architecture enables scalability, flexibility, and rapid deployment of updates and innovations.

Open integration is a key characteristic, as SAP S/4HANA provides robust integration capabilities with other SAP solutions (SAP SuccessFactors, SAP Ariba, etc.) and third-party applications through open APIs. This ensures seamless data exchange and process integration across the enterprise ecosystem.

SAP HANA is a fast and powerful database that provides real-time, multimodel data analytics, which offers many more benefits than just storing data, serving it, and providing a single source of truth. SAP HANA serves as a platform for ERP software and other business applications in the following ways:

- **Real-time analytics**
  SAP S/4HANA enables real-time analytics and reporting directly on transactional data stored in SAP HANA. This capability provides instant insights into business operations, enhancing decision-making agility and responsiveness.
- **Advanced machine learning**
  Embedded machine learning capabilities in SAP S/4HANA leverage SAP HANA's in-memory computing power to perform complex predictive analytics, automate routine tasks, and optimize business processes based on data-driven insights.
- **Modern user experience**
  SAP S/4HANA adopts SAP Fiori as its user interface, offering a modern and intuitive experience across devices. SAP Fiori enhances user productivity with role-based access, simplified navigation, and personalized dashboards.
- **Scalability and flexibility**
  The architecture of SAP S/4HANA supports scalability and flexibility, allowing organizations to scale resources dynamically based on business needs. Cloud deployment options further enhance scalability by providing elastic computing resources.
- **Security and compliance**
  SAP S/4HANA incorporates robust security features to protect data integrity and confidentiality. It complies with industry-specific regulations and standards, ensuring data privacy and regulatory compliance.

SAP S/4HANA's architecture and characteristics represent a significant evolution from traditional ERP systems. By leveraging in-memory computing, simplified data models, a modular structure, and a cloud-first approach, SAP S/4HANA enables organizations to achieve real-time insights, operational efficiency, scalability, and enhanced user experiences.

Integrating SAP S/4HANA with EDI involves leveraging their respective architectures (to be discussed in Chapter 3) and characteristics to streamline data exchange, enhance operational efficiency, ensure compliance, and support informed decision-making across the organization. SAP S/4HANA is designed to be scalable, accommodating growing transaction volumes and business needs. This scalability supports organizational growth and operational expansion. Integration between SAP S/4HANA and EDI includes robust security measures to protect sensitive data during transmission and storage. This ensures compliance with data protection regulations and builds trust with trading partners.

Figure 1.3 shows the architecture and characteristics of SAP S/4HANA. The architecture is designed in such a way that it can be integrated with any application and retrieve the data for any purpose in near real time, by robust design, governance, and standards. The key characteristics of SAP S/4HANA include a smaller total data footprint, faster analytics and reporting, unlimited workload capacity, and multitenancy.

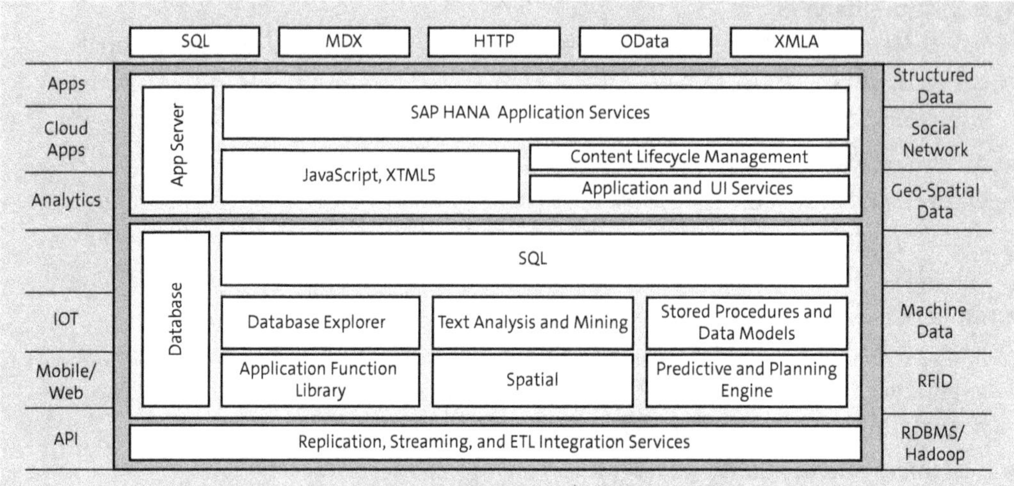

Figure 1.3 High-Level Architecture and Characteristics of SAP S/4HANA

1.2.2 Modern Technology

SAP S/4HANA uses lean architecture, a simplified data model, and a new user experience. SAP S/4HANA is integrated with several advanced technologies, including AI, machine learning, IoT, and advanced analytics. The SAP HANA in-memory database architecture and the integration of advanced technologies allow SAP S/4HANA to help solve complex problems in real time and analyze data faster.

The modern technology enables customers to undergo a digital transformation, where customers can modify existing business processes and models or create new ones. This allows the customers to be more flexible, responsive, and resilient to changing business requirements, customer demands, and environmental conditions.

An intuitive user interface makes it simple for customers to access necessary insights and collaborate across departments. A single source of truth for the entire enterprise and robust in-memory architecture allows organizations to access crucial statistics and analytics in real-time.

SAP S/4HANA can be integrated with other modern, cutting-edge technologies like IoT and AI. This allows businesses to connect with sensors and devices to collect real-time data and leverage AI for automation and further improve operational efficiency.

Using the following modern technology, SAP S/4HANA seamlessly integrates with other applications, cloud apps, public authorities, mobile apps, and on-premise systems:

- **API Management**
  API Management allows organizations to design, publish, and manage application programming interfaces (APIs). It enables you to expose and manage APIs for both internal and external consumption, making it easier to connect with third-party applications.

- **SAP Data Services**
  SAP Data Services is an extract, transform, load (ETL) tool that allows you to extract data from various sources, transform it, and load it into SAP S/4HANA. It's often used for data migration and data integration projects.
- **SAP Gateway**
  SAP Gateway enables the creation of OData services that provide access to SAP S/4HANA data and processes. It simplifies the development of mobile and web applications that need to interact with SAP data.
- **Remote function call (RFC)**
  RFC is a protocol used to call functions and methods in SAP systems remotely. It's commonly used for synchronous communication between SAP systems and non-SAP systems.
- **SOAP and REST web services**
  SAP S/4HANA supports both SOAP and REST web services for integration with external systems. These are commonly used for real-time and web-based integrations.
- **IDocs**
  IDocs are standard data structures used for asynchronous communication between SAP systems and non-SAP systems. They are often used for batch processing and data exchange.
- **SAP Business Technology Platform (SAP BTP)**
  SAP BTP offers a range of tools and services for integration, including event-driven architecture, API management, and data integration. It's a platform that complements SAP S/4HANA for comprehensive integration capabilities.

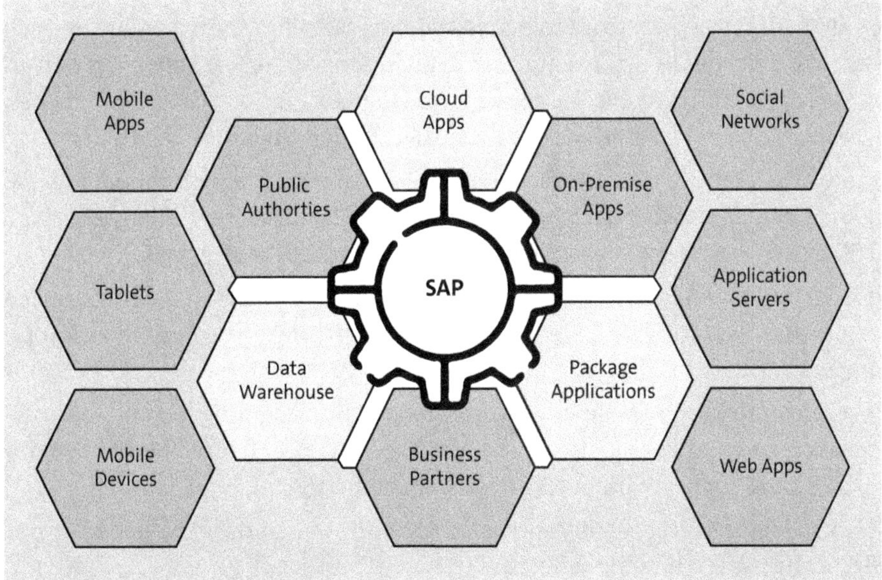

Figure 1.4 SAP Modern Technology

SAP S/4HANA with EDI enhances efficiency, customer responsiveness, and overall business agility. This provides a competitive edge in the marketplace by improving service levels and operational excellence. Integrating SAP S/4HANA with EDI offers comprehensive features and benefits that streamline business operations, improve data accuracy, ensure compliance, and drive cost savings and efficiency gains across the organization.

Figure 1.4 shows examples of how SAP S/4HANA can integrate with the multiple business functions that help customers manage complex business processes effectively in real time.

1.2.3 User Experience

SAP Fiori is a new user experience (UX) for SAP S/4HANA applications used in a variety of applications that used to be covered by multiple other technologies (e.g., SAPUI5, Web Dynpro ABAP, SAP GUI, Web client UIs and for arbitrary URLs). With SAP Fiori, business and technology come together to help users get their regular business functions done more easily: work approvals, financial tasks, calculation tasks, and various self-service tasks.

SAP Fiori was first introduced in 2013, with its primary focus being a mobile-first-display approach. SAP Fiori serves as an entry point to various applications, including custom-built and third-party applications that can be viewed or accessed on various devices with an adaptive frontend. It also provides services such as navigation, personalization, bookmarking, app configuration, notifications, usage analytics, and user information.

SAP Fiori offers the following benefits:

- The SAP Fiori interface utilizes a role-based design that tailors the experience to each user's role. This means, for example, that finance personnel see apps relevant to their tasks, while salespeople see apps specific to managing customers and quotes. This reduces clutter and makes it easier for users to find what they need quickly.
- The intuitive interface uses clear icons and labels, making it easy for users to navigate and understand what each function does. In addition, features like search bars and filters allow users to quickly find the information they're looking for.
- The SAP Fiori design interface is responsive, meaning it adapts to different screen sizes and devices. This allows users to access SAP S/4HANA from desktops, laptops, and even tablets for on-the-go access.
- SAP S/4HANA streamlines and improves workflows by presenting relevant information and actions within a single app. This reduces the need to jump between multiple screens and applications, making tasks more efficient.
- SAP Fiori adheres to accessibility standards, ensuring that users with disabilities can navigate and use the interface effectively.

- The interface supports compliance with regulatory requirements related to data privacy and security, enhancing trust and reliability.
- SAP Fiori allows users to personalize their interface by setting preferences for commonly used applications, layouts, and themes. This customization enhances user comfort and efficiency.
- Organizations can customize SAP Fiori applications to align with specific business processes and branding requirements, ensuring a tailored user experience.

SAP Fiori apps can display EDI document statuses, facilitate exception handling, and provide analytics on transaction volumes. SAP Fiori also can integrate with monitoring tools to track EDI document flows and provide alerts for issues such as failed transmissions or delays. Users can take immediate action through SAP Fiori apps to resolve issues and ensure smooth operations.

SAP Fiori's responsive design allows users to access EDI-related tasks and analytics on mobile devices, enhancing flexibility and productivity for users who need to manage EDI processes remotely, and its analytical capabilities complement EDI by providing insights into transaction trends, processing times, and compliance metrics. Users can monitor EDI performance and optimize processes based on real-time data.

Figure 1.5 illustrates the SAP Fiori launchpad system.

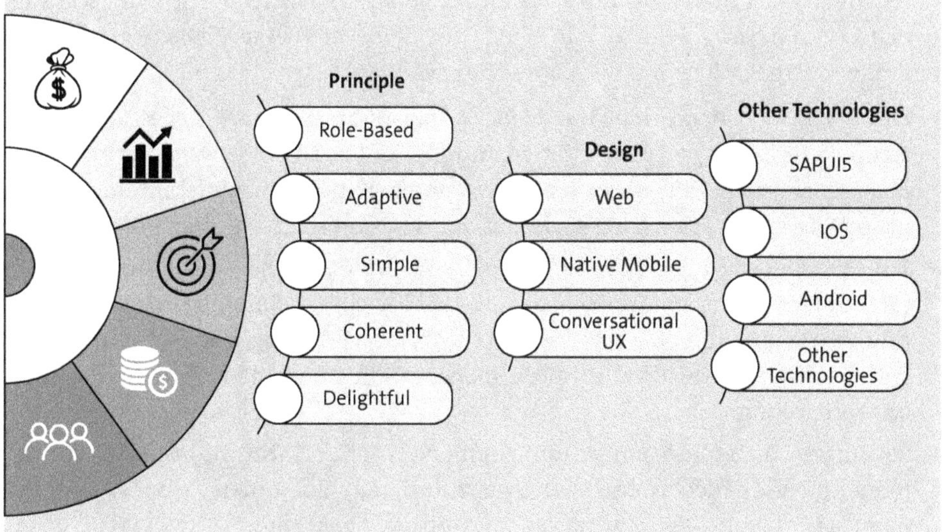

Figure 1.5 SAP Fiori User Interface

SAP Fiori Launchpad on SAP S/4HANA and in SAP BTP

Let's review the difference between SAP S/4HANA and SAP BTP in relation to SAP Fiori launchpad:

- **SAP Fiori launchpad on SAP S/4HANA**
  Implemented on an SAP Fiori frontend server and dependent on SAPUI5 and OData services; uses SAP Fiori launchpad designer
- **SAP Fiori launchpad on SAP BTP**
  Implemented as a software-as-a-service (SaaS) solution as a launchpad service; uses SAP Fiori Launchpad site manager

1.2.4 Business Processes

An SAP S/4HANA business process is a set or a group of activities that must be performed to complete an end-to-end business scenario. A typical business process involves several steps, such as creating a document, posting it, reviewing it, and finally releasing it. Each step can have multiple levels of approval and can involve different stakeholders, such as managers, accountants, and production supervisors.

SAP S/4HANA emphasizes breaking down the business divisions that operate independently and processes are designed to flow seamlessly across functions, improving collaboration and efficiency. For example, posting a sales order automatically triggers actions in procurement and manufacturing.

SAP S/4HANA encompasses a wide range of business processes across various functional modules, each designed to streamline operations, enhance efficiency, and support decision-making with real-time insights. Some of the key business processes typically covered by SAP S/4HANA are as follows:

- SAP S/4HANA Finance handles financial transactions, general ledger accounting, accounts receivable/payable, asset accounting, and financial reporting in real time. Management accounting supports cost and profitability analysis, internal orders, product costing, and budgeting to provide insights into cost structures and profitability.
- The sales and distribution channel handles sales order management and processes sales orders, quotations, pricing, and availability checks. Shipping and transportation will manage shipping activities, transportation planning, and execution. Billing and invoicing will generate invoices, manage billing documents, and handle revenue recognition.
- For supply chain management (procurement), SAP S/4HANA will manage procurement processes such as purchase requisitions, purchase orders, supplier management, and invoice verification. For inventory management, it will track goods movements, manage inventory levels, and optimize warehouse processes. Production planning will also be streamlined as it will plan and schedule production activities, manage bills of materials (BOMs), and monitor shop floor operations.
- In the manufacturing sector, with shop floor control, production activities can be monitored, work orders scheduled, and work-in-progress (WIP) inventory can be tracked. Quality management manages quality inspections, control plans, and quality notifications to ensure product quality and compliance.

- In human resources, SAP S/4HANA handles employee master data, payroll processing, time management, and benefits administration. With talent management, it will manage recruitment, training, performance evaluations, and career development.
- Asset management in SAP S/4HANA focuses on managing and maintaining an organization's assets throughout its lifecycle. It employs systematic approaches such as asset tracking, planning, maintenance, and analysis to ensure optimal asset utilization and performance.
- For sales and services, as a customer relationship management tool, SAP S/4HANA manages customer interactions, sales orders, service requests, and customer complaints. In relation to marketing, SAP S/4HANA plans and executes marketing campaigns, tracks campaign performance, and analyzes customer segmentation.

SAP S/4HANA integrates seamlessly with other SAP offerings (e.g., SAP Fieldglass, SAP SuccessFactors, SAP Ariba) and third-party applications through open APIs for end-to-end business process and provides real-time insights into key performance indicators (KPIs), financial statements, and operational metrics. It uses machine learning algorithms to predict future trends, customer behaviors, and demand patterns.

For example, the travel-to-reimburse business process, shown in Figure 1.6, is executed across various systems for several activities to determine what needs to be done to complete the travel-to-reimburse business scenario. SAP Concur is used in the front-end, and it works in conjunction with SAP SuccessFactors and SAP S/4HANA solutions to get the businesses to manage end-to-end processes for travel booking and reimbursement to the employee.

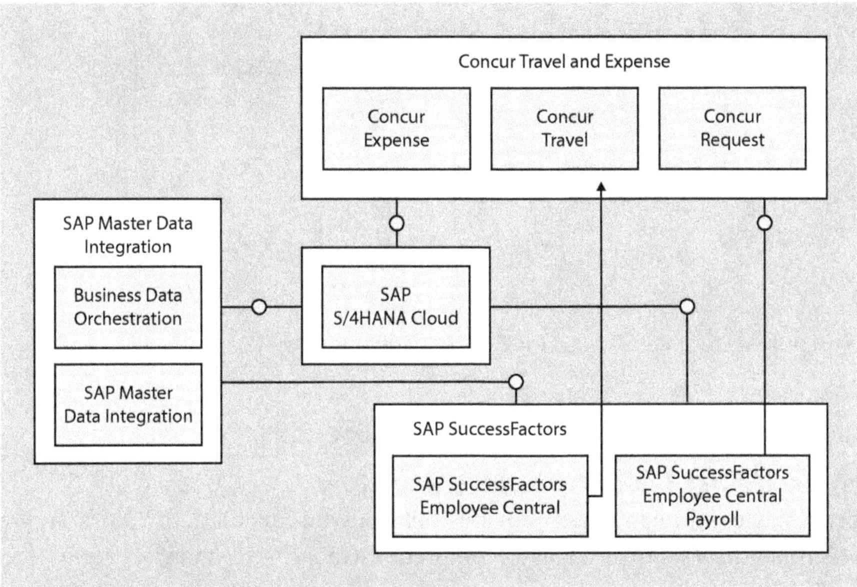

Figure 1.6 High-Level Business Process for Travel to Reimburse

The business process is comprised of solution value flow diagrams that describe its implementation from a business perspective and identify how value is generated by the organization. In this example, the holistic end-to-end solution for travel and expenses has a seamless integration through SAP SuccessFactors Employee Central to SAP Concur and SAP S/4HANA Cloud to consolidate employee and financial information across SAP.

Integrating SAP S/4HANA with EDI significantly enhances various business processes across organizations. Figure 1.7 shows a brief look at how these two technologies work together to streamline operations in a procurement process.

The purchase order is created in SAP S/4HANA by a procurement team for goods or services. The purchase order data is automatically converted into an EDI format (e.g., ANSI X12 850) and transmitted to suppliers electronically. Suppliers receive the EDI purchase order and process it directly in their ERP systems, triggering a fulfillment processes. As an order acknowledgment, the supplier sends an EDI 855 document confirming the receipt of the purchase order, providing order details, and indicating shipment dates. The buyer's SAP S/4HANA system then processes the EDI 855 document, updating the purchase order status and initiating subsequent steps in the procurement cycle.

Figure 1.7 shows a high-level business process for procurement involving EDI messages, illustrating a series of activities that an organization undertakes to achieve specific objectives in relation to a procurement process between SAP S/4HANA and an EDI partner.

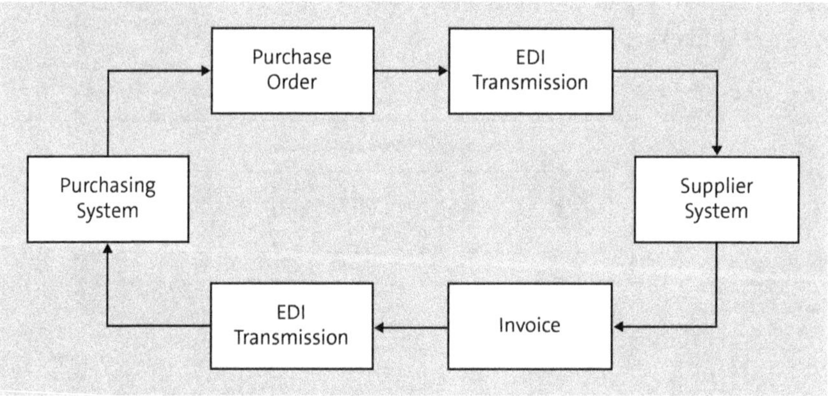

Figure 1.7 High-Level Business Process for Procurement Involving EDI

1.2.5 Intelligence and Analytics with SAP Analytics Cloud

SAP Analytics Cloud is a SaaS business intelligence (BI) platform. SAP Analytics Cloud is made specifically with the intent of providing all analytics capabilities to users in one place. SAP Analytics Cloud provides a comprehensive suite of analytics tools that enable businesses to collect, analyze, transform, and enhance the data, turning it into actionable insights by adding the power of operational analytics and business intelligence, thus helping decision makers to make confident decisions.

SAP Analytics Cloud allows users to create interactive dashboards and reports using various data visualization techniques like charts, graphs, and maps. With ad hoc analysis, users can explore data interactively, drill down into details, and uncover insights.

With augmented analytics, SAP Analytics Cloud uses machine learning and AI to provide smart insights and recommendations based on data patterns and trends. Users can perform predictive analytics and forecasting using built-in algorithms without needing deep technical knowledge.

In financial planning, SAP Analytics Cloud supports financial planning and budgeting processes, enabling organizations to create models, perform what-if analyses, and collaborate on financial plans. It also includes workforce planning capabilities for organizations to plan and optimize their human resources.

SAP Analytics Cloud integrates with various SAP and non-SAP data sources, including SAP HANA, SAP S/4HANA, SAP BW, SAP BusinessObjects, and third-party systems. SAP Analytics Cloud also includes robust security features such as role-based access control (RBAC), encryption, and data masking to ensure data security and compliance with regulations like GDPR.

SAP Analytics Cloud and EDI serve different but complementary roles within an organization's IT infrastructure, particularly when integrated to enhance data analysis and decision-making capabilities. Integration of SAP Analytics Cloud with EDI will add more value in data integration, where SAP Analytics Cloud can connect to EDI systems and integrate EDI transaction data with other enterprise data sources. This allows organizations to combine transactional data with operational and financial data for comprehensive analysis. For real-time analytics, SAP Analytics Cloud provides real-time data connectivity, enabling organizations to analyze EDI transactions as they occur. This capability supports proactive decision-making and immediate response to business events. SAP Analytics Cloud's visualization capabilities help transform EDI transaction data into meaningful insights. Companies can create dashboards and reports to monitor transaction volumes, processing times, and compliance metrics by analyzing historical EDI data within SAP Analytics Cloud, and they can apply predictive analytics to forecast transaction volumes, anticipate supply chain demands, and optimize inventory levels.

In a few words, integrating SAP Analytics Cloud with EDI enhances data analysis, operational efficiency, and decision-making capabilities. This integration allows organizations to leverage real-time EDI data alongside other enterprise data sources, transforming raw data into actionable insights for strategic advantage.

Integrating SAP Analytics Cloud with EDI offers numerous benefits for businesses. First, it enables enhanced decision-making by providing real-time access to EDI data within the cloud, allowing for timely insights that support informed strategic decisions. Second, the integration streamlines operations by automating data exchange processes and reducing reliance on manual input, thereby boosting operational efficiency and data accuracy. Third, SAP Analytics Cloud's robust analytics capabilities delve deep into EDI transactions, supply chain operations, and overall business performance, offering comprehensive insights that drive improvements across various facets of the organization. SAP Analytics Cloud also fosters collaboration by enabling seamless sharing of actionable insights and reports across departments and with external stakeholders, enhancing communication and alignment. Finally, the integrated solution is scalable, adapting flexibly to organizational growth and evolving analytics needs, ensuring continued support for expanding business requirements.

Figure 1.8 gives a high-level overview of SAP Analytics Cloud.

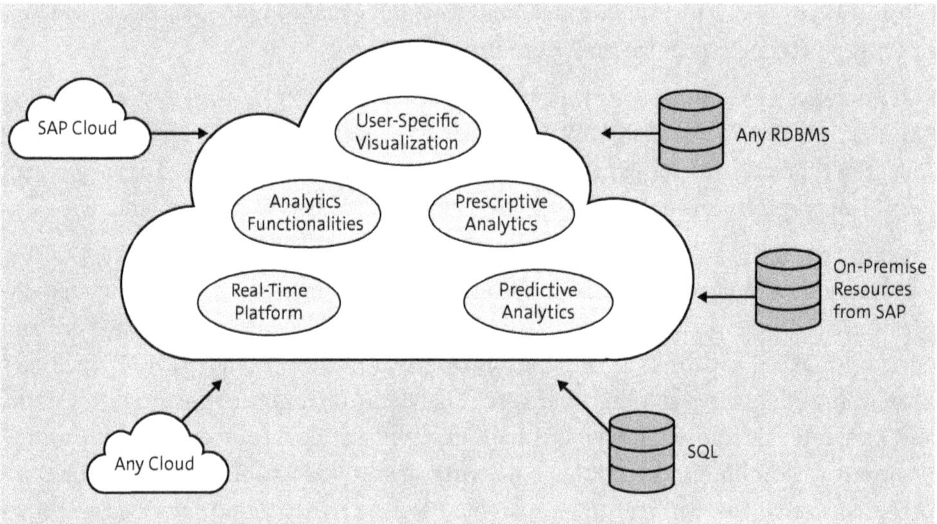

Figure 1.8 SAP Analytics Cloud Overview

1.3 Deployment Options

SAP offers the following deployment options for the current SAP Business Suite and for SAP S/4HANA. Each deployment option has its own benefits, so you should evaluate the options to determine the best course of action for your business needs. For all the options, SAP provides migration tools and services to help customers switch between deployment options and migrations:

- **On-premise**
  In this common model, a company acquires a license and installs the software in its own data center, where SAP S/4HANA system is installed and maintained on the company's own IT infrastructure. It offers a traditional licensing model and has the highest level of control and customization. The planning, testing, security, access,

system availability, controlling, and upgrading of the system are the complete responsibility of the customer.

- **Cloud**
  In this model, the customer rents the software. SAP deploys the software in an SAP-owned data center, where the SAP S/4HANA system is hosted and maintained by SAP or a third-party cloud service provider like Amazon Web Services (AWS), Google Cloud Platform (GCP), or Microsoft Azure. If it is in the SAP cloud, the infrastructure and maintenance are under SAP's control, with the automatic latest innovation upgrades occurring quarterly without a company's direct participation.
- **Hybrid**
  In this model, required applications are operated partially by the customer and partially by SAP in the cloud. Both parts can be linked and integrated with each other. This option is a combination of on-premises and cloud deployment. Some components of the SAP S/4HANA system are deployed on-premises while others are hosted in the cloud. This option provides the benefits of both on-premise and cloud deployment, such as control and customization as well as scalability and flexibility.

Figure 1.9 shows the various deployment options for SAP. There are two main SAP S/4HANA deployment options: cloud and on-premise. When choosing a deployment option, it's about balancing standardization versus flexibility. Choosing the cloud and standardization helps increase agility, speed, innovation, and flexibility, whereas on-premise deployment brings your own license and infrastructure subscription, allowing more freedom and flexible service offerings. In this scenario, the company is responsible for managing the software and infrastructure.

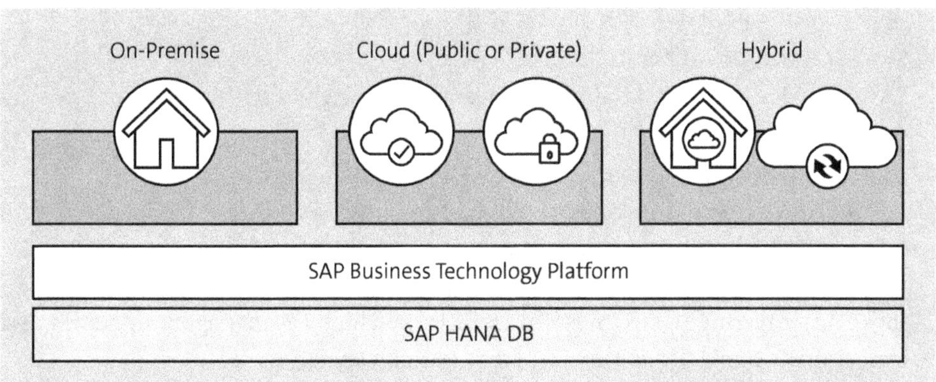

Figure 1.9 High-Level Overview of Deployment

Choosing the right deployment option depends on factors like budget, IT capabilities, scalability needs, data security requirements, and desired level of control over customization and management. The SAP S/4HANA deployment options currently available are as follows:

- **SAP S/4HANA Cloud**
  SAP S/4HANA Cloud is the SaaS edition of the SAP S/4HANA ERP system designed for the SAP HANA in-memory database. SAP S/4HANA Cloud is a shared public cloud edition hosted and operated by SAP. It is shared with other customers and is provided through subscription; mandatory upgrades are managed by SAP. Automated and manual regression tests are also available.
- **SAP S/4HANA Cloud, Extended Edition**
  Formerly called the single-tenant edition, the extended edition of SAP S/4HANA Cloud has all the advantages of the cloud with more flexibility and control. Like SAP S/4HANA Cloud, the extended edition uses a subscription service, and its dedicated landscape infrastructure is operated by SAP. This deployment also relies on the greenfield method. Unlike SAP S/4HANA Cloud, however, the extended edition needs to be upgraded by SAP and the customer every two years. SAP?S/4HANA Cloud Extended Edition has the same SAP S/4HANA functional scope as the on-premise version. In-app extensibility includes key user extensibility with some classic ABAP extensibility, plus further extensions through SAP Extension Suite.
- **SAP S/4HANA Cloud Private Edition**
  SAP S/4HANA Cloud Private Edition was released in 2021 as the latest deployment option for migration. It is simply the SAP S/4HANA on-premise software solution stack offered in a private cloud and managed by hyperscalers. Available under a RISE subscription contract, the private cloud enables customers to create a deployment closely suited to their needs but must stay in mainstream maintenance. The scope, extensibility, and configuration are the same as for SAP S/4HANA Extended Edition, and implementation can be done through a greenfield, system conversion (brownfield), or selective data transition (hybrid) approach.
- **SAP S/4HANA Cloud Public Edition**
  This provides a full cloud approach, resides on the provider's server, and is fully operated and maintained by SAP.
- **SAP S/4HANA on-premise managed by SAP HANA Enterprise Cloud**
  SAP S/4HANA on-premise offers individual and private hardware and infrastructure hosted by SAP, by a hyperscaler, or in a data center. It has a bring-your-own-license and infrastructure subscription, allowing more freedom and flexible service offerings. The company using the software is responsible for managing upgrades and choosing the speed of adoption. Upgrades should be performed annually.

  SAP S/4HANA managed by SAP HANA Enterprise Cloud has a full SAP S/4HANA industry and country scope, with customization, modification, and extensibility all possible as with SAP ERP. Like SAP S/4HANA Cloud Private Edition, SAP's on-premise deployment option can be implemented using the new implementation method, a system conversion, or selective data transition. This approach provides tools that selectively transfer configuration and custom code from older ERP systems into SAP S/4HANA by migrating the database, applying software updates, and doing a data conversion from third-party data models to the new, SAP-specific SAP S/4HANA data model. Companies can use SAP Solution Manager to manage their project design.

- **SAP S/4HANA on-premise**
  SAP S/4HANA on-premise or managed by a cloud provider hyperscaler (also known as AnyPremise) is a deployment option managed by a customer database or third party (a hyperscaler such as Azure, AWS, or GCP). The license is either perpetual or a bring-your-own model, and upgrades are done by the company running the software. Customization, modification, and extensibility are possible as with SAP ERP, and configuration uses the full IMG based on a fit-to-standard approach in the explore phase.

As stated previously, SAP offers public and private cloud models for deployment. SAP S/4HANA Public Cloud Edition is a SaaS solution in which all companies using the public SAP cloud share the same cloud space. It typically runs in a multitenant server where all the companies share the resources of the server. In a public cloud, each company's data and applications are hidden from the others, but because they are sharing certain resources and the maintenance of the solution, the company's data is taken care of by the cloud provider. SAP S/4HANA Private Cloud Edition typically runs in a single-tenant server, where only a single tenant (company) uses the resources of the server. Software runs in a private network protected by a firewall, similar to an on-premise system. The main differences between a private cloud and an on-premise installation are who has responsibility for maintaining the server and the license for the software installed on the server.

For example, with SAP S/4HANA Cloud Private Edition, SAP is responsible for maintaining the business software, but a company can choose to have its software on a server in an SAP data center or a server in a data center from one of SAP's hyperscaler partners. In this case, the hyperscaler partner is responsible for maintaining the server.

Table 1.2 shows the different features in SAP S/4HANA Cloud Public and Private Editions.

Features	S/4HANA Cloud Public Edition	SAP S/4HANA Cloud Private Edition
Ownership and control	Infrastructure isn't owned or controlled by the company using the software. It relies on the cloud provider for maintenance and updates.	Companies have more control over the infrastructure, whether hosted on premise or with a cloud provider.
Upgrades	Up to four times a year and managed by SAP.	Company-driven and managed.

Table 1.2 Features of SAP S/4HANA Cloud Public Edition versus SAP S/4HANA Cloud Private Edition

Features	S/4HANA Cloud Public Edition	SAP S/4HANA Cloud Private Edition
Type of implementation	Greenfield approach. Could be a component of selective data transition.	Brownfield and greenfield approach and selective data transition.
Deployment	Infrastructure as a service (IaaS; SAP Data Center, or GCP, AWS, Azure, or Alibaba Cloud).	IaaS, HEC, customer data center, hosting providers.
Customization	Standardized solution and limiting customization options.	Customization options and tailoring infrastructure are possible to meet requirements.
Integration	Integration with other systems may require extra effort.	Integration with other systems is easy when compared to public cloud.
Security and compliance	Security and compliance are primarily the responsibility of the cloud provider.	Security and compliance are the responsibility of the company.
Operations	Offers 99.9% availability and high availability.	Company and private cloud provider define the availability and high availability.
Scalability and performance	An on-demand approach offers scalability and elasticity and performance is a responsibility of SAP.	Scalability is determined by the company and involves effort and time. Performance can be optimized based on specific requirements.

Table 1.2 Features of SAP S/4HANA Cloud Public Edition versus SAP S/4HANA Cloud Private Edition (Cont.)

Choosing between the public and private cloud options depends on your organization's unique requirements, priorities, and constraints. While public clouds offer scalability and convenience, private clouds provide more control over security, compliance, and customization.

Before jumping into public cloud, private cloud, or hybrid options, it's worth considering the two-tier ERP deployment option.

In a two-tier SAP ERP strategy, an organization implements different SAP ERP systems for separate parts of the business. By combining two different deployments of SAP S/4HANA, a business can take advantage of the fast innovation in the public cloud while still allowing for a high degree of customization for strategic processes in the private cloud or on premise. The two-tier ERP strategy can be divided into three model types:

- **Headquarters and subsidiaries model**
  The headquarters runs a highly customizable on-premise or private cloud ERP system, and subsidiaries run a standardized public cloud ERP with a specific scope (see Figure 1.10).

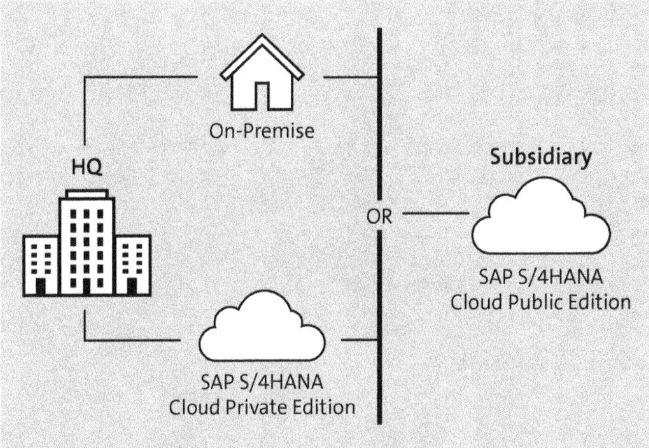

Figure 1.10 High-Level Diagram of Headquarters and Subsidiaries Model

- **Central services model**
  The organization has a line-of-business spin-off running a standardized public cloud ERP system (see Figure 1.11). For example, finance would be a separate legal entity and run in the public cloud as *shared services*, with the rest of the business running on a highly customizable on-premise or private cloud ERP system.

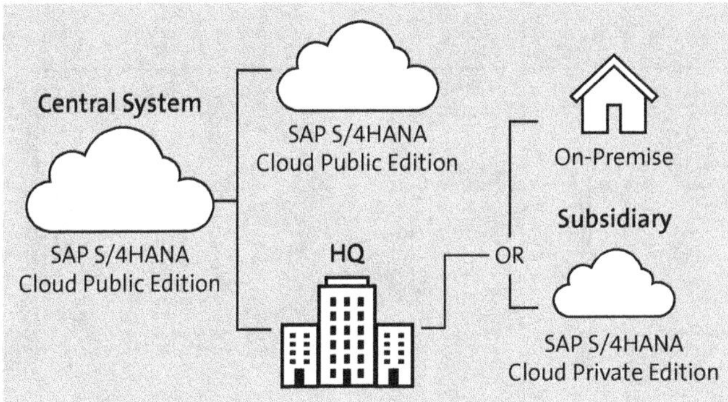

Figure 1.11 High-Level Diagram of Central Services Model

- **Ecosystem model**
  The headquarters runs a highly customizable on-premise or private cloud ERP system and brings its subcontractors or dealers into a standardized public cloud ERP system (see Figure 1.12).

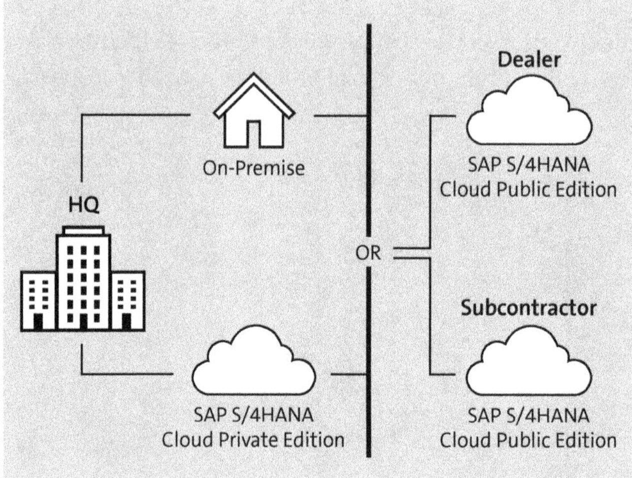

Figure 1.12 High-Level Diagram of Ecosystem Model

1.4 Implementation Options

SAP S/4HANA implementation options can vary depending on the specific needs and requirements of a business. The options are greenfield, brownfield, and hybrid. These options can be implemented by the SAP best practices methodology called *SAP Activate*.

Figure 1.13 gives a brief overview and comparisons of the greenfield, brownfield, and hybrid approaches. We'll further discuss each of these, along with the SAP Activate methodology, in the following sections.

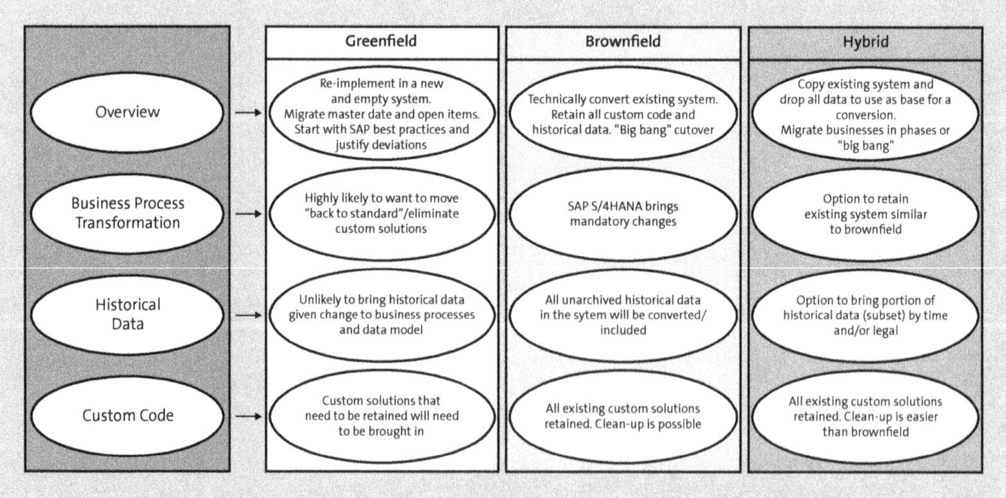

Figure 1.13 Major Types of SAP Implementation

1.4.1 SAP Activate

SAP S/4HANA project implementation is ongoing with a simple, flexible, modular, and agile methodology called SAP Activate (formerly the ASAP methodology). The SAP Activate methodology is a combination of best practices, guided configuration, and agile methods to simplify an SAP S/4HANA implementation.

The SAP Activate methodology is the next-generation agile approach for SAP S/4HANA implementation and project management, building on the foundation of the ASAP methodology used to implement SAP on-premises solutions. SAP Activate delivers ready-to-run business processes optimized with a reference solution designed for continuous innovation. It provides best practices for migration, integration, and configuration for SAP S/4HANA. It supports different types of deployments, such as new implementations, system conversion, and landscape transformation.

SAP Activate includes six phases in its implementation methodology:

1. **Discover**
   This first phase helps to develop a roadmap and the strategies for the implementation by defining goals based on SAP journey. This evaluates various business processes and technical research across the enterprise functions, and it enables the customer to prepare a business case for SAP S/4HANA adoption.
2. **Prepare**
   In this phase, the project team conducts the initial planning and preparation activities to get the project started. The team defines the project goals, the scope, and a project plan, and a migration strategy and approach are defined.
3. **Explore**
   In this phase, the project team verifies the business requirements, confirms business processes, performs solution validation, reviews details, and prepares for implementation and system conversion.
4. **Realize**
   The project team uses a series of iterations to build and test a complete business and system environment. Based on the inputs from the previous phases, the development environment builds and tests the configuration and migration process while finalizing end user training materials and documentation.
5. **Deploy**
   The project team prepares and executes the system for production release (go live). The process includes setup of the cutover plan, transitioning business operations to the new production system, and completing all end user training.
6. **Run**
   In the final phase in the SAP Activate methodology, the project team winds up all activities, fine-tunes the software now that it's deployed, and completes a handover of all details to the support team.

While SAP Activate primarily focuses on the implementation methodology and lifecycle management of SAP S/4HANA, if the customers have EDI integration, it can be an important aspect of the deployment strategy, especially for businesses that rely heavily on electronic document exchange with their trading partners. During the realize phase of SAP Activate, integration points for EDI transactions can be identified and mapped out. This involves understanding which business processes will involve EDI and configuring SAP systems to accommodate these interactions. As part of the deployment process (deploy phase), data migration strategies might include converting existing EDI formats or integrating EDI data into the new SAP environment. This ensures continuity of business operations without disruption in electronic document exchange.

EDI transactions often require rigorous testing to ensure that data is accurately exchanged between systems. SAP Activate provides methodologies for testing and validation during the deploy phase, ensuring that EDI integrations meet business requirements and are error-free. In the run phase of SAP Activate, ongoing support and maintenance activities include monitoring EDI transactions, resolving any issues that arise, and optimizing performance to ensure seamless operation of business processes.

Figure 1.14 shows the high-level framework from SAP for implementing SAP S/4HANA and other SAP solutions.

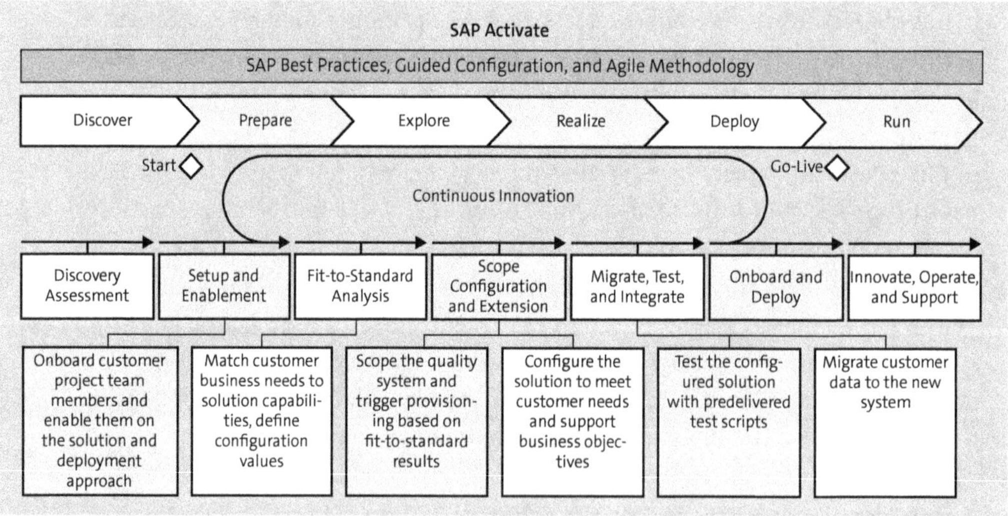

Figure 1.14 High-Level Diagram for SAP Activate

1.4.2 Greenfield Approach

Greenfield is a term used to describe a brand-new implementation of the SAP S/4HANA system. It involves starting from scratch with a fresh implementation of the system, allowing the customer to take full advantage of the latest features and capabilities of the SAP S/4HANA system. This includes the ability to streamline business processes, improve data analytics and reporting, and enable real-time decision-making and an amazing opportunity to eliminate outdated or unnecessary business processes and streamline operations.

Greenfield implementation can be the best option for organizations looking to modernize their IT infrastructure and take advantage of the latest technologies and capabilities of the SAP S/4HANA system. Implementing SAP S/4HANA and integrating EDI involves careful planning to ensure seamless digital transformation and efficient business processes. Starting fresh with SAP S/4HANA allows organizations to adopt standardized processes and best practices. EDI integration can be planned from the outset to ensure new business processes are optimized for electronic document exchange. It requires mapping out EDI integration points early in the implementation planning phase. Data migration planning may include mapping EDI data formats to SAP S/4HANA structures.

1.4.3 Brownfield Approach

Brownfield describes an approach to upgrading an existing SAP ERP system to SAP S/4HANA while retaining most of the existing configuration, data, and customizations. It is an upgrade that minimizes the impact on the existing system landscape and aims to preserve as much of the existing investment as possible.

The brownfield approach lets the customer reevaluate and edit their existing processes and port over the ones that are working in the legacy SAP system to SAP S/4HANA. The customer can get SAP S/4HANA up and running, and at the same time, migrate the existing SAP workflows and systems over to SAP S/4HANA. This approach is considered less disruptive and lower risk compared to other migration approaches as it allows organizations to leverage their existing investments and minimize disruption to their business operations.

Converting from an existing SAP ERP system to SAP S/4HANA will allow the customer to retain historical data and configurations. EDI integrations can be maintained or enhanced during the conversion process, ensuring compatibility of existing EDI setups with SAP S/4HANA. Customizations and integrations with EDI partners may need adjustments to align with new SAP S/4HANA capabilities.

1.4.4 Hybrid Approach

A *hybrid* approach is a migration strategy that involves a combination of the greenfield and brownfield approaches to migrating to SAP S/4HANA. This approach is particularly suitable for organizations that have complex system landscapes, including multiple SAP and non-SAP systems. It starts with a duplicate copy of the customer SAP system, including configuration settings and customizations, but master data and transactional data are removed.

In the hybrid approach, some systems are migrated to SAP S/4HANA using a greenfield approach, which involves a new installation of SAP S/4HANA with a completely new data model and configuration. At the same time, other systems are migrated using a brownfield approach, which involves upgrading existing SAP ERP systems to SAP S/4HANA while retaining most of the existing configuration, data, and customizations.

This approach allows organizations to leverage the benefits of both approaches. The greenfield approach can be used for systems that are relatively new or have fewer complex configurations, while the brownfield approach can be used for systems that have more complex configurations or are critical to the organization's operations.

The hybrid approach will allow for migration of selected data and configurations from existing systems to SAP S/4HANA. EDI integrations can be selectively migrated or redesigned to align with the new system landscape and there will be detailed mapping and validation of EDI data objects and formats. Integration testing will ensure seamless communication with trading partners after migration.

1.5 Summary

In this chapter, you learned about SAP S/4HANA, including a high-level description of its functions and properties, how it fits within SAP HANA, and how it is a next-generation ERP solution.

We also looked at the architecture and characteristics of SAP S/4HANA, how it excels as modern technology, and its user experience, business processes, and intelligence and analytics.

We discussed the three deployment options and the implementation methodology as well as migration strategies. We also looked into integrating SAP S/4HANA with EDI, which involves leveraging modern ERP capabilities to streamline electronic document exchange, enhance operational efficiency, and support business growth in a digital economy. It strengthens relationships with trading partners by improving collaboration and responsiveness in supply chain and financial transactions.

In the next chapter, we will look at the integration technologies that are available with SAP S/4HANA.

Chapter 2
Integration Technologies

This chapter provides information about the integration technologies that are available within SAP S/4HANA and the integration services that are a part of SAP BTP. We will look at the strategy and recommendations for integration.

SAP S/4HANA offers a variety of integration technologies to connect with other applications and data sources. These integrations are crucial for modern businesses that rely on seamless information flow across their entire ecosystem. SAP S/4HANA supports integration through a mix of legacy and modern integration technologies. Legacy technologies include RFC, Business Application Programming Interfaces (BAPIs), and Application Link Enabling (ALE) and IDocs. While still functional, these might not be the most efficient for new integrations. For modern integration technologies, SAP S/4HANA introduces new options like OData APIs, SOAP APIs, and SAP Integration Suite. These offer greater flexibility, real-time data exchange, and easier development compared to legacy methods.

The best integration technology depends on customer needs. Some factors to consider are the type of integration; whether connecting to an on-premise application, a cloud application, or a data lake; the data volume and complexity; how much data needs to be exchanged; the structure of the message; and whether the exchange has to be in real time or handled via batch processing.

In the context of SAP S/4HANA and EDI, *integration technologies* typically refers to the methods and tools used to connect SAP S/4HANA with external systems for EDI transactions, where middleware solutions facilitate communication between SAP S/4HANA and external EDI systems. These integration technologies and capabilities help organizations leverage the full potential of SAP S/4HANA by ensuring seamless data flow, process automation, and connectivity across the enterprise landscape to meet an organizations needs.

This chapter provides an extensive exploration of integration technologies within SAP S/4HANA, emphasizing SAP's strategic recommendations for seamless integration. It begins by outlining SAP's integration strategy, emphasizing the importance of leveraging modern technologies and services within SAP S/4HANA environments. The chapter compares integration technologies inherited from SAP ERP with new and improved capabilities in SAP S/4HANA, such as OData Services and APIs, SAP Application Interface Framework, enhanced output control, and Business Rules Framework plus (BRFplus).

Furthermore, this chapter introduces SAP Business Technology Platform (SAP BTP), highlighting its role in enhancing integration capabilities. Key components discussed include the SAP Integration Suite, which offers services like Integration Assessment, Cloud Integration, Integration Advisor, Trading Partner Management, API Management, SAP Business Accelerator Hub, Open Connectors, and Event Mesh. The cloud connector is also detailed as a crucial tool for securely connecting on-premise systems with cloud solutions.

2.1 SAP Strategy and Recommendations for Integration

The SAP S/4HANA integration strategy focuses on leveraging its architecture and capabilities to connect seamlessly with other systems, both within and outside the organization. Some of the key integration principles to be considered are as follows:

- **Predefined integration**
  SAP prioritizes seamless out-of-the-box integration between its intelligent suite applications. This is achieved through "suite quality" features like aligned domain models, harmonized APIs, and event-based architecture.
- **Open integration**
  SAP provides feature-rich, prebuilt connectors for more than 100 non-SAP applications.
- **Holistic integration**
  SAP offers a comprehensive integration technology portfolio catering to diverse integration needs in cloud and hybrid landscapes. This includes legacy technologies for existing systems and modern options for real-time data exchange.
- **AI-driven integration**
  Looking ahead, SAP is incorporating AI to automate integration tasks, simplify development, and optimize data flow.

In today's rapidly advancing technological landscape, organizational systems have expanded beyond limited applications and solutions. Traditional integration middleware solutions are struggling to keep up with the complex demands faced by organizations in the modern era. This poses a significant challenge across processes, applications, landscapes, security, and authorization. Addressing this issue is crucial for organizations.

To tackle these challenges, SAP has introduced the concept of intelligent enterprises, aiming to centralize digital interactions. The focus is on simplifying the integration process, thereby reducing project costs and timelines compared to traditional middleware tools. This initiative has resulted in SAP Integration Suite, a versatile, dynamic, and enterprise-grade platform available as part of SAP BTP. This platform is designed to streamline integration for individuals with varying skill sets, offering a range of integration approaches.

SAP's recommendation for integration is to use the cloud-based SAP Integration Suite offering. It is a middleware solution that supports process integration between SAP and non-SAP systems using open standards and modern integration technologies (web services, Groovy, REST, XML, JSON, etc.).

SAP Integration Suite is an integration platform as a service (iPaaS) that helps customers integrate on-premise and cloud-based processes, services, applications, events, and data. SAP Integration Suite is an industry-leading and enterprise-grade offering that helps businesses seamlessly connect and integrate their applications, data, and processes within their organization and beyond.

As shown in Figure 2.1, SAP Integration Suite includes prebuilt integrations with integration practice and patterns, architecture blueprints, and best practices for cloud and hybrid landscapes. It offers harmonized access to popular third-party cloud applications, which benefits all integration styles. These applications include process, data, analytics, user, and asset integration offerings to deliver significant value to your business. Features also include tools for designing, publishing, and managing APIs, and extending and adapting existing integration methodology with an open interface to address your unique business needs and requirements. We'll discuss SAP Integration Suite in more detail in Section 2.4.1.

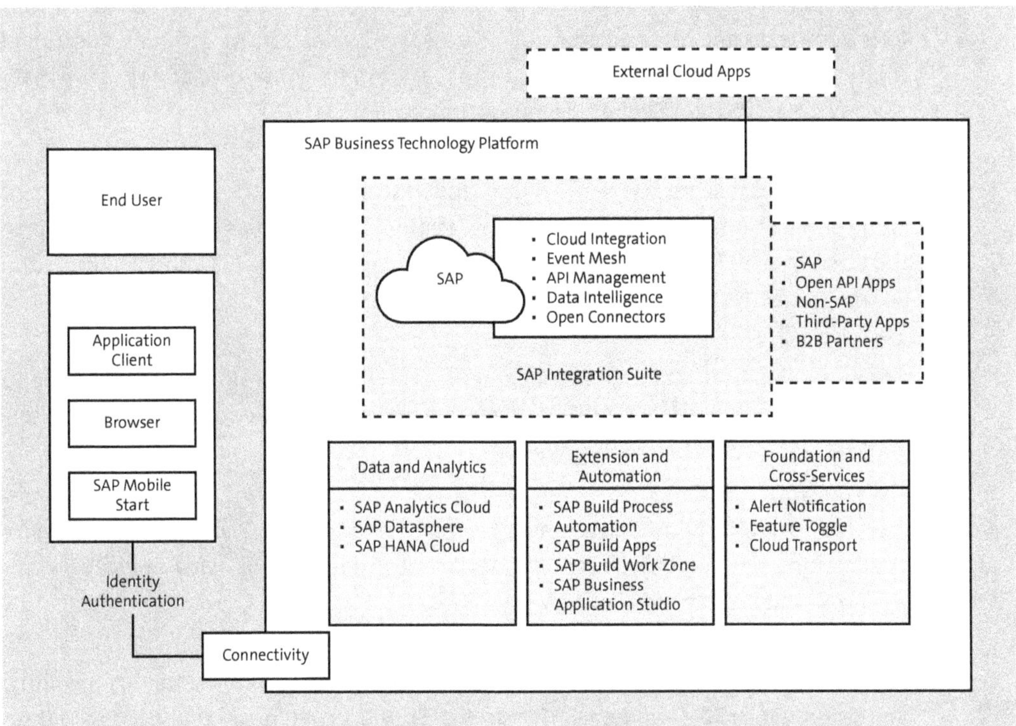

Figure 2.1 SAP BTP and SAP Integration Suite

SAP Integration Suite provides powerful tools and capabilities for integrating EDI within EDI partners through its API Management capabilities that enable organizations to expose EDI-related APIs securely. This allows external partners and systems to integrate seamlessly with SAP applications via governed APIs and also supports event-driven architectures for real-time integration scenarios, enhancing responsiveness and agility in EDI transactions and business processes.

SAP Cloud Platform and SAP BTP

SAP Cloud Platform and SAP BTP are the same; the latter is a rebranding of the former. SAP Integration Suite is one of the SaaS capabilities available in SAP BTP.

2.2 Technologies from SAP

SAP offers a variety of integration technologies to facilitate seamless data exchange and communication between SAP systems and other applications. Here's an overview of the key SAP integration technologies:

- **SAP Process Integration/SAP Process Orchestration**
  SAP Process Integration is used for integrating SAP and non-SAP systems using various protocols and data formats. SAP Process Orchestration includes additional orchestration capabilities, allowing for more complex workflows and process management. SAP Process Orchestration is supported till 2030.
- **Cloud Integration and SAP Integration Suite**
  A cloud-based integration service that enables the integration of on-premise and cloud applications. Provides prebuilt integration content and supports various protocols such as REST, SOAP, and OData. Cloud Integration and SAP Integration Suite are the same; the latter is an evolution of the former.
- **SAP BTP**
  Combines database and data management, analytics, application development, and integration into a single platform. Facilitates building and extending applications while integrating with existing systems.
- **SAP Gateway**
  A framework that allows developers to create and expose OData services for SAP applications. Enables easy access to SAP data via RESTful APIs, allowing integration with web and mobile applications.
- **SAP Data Services**
  A data integration and transformation tool that allows for data cleansing, profiling, and loading into SAP and non-SAP systems. Supports batch and real-time data integration.
- **IDocs**
  A standard data structure used in SAP for transferring data between SAP systems and external systems. Facilitates asynchronous communication and is commonly used for batch processing.

- **SAP Fiori and SAPUI5**
  SAP Fiori is a user experience (UX) design approach, and SAPUI5 is its development toolkit. They can be integrated with backend systems through OData services for real-time data retrieval and updates.
- **Event Mesh (formerly SAP Cloud Platform Messaging)**
  A messaging service that enables event-driven architecture and real-time communication between applications. It supports various messaging patterns like publish/subscribe.
- **Application Link Enabling (ALE)**
  A technology for connecting distributed SAP systems using IDocs for asynchronous data exchange. Suitable for integrating different SAP environments.
- **SAP Analytics Cloud**
  Allows integration of analytics capabilities into SAP applications and connects with various data sources. Facilitates real-time data access and reporting.

These technologies provide flexibility in integrating various systems, whether they are on-premise, in the cloud, or in hybrid environments, ensuring that businesses can maintain efficient processes and data flow.

Table 2.1 provides an overview of SAP ERP middleware.

Product	Product Release
SAP Exchange Infrastructure 2.0	2003
SAP Exchange Infrastructure 3.0	2005
SAP Process Integration 7.0	2006
SAP Process Integration 7.1	2008
SAP Process Integration 7.3	2011
SAP Process Orchestration 7.4	2013

Table 2.1 Product Names and Release Dates for SAP ERP Middleware

The middleware listed in Table 2.1 is out of support from SAP. SAP Process Orchestration version 7.5, which was released in 2015, is still in use and will be supported until the end of 2030. SAP Process Orchestration is a combination of different tools including SAP Process Integration, enabling not just system integration but also business process management, business rules management, and other technical components like SAP Composite Application Framework (supports the creation of persistent data) and SAP Extended Enterprise Content Management by OpenText. For companies that are looking to move their applications and integrations to the cloud, SAP Integration Suite is the logical successor to SAP Process Orchestration. We'll discuss SAP Integration Suite in more detail in Section 2.4.1.

Let's look at SAP Process Orchestration 7.5 and its three components—SAP Process Integration, SAP Business Process Management (SAP BPM), and SAP Business Rules Management (SAP BRM):

- **SAP Process Orchestration 7.5**
  SAP Process Orchestration is a strategic integration offering for an on-premise deployment from SAP. It covers the full functional range of an integration platform, including message processing, connectivity, and business process and rules management. The current version of SAP Process Orchestration is 7.5 and is a solution bundle running purely on a Java application server which consists of SAP Process Integration on Java (an Advanced Adapter Engine Extended [AEX]), SAP Business Process Management, and SAP Business Rules Management.
- **SAP Process Integration**
  This is an Enterprise Service Bus (ESB) that traditionally uses both ABAP and Java stacks (up to SAP Process Integration 7.4) to enable message-based integration between IT systems, using a number of adapters to send messages to and receive them from these systems. It is possible to implement SAP Process Integration so that it runs only on the Java stack that can activate and use the other SAP Process Orchestration components (SAP BPM and SAP BRM). As of version 7.5, it is only a Java stack in SAP Process Orchestration 7.5.
- **SAP Business Process Management**
  This is a business process management offering to create processes and workflows among people, systems, and business partners. It supports joint modeling of processes, central process execution via a Java-based engine, provision of interfaces for users, and integration of business rules into processes.
- **SAP Business Rules Management**
  This is used to set user-defined business rules that are required in business processes. For example, SAP Process Orchestration includes all the tools you need to create business logic, applications, and integration to close gaps between your IT systems.

  To enable automatic decisions by using business rules, the business users participate in and control rule definition, while the business process experts model, validate, deploy, update, and archive business rules through the lifecycle. The business rules describe the operation of, definition of, and constraints on the behavior of a business and enable decision automation.

Figure 2.2 shows the components of SAP Process Orchestration, where SAP Process Integration, SAP BPM, and SAP BRM are the components of SAP Process Orchestration.

In SAP Process Orchestration, the Advanced Adapter Engine (AAE) is meant to connect SAP systems and non-SAP systems together by the various adapters in the Advanced Adapter Engine to convert XML and HTTP-based messages to the specific protocol and format required by the systems.

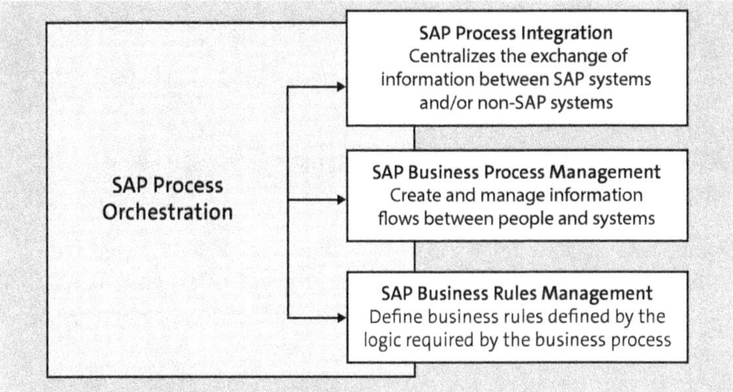

Figure 2.2 SAP Process Orchestration and Its Components

Advanced Adapter Engine Extended (AEX) provides the connectivity capabilities of the AAE as well as the design and configuration tools (ES Repository and the Integration Directory) to set up scenarios based on the AAE. Figure 2.3 shows the AAE, which is the part of SAP Process Orchestration. AEX connects to the AAE.

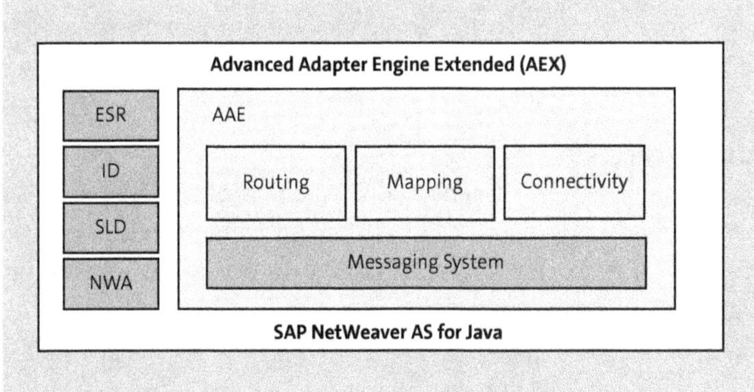

Figure 2.3 AEX Provides Connectivity Capabilities for AAE

Table 2.2 differentiates the key features between SAP Integration Suite in SAP BTP as a component and SAP Process Orchestration. We'll discuss SAP Integration Suite further in Section 2.4.1.

Features	SAP Integration Suite	SAP Process Orchestration
Deployment	Cloud-based service hosted by SAP. This IPaaS solution can be hosted on AWS, Azure, or GCP and will be handled by SAP.	On-premises installation in company landscape and handled by the company.

Table 2.2 Key Differentiating Features between SAP Integration Suite and SAP Process Orchestration

Features	SAP Integration Suite	SAP Process Orchestration
License model	Monthly subscription fee.	Core-based product license and annual fee.
Business-to-business (B2B) support	Supported via Trading Partner Management.	Supported via B2B add-on.
Architecture	Multitenancy, data isolation, and virtualization are possible. Automated failover and rolling software updates.	Single-tenant solution with no built-in virtualization capabilities. Failover to be configured by company.
Software updates	SAP periodically updates.	Needs to be installed by the company.
Scalability	Highly scalable as it is a cloud-based solution managed by SAP.	Company responsible for additional nodes/servers.

Table 2.2 Key Differentiating Features between SAP Integration Suite and SAP Process Orchestration (Cont.)

Integration technologies with EDI and SAP offer several significant benefits that contribute to improving efficiency, accuracy, and overall business performance. Here are some key advantages:

- **Automation of business processes**
  Integration technologies automate the exchange of business documents such as purchase orders, invoices, and shipping notices between SAP and external systems using EDI standards. This automation reduces manual intervention, speeds up transaction processing, and minimizes errors.
- **Improved data accuracy**
  By eliminating manual data entry and reducing the likelihood of human error, integration technologies ensure that data exchanged between SAP and EDI systems is accurate and consistent. This enhances the reliability of information across business processes.
- **Enhanced efficiency**
  Automating data exchange and streamlining business processes lead to improved operational efficiency. Organizations can reduce processing times, shorten cycle times for order fulfillment and invoicing, and optimize resource utilization.
- **Real-time visibility and tracking**
  Integration technologies provide real-time visibility into transaction statuses and process flows across the supply chain. This visibility enables businesses to track the movement of goods, monitor order statuses, and respond quickly to changes or issues, enhancing overall supply chain management.

- **Compliance with EDI standards**
  Integration technologies ensure compliance with EDI standards (such as EDIFACT or ANSI X12) required by trading partners and regulatory authorities. Adhering to these standards facilitates seamless communication and collaboration with partners, avoiding penalties and improving business relationships.
- **Scalability and flexibility**
  SAP's integration capabilities, such as SAP Integration Suite, support scalability and flexibility. Businesses can easily integrate new trading partners, add additional EDI transactions, or expand their business operations without significant IT overhead.
- **Analytics and insights**
  Integration technologies enable the aggregation and analysis of data from EDI transactions within SAP systems. This data can be leveraged for business intelligence, trend analysis, and decision-making, driving continuous improvement and strategic initiatives.

Integration technologies in SAP and EDI are instrumental in modernizing business operations by automating and enhancing the exchange of critical business data. By leveraging SAP's robust capabilities and integration tools, organizations can achieve significant improvements in efficiency, accuracy, and compliance, ultimately driving competitive advantage and operational excellence. They play a crucial role in achieving digital transformation goals and enhancing overall business efficiency and effectiveness.

Figure 2.4 illustrates how an SAP system integrates with EDI systems. The SAP sender system sends a purchase order as an IDoc to SAP middleware, where it is converted from an IDoc to an EDI purchase order in 850 format and sent to the SAP receiver system via VPN and an EDI hub. The SAP receiver system processes the PO and sends the PO acknowledgement in IDoc format to the middleware, where is it converted to an EDI message and sent across VPN to the SAP sender system.

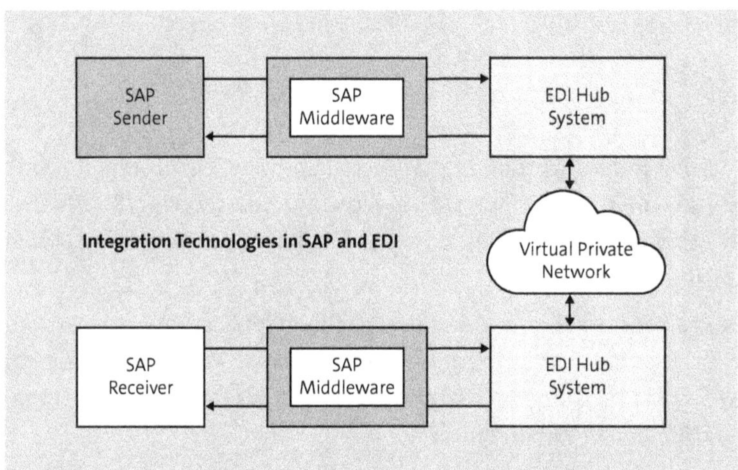

Figure 2.4 Integration Technologies in SAP and EDI

2.3 New and Improved Technologies

Although there are many new and improved technologies in SAP, here we will discuss OData, SAP Application Interface Framework, new output control, and BRFplus.

OData is a standardized protocol for building and consuming RESTful APIs, making it easier to access and manipulate data over the web. OData is an open protocol that allows clients to access and manipulate data using standard HTTP methods (GET, POST, PUT, DELETE). It supports querying, filtering, and pagination of data, providing a flexible way to interact with RESTful services. SAP leverages OData to expose its data and business logic, allowing developers to create applications that can easily integrate with SAP systems.

SAP Application Interface Framework is a powerful tool that simplifies the management, monitoring, and integration of application interfaces within SAP and with external systems. By providing a structured approach to interface development, error handling, and process integration, SAP Application Interface Framework enhances data quality and operational efficiency, enabling organizations to respond swiftly to changing business requirements. Its flexibility, combined with robust monitoring and error handling features, makes it an essential component of SAP's integration landscape.

New output control in SAP S/4HANA represents a significant advancement in how organizations manage document output. With its centralized configuration, flexible output options, and integration with SAP Fiori, it provides a streamlined, user-friendly approach to document processing.

BRFplus is a powerful tool for managing business rules within SAP applications. By separating business logic from application code and providing a user-friendly interface, BRFplus empowers organizations to respond quickly to changing business needs. Its integration capabilities, versioning, testing, and compliance features make it a vital component for organizations looking to enhance their decision-making processes and improve operational efficiency.

2.3.1 OData

OData is a widely used standard for building and consuming RESTful APIs. In SAP, OData is used to expose data from various SAP applications and services as APIs that can be consumed by other applications or services. The following are some key points about OData services and APIs:

- OData services can be consumed by a wide range of client applications, such as web applications, mobile apps, and other third-party applications. SAP provides several tools and frameworks for consuming OData services, such as the SAPUI5 development toolkit for HTML5 and Cloud Integration.
- OData services in SAP support a range of operations, such as read, create, update, and delete (CRUD) operations, as well as filtering, sorting, and pagination. OData ser-

vices in SAP are secured using various authentication and authorization mechanisms, such as OAuth 2.0 and SAML.

- SAP provides several prebuilt OData services for common business scenarios, such as the Sales Order Processing OData service for sales and distribution or the Purchase Order Approval OData service for materials management.

APIs play a crucial role in enabling EDI integration by providing a structured and secure way for EDI systems to exchange data with SAP applications and leverage OData services. APIs in SAP for EDI integration enable organizations to achieve seamless data exchange, improve operational efficiency, and support digital transformation initiatives effectively. These technologies provide the flexibility, scalability, and security required to integrate SAP systems with EDI ecosystems and meet evolving business demands.

OData is an International Organization for Standardization/International Electrotechnical Commission (ISO/IEC)-approved Organization for the Advancement of Structured Information Standards (OASIS) standard that defines a set of best practices for building and consuming RESTful APIs. ISO/IEC is an international standard that manages information security. OData assists with building RESTful APIs via various approaches to define request and response headers, status codes, HTTP methods, URL conventions, media types, payload formats, query options, and so on. OData also provides guidance for tracking changes, defining functions for reusable procedures, and sending asynchronous/batch requests. Figure 2.5 illustrates these concepts.

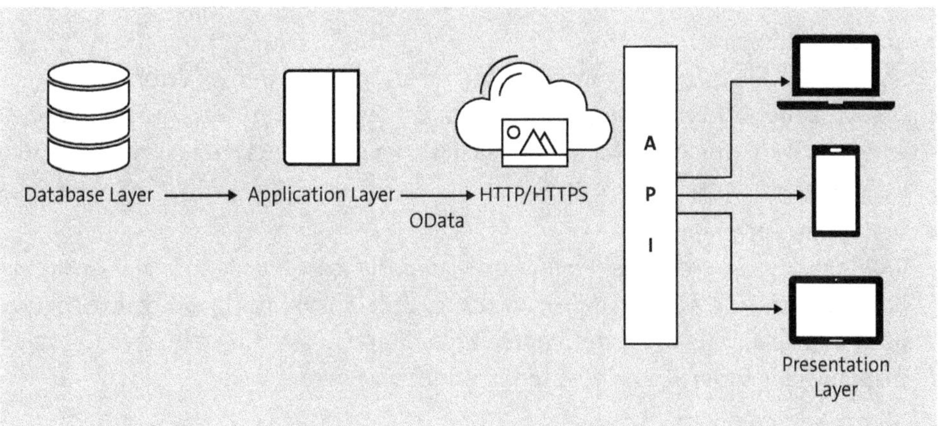

Figure 2.5 OData Services and APIs

2.3.2 SAP Application Interface Framework

There are various technologies through which customers can exchange data between an SAP application system and the rest of the world. A few of the options can use RFC and call a remote function in an external system or let the external system call a function in the SAP system. The legacy approach can use a file as an exchange medium between the applications and can choose between the SOAP protocol, OData calls, and EDI communication by sending an IDoc message directly from one system to another with the help of a combined RFC connection. As we have numerous options, it is a challenge to have control over all those interfaces, in terms of both their consistent development and their monitoring. SAP Application Interface Framework can be a solution to those challenges. SAP Application Interface Framework is an add-on that can be installed in any SAP application system. In SAP S/4HANA, it is even easier because it is already one of the standard components.

SAP Application Interface Framework is a framework that provides a set of tools and functionalities for managing interfaces inbound to or outbound from SAP systems. It enables businesses to monitor, control, and manage the exchange of data between different systems in real time.

The following are some key features of SAP Application Interface Framework:

- **Interface development**
  SAP Application Interface Framework provides tools for developing interfaces in an SAP system. It supports various exchange technologies like IDoc, SOAP, file upload, and so on.
- **Message monitoring**
  SAP Application Interface Framework enables businesses to monitor messages exchanged between SAP and external systems in real time. It provides a central monitoring cockpit that displays the status and progress of each interface.
- **Error handling**
  SAP Application Interface Framework provides tools for handling errors and exceptions that occur during interface processing. It enables businesses to define rules and workflows for error handling, such as sending alerts, triggering corrective actions, or reprocessing messages.
- **Validation and mapping**
  SAP Application Interface Framework supports data validation and mapping between different systems. It provides a structural mapping capability that enables users to map data fields between different interfaces. It also allows for creating value mapping tables for external to internal value conversions.
- **Alerting and reporting**
  SAP Application Interface Framework enables businesses to set up alerts and notifications for interface errors or exceptions. It also provides reporting tools that enable businesses to analyze interface performance and identify bottlenecks.

SAP Application Interface Framework is available in cloud and on-premise systems. For on-premise systems, SAP Application Interface Framework is available as a part of the SAP S/4HANA on-premise edition.

The on-premise version of SAP Application Interface Framework offers the flexibility to use a variety of user interfaces depending on an organization's preference. You have the option of working with SAP GUI with its customizable properties or with a ready-to-use web interface based on Web Dynpro technology.

There are three user personas that are relevant for on-premise SAP Application Interface Framework:

- **Business user**
  For a business user, the biggest advantage that you have with SAP Application Interface Framework is a feature-rich monitoring and error handling tool that needs little setup time and offers a very capable framework that monitors all interfaces. In addition, it can also allow for fixing messages without the need to involve IT specialists. This saves a lot of time in the end-to-end processing of messages and will significantly improve the overall efficiency of the error-handling capabilities of the implementation. It also can maintain specific hints for errors, customer functions, and data links, which will give an easy-to-use point of reference in case a customer encounters the same error in subsequent error handling scenarios.
- **Administrator**
  For an administrator, SAP Application Interface Framework offers tools to manage all of SAP Application Interface Framework and access advance reporting and tracing tools.
- **Developer**
  For a developer, SAP Application Interface Framework reduces the overall time spent in implementing new interfaces by providing customizations for easy interface implementation. These customizations eliminate the need for extensive coding by providing automated and step-by-step guidelines that enable implementing and reusing interface components across systems. Components like mappings, actions, and checks can be reused across interfaces, improving the efficiency of developing and deploying interfaces.

In the cloud, SAP Application Interface Framework is available as a set of apps under the **Message Monitoring** section:

- **Message Dashboard**
  This app offers a consolidated view of all the messages passing through the interfaces that support monitoring and allows you to fix errors and download the message payload to get an idea of the structure to make better decisions while handling errors. You can change the payload and restart or cancel messages based on how you want to handle the messages that are stuck in processing. In addition, if there is a dependency on any of the other apps in the SAP S/4HANA Cloud catalog that impact your messages, the dashboard app offers forward navigation to help navigate to the required app with a single click.

- **Assign Recipients to Users**
  For the administrator, the Assign Recipients to Users app enables finding and adding users to predefined recipients easily. It can also define if a technical user can see messages that are in process or that are only in an error state, and basically control which messages are shown to or hidden from specific users. You can also control the type of log messages that are displayed to the user in the message dashboard, giving you a fine level of control over the monitoring view for each user.
- **Configure Value Mapping**
  The Configure Value Mapping app offers the ability to configure value mapping for interfaces and map a single external value to multiple external values or one internal value. You also have the option of defining validity periods in case the mappings are time-bound. For example, if you need to match the data you have to an existing template from your business partner, value mapping can help configure that, letting you can match the fields from your template to your business partner's template.

Figure 2.6 shows how SAP Application Interface Framework has a local monitoring tool integrated that runs and monitors all the application interfaces that a business needs. When a message comes to SAP S/4HANA from either an SAP or non-SAP system via SAP middleware (SAP Integration Suite or SAP Process Orchestration), it is passed through SAP Application Interface Framework, which is a part of the SAP S/4HANA core component where the message is monitored. SAP integration developers can configure, for example, value mapping and data validation. Business users can look at the error messages and respond accordingly.

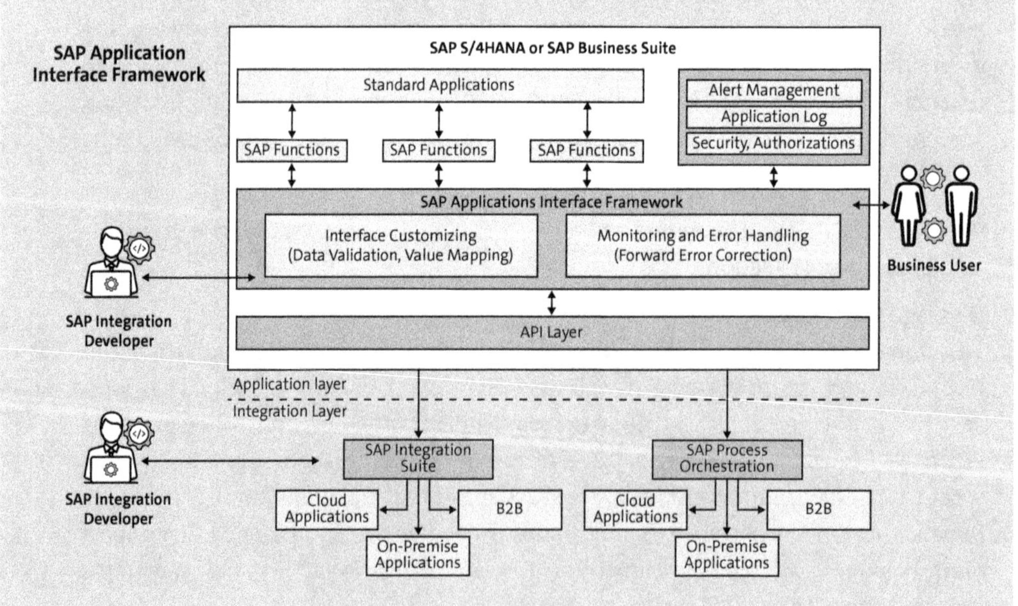

Figure 2.6 SAP Application Interface Framework: Centralized Monitoring Tool

Table 2.3 gives an overview of typical integration patterns best implemented in either the integration layer or the application layer. The combination of SAP Application Interface Framework with SAP Integration Suite has great potential. It is recommended that the integration patterns be implemented in the integration layer via the Cloud Integration capability of SAP Integration Suite.

SAP Application Interface Framework in the application layer complements SAP Integration Suite, which is the integration layer, as per the integration patterns. You can select the best fit for your implementation.

Integration Pattern	Integration Layer	Application Layer
Integration technology	SAP Integration Suite	SAP Application Interface Framework
Structure mapping	Good fit	Partial fit
Routing	Good fit	Not a fit
Splitter	Good fit	Not a fit
Aggregator	Good fit	Not a fit
Protocol adaptation	Good fit	Not a fit
Value mapping—static code list	Good fit	Partial fit
Value mapping—dynamic code list	Partial fit	Good fit
Data validation	Good fit	Good fit
Forward error correction (business user)	Partial fit	Good fit

Table 2.3 Samples of Integration Patterns

Let's take a closer look at some of these integration patterns:

- *Structure mappings* can mediate between different structures of the same format (e.g., two different XML structures) or even change the format (e.g., from JSON to XML).
- The *routing* pattern is used to determine for a given message a set of receivers based on static settings (technical routing) or message payload data (content-based routing). Logical receiver destinations define a message receiver within the sending application. The determination of logical receiver destinations within an application is usually tightly coupled with the application data and process model and, hence, is part of the application layer. An example is the configuration of the distribution model in the ABAP programming language.

- The *splitter* pattern creates multiple messages out of a received message by fragmenting the payload by a splitter rule. By then applying the routing pattern, split message parts might be routed to different receivers.
- The *aggregator* pattern collects multiple inbound messages until a modeled threshold is reached. A single payload is then constructed out of all aggregated payloads by applying an aggregation rule. Finally, this single message is forwarded to one or more receivers.
- The *protocol adaptation* pattern handles the mediation between two different protocols used by the sending and receiving applications (e.g., Java Message Service, REST, or SOAP4).

The following patterns can be implemented either in the integration layer or the application layer:

- The *value mapping* pattern transforms field values that are exchanged between two applications. Because this pattern is tightly coupled with business application data, business users must be able to maintain value mappings in a dynamic way. The value mapping pattern is typically covered by SAP Application Interface Framework as business users may detect application errors, such as incomplete value mappings, due to unforeseen values coming up. They then must extend these mappings in an agile way close to the application. As opposed to dynamic value mappings, static code-list-value mappings (e.g., country code mappings) can be best implemented in the integration layer, as they can be reused across multiple integration scenarios between various applications.
- The *data validation* pattern checks a received message with respect to schema correctness or data correctness. This pattern can be applied in the integration layer as well as in the application layer. SAP Application Interface Framework particularly supports this pattern, as it provides full access to application data and functions.
- The *forward error correction* pattern handles an application error during message processing by forwarding it to the receiver. One key capability of SAP Application Interface Framework (in the application layer) is this exact type of error correction and resolution by business users.

2.3.3 New Output Control

In SAP, *output control* refers to the process of defining, managing, and distributing various types of documents and messages, such as purchase orders, invoices, shipping documents, and payment advice, to customers, vendors, and other business partners. The primary goal of output control in SAP is to streamline the communication and document exchange process among various business entities, improve operational efficiency, and reduce costs.

The new output control functionality in SAP S/4HANA provides a modern and flexible approach to output control by introducing a new architecture that enables users to define, configure, and distribute output using a single platform. Some of the key features of new output control include the following:

- **Centralized output control**
  New output control allows users to define and manage output for various business processes centrally. This eliminates the need for multiple output types, simplifies configuration, and improves consistency across the organization.
- **Template-based output**
  With new output control, users can create templates for various output types, such as email, fax, or print. These templates can be customized to meet specific business requirements and can be reused across different business processes.
- **Interactive forms**
  New output control supports interactive forms, which enable users to create dynamic forms that can be filled out electronically. These forms can be integrated with backend systems and can be distributed via email or other channels.
- **Digital signatures**
  New output control supports digital signatures, which enable users to sign documents electronically. This improves the security of the document exchange process and eliminates the need for physical signatures.

In SAP ERP, output control uses the condition technique and is based on table NAST, containing the properties in a specific context. In SAP S/4HANA, new output control is based on BRFplus, which includes cloud qualities such as extensibility engagement and multitenancy engagement.

Table 2.4 describes the relevant form technology in relation to the output channel.

Output Channel	Form Technology
Printout	Adobe Forms using Fragments
Email	Adobe Forms
XML	SmartForms, SAPScripts

Table 2.4 Output Channels and Associated Form Technologies

Overall, new output control in SAP provides a modern and flexible approach to output control, which helps organizations streamline their communication and document exchange processes, improve efficiency, and facilitate the transformation of business documents into formats (EDIFACT, ANSI X12) required for EDI.

Figure 2.7 details the output control options available. In SAP ERP, the message determination was based on table NAST. BRFplus replaces this in SAP S/4HANA. The target architecture is now based on Adobe Document Server and Adobe Forms only.

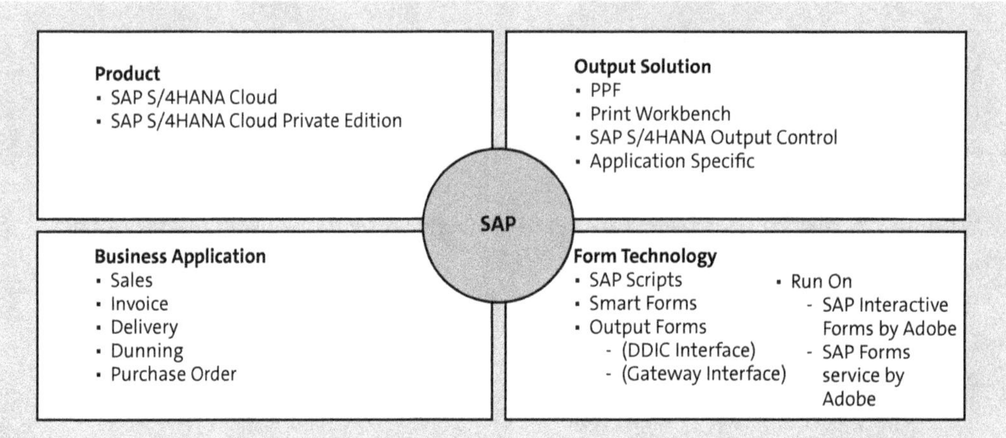

Figure 2.7 Output Control: Available Options across All SAP S/4HANA Editions

Can I Use Both the Old and New SAP S/4HANA Output Control Frameworks Simultaneously?

You cannot use both frameworks simultaneously for an application object, but you can use the old output control framework for maintenance orders and the new SAP S/4HANA output control framework for maintenance notifications.

2.3.4 Business Rules Framework Plus

BRFplus is a powerful tool in SAP S/4HANA that enables organizations to create and manage business rules in a flexible and efficient manner. BRFplus provides a central repository for business rules that can be used across different business processes and applications. It acts as the interface between the caller program and the business rules logic implemented within the rule.

BRFplus can be used as a simple rules engine handling value mappings or verification based on business logic. In such scenarios, it can be provided with one or a couple of values to determine a single output value, usually by transforming external to internal values (or the other way around). It can also perform basic checks like verifying the length, prefix, or structure of a value provided in the BRFplus function input. An important fact to remember is that BRFplus objects are reusable and can call one another. To achieve a simple transformation or complex logic, the same objects in BRFplus are used. There will always be functions and expressions. A *function* is an interface provider for BRFplus logic, and *expressions* are the heart of BRFplus, running all types of logic required by the business. To make things simple for simpler use cases, BRFplus introduced top expressions for functions running in functional mode. In this mode, a function is focused only on providing an interface to a single expression. As mentioned, BRFplus objects can also call one another, so functional mode is also a very efficient way to encapsulate some standalone logic to be reused by many other BRFplus objects.

Figure 2.8 gives a glimpse of BRFplus creating a function. Creating a function in BRFplus involves defining the business logic in a structured manner using input and output parameters, implementing the logic through various constructs, testing, and finally activating the function for use within SAP applications. This framework allows organizations to manage business rules effectively, enhancing agility and responsiveness to changing business needs.

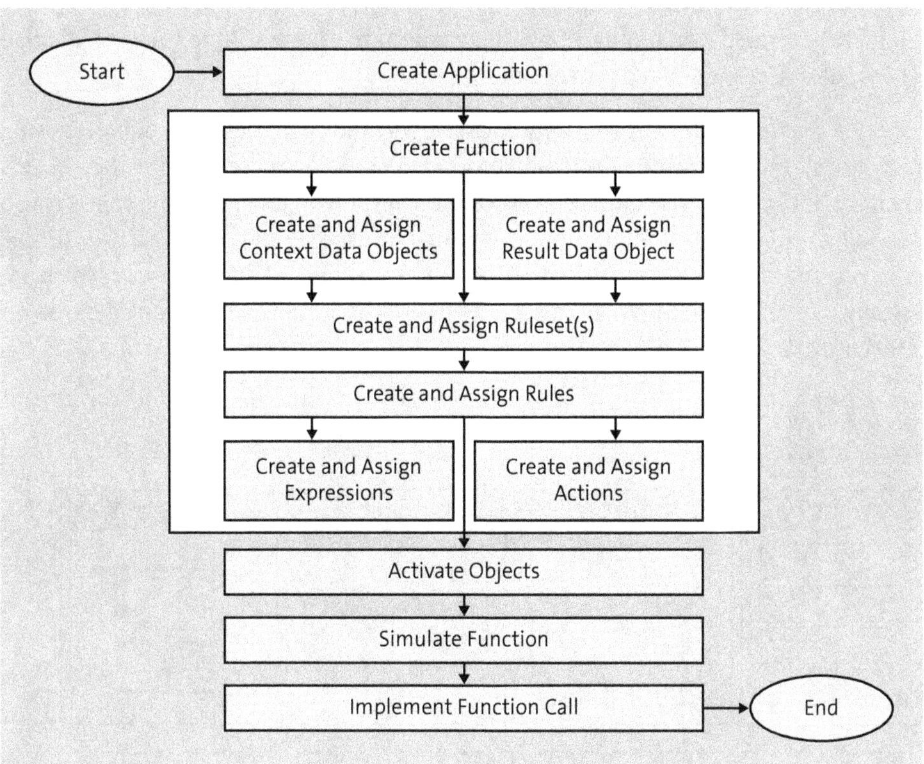

Figure 2.8 BRFplus Creation Steps

Some of the key features of BRFplus in SAP S/4HANA include the following:

- **User-friendly interface**
  BRFplus provides a user-friendly interface that allows business analysts and subject matter experts to create and manage business rules without the need for programming skills.

- **Flexible rule modeling**
  BRFplus supports various types of rule modeling, including decision tables, decision trees, formulas, and expressions. This enables users to choose the best modeling approach for their business requirements.
- **Integration with SAP systems**
  BRFplus can be integrated with other SAP systems, such as SAP ERP and SAP CRM, enabling users to reuse business rules across different applications and business processes.
- **Customizable rule execution**
  BRFplus enables users to customize the execution of business rules based on their specific requirements. This includes the ability to define rule sets, rule chains, and rule priorities.
- **Support for versioning**
  BRFplus supports versioning, enabling users to manage multiple versions of business rules and track changes over time.

Figure 2.9 shows BRFplus with a comprehensive API and UI for defining and processing business rules. The figure shows that decision tables are going to hold the data, which consists of all possible inputs and corresponding output values. A decision tree can also be used instead of decision tables, but this would be used when you have a very simple yes or no type of validation. Inside the functions/formulas, we have *rule sets*, which are basically containers that hold all the sets of rules. These rules are processed based on the decision tables.

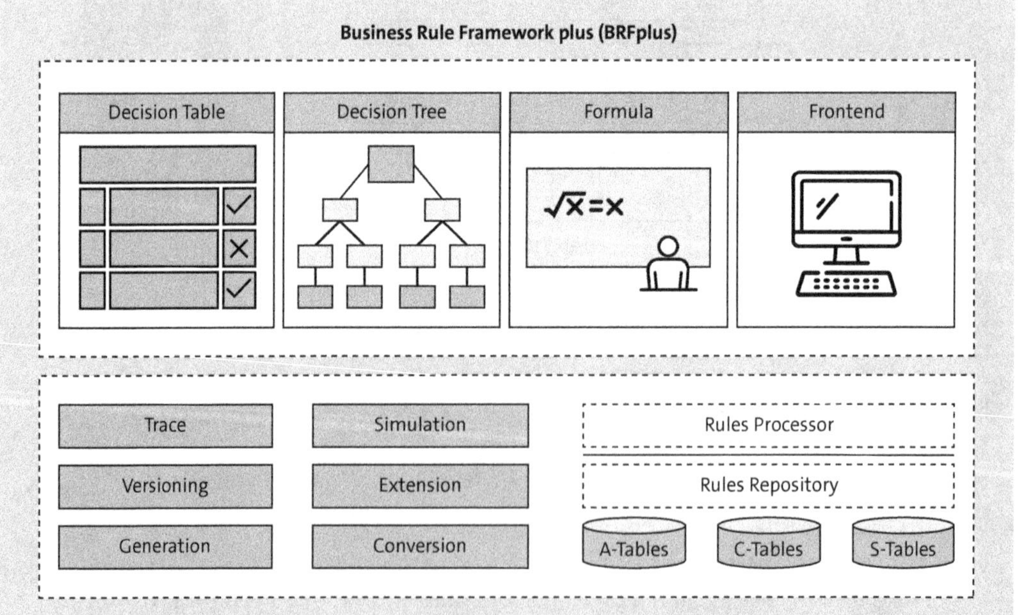

Figure 2.9 Components of Business Rules Framework Plus

2.4 SAP Business Technology Platform

SAP BTP is a cloud-based platform offered by SAP that enables organizations to build, integrate, and extend business applications and services. It provides a range of integration services that enable users to connect different systems and data sources, including SAP and non-SAP systems.

Integration is a key pillar of SAP BTP, so SAP BTP offers SAP Integration Suite as an integration tool for seamlessly integrating SAP and non-SAP solutions. SAP Integration Suite provides prebuilt adapters for EDI standards such as EDIFACT, ANSI X12, and AS2, simplifying connectivity and data exchange among systems. The API Management capabilities in SAP Integration Suite enable secure and governed access to APIs, facilitating integration with EDI systems and external partners. (For more details on configuring SAP Integration Suite, see Chapter 9.)

In this section, we'll provide a high-level overview of SAP BTP's integration technologies: SAP Integration Suite and the cloud connector.

2.4.1 SAP Integration Suite

The modern era's IT landscapes are increasingly distributed across corporate data centers and multiple clouds with business processes supported by heterogeneous applications and composable modules. These must all exchange data with other systems, making automation and interactions challenging. Plus, there is a growing need to collaborate with customers and partners across value chains to avert supply chain disruption, increase efficiency, and improve sustainability.

These business objectives make connecting internal and external systems a top priority. With SAP Integration Suite, you can integrate solutions inside and outside your data center and can accelerate connectivity with prebuilt integration flows, connectors to hundreds of systems, thousands of APIs, and other capabilities that simplify integrations with SAP and third-party applications. As shown in Figure 2.10, connecting people in real time becomes a reality with event-driven architectures and intelligent technology that helps organizations automate comprehensive business processes to gain efficiency and visibility.

SAP Integration Suite is a cloud-based solution that provides a set of integration tools and services to connect SAP systems with other applications and services. It enables seamless integration between cloud-to-cloud, cloud-to-on-premise, and on-premise-to-on-premise applications, allowing organizations to streamline their business processes and data exchange. It is like a proxy server that supports the integration of processes between SAP and non-SAP systems, using open standards and modern integration technologies (web services, Groovy, REST, XML, JSON, etc.)

Figure 2.10 shows that SAP Integration Suite is the strategic integration service within SAP BTP. It facilitates the integration of business processes that span different departments, organizations, or companies.

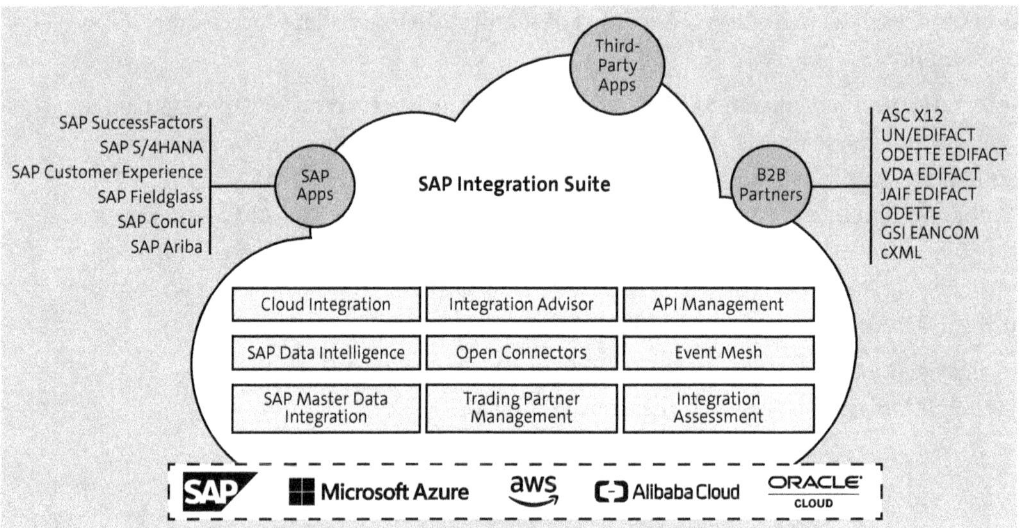

Figure 2.10 SAP Integration Suite, Part of SAP BTP

In the following sections, we'll walk through the key components, use cases, and integration approaches of SAP Integration Suite.

Key Components and Use Cases

SAP Integration Suite consists of the following key components, as shown in Figure 2.11:

- **Cloud Integration**
  Cloud Integration is one of the central components of SAP Integration Suite. As businesses move to the cloud, it is crucial for most elements to be unified. The organization must integrate different applications and systems into their hybrid landscape, maintained and developed on diverse technology stacks with varying standards of security and interface specifications. Cloud Integration allows customers to connect the applications seamlessly and quickly, irrespective of the landscape.
- **API Management**
  This tool enables you to plan, develop, and manage your application's interfaces from a central point. In principle, you can use application interfaces without this tool, but when implementing comprehensive security policies, monitoring data traffic, monetizing calls, or handling errors, you need the support of API Management.

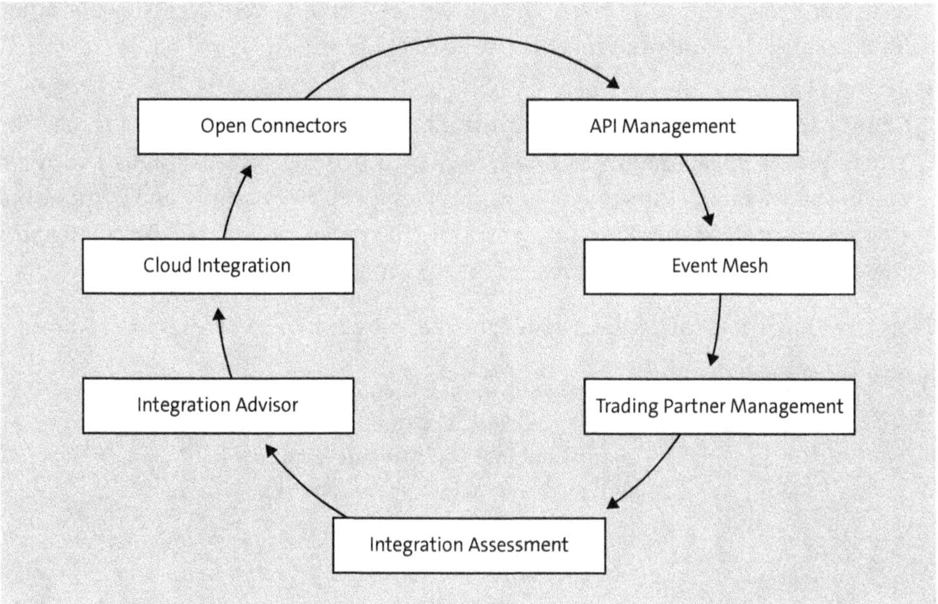

Figure 2.11 SAP Integration Suite and Its Components

- **Open Connectors**
  Cloud-based applications have APIs defined by their providers. The APIs of applications will differ significantly in their supported technologies and authentication methods, among other parameters. Open Connectors make the APIs of diverse cloud-based applications available for the organization uniformly via a combined REST API layer. With this, you will have normalized error handling, connectivity mechanisms, and authentication for your applications.
- **Integration Advisor**
  Most of the work in integration relates to the mapping of business interfaces executed by a technical developer and functional consultant. This is a lengthy and time-consuming process. The Integration Advisor tool in SAP Integration Suite accelerates the mapping and development of business interfaces through its use of EDI mappings. When developers create a new mapping, the knowledge from all previously created mappings is used, so the new mapping is automatically generated.
- **Integration Assessment**
  This tool focuses on data orchestration and integration. The addition of machine learning in the latest versions of SAP Integration Suite enables more focus on the operationalization of machine learning and data science to ease data clustering, insights, and regression. Data Intelligence also supports SAP Analytics Cloud and SAP Data Warehousing Cloud in data integration.
- **Event Mesh**
  This tool handles all event-driven communication in SAP Integration Suite. In event-driven communication, applications will send brief messages to event brokers when

a specific event occurs. The event broker then makes this event accessible for other applications to decide which events affect them.

- **Trading Partner Management**
  Trading Partner Management in SAP Integration Suite is a feature designed to facilitate seamless collaboration and data exchange between organizations and their trading partners. SAP Integration Suite provides the necessary tools and frameworks to manage interactions effectively, ensuring that organizations can easily integrate their business processes with those of their partners.

Table 2.5 summarizes the services and high-level use cases of SAP Integration Suite.

Cloud Integration	End-to-end application-to-application (A2A) process integrations, AI-accelerated B2B and EDI integrations, and business to-government (B2G) integrations, with over 2,000 prebuilt integrations delivered as integration packs
API Management	Full lifecycle management of enterprise APIs (design, develop, manage, and developer engagement), with support for customers launching their own API domains and running enterprise API programs
Integration Advisor	AI-based mapping proposals to accelerate onboarding of trading partners through accelerated mappings, with support for EDI metadata management and B2B library support
Open Connectors	Over 150 open connectors to simplify third-party connectivity through one to-many integrations with various integration hubs such as SAP CRM, SAP ERP, cloud storage, and social media
Event Mesh	Event-based integrations with prepackaged events from SAP applications such as SAP S/4HANA and the SAP Customer Experience and SAP SuccessFactors solutions
Trading Partner Management	Creation and maintenance of B2B scenarios
Integration Assessment	Shaping and execution of your integration technology strategy following the SAP Integration Solution Advisory Methodology

Table 2.5 SAP Integration Suite: Services and Use Cases

The robust functionality in SAP Integration Suite means customers can confidently integrate and manage their mission-critical processes across SAP and third-party applications in complex and distributed landscapes. Enterprise-grade capabilities include 99.95% uptime, multizone availability, failover prevention, and elastic scaling for high performance and throughput.

Innovating with modular functionalities becomes easier as SAP Integration Suite offers capabilities for composability, real-time event management, and API management.

Organizations can also use a centralized hub (SAP Business Accelerator Hub) to access over 3,800 APIs from SAP and its partners that help companies innovate faster. SAP Integration Suite offers over 3,400 ready-to-go integrations and connectors for both SAP and third-party solutions

Real-world problems often span multiple lines of business, services, and APIs, and it can be a challenge for even the most experienced developers to understand the relationships among them. With SAP Integration Suite, companies can take advantage of data semantics in SAP One Domain Model defined for business objects such as the product, customer, or order information. Organizations can also understand the business context in which they are used. Meanwhile, with other capabilities in SAP Integration Suite, such as Graph, and SAP Business Accelerator Hub, you can quickly explore, test-drive, and document APIs to access SAP-managed data, regardless of where it resides.

Integration Approaches

SAP Integration Suite also includes various prebuilt integration content packages, such as integration flows, templates, and adapters, which can be used to accelerate integration projects and reduce the time and effort required to build integrations. A few of the integration approaches from SAP Integration Suite are as follows:

- *Application-to-application integration* integrates the end-to-end processes across on-premise or cloud and non-SAP/SAP solutions. A2A integration uses prepackaged integrations and over 100 open connectors. These prebuilt connectors can be used to connect to systems and services without the need for custom coding.
- *B2B integration* is a traditional EDI and API-based real-time integration to make collaboration between businesses possible. SAP Integration Suite will support emerging settings with blockchain to facilitate trustful business network relationships.
- *B2G integration* is meant for businesses that work electronically with government agencies. Many government agencies enact e-government initiatives to drive digital interactions between the government and its users. SAP Integration Suite has prepackaged integrations for organizations to ease compliance with regulations.
- Real-time *API-based integration* allows businesses using APIs to generate omnichannel access while hiding the complexity of heterogeneous landscapes. A real-time API-based integration will dramatically improve customer experiences while supporting channel access as it opens processes and data for a beneficial experience on multiple channels.
- *Event-driven integration* supports decoupled connections across different applications, people, and IoT devices. Event-driven integration will enable a sense-and-respond pattern with enterprise messaging and a function-as-a-service without a server.

- *IoT integration* connects machines, assets, and devices to your core business processes to support real-time supply chains, digital manufacturing, and predictive maintenance. SAP Integration Suite has device connectivity, onboarding, and edge processing along with a well-defined digital twin foundation to support IoT integration.
- *Orchestration* is a key integration approach that builds process-based digital applications using business rules and workflows. It supports custom process applications and integration scenarios seamlessly.

2.4.2 Cloud Connector

The cloud connector is a software component that enables secure and reliable communication between on-premise systems and SAP BTP or other cloud-based applications. It establishes a secure tunnel between the on-premise system and the cloud-based application, allowing the cloud-based application to access data and services from the on-premise system.

The cloud connector is designed to work with SAP BTP, but it can also be used to connect on-premise systems to other cloud-based applications, such as Salesforce or Workday. It provides a secure connection without the need to open additional ports in the firewall, reducing the risk of unauthorized access and data breaches.

The cloud connector runs as a lightweight agent on a computer inside the organization's firewall and establishes a secure connection to SAP BTP over the internet. It also supports different protocols and standards, such as HTTPS, TLS, and OAuth, to ensure secure communication.

Overall, the cloud connector provides a secure, reliable, and scalable way to connect on-premise systems with cloud-based applications, enabling organizations to leverage the benefits of cloud-based solutions while protecting their valuable data and assets.

Figure 2.12 illustrates how the cloud connector can propagate the identities of cloud users to on-premise systems in a secure way. The cloud connector serves as a link between SAP BTP applications and on-premise systems. It combines easy setup with clear configuration of the systems that are exposed to SAP BTP.

Cloud Connector

The cloud connector propagates the identities of cloud users to on-premise systems in a secure way.

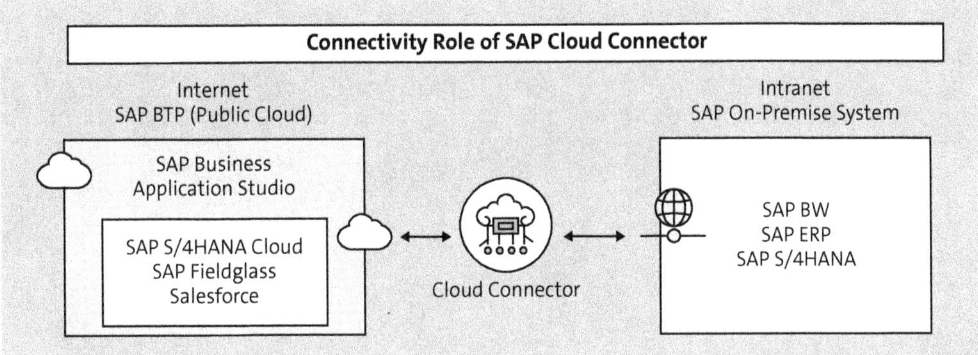

Figure 2.12 Cloud Connector Linking Internet and Intranet Systems

2.5 Summary

In this chapter, you learned about the integration technologies that are available with SAP S/4HANA and the integration strategy, emphasizing the use of modern technologies with enhanced capabilities such as OData Services and APIs, SAP Application Interface Framework, the new output control, and BRFplus. We also looked at integration capabilities with SAP Integration Suite such as Integration Assessment, Cloud Integration, Integration Advisor, Trading Partner Management, API Management, SAP API Business Hub, Open Connectors, and SAP Event Mesh. Finally, we discussed the cloud connector and its benefits when linking on-premise systems with cloud solutions.

In the next chapter, we will introduce electronic data interchange and explore standards for messages and communication.

3

Chapter 3
Electronic Data Interchange

This chapter introduces EDI. It begins by outlining the EDI concept and characteristics and exploring the EDI standards for messages and communication, with an emphasis on EDI messages such as UN/EDIFACT, ASC X12, and GS1 XML, involving EDI communications such as VAN, SFTP, and AS2. We also delve into the characteristics of the EDI integration pattern and security.

Electronic data interchange (EDI) is a system that allows the exchange of business documents and information electronically, between different computer systems or applications, without human intervention. EDI is a fast and secure method of exchanging business information, reducing the need for paper-based processes and manual data entry.

This chapter serves as an introductory guide to EDI. It begins by defining EDI and outlining its key components, emphasizing its role in facilitating electronic communication of business documents between trading partners. The chapter explores the benefits and limitations of EDI, highlighting its efficiency in reducing manual processes while addressing potential challenges such as implementation complexity.

A significant portion of the chapter is dedicated to EDI standards for messages and communication protocols. It details prominent message standards including UN/EDIFACT, ASC X12, GS1 XML, and their common elements. Special focus is placed on UN/EDIFACT, delving into its syntax, semantics, and specific transaction sets relevant to sales and distribution processes.

This chapter also discusses EDI standards for communication, covering the value-added network (VAN), SSH File Transfer Protocol (SFTP), and Applicability Statement 2 (AS2). It explores AS2 in detail, highlighting its role in secure and reliable data transmission between business partners, along with additional security considerations crucial for EDI implementations. The chapter concludes by summarizing the characteristics of the EDI integration pattern, underscoring its importance in enabling seamless electronic data exchange across different organizational systems.

3.1 Concept and Characteristics of EDI

EDI streamlines the business document process electronically, without the need to print, sign, fax, or email paperwork. Instead documents flow through the system from the buyer to the supplier and back again; for example, a purchase order from the buyer goes to the supplier, and then an invoice is sent back to the buyer.

In this section, we'll discuss this process flow and look at some of the key components, benefits, and limitations of EDI, as well as how it integrates with SAP S/4HANA.

3.1.1 EDI Process Flow

Let's start with an overview of how EDI works. Some of the key characteristics of EDI include the following:

- **Standardization**
  EDI uses standardized message formats, such as ANSI X12, EDIFACT, or XML, which enable businesses to exchange information in a uniform way, regardless of the systems or applications used.
- **Automation**
  EDI enables the automation of business processes, including the creation, transmission, and receipt of business documents, reducing the need for manual data entry and paper-based processes.
- **Speed and accuracy**
  EDI enables the fast and accurate exchange of business information, reducing the time requirements and errors associated with manual data entry and paper-based processes.
- **Security**
  EDI provides secure transmission of business information, using encryption and other security measures to protect data confidentiality and integrity.
- **Cost savings**
  EDI can help organizations save costs associated with paper-based processes, such as printing, mailing, and storage, as well as reduce errors and improve operational efficiency.

Typically, the most common business documents exchanged through EDI are purchase orders, invoices, and advance shipping notifications (ASNs), but there are many other business documents that can also be exchanged, such as paperwork relating to logistics.

As EDI documents are processed by computers, there must be a standardized format so that the computer can understand the language and read the documents. This is standardized across the EDI system, rather than being set by each company, and it describes each piece of information and what the format is. When businesses exchange documents through EDI, they must first agree on which standard and version they will use—for example, ANSI 5010 or EDIFACT version D12. To allow the processing of the business documents, usually either in-house software or a service provider allows for translation of the EDI documents. The exchange of these EDI documents occurs via two different companies, which are called *business partners* or *trading partners*.

Before we go further into detail about EDI, let's review the traditional way of processing a purchase order and invoice between a customer and supplier, as shown in Figure 3.1. Here, the purchase order is created manually via a phone call and clarifying if any document is sent. No standard purchase order template is maintained, apart from the customer and seller's own systems and data formats to process their requirements, so manual intervention is required in this time-consuming process.

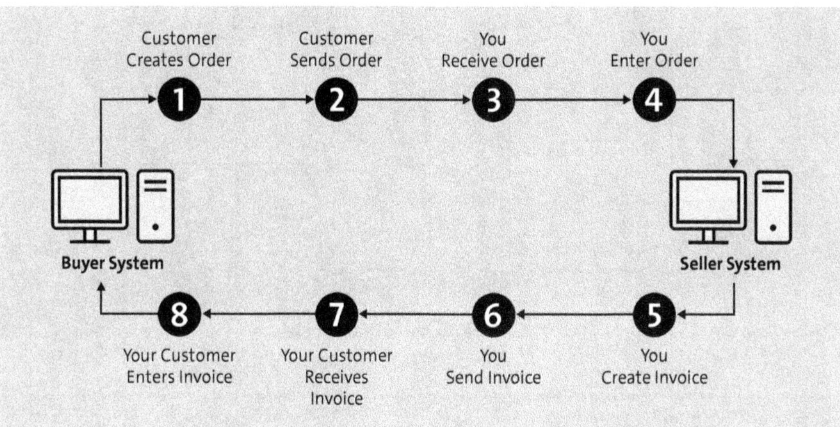

Figure 3.1 High-Level Diagram: Traditional Way of Buying and Selling

In contrast to Figure 3.1, Figure 3.2 shows the basic EDI process flow, where the buyer sends the purchase order to the supplier, indicating types, quantities, and agreed-upon prices for products or services required. In return, an invoice is sent from the supplier to a buyer for a sales transaction, indicating the products, quantities, and agreed-upon prices for products or services.

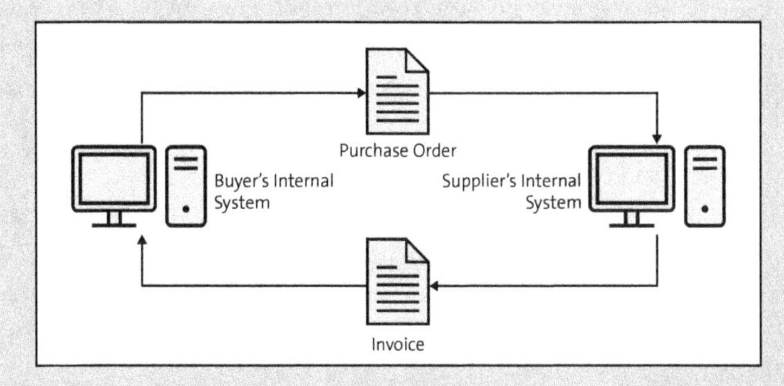

Figure 3.2 Purchase Order and Invoice Exchange between Business Partners

Figure 3.3 shows where the EDI transmission is involved during the purchase order and invoice process between the buyer and supplier. A buyer creates a purchase order in their format and sends it across to the EDI translator to convert into EDI purchase order format 850, which is then sent across the EDI network to the seller, where the EDI translator converts the purchase order into their own format that their internal system can accept. After processing the purchase order, the seller sends the invoice in their format, which is converted into EDI format and sent across to the buyer; the EDI translator on the receiver side then converts invoice 810 into an acceptable internal format in the buyer's system.

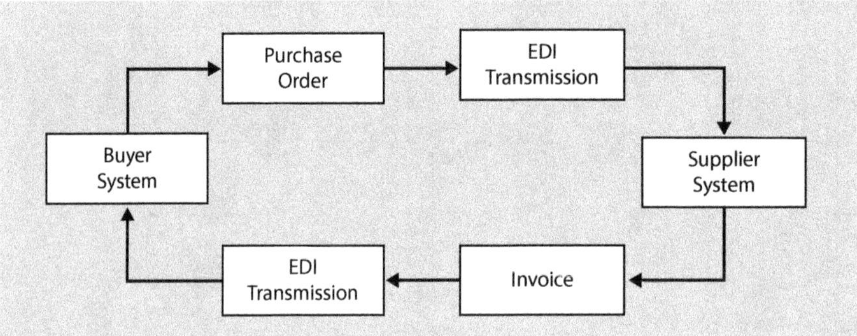

Figure 3.3 Business Document Exchange Using EDI Transmission

EDI Basic Structure

An EDI data structure resembles layers of envelopes. The envelopes separate different types of data and carry the sender and receiver address information. A receiver uses these envelopes to determine if all of the data has been received.

There are three steps to send EDI documents—preparing the document, translating the document into EDI format, and transmitting the EDI document to respective partners:

1. **Prepare the document**
   The first step is to collect and organize the data. The sources of data and the methods available to generate the electronic documents can be shared in different ways, like entering the data manually in the system, exporting the data, or using any application for the file.
2. **Translate the document into EDI format**
   The second step is to convert the internal data format into the EDI standard format using the appropriate segments and data elements. This can be achieved by EDI translation software or by mapping to the EDI data. Alternatively, there are translation services for EDI format that handle translation to and from the EDI format on an organization's behalf.

3. **Transmit the EDI document to respective partners**
   After the data is converted to the appropriate EDI format, it is ready to be transmitted to business partners. For this a channel is required, and there are several options: to connect directly using AS2 or another secure internet protocol; to connect to an EDI network provider (also referred to as a VAN provider) using your preferred communications protocol and relying on the network provider to connect to your business partners using whatever communications protocol your partners prefer; or a combination of both, depending on the particular partner and the volume of transactions you expect to exchange.

Figure 3.4 shows the buyer using a EDI translator to convert his internal format document to standard EDI. By the assistance of EDI transmission, the purchase order in EDI format is sent to the seller, where the seller uses an EDI translator to convert the document into his required format using an EDI translator so that the seller can process the received purchase order in his own format.

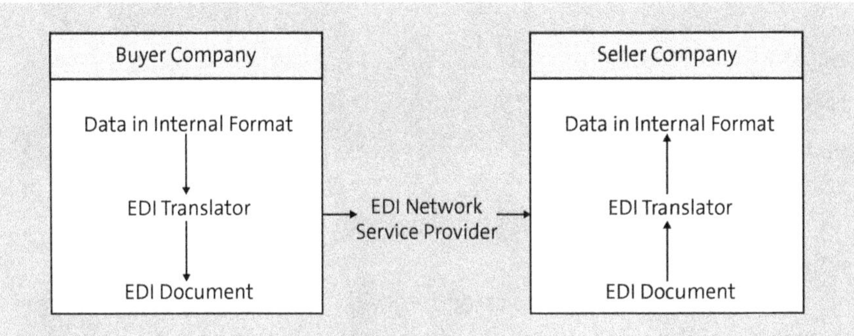

Figure 3.4 Logical Flow between Buyer and Seller Using EDI Network Service Provider

3.1.2 Key Components

The key components of EDI include the following:

- **Standards**
  EDI standards define the format and structure of electronic documents exchanged between trading partners. Some of the commonly used EDI standards include ANSI X12, EDIFACT, TRADACOMS, and XML.
- **Communication protocols**
  EDI requires communication protocols to enable the exchange of electronic documents between different computer systems or applications. Some of the commonly used communication protocols for EDI include AS2, FTP, and VAN.
- **Translation software**
  EDI documents must be translated from the standard format used by the sender to the standard format used by the receiver. This is typically done using EDI translation software, which converts the data into the appropriate format for each trading partner.

- **Mapping software**
  EDI mapping software is used to transform data between different formats, allowing trading partners to exchange information using their own internal formats.
- **Security software**
  EDI documents contain sensitive business information and require secure transmission. EDI security software, including encryption and digital signatures, is used to protect the confidentiality and integrity of the data being exchanged.
- **Business process automation**
  EDI enables the automation of business processes, including the creation, transmission, and receipt of electronic documents, reducing the need for manual data entry and paper-based processes.

Overall, the key components of EDI work together to enable the electronic exchange of business documents and information, providing standardization, secure communication, translation and mapping, and business process automation.

3.1.3 Benefits and Limitations

EDI offers several benefits for organizations, including the following:

- **Improved efficiency**
  EDI streamlines business processes by automating data entry and document processing, reducing the time and errors associated with manual processing.
- **Cost savings**
  EDI can help organizations save costs associated with paper-based processes, such as printing, mailing, and storage, as well as reduce errors and improve operational efficiency.
- **Improved accuracy**
  EDI eliminates manual data entry, reducing errors and improving data accuracy.
- **Faster response times**
  EDI enables faster communication and response times between trading partners, leading to improved customer satisfaction and operational efficiency.
- **Increased visibility**
  EDI provides greater visibility into the status of orders and transactions, enabling better supply chain management and forecasting.
- **Scalability**
  EDI systems are scalable and can handle large volumes of transactions efficiently. As businesses grow, EDI can easily accommodate increased data exchange requirements without significant upgrades or changes.
- **Compliance**
  Many industries and trading partners require compliance with a standard for business transactions. Implementing EDI ensures organizations meet regulatory and industry standards, ensuring smooth business operations.

- **Strategic benefits**
  All transaction statuses are available in real time via EDI, enabling customers to adjust to changing consumer demands and market trends. This also gives companies the ability to speed up the supply of new products and enhance existing products.

However, EDI also has some limitations:

- **High initial investment**
  Implementing EDI requires significant upfront investment in hardware, software, and training.
- **Complexity**
  EDI can be complex, requiring specialized knowledge and skills to implement and maintain.
- **Limited flexibility**
  EDI requires a standardized format for data exchange, which may limit the flexibility of organizations to customize their processes.
- **Security concerns**
  EDI documents contain sensitive business information and require secure transmission. Security risks such as data breaches or cyberattacks can compromise the confidentiality and integrity of the data.
- **Limited adoption**
  Despite the benefits, EDI adoption can be limited, particularly among small and medium-sized enterprises that may lack the resources and expertise to implement and maintain an EDI system.

Organizations need to weigh the benefits against the potential limitations and risks before implementing an EDI system.

3.1.4 EDI Integration with SAP S/4HANA

EDI is a multitude of standards for electronic communication between organizations that enable them to not only integrate but also automate their business processes through the use of various EDI transactions, such as purchase orders, purchase order acknowledgments, purchase order changes, ASNs, invoices (810), and so on. These EDI transactions are integrated with SAP S/4HANA to automate business processes in an organization.

Let us consider a scenario where a business partner would like to send an EDI message to SAP S/4HANA for a business process. Per SAP, the best practice is integrating SAP S/4HANA via SAP Integration Suite, which is a service in SAP BTP.

A *business partner* is a non-SAP system that receives or sends an EDI message. To set up a connection with a business partner is to engage with the partner to understand three main things. The first is how to connect to their organization. This is the connectivity aspect that leverages SFTP, FTP, FTPS, VAN, or a direct encrypted connection using AS2.

Second, you need to understand the EDI standard that your business partners require or would like to exchange. Finally, you need to understand and agree to the EDI message types, also known as EDI transactions, with your partner. The key is understanding which partner dictates what EDI transactions, protocols, and standards to use. EDI allows businesses to exchange business documents such as purchase orders, invoices, ASNs, and other transactional information electronically in a standardized format.

Figure 3.5 shows how a non-SAP business partner sends an EDI message to a business running SAP via standard communication channels to SAP integration tools, where the EDI messages are transformed to the IDoc format, SAP's standard document type, to be processed internally in the SAP system.

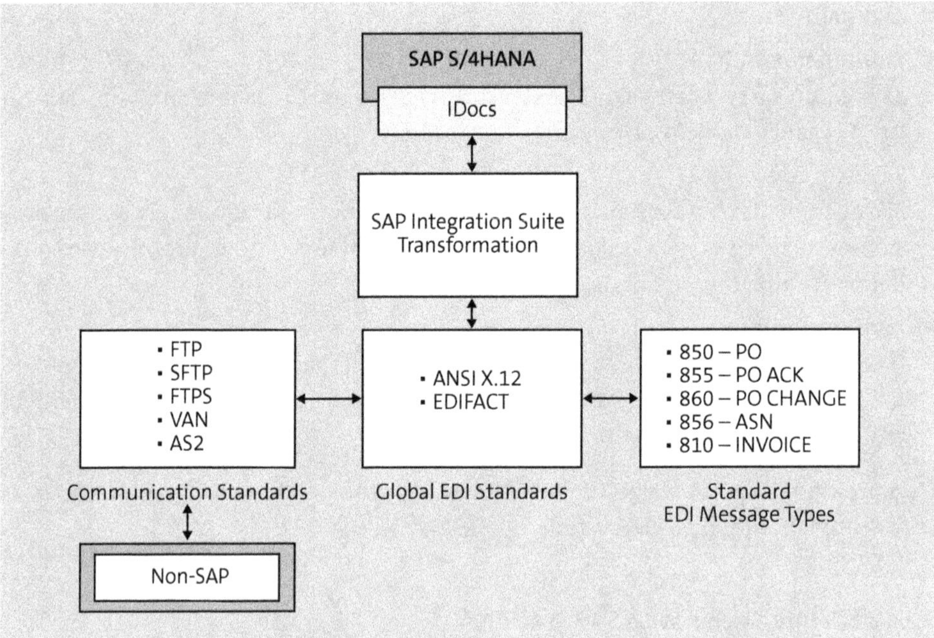

Figure 3.5 Integrating SAP S/4HANA with EDI Business Partners

API Options for EDIFACT and ANSI X.12

The option of integrating via APIs directly in SAP S/4HANA isn't as effective and does not perform as well when we look at it in terms of transactional volume.

3.2 EDI Standards for Messages

There are several EDI standards for messages, which define the format and structure of electronic documents exchanged between trading partners. These standards ensure that data is exchanged accurately, efficiently, and consistently across various industries and ensure all parties involved understand the data being transmitted, thus streamlining processes and reducing errors.

An EDI standard builds on the following four basic principles:

- **Syntax rules**
  The syntax rules define the allowed characters and the allowed order in which the individual characters may be used.
- **Codes**
  Within an EDI file, most information is accurately identified using codes—for example, currency codes, country codes, codes for identifying a particular date format, and so on.
- **Message design**
  The message design defines the structure of a particular message type. Message types include, for example, purchase orders, delivery notes, invoices, and so on.
- **Identification of values in an EDI message**
  Depending on the standard used, there are three ways in which a value can be identified in an EDI file:
  - **Implicitly by the position in the message**
    This technique is used in flat file formats and CSV files. The exact position and semantics of a given value is defined in the accompanying documentation. For example, a line beginning with the characters 100 stands for the header line, the characters from positions 4–17 in a line beginning with 100 represent the GLN of the sender, and so on.
  - **Implicitly through the use of separators**
    Using a set of predefined separator characters, a message structure is explicitly defined. The message structure defines which building blocks are to be used and in which order the building blocks must be aligned to form a correct EDI message, such as an orders message. This technique is used in EDIFACT files.
  - **Explicitly through the application of metadata**
    With the help of additional data, the meaning of the individual information fragments in a file is specified more precisely. This technique is used in XML, for example, where the actual information is enclosed by means of markup elements and attributes. Some of the commonly used EDI standards for different types of messages include ANSI X12, EDIFACT, TRADACOMS, HIPAA, SWIFT, GS1 XML, and Common Parts EDI. The choice of EDI standard depends on factors like the industry, trading partners, and geographical location.

In the following sections, we will look at the main standards for UN/EDIFACT, ASC X12, and GS1 XML.

3.2.1 Main Standards

There are different EDI standards that govern the electronic exchange of business documents between trading or business partners. Let's walk through each of the main standards using the example in Figure 3.6, which shows a basic sample purchase order in a paper format.

PURCHASE ORDER

XYZ Company
1 Carole Drive,
Sydney, Australia, 2110

PO # 1983
PO Date: 08-01-2017

Item #	Quantity	Unit of Measure	Price	Product ID
1	10	EA	30.00	2061979B

Total Items : 1

Total Quantity :10

Figure 3.6 Purchase Order in Paper Format

This purchase order can be converted into the following standard formats:

- **UN/EDIFACT**
  Developed by the United Nations, EDIFACT is a widely used EDI standard in Europe, used for the exchange of various types of business documents such as orders, invoices, and shipping notices. The standardization organization behind UN/EDIFACT is the United Nations Center for Trade Facilitation and Electronic Business (UN/CEFACT). UN/EDIFACT is one of the most widely used EDI standards today. Figure 3.7 shows how the paper format purchase order looks converted into an EDI purchase order format.

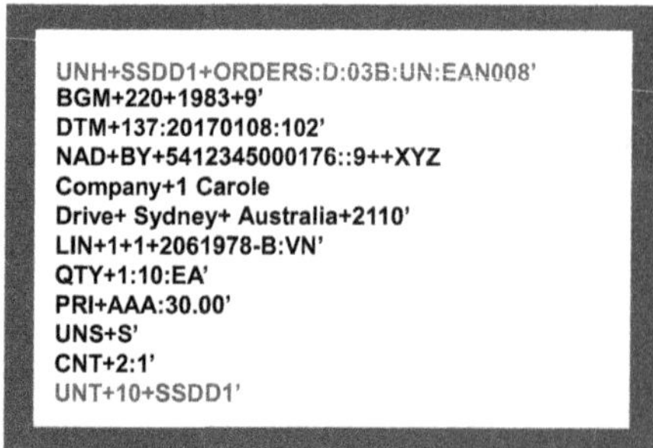

```
UNH+SSDD1+ORDERS:D:03B:UN:EAN008'
BGM+220+1983+9'
DTM+137:20170108:102'
NAD+BY+5412345000176::9++XYZ
Company+1 Carole
Drive+ Sydney+ Australia+2110'
LIN+1+1+2061978-B:VN'
QTY+1:10:EA'
PRI+AAA:30.00'
UNS+S'
CNT+2:1'
UNT+10+SSDD1'
```

Figure 3.7 Purchase Order in EDI Format

- **ASC X12**
  Developed by the American National Standards Institute (ANSI), ASC X12 is a widely used EDI standard in North America, used for the exchange of various types of business documents such as purchase orders, invoices, and shipment notices. (The ASC X12 standard is sometimes called EDI ASC X12, ANSI X12, ANSI ASC X12, or simply the X12 standard. X12 is a set of standards and rules that determine a specific syntax for structuring and transferring electronically business documents between partners.)

  The ASC X12 standard defines the sequence of segments in the transaction set and also the sequence of elements within each segment. The relationship between segments and elements can be compared to the relationship between records and fields in a database environment. Each transaction set contains a transaction set header (designated ST), a transaction set trailer (designated SE), and a single message, enveloped within the header and trailer. Figure 3.8 shows how the paper format purchase order looks converted into an ASC X12 purchase order format.

```
ST*850*540001
BEG*00*SA*1983*65*20170108
NI*SO*XYZ Company
N3*1 Carole Drive
N4*Sydney*Australia*2110
PO1*1*10*EA*30.00**VN*2061979-B
CTT*1*10
SE*8*54001
```

Figure 3.8 Purchase Order in ASC X12 Format

- **GS1 EDI**
  GS1 EDI provides global standards for electronic business messaging that allow automatic electronic transmission of agreed-upon business data between trading partners. This automation ensures that the exchange is done in a rapid, efficient, and accurate manner.

  In the GS1 system, GS1 EDI is composed of three components: GS1 EANCOM, GS1 XML, and UN/CEFACT XML. Let's look at these in more detail:

  - GS1 EANCOM takes UN/EDIFACT standards, which are internationally agreed-upon standards to facilitate the electronic interchange of data, and focuses them on the trade and transport sectors. It provides clear definitions, explanations, and examples so that trading partners can exchange commercial documents in a simple, accurate, and cost-effective manner. As well as using UN/EDIFACT standards, GS1 EANCOM also incorporates GS1 standards, like the physical identification of trade items (GTINs), logistics units (SSCC), and Global Location Numbers (GLNs), to further facilitate the exchange of information in the supply chain.
  - GS1 XML is a standardized electronic data exchange format developed by GS1. It is designed to enable seamless electronic communication between trading partners in the supply chain, including manufacturers, retailers, distributors, and logistics providers. GS1 XML supports the exchange of a wide range of business docu-

ments, including purchase orders, invoices, shipping notices, and product information. The format is based on XML, which is a widely used format for structuring and exchanging data over the internet. One of the key benefits of using GS1 XML is that it helps to improve the efficiency and accuracy of the supply chain by enabling automated data exchange between trading partners. This can help to reduce the errors, delays, and costs associated with manual data entry and processing.

- UN/CEFACT's goal is simple, transparent, and effective processes for global commerce. It aims to help business, trade and administrative organizations from developed, developing, and transitioning economies to exchange products and services effectively.

  UN/CEFACT focuses on two main areas of activity to make international trade processes more efficient and streamlined—namely, trade facilitation and electronic business:

  - *Trade facilitation* involves the simplification of trade procedures to standardize and harmonize the core information used in trade documents, to ease the flow of information between parties by relying on appropriate information and communication technology, and to promote simplified payment systems to foster transparency, accountability, and cost-effectiveness.
  - *Electronic business*, in the UN/CEFACT context, focuses on harmonizing, standardizing, and automating the exchange of information that controls the flow of goods along the international supply chain.

Figure 3.9 shows a template of the GS1 electronic data exchange format.

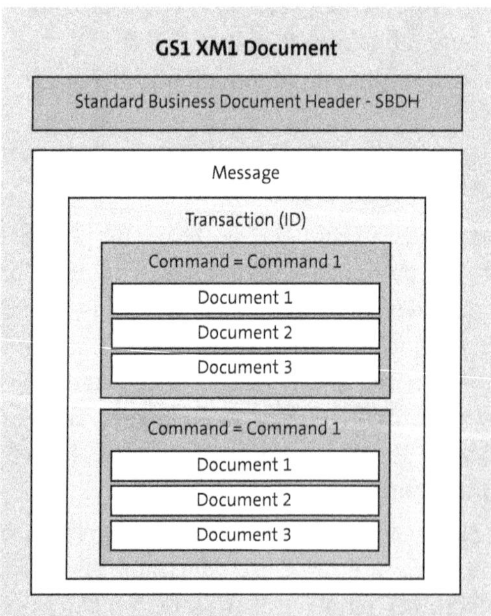

Figure 3.9 GS1 XML Format

Other EDI message standards are as follows:

- **TRADACOMS**
  Developed by the UK Article Numbering Association, TRADACOMS is a widely used EDI standard in the UK, used for the exchange of various types of business documents such as orders, invoices, and shipping notices.
- **VDA**
  Developed by the German Association of the Automotive Industry, VDA is a standard used for the exchange of electronic business documents in the automotive industry.
- **HIPAA**
  Developed by the US Department of Health and Human Services, HIPAA is a standard used for the exchange of electronic health care information.
- **ODETTE**
  The automotive industries in Europe developed the ODETTE EDI standard to facilitate the exchange of EDI documents with their trading partners. Communication standards such as OFTP and OFTP2O arose due to the popularity of ODETTE.
- **SWIFT**
  Developed by the Society for Worldwide Interbank Financial Telecommunication (SWIFT), SWIFT is a standard used for the exchange of financial messages between banks.

Overall, the choice of EDI standard used depends on the industry, geographic location, and trading partner requirements. Figure 3.10 shows the EDI message standards used commonly in various regions.

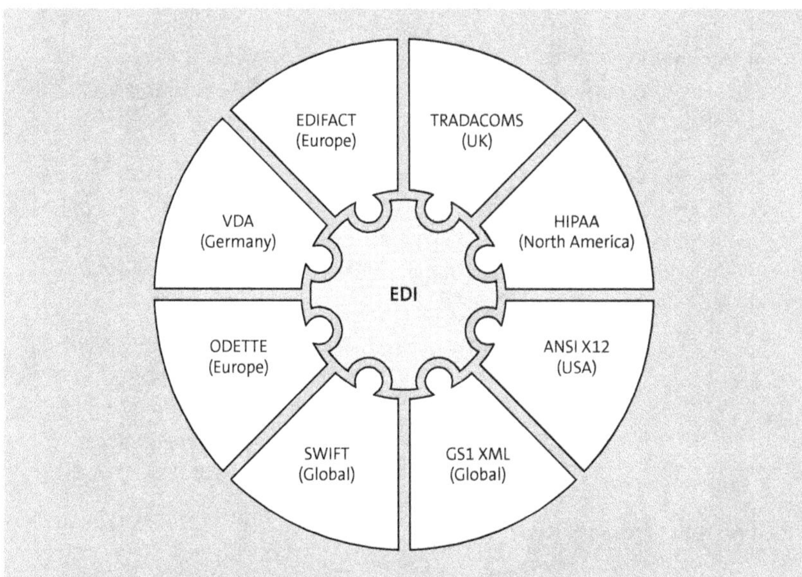

Figure 3.10 Commonly Used EDI Message Standards

SAP has added 22 messages in the message implementation guidelines (MIGs) in SAP Integration Suite to assist with EDI integration. MIGs are created by choosing from a variety of predefined message definitions offered by the Integration Advisor's type systems. The key benefit of this approach is that you receive a ready-made design template that adheres to one of the widely accepted standards for message integration in the industry.

In SAP Integration Suite, follow the menu path **Design • MIGs** to find a few standard EDI messages. Under **Create Options**, you can create MIGs for 22 messages, as described in Table 3.1.

Icon	Standard Message	Description
X12	**ASC X12: Accredited Standards Committee X12**	ASC X12 is an EDI standard and related message types created and maintained by the Accredited Standards Committee (ASC). It was chartered by ANSI in 1979. More than 315 X12 EDI standards for national and global markets cover multiple business domains, including communications, finance, government, insurance, supply chain, and transportation.
O	**Automotive CL: Error—Automotive Codelists: Error Codelists**	Automotive codelists for error situations defined jointly by ODETTE and VDA—to be used in message guidelines based on JAIF EDIFACT, ODETTE EDIFACT, and VDA EDIFACT.
JAIF	**Automotive CL: JAIF—Automotive Codelists: JAIF**	Automotive codelists defined by JAIF—to be used in message guidelines based on JAIF EDIFACT, ODETTE EDIFACT, and VDA EDIFACT.
O	**Automotive CL: ODETTE: Automotive Codelists: ODETTE**	Automotive codelists defined by ODETTE—to be used in message guidelines based on JAIF EDIFACT, ODETTE EDIFACT and VDA EDIFACT.
UN/EDIFACT	**Automotive CL: UNC, Automotive Codelists: Adopted UN/EDIFACT Codelists**	In the automotive industry, ODETTE EDIFACT, VDA EDIFACT, and JAIF EDIFACT are three subsets of UN/EDIFACT. This type system contains the joint collection of adopted UN/EDIFACT codelists for these subsets. Note that the codelists are based on UN/EDIFACT and use the same codelist identifiers as in UN/EDIFACT but the code values itself might be extended or modified. Moreover, all versions of the three subsets always refer to the latest version of these codelists.

Table 3.1 Twenty-Two Messages in MIGs

Icon	Standard Message	Description
VDA	**Automotive CL: VDA—Automotive Codelists: VDA**	Automotive codelists defined by VDA—to be used in message guidelines based on VDA EDIFACT.
cXML	**cXML—Commerce eXtensible Markup Language**	cXML is a streamlined protocol intended for consistent communication of business documents among procurement applications, e-commerce hubs, and suppliers. The protocol does not include the full breadth of interactions some parties may wish to communicate. However, through the use of extrinsic elements and newly defined domains for various identifiers, the protocol is easily expanded. This expansion is the limit of point-to-point configurations necessary for communication.
GS1	**GS1 Codelists: Code lists defined and maintained by GS1**	Code lists are the standardized lists of values allowed at the given location of the GS1 EDI message. This is the collection of all code lists defined and maintained by GS1. GS1 develops its own code lists only if no suitable international standard exists. Whenever possible, the international standard code lists are used.
GS1	**GS1 EANCOM: GS1 EANCOM**	GS1 EANCOM is a GS1 subset of the UN/EDIFACT standard. It contains only the message elements required by business applications especially for domains supported by GS1 organizations such as retail, consumer products, transport, and logistics. Omitted are optional elements that are not relevant for these specific use cases. Currently, Integration Advisor provides the so-called subset message standard (but not the MIGs that are defined on top of the message standard).
GS1	**GS1 XML: GS1 XML Standards**	GS1 uses XML to create a set of standard messages for the GS1 EDI. GS1 XML messages can be exchanged using any technical solution or internet transport protocol.

Table 3.1 Twenty-Two Messages in MIGs (Cont.)

Icon	Standard Message	Description
ISO	**ISO: International Standardization Organization—Codelists and Schemes**	The International Standardization Organization (ISO) creates documents that provide requirements, specifications, guidelines or characteristics that can be used consistently to ensure that materials, products, processes, and services are fit for their purpose. In this scope, ISO also develops and maintains international standardized codelists and identifier schemes. These codelists and identifier schemes can be generally used across all type systems (A2A/B2B standard libraries).
JAIF	**JAIF EDIFACT: JAIF EDIFACT Messages**	JAIF EDIFACT is a subset of the UN/EDIFACT standard from the Joint Automotive Initiative Forum (JAIF) organization and is used in the global automotive industry. Currently, Integration Advisor provides the so-called subset message standard (but not the MIGs that are defined on top of the message standard).
O	**ODETTE: ODETTE Messages (old-style)**	The ODETTE type system provides the early ODETTE messages and codelists that were defined by ODETTE in the 1980s and 1990s and were used in the European automotive industry. Although these messages are based on the EDIFACT syntax (ISO 9735), they are not derived from the EDIFACT messages but are independent message type definitions.
O	**ODETTE EDIFACT: ODETTE EDIFACT Messages**	ODETTE EDIFACT is a subset of the UN/EDIFACT standard from the ODETTE organization and is used in the European automotive industry.
SAP S/4 HANA	**SAP S/4HANA Cloud OData**	OData is a web protocol for querying and updating data. It applies and builds upon web technologies such as HTTP, Atom Publishing Protocol (AtomPub), and JSON to provide access to information from a variety of applications.
SAP S/4 HANA	**SAP S/4HANA Cloud SOAP**	SOAP is a standardized, messages-based protocol that is used by web services for communication between the service consumer and the service provider. A SOAP message comprises a header, which contains communication control parameters, and a body, which contains the payload.

Table 3.1 Twenty-Two Messages in MIGs (Cont.)

Icon	Standard Message	Description
	SAP S/4HANA on-premise IDoc	Standard SAP format for electronic data interchange between systems (Intermediate Document or IDoc). Different message types (such as delivery notes or purchase orders) generally correspond to different special formats, known as IDoc types. Multiple message types with related content can, however, be assigned to a single IDoc type.
	SAP S/4HANA on-premise SOAP	SOAP is a standardized, messages-based protocol that is used by web services for communication between the service consumer and the service provider. A SOAP message comprises a header, which contains communication control parameters, and a body, which contains the payload.
	TRADACOMS: TRADACOMS EDI Standard	TRADACOMS was the core UK message standard from the early days of EDI in the UK retail industry. It was widely used by retailers and wholesalers for EDI order and invoice processes between buyers and their mainly UK-based suppliers. However, since 1998, existing users have only received basic TRADACOMS support from GS1 UK for current applications and there has been no development of any new message types. Since July 1, 2017, GS1 UK has no longer provided any support for TRADACOMS. All users have been advised to adopt the recognized GS1 EDI standards: EANCOM, GS1 XML, and the GS1 UN/CEFACT XML profile.
	UN/CEFACT: Trade Facilitation Recommendations	UN/CEFACT's trade facilitation recommendations are made to improve the ability of business, trade, and administrative organizations from developed, developing, and transitional economies to exchange products and relevant services effectively. Its principal focus is on facilitating national and international transactions, through the simplification and harmonization of processes, procedures, and information flows, and so contribute to the growth of global commerce.

Table 3.1 Twenty-Two Messages in MIGs (Cont.)

Icon	Standard Message	Description
UN/EDIFACT	**UN/EDIFACT: United Nations Electronic Data Interchange for Administration, Commerce and Transport**	UN/EDIFACT is the international EDI standard developed under the United Nations. The work of maintenance and further development of this standard is done through UN/CEFACT under the UN Economic Commission for Europe, in the finance domain working group, UN CEFACT TBG5. EDIFACT has been adopted by ISO.
VDA	**VDA EDIFACT: VDA EDIFACT Messages**	VDA EDIFACT is a subset of the UN/EDIFACT standard from the VDA organization (Verband der Automobilindustrie Germany) and is used in the German automotive industry.

Table 3.1 Twenty-Two Messages in MIGs (Cont.)

In addition to these standard MIGs, SAP allows you to create custom types.

3.2.2 UN/EDIFACT Standard in Detail

UN/EDIFACT is a set of international EDI standards developed by the United Nations. It is used for the electronic exchange of business documents between trading partners, including orders, invoices, and shipment information. UN/EDIFACT provides a standardized syntax and message structure for electronic data interchange, ensuring that trading partners can understand and interpret each other's messages. The standard defines the structure and content of messages, as well as the rules for transmitting and receiving them.

UN/EDIFACT messages are made up of *segments*, which are groups of related data elements. Each segment has a three-letter code that identifies its purpose, such as identifying the sender or receiver, specifying the message type, or providing details about a particular transaction. The data elements within each segment are identified by a code or tag, and each has a specific meaning and format. The standard includes a wide range of message types, covering everything from simple messages like purchase orders and invoices to more complex messages like customs declarations and transport instructions. Each message type is identified by a unique message type identifier (UNH), which specifies the type of message and the version of the standard used.

UN/EDIFACT is used by businesses and organizations around the world for EDI, particularly in industries like transportation, logistics, and finance. It helps to streamline business processes, reduce errors, and improve the overall efficiency of the supply chain. However, implementing and using the standard can require significant investment in technology and training.

An EDIFACT message is a single business document. Each EDIFACT message is always identified by a six-character name. Typical EDIFACT messages include the following:

- ORDERS, purchase order messages
- ORDCHG, purchase order change request messages
- ORDRSP, purchase order response messages
- DESADV, dispatch advice messages
- INVOIC, invoice messages

EDIFACT messages are made up of a collection of sequenced segments within defined areas. Some segments may be used in more than one area. Messages begin with the message header segment (UNH) and end with the message trailer segment (UNT). Let's look more closely at the EDIFACT segment, EDIFACT data element, and EDIFACT qualifier:

- **EDIFACT segment**
  An EDIFACT segment is a collection of logically related data elements in a fixed, defined sequence. An EDIFACT segment contains a three-character alphanumeric code that identifies the segment (a *segment tag*) and a variable length data element, which can be either simple or composite.

  EDIFACT segments must be separated by a data element separator (a *delimiter*), which is normally + or :, and terminated by a segment terminator. All segments are fully documented in the United Nations Trade Data Interchange Directory (UNTDID). These tables list the segment position, segment tag, and segment name. Segment tables also specify if a segment must appear in a message using the requirements designator, M (mandatory) or C (conditional), and how many times a particular segment may repeat.

  In EDIFACT, there are two kinds of segments: service segments and generic segments. Service segments include the following:
  - Envelopes (UNB-UNZ, UNG-UNE, UNH-UNT)
  - Delimiter string advice (UNA)
  - Section separator (UNS)

  Generic segments include the following:
  - BGM for the beginning of a message, which contains the document number
  - DTM for dates and times
  - NAD for name and address data
  - MOA for monetary amounts
- **EDIFACT data element**
  A simple data element contains one piece of information, whereas a composite data element contains more than one piece of information. In EDIFACT, all mandatory data elements must contain data. Conditional data elements may or may not contain data, depending on the requirements of the particular transmission.

Because data elements must be accounted for by their position in the segment, if an optional or conditional data element does not have data, then that data element must still be accounted for in its position within the segment by using the appropriate number of data element separators to "skip over" the empty field. Data elements can have different types: numeric, alphabetical, and alphanumeric.

- **EDIFACT qualifier**
  A composite data element is made up of two or more pieces of data (known as *components*), which form a single data unit. Typically, the first data element is the value that is being qualified. The second data element is typically the qualifier. These are typically ID (code value) fields. The qualifier gives additional definition to the value.

Figure 3.11 shows a detailed explanation of the EDI sample purchase order using EDIFACT.

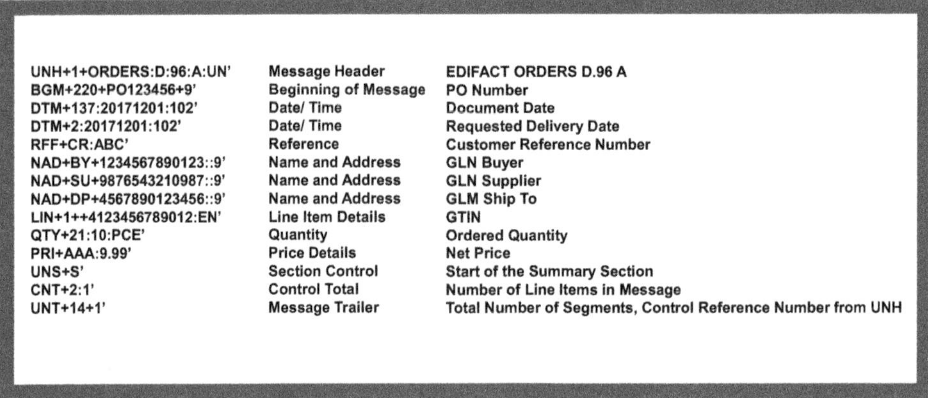

UNH+1+ORDERS:D:96:A:UN'	Message Header	EDIFACT ORDERS D.96 A
BGM+220+PO123456+9'	Beginning of Message	PO Number
DTM+137:20171201:102'	Date/ Time	Document Date
DTM+2:20171201:102'	Date/ Time	Requested Delivery Date
RFF+CR:ABC'	Reference	Customer Reference Number
NAD+BY+1234567890123::9'	Name and Address	GLN Buyer
NAD+SU+9876543210987::9'	Name and Address	GLN Supplier
NAD+DP+4567890123456::9'	Name and Address	GLM Ship To
LIN+1++4123456789012:EN'	Line Item Details	GTIN
QTY+21:10:PCE'	Quantity	Ordered Quantity
PRI+AAA:9.99'	Price Details	Net Price
UNS+S'	Section Control	Start of the Summary Section
CNT+2:1'	Control Total	Number of Line Items in Message
UNT+14+1'	Message Trailer	Total Number of Segments, Control Reference Number from UNH

Figure 3.11 Sample EDI Purchase Order Message with Explanation

3.2.3 Syntax and Semantics

In the context of UN/EDIFACT, *syntax* refers to the rules governing the structure and composition of messages exchanged between trading partners. The syntax rules specify the format and sequence of the message segments and the allowable data elements and codes within each segment. The UN/EDIFACT syntax rules ensure that messages are consistent, complete, and interpretable by trading partners.

For example, a UN/EDIFACT purchase order message may consist of segments like UNH (message header), BGM (beginning of message), DTM (date/time/period), and NAD (name and address). Each segment is made up of one or more data elements, each of which has a specific meaning and format. The syntax rules specify the allowable codes and formats for each data element, as well as the mandatory and optional data elements for each segment.

Semantics in UN/EDIFACT refers to the meaning of the message content and how it relates to business processes. The semantics of a message define the intended business process and the rules for interpreting the message content to execute that process. For example, the semantics of a UN/EDIFACT purchase order message specify the intended purchase transaction, including the buyer, seller, items to be purchased, quantities, prices, and delivery requirements. UN/EDIFACT provides a set of standardized message types that are widely used for electronic data interchange between trading partners. The syntax and semantics of these message types are defined in UN/EDIFACT documentation, which specifies the allowable segments, data elements, and codes for each message type. Trading partners must comply with these standards to ensure interoperability and successful electronic communication.

3.2.4 Transaction Messages for Sales and Distribution

The following are some of the key message types used in the sales and distribution area of UN/EDIFACT:

- ORDERS
  Purchase order message used to request goods or services from a supplier. It includes information on the buyer, seller, item details, quantities, and delivery requirements.
- ORDRSP
  Order response message used to acknowledge receipt of an order or to provide a response to a buyer's purchase order. It includes information on the seller's acceptance or rejection of the order, delivery dates, and pricing.
- DESADV
  Dispatch advice message used to provide notification of a pending delivery or shipment. It includes details of the shipment, such as the delivery date, the carrier, and the items being shipped.
- ASN
  An ASN is an EDI message sent from the shipper to the receiver prior to the departure of the shipment from the shipper's facility. The message includes complete information about the shipment and its contents.
- INVOIC
  Invoice message used to request payment for goods or services provided. It includes information on the buyer, seller, pricing, and payment terms.
- REMADV
  Reminder message used to request payment for goods or services that have not been paid for within the agreed-upon time frame.
- CREDIT
  Credit note message used to provide a credit for goods or services that have been returned or for which a refund has been issued.
- CREADV
  Credit advice message used to provide notification of a credit issued to a buyer.

These transaction messages are used to support a range of business processes related to sales and distribution, including purchase orders, order acknowledgments, delivery notifications, invoicing, and credit processing. By standardizing the format and content of these messages, UN/EDIFACT helps to streamline business processes, reduce errors, and improve overall efficiency in the sales and distribution area.

Figure 3.12 shows a brief explanation of the EDI messages involved in sales and distribution. The EDI 850 is an electronic data interchange transaction set that provides details about an order. In response, sellers send an EDI 855 purchase order acknowledgement to buyers. The EDI 856, or ASN, notifies the receiving party about a shipment and describes its contents. Finally, the EDI 810 invoice is sent by the seller to the buyer to indicate charges due and request payment, serving as a digital alternative to traditional invoices.

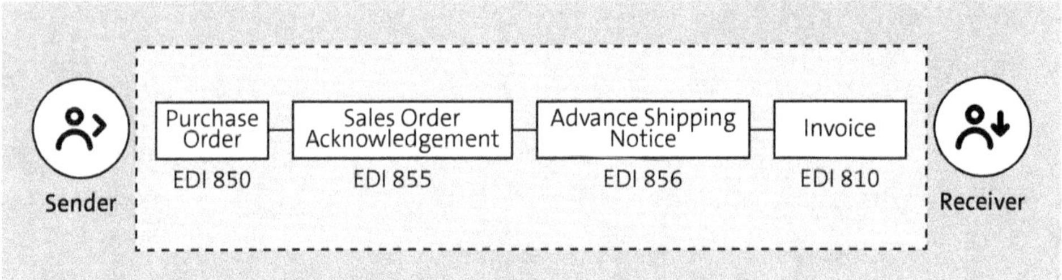

Figure 3.12 Basic Flow of Sales and Distribution with EDI Messages

3.3 EDI Standards for Communication

EDI standards for messages focus on the content and format of the data being exchanged, while EDI standards for communication deal with the methods and protocols for transmitting that data securely and reliably. Understanding both aspects is essential for successful EDI implementation. There are several EDI standards for communication, each with its own syntax, data elements, and message types that must be adhered to in order to successfully exchange data between trading partners. By standardizing the format and content of EDI messages, these standards help to improve the efficiency and accuracy of business processes, reduce errors, and simplify communication between trading partners. This section will cover some of the most widely used EDI standards.

3.3.1 Value-Added Network

A VAN is a private, hosted service that provides a secure way to exchange electronic messages and business documents between companies. It is a secured electronic postal service for trading partners. In addition to basic data transmission services, VANs provide value-added services to improve the efficiency and security of EDI transactions. The following are some of the most common value-added services offered by VANs:

- **Data translation**
  VANs can provide data translation services to convert EDI messages from one format to another to ensure that trading partners can communicate in a common format.
- **Document tracking and management**
  VANs can track the status of EDI documents throughout the entire transaction process, from the sender to the receiver. They can also provide document management services to store, archive, and retrieve EDI documents as needed.
- **Trading partner management**
  VANs can manage the communication and relationship between trading partners by providing services such as partner onboarding, compliance checking, and message routing.
- **Error detection and correction**
  VANs can perform error detection and correction services on EDI documents to ensure the accuracy and completeness of the data being exchanged.
- **Data security**
  VANs can provide data security services to protect EDI documents from unauthorized access, data breaches, and cyberattacks. These services may include encryption, digital signatures, and secure data storage.
- **Value-added information**
  VANs can provide value-added information services such as tracking and monitoring of industry news, regulatory updates, and supply chain trends.

By providing these value-added services, VANs help to streamline the EDI communication process, reduce errors and delays, and improve the overall efficiency and security of electronic business transactions.

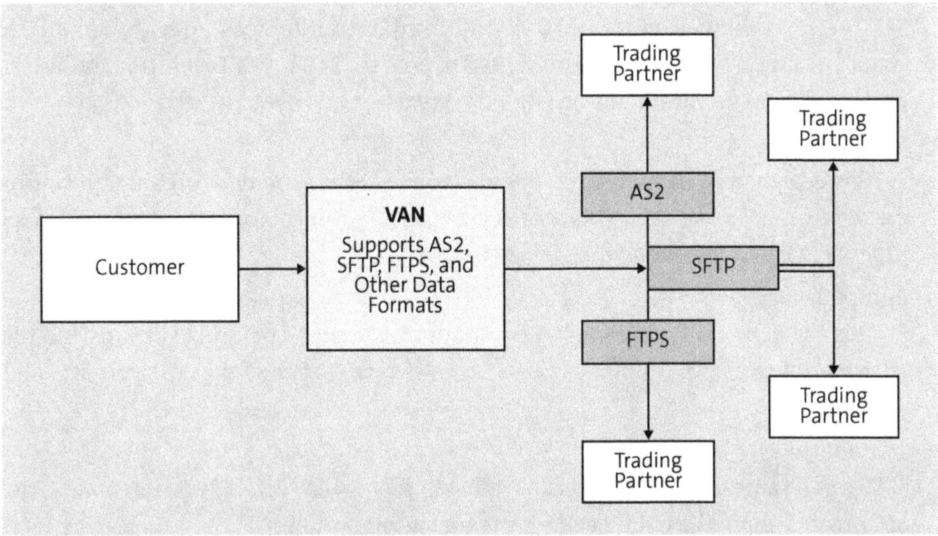

Figure 3.13 Customer System Interacting with Trading Partners via VAN

A VAN infrastructure is a specialized system designed to enhance the exchange of electronic documents between trading partners. It provides secure and reliable data transmission methods; supports multiple communication protocols like AS2, FTP, and HTTP(S) for compatibility; and includes message formatting services for various EDI standards such as ANSI X12 and EDIFACT. The infrastructure also emphasizes data security through encryption and authentication, offers value-added services like data archiving and error handling, and acts as an intermediary to facilitate seamless communication among trading partners using different systems and technologies. Figure 3.13 shows the VAN infrastructure to connect trading partners for the exchange of business documents.

3.3.2 SSH File Transfer Protocol

SFTP is a secure protocol that enables the transfer of files over an encrypted connection, functioning as an extension of the Secure Shell (SSH) protocol to ensure data confidentiality and integrity. The process begins with the client establishing a connection to the server through SSH, which requires authentication via passwords, public keys, or certificates. Once authenticated, an encrypted session is created, safeguarding data from eavesdropping and tampering. SFTP allows users to perform various file operations, such as uploading, downloading, renaming, and deleting files, all within this secure session, which encrypts both file content and metadata during transmission. Finally, the session is securely closed after completing the file operations.

SFTP is a secure and reliable protocol for transferring files between computers over a network. It is commonly used in EDI to transfer large files, such as EDI documents, between trading partners. Here are some of the advantages of using SFTP for EDI:

- **Security**
  SFTP uses a combination of encryption and authentication techniques to secure the transfer of data between systems. This ensures that the data being transmitted is protected from unauthorized access, data breaches, and cyberattacks.
- **Reliability**
  SFTP is designed to handle large files and can recover from network disruptions or errors without losing data. This makes it a reliable protocol for transferring EDI documents that can be several megabytes in size.
- **Compatibility**
  SFTP is a widely used protocol and is supported by most operating systems and file transfer software. This makes it easy to integrate into existing EDI systems and workflows.
- **Automation**
  SFTP can be automated using scripts or batch files, which can help to streamline the EDI process and reduce the need for manual intervention.

- **Auditing and logging**
  SFTP provides detailed logging and auditing features, which can help to track the transfer of EDI documents and ensure compliance with regulatory requirements.

3.3.3 Applicability Statement 2

AS2 is a popular standard for secure and reliable data transmission over the internet. It is widely used in B2B transactions and supply chain management to exchange electronic documents, such as purchase orders, invoices, and shipping notices, between trading partners. The special feature of the communication protocol AS2 is that the protocol has an acknowledgement of receipt—the so-called message disposition notification (MDN). It serves as proof that the files have been delivered properly and on time. In addition, the AS2 protocol offers the possibility to sign or encrypt files.

AS2 provides a set of rules and guidelines for the exchange of electronic documents between trading partners using the HTTP(S) protocol. It uses digital certificates and encryption to provide confidentiality and authentication of the parties involved in the exchange. AS2 also provides reliable message delivery, including message acknowledgment and error reporting, to ensure the successful transmission of data.

AS2 enables a secure and efficient exchange of data between trading partners, regardless of their location or technology infrastructure. It provides a standardized and widely adopted protocol for electronic document exchange, enabling interoperability between different systems and applications. AS2 is widely used in industries such as retail, healthcare, and logistics to enable secure and efficient exchange of electronic documents between trading partners. It is supported by many software vendors and is often used in conjunction with other B2B standards, such as EDI and XML.

AS2 is a protocol used to securely exchange files through the internet, using the HTTP or HTTPS web protocol as the transport mechanism. It is used extensively for the exchange of business documents such as EDI messages like EDIFACT and X12 and document types like XML, JSON, CSV, binary, or flat files. The AS2 protocol is primarily a secure file transfer protocol utilizing S/MIME capabilities for digital encryption, signature, and nonrepudiation. Unlike FTP, SFTP, or FTPS, it operates over the HTTP and HTTPS web protocols and is used extensively for the exchange of EDI files. The AS2 protocol secures the files with encryption and uses digital signatures to guarantee authenticity. Signed receipts issued by the receiver provide legally valid nonrepudiation.

AS2 is also referred to as *EDIINT AS2*, with EDIINT short for Electric Data Interchange over the Internet. EDIINT is a working group of the Internet Engineering Task Force (IETF), tasked with the responsibility to create secure protocols for business communications over the internet. The EDIINT working group has created the AS1, AS2, and AS3 protocols for the exchange of EDI documents between businesses.

Figure 3.14 illustrates a message flow for AS2.

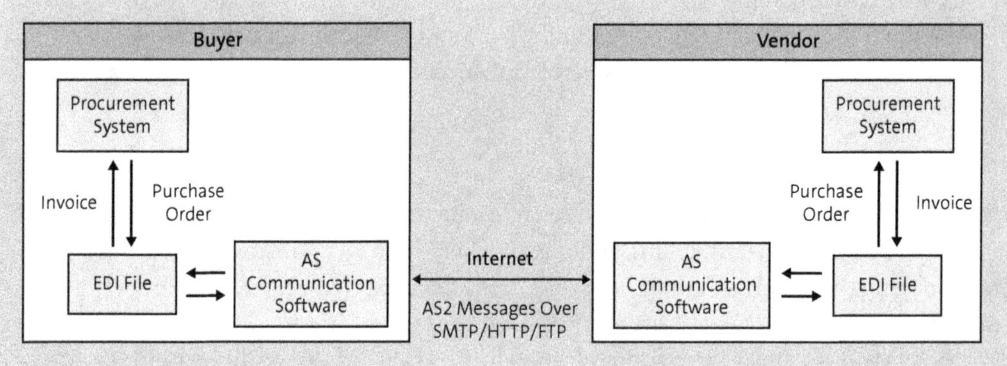

Figure 3.14 Brief AS2 Message Flow

Due to its simplicity and adoption by many industries across the world, AS2 remains the most widely used and preferred mechanism today for the B2B exchange of EDI and other documents. While AS1 became obsolete with the adoption of AS2, AS3 and AS4 have not gained wide adoption due to the popularity and use of AS2.

Table 3.2 lists the AS protocols.

Protocol	Transport	Security
AS1	SMTP and POP3	PGP/MIME and S/MIME
AS2	HTTP/S	S/MIME
AS3	FTP	S/MIME
AS4	Web services	WS-Security

Table 3.2 Protocols and Their Associated Transport and Security for AS2

AS2 was defined for the exchange of structured business documents, such as EDI files. However, the AS2 specification is agnostic in terms of file content. Businesses of all types and sizes use AS2 to exchange files such as XML, JSON, CSV, flat files, text files, and images in addition to EDI files.

Software capable of communicating over the AS2 protocol is called an AS2 server. Similar to having a mail server or website hosted within a corporate network, an AS2 server can also be deployed within an internal corporate network. AS2 servers can be deployed outside a corporate network, such as on the cloud. Another common option is to use AS2 as a service in the common SaaS model. This requires no hardware or software to be purchased, installed, or maintained, much like using a web-based email service. Files provided to the AS2 solution will be transmitted to the destination partners over AS2, while files received over AS2 will be accepted and stored for later consumption, much like store-and-forward email.

The file to be sent is signed using the private key of the sender system and then encrypted using the certificate (public key) of the receiver system. Then the file is encrypted using the private key of the sender system. The signed and encrypted file—now unreadable to third parties—is then sent via HTTP or HTTPS. If the header of the message—the information about who sends something to whom—is to be protected against access by third parties, then HTTPS should be used to encrypt the communication channel information.

After the file has been received by the receiver system, the file is decrypted using its own private key. In case of problems decrypting the message, a negative MDN will be sent with the "Error decrypting" message (synchronous or asynchronous depending on configuration). If the file has been successfully decrypted, the signature will now be verified (message integrity). Also in this case, a negative MDN is sent to the sender system in case of an error: "Signature error" (depending on configuration synchronous or asynchronous).

If the file was successfully decrypted and the signature check was successful, then an MDN with the indicator of successful receipt is sent at the end.

Figure 3.15 shows an AS2 transaction flow with a procedure involving signature and encryption. Exchanging files using the AS2 protocol with digital signatures, encryption, and MDNs is called AS2 communication. The AS2 protocol enables secure and reliable electronic data exchange between systems over the internet by utilizing digital certificates, secure communication channels, and encryption algorithms. The process begins with the encryption of data using algorithms like AES to ensure confidentiality.

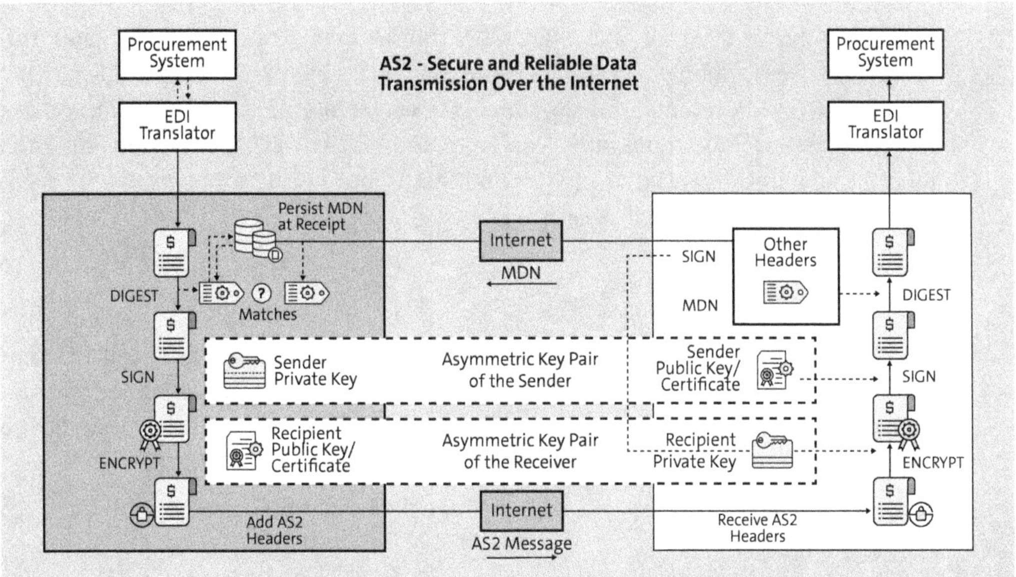

Figure 3.15 AS2: Popular Standard for Secure and Reliable Data Transmission with Acknowledgment

A digital signature is then generated using a private key, which is verified with the recipient's public key, ensuring message integrity. The encrypted and signed message is transmitted using HTTP or HTTPS protocols for secure delivery. Upon receipt, the recipient generates an acknowledgment (MDN) to confirm successful delivery back to the sender/procurement system. Finally, the sender/procurement system decrypts the message and verifies the digital signature to ensure that the data has remained intact during transmission. This is provided by AS2 software deployed on premise or in the cloud, or utilized as a service (SaaS).

How Does AS2 Differ from FTP and SFTP?

AS2 differs from traditional FTP or SFTP file transfer mechanisms due to its use of payload hashing, digital signatures, and encryption. In addition, the digitally signed receipt called the MDN, issued by the recipient, serves the purpose of legally valid nonrepudiation, which is not available in basic SFTP or FTP.

One of the key features of AS2 is secure data transmission. AS2 uses digital certificates and encryption (typically using Secure Sockets Layer [SSL]/Transport Layer Security [TLS]) to ensure data confidentiality and integrity during transmission over the internet. It supports both synchronous (real-time) and asynchronous (store-and-forward) messaging modes, enabling flexibility based on business needs. For message integrity and non-repudiation, AS2 employs digital signatures to verify the authenticity of messages and ensure that messages cannot be altered without detection. AS2 does support MDNs acknowledging the successful receipt and processing of messages, which add values in reliability and acknowledgment.

AS2 is commonly used for exchanging EDI transactions (like EDI 850, 856, and 810) securely between trading partners. It ensures compliance with EDI standards (e.g., ANSI X12, EDIFACT) while maintaining the benefits of secure internet-based communication. AS2 provides a robust framework for secure and efficient EDI transactions over the internet, enabling organizations to streamline B2B communication, comply with regulatory requirements, and enhance operational efficiency.

3.3.4 Additional Security Aspects

EDI is a standard for exchanging business documents electronically between different organizations. While EDI provides many benefits, including increased efficiency, accuracy, and speed of document exchange, it also introduces security risks that need to be addressed to ensure the confidentiality, integrity, and availability of data.

Some additional security aspects of EDI include the following:

- **Access control**
  Access to EDI systems and data should be restricted only to authorized personnel. This can be achieved by strong authentication mechanisms, such as passwords or multifactor authentication, and by limiting access based on job role and responsibility.

- **Encryption**
  EDI data should be encrypted during transmission to protect against interception and unauthorized access. This can be achieved using technologies such as SSL/TLS or S/MIME.
- **Nonrepudiation**
  EDI messages should provide *nonrepudiation*, meaning that the sender cannot deny sending the message and the recipient cannot deny receiving the message. This can be achieved using digital signatures or message encryption.
- **Monitoring and auditing**
  EDI systems should be continuously monitored and audited to detect and prevent security breaches. This includes logging and analyzing system activity, reviewing access logs, and performing periodic vulnerability assessments and penetration testing.
- **Disaster recovery and business continuity**
  EDI systems should have robust disaster recovery and business continuity plans in place to ensure that critical business functions can continue in the event of a system failure or disaster.

Addressing these additional security aspects of EDI can help organizations mitigate security risks and ensure the secure and reliable exchange of electronic business documents.

3.4 Characteristics of the EDI Integration Pattern

The EDI integration pattern is a popular approach to integrating EDI systems with other enterprise applications and systems. Sensitive digital documents from various systems and trading partners can be exchanged in a single, standardized format. This process simplifies data exchange and reduces data entry errors. Some of the key characteristics of the EDI integration pattern include the following:

- **Message transformation**
  EDI messages often use a different format and structure than ERP systems. The EDI integration pattern involves transforming EDI messages into a format that can be consumed by other systems, such as XML or JSON, and vice versa.
- **Standards-based approach**
  EDI systems use standardized formats and protocols to ensure interoperability between different systems and trading partners. The EDI integration pattern leverages these standards and protocols to enable seamless integration between EDI systems and enterprise systems.
- **Routing and message delivery**
  EDI messages may need to be routed to different systems and applications based on their content or destination. The EDI integration pattern involves setting up routing

rules and workflows to ensure that messages are delivered to the right destination in a timely and efficient manner.

- **Message validation and error handling**
  EDI messages must be validated to ensure that they meet the required standards and comply with the format and structure defined by the trading partner. The EDI integration pattern involves implementing validation rules and error handling mechanisms to ensure that invalid messages are rejected or corrected before they are processed.
- **Integration with business processes**
  EDI messages often represent business transactions, such as purchase orders, invoices, and shipping notices. The EDI integration pattern involves integrating these transactions with the organization's business processes, such as order fulfillment, invoicing, and inventory management, to enable end-to-end process automation.

3.5 Example Use Case for EDI

Now that you understand how EDI works, let's take a look at a practical use case. In the healthcare industry, EDI allows institutions, insurers, and patients to securely exchange data in standardized formats. It simplifies document transmission for organizations using protocols like HIPAA EDI or ASC X12, benefiting health plans and care providers. EDI has effectively reduced IT costs, interoperability issues, conversion errors, and security risks in the industry.

Healthcare providers, like hospitals and clinics, frequently submit claims to insurers for reimbursement of medical services provided to patients. Traditionally, this involved paper forms or electronic submissions through web portals. Many providers have adopted EDI to streamline this process, using standards like ASC X12 or HIPAA transactions. EDI formats and transmits claims data electronically, including patient demographics, diagnosis codes (ICD-10), procedure codes (CPT/HCPCS), service dates, provider details, and billing specifics. This structured approach ensures consistency and compatibility across systems.

EDI automation eliminates manual data entry and reduces paperwork, allowing providers to submit claims directly from their practice management or electronic health record (her) systems. This speeds up submissions, reduces errors, and enhances efficiency. Insurance payers receive EDI claims, process them electronically, and respond with remittance advice or claim status updates. This ensures faster reimbursement and allows providers to track claim status in real time.

EDI transactions comply with HIPAA regulations, ensuring patient data confidentiality through encryption and secure protocols. Implementation of EDI also reduces administrative costs associated with printing, postage, and manual labor, while minimizing errors and improving operational efficiency.

Some of the benefits of EDI in healthcare claims processing include the following:

- **Efficiency**
  Streamlined claims processing reduces administrative burden and speeds up revenue cycle management.
- **Accuracy**
  Electronic data exchange reduces errors from manual entry, ensuring data accuracy and completeness.
- **Speed**
  Quicker claims processing results in faster reimbursement, enhancing cash flow for healthcare providers.
- **Compliance**
  Adherence to HIPAA EDI standards ensures regulatory compliance with data security and privacy requirements.
- **Improved relationships**
  EDI promotes better communication and collaboration between healthcare providers and insurance payers, improving operational efficiency and service delivery.

Figure 3.16 shows that EDI messages allow institutions, healthcare partners, insurers, and patients to securely exchange data in standardized formats. Hospitals and clinics will use the EDI message format to hold information such as patient demographics, service dates and provider details, diagnostic codes, and bill specifics in messages that can be read by the claimer or insurer to process the claims.

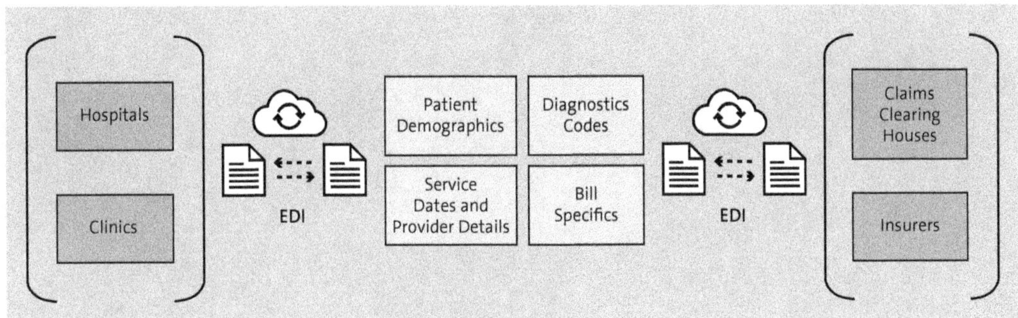

Figure 3.16 EDI Messages Flow in Healthcare

3.6 Summary

In this chapter, you learned about electronic data interchange and its key components, including EDI standards for messages and communication protocols. You also learned about the characteristics of the EDI integration pattern.

Overall, the EDI integration pattern is a robust and proven approach to integrating EDI systems with enterprise systems, enabling organizations to streamline their business processes, reduce errors, and improve the efficiency and accuracy of data exchange.

In the next chapter, we will discuss EDI processing in the SAP S/4HANA landscape.

Chapter 4
EDI Processing in the SAP S/4HANA Landscape

4

The first part of this chapter gives basic information on the business processes for which we enable the EDI scenario. The second part is a close look at the technical solution design. Both parts together will let you fully understand the overall context.

In the previous chapters, you were introduced to SAP S/4HANA and the SAP integration components. There was also a brief introduction to the B2B EDI integration pattern and its standards. In this chapter, we will take a step forward and start defining an EDI processing environment for our use case. We will go through each component included in the overall technical design, explaining its role and the motivation behind the choice. But before we go into the technical details, we start with an introduction to the sales and distribution process in the context of SAP S/4HANA. This basic understanding is a prerequisite to fully understand the overall functional and technical process flow and the unique EDI transactions that we will implement later on in Chapter 5 through Chapter 9.

4.1 Business Process

A *business process* is a set of related activities that are performed by an organization to achieve a specific goal. Business processes are designed to be efficient, effective, and repeatable and they are critical to the success of any organization. A business process typically involves multiple steps and may involve the use of technology, human resources, and other resources.

Business processes in SAP are the set of activities that are performed within the system to achieve a specific business goal. Business processes in SAP S/4HANA are like those in traditional SAP ERP systems, but they are streamlined and simplified. SAP S/4HANA offers a range of preconfigured business processes, and these can be customized to meet the specific needs of each organization. They integrate various business functions and processes into a single system, enabling organizations to streamline their operations and improve efficiency.

SAP offers a wide range of modules that cover different business processes and functional areas, such as finance and accounting, human resources, supply chain management, sales and distribution, and production planning.

When we look at SAP's intelligent enterprise strategy, which represents the core solutions, we can distinguish six main business processes:

- Source-to-pay
- Design-to-operate
- Lead-to-cash
- Request-to-service
- Record-to-report (finance)
- Hire-to-retire (people)

Let's take a closer look at lead-to-cash as this is the area where we will implement the EDI solution. Lead-to-cash consists of all activities that deliver an end-to-end customer experience. From this main business process, we could extract subprocesses like quote-to-order, sell-from-stock, and invoice-to-cash. The starting point could be presales activities related to the contact, lead, opportunity, or quote. The next steps are related to the actual customer sales order management and fulfillment. The last steps are related to the goods delivery and invoicing, and at the very end we may have some additional analytics and reporting. It is also important to note that although SAP S/4HANA is in the center of this process, there are also other SAP products to complement and extend the scenario. For the lead-to-cash process, this can run on SAP Sales Cloud, SAP S/4HANA, and SAP Analytics Cloud. In this sample setup, SAP Sales Cloud would be responsible for the presales activities, SAP S/4HANA would execute the sales and distribution processing, and SAP Analytics Cloud would capture the information from both previously mentioned systems and deliver real-time KPIs and reports.

The order-to-cash process is a component of the sales and distribution application. We'll take a closer look at sales and distribution within SAP S/4HANA, which takes care of order fulfillment and billing. Having some functional knowledge will help with understanding the overall EDI scenario. The starting point of our introduction will be Figure 4.1, which shows the key elements of the order-to-cash process within SAP S/4HANA. The process goes from the left-hand side to the right-hand side, and in this simplified diagram, each box represents not only a specific action but also a specific business document behind it. We will limit our scope to sales order management, shipping, and billing. There are some additional activities or even documents that come into play before the sales order itself, but they are not part of our demo scenario, which will start with the sales order. What happens after invoice creation is critical for each organization, including from a legal perspective, but this is also not part of the demo. The invoice will be the last document in our simplified process.

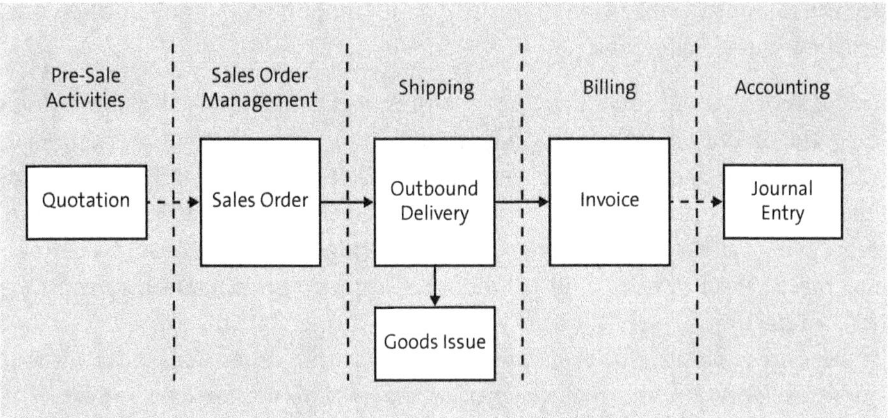

Figure 4.1 Key Elements of Order-to-Cash Process in SAP S/4HANA

Before we go into the details of these activities and documents, we need an introduction to the basic aspects of the organizational structure and master data in SAP S/4HANA.

4.1.1 Organizational Structure in Sales and Distribution

There are dozens of different objects that define organizational structures within SAP ERP systems. Among these objects, we can distinguish those which are used and shared across multiple functional modules like company codes or plants, but there are also many objects that are dedicated and unique to specific functional modules. Looking at sales and distribution and its unique objects, we can distinguish the sales organization, distribution channel, division, and a few others that are less relevant for our case. The listed objects are mandatory prerequisites for processing sales transactions. Understanding their concept and relations among them is also necessary to fully understand the configuration activities performed in the following chapters.

Let's look at short definitions of the key objects:

- **Sales organization**
  This is a four-character alphanumeric key, and in most cases it represents the legal entity responsible for the sales and distribution. For example, a company that operates globally can create unique sales organizations for each country in which it sells products.
- **Distribution channel**
  This represents how goods or services are distributed to the customers. These channels could include retail, online sales, wholesale, direct sales, and so on.
- **Division**
  If a company sells multiple products or services that have similar characteristics or belong to the same line of business, it may consider grouping them to specialize and

make the sales processes more efficient. The different divisions could include cars, bikes, food, computers, and so on.

The combination of a sales organization, a distribution channel, and a division forms a *sales area*, the definition of the exact organizational unit and distribution channel in which the company sells goods and performs all sales-related activities. Each document in sales and distribution must be created in a specific sales area. This also counts for master data objects specific for sales and distribution, such as the business partner, material master data, and pricing conditions. These have elements that must be defined on a sales area level. This means that some master data objects are active for unique sales areas only and they cannot be reused within other sales areas without conscious configuration and maintenance of the specific master data record. That makes sense: if you look back at the sample setup where sales organizations are created per country, it would be necessary to have the possibility to maintain different price conditions among them. This would even come as a legal requirement as each country may have different taxes. Figure 4.2 shows the relations among the sales organization, distribution channel, and division and the sample sales areas defined from these.

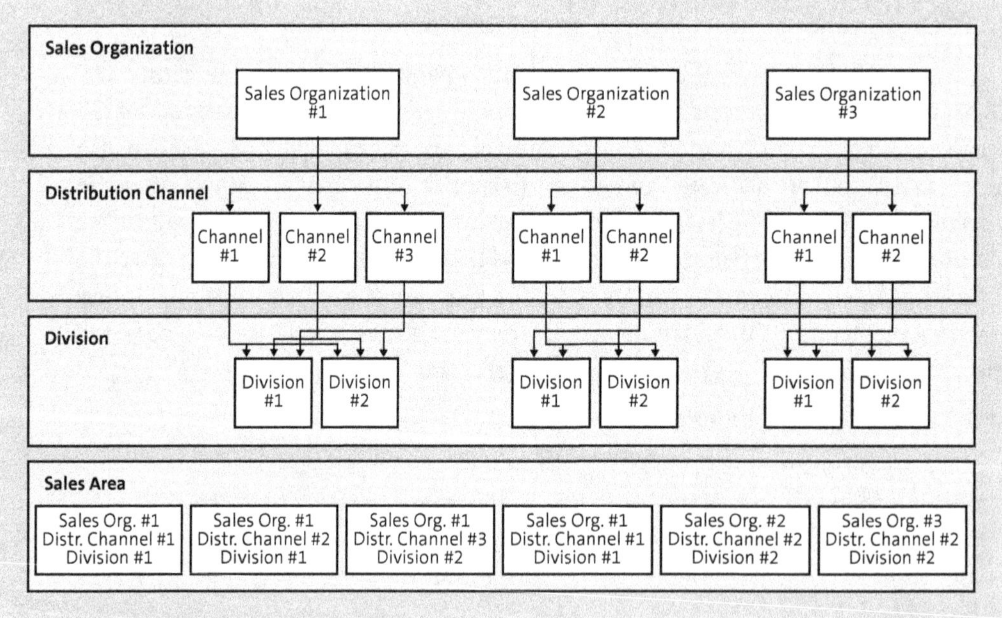

Figure 4.2 Organizational Units in Sales and Distribution Area

Now that you're familiar with the organizational structure in sales and distribution and its main elements, let's take a short look at the two master data objects: the material master record and the business partner, starting with the latter.

4.1.2 Business Partners

If you have worked with older SAP ERP systems, then you probably recall that the customer master data and vendor master data had to be maintained separately. SAP S/4HANA introduces the *business partner* concept, which provides a centralized repository and approach for managing all types of business partners. This improves data quality and streamlines business processes and reporting. It increases the overall efficiency of the data maintenance as well as relationships with business partners.

The business partner is the main entity with which an organization has any kind of relationship. The different relationships are roles like customer and vendor, as well as employee and many more. The starting point is the general data maintained in Transaction BP; all the transactions which you may know from SAP ERP and which were related to direct maintenance of customers, vendors, and others are obsolete in SAP S/4HANA. In SAP S/4HANA, you will not use transactions like XK01, XD01, VD01, FD01, FK01, and dozens of others. Each business partner has a unique identifier and can be grouped as follows:

- Business partner categories (person/organization/group)
- Grouping (number range)
- Roles (customer, vendor, bill recipient, goods supplier, ship-to party, etc.)

Figure 4.3 shows a sample business partner from the organization category. It has multiple roles created: business partner (general), collections management, customer, customer (financial accounting), and financial services BP.

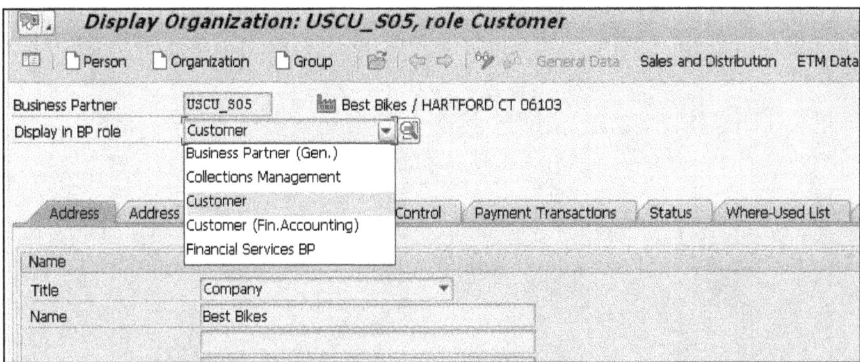

Figure 4.3 Sample Business Partner and Its General Data

Figure 4.4 shows the same business partner, but this time in the view of its sales and distribution data. This business partner has a customer role. It is active for a specific sales area, and within this sales area it has multiple partner functions like sold-to, ship-to, payer, and bill-to party.

Figure 4.4 Sample Business Partner and Its Sales and Distribution Data

4.1.3 Material Master Record

The material master record is the next critical element of each sales-related document. We cannot trigger any sales activities without having properly defined master data for our products.

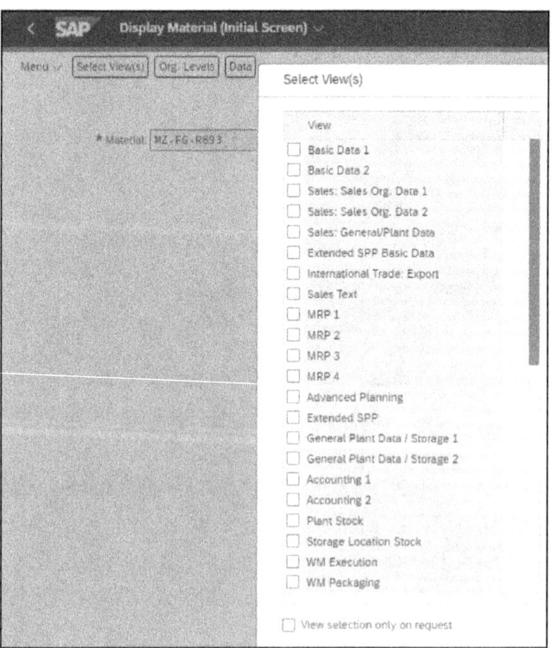

Figure 4.5 Different views Available when Displaying Sample Material

Like in case of a business partner, each material also has a unique identifier. Another similarity is that material master record is also maintained in multiple areas (views): basic data, sales and distribution data, purchasing data, accounting, costing, and many more. The basic data is a mandatory element, and it is complemented by the additional elements depending on the requirements, use, and purpose of the material. For our scenario, we will need a product that has material master data maintained on the sales organization level. This is the prerequisite to use the material in sales-related processes. Figure 4.5 shows the different views available when displaying sample material.

Figure 4.6 shows sales-related data of a sample material.

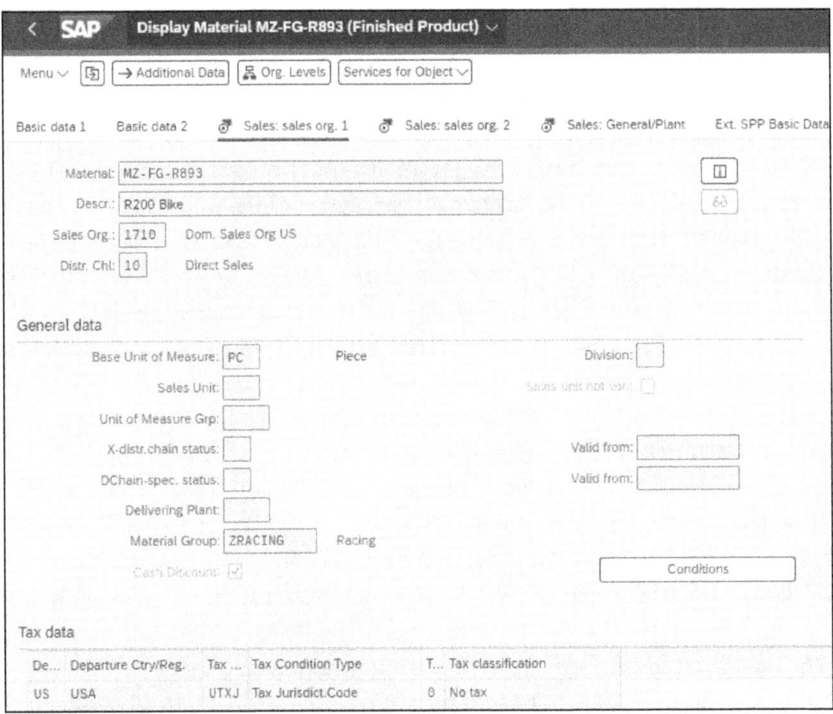

Figure 4.6 Sample Material and Its Sales-Specific Data

In Chapter 5, Section 5.1.2, you will find additional information on the exact sales area, business partner (customer), and material used in our EDI scenario.

4.1.4 Sales and Distribution Processes

In this section, we will explain the foundation of the sales and distribution process. This will be a simplified explanation of the key aspects, but very helpful for proper understanding of our overall EDI scenario. Let's look again at the diagram shown in Figure 4.1.

The first step mentioned in the diagram is related to presales activities. This is something we will ignore in our case, but you need to be aware that the sales order document itself can be created with a reference to some presales documents like, for example, a quotation or sales contract.

The second step in the diagram is sales order management. This is where we start: we will create a standard sales order document based on an EDI message from our trading partner (via an interface). We will have to make sure that the document is created within correct sales area and with correct master data elements. While the trading partner may know our products and their identifiers, it is unlikely that they know the exact identifiers of business partners existing in our SAP S/4HANA system. The same counts for the sales area; this is our internal configuration, specific to SAP S/4HANA, and we cannot expect any third party to specify the exact sales area in which they want to buy goods. These are topics that need to be handled during the functional and technical configuration of the EDI transactions. The moment our sales document is created within SAP S/4HANA, we automatically trigger a sales order response message for the business partner to acknowledge receipt of their request and its acceptance (or rejection). While discussing the process with the trading partner, you may hear about a *purchase order* instead of *sales order*, and this is also correct: the trading partner purchase goods from our company which acts as supplier for them. In the ERP system of our trading partner, this document exists as a purchase order, which is what a buyer issues to a seller. When we look at the sales order document, it is also important to understand its structure. All the elements of the sales order document can be grouped into three levels: sales document header, sales document item, and schedule lines belonging to the subsequent item. There is always one header section and usually a few or several line items, one per ordered material, but it is not a problem for SAP S/4HANA to handle sales order documents with hundreds of items. Each item line can also have several schedule lines, this is where delivery-related data like quantity and delivery dates are stored. In case of partial delivery of an item, there can be several schedule lines on that item. The data located on the document header is applicable to the whole document; the data located on the item level is applicable to the specific item. The SAP Fiori applications and older SAP GUI transactions offer multiple views on the sales document, all of which distinguish the document structure based on the mentioned levels. You can see the key elements of the sales order document in Figure 4.7.

From the elements mentioned previously, we will pay some extra attention to the header partners as these are always an important aspect of any EDI implementation and processing. As we mentioned earlier, SAP S/4HANA follows on the concept of business partners but we will still have parties with different functions involved in the sales process. There are four mandatory partner functions for sales order processing:

- Sold-to party—places the order
- Ship-to party—receives goods
- Bill-to party—receives the invoice
- Payer—pays the invoice

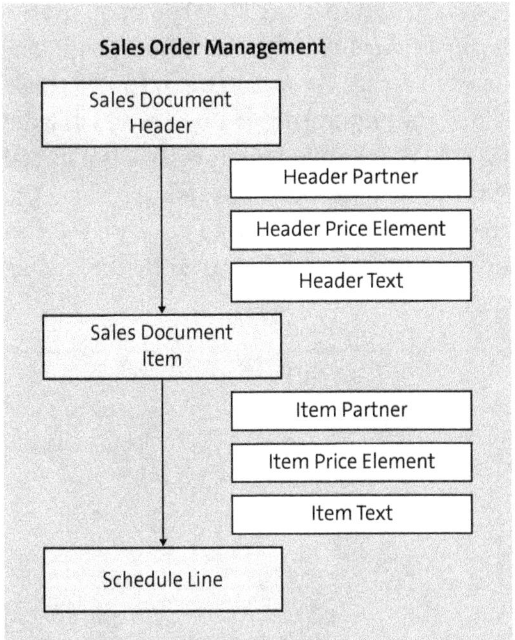

Figure 4.7 Structure of the Sales Order Document

When creating sales order documents, it is not necessary to enter all four partner functions. The key one is the sold-to party, and it is possible to configure the system and maintain the master data in such a way that all the others will be determined automatically based on the sold-to value. This also is true for EDI processing: it is not necessary to have all these parties in the inbound EDI sales order message. You may notice that there is nothing about the vendor being a mandatory party. That's correct as the vendor is covered by the sales area, so instead of having a mandatory vendor party there is a requirement that in each sales document it is mandatory to specify the sales area. Looking at the sales area defined within SAP S/4HANA, we can determine what the sales organization is and at the end also the exact legal entity that sells and delivers the goods.

Figure 4.8 shows a sample sales order displayed in an SAP Fiori application. You can clearly see the header information located at the top part of the screen, like **Sales Order Number**, **Sold-to Party**, **Ship-to Party**, and more. In the middle part of the screen, in the **General Information** section, you will find additional header-related data. At the bottom of the screen, you can see the **Items** section, which contains item-related details; in this example, there is just one line item.

Shipping is the third step in the simplified order-to-cash process. The outbound delivery document will be created with a reference to the sales order document, and it will automatically pull the necessary data from the sales order header and item lines. When executing the EDI scenario, we will simulate warehouse-related activities by maintaining a picking quantity in the outbound delivery document as if we had prepared the goods for shipping. You can imagine this step as moving the goods from storage location in the warehouse to a loading ramp or even a truck. After that, we will post a goods issue for the outbound delivery. This would be the physical departure of products from the warehouse. The goods issue is the last step of the shipping process. This material movement automatically posts a stock removal from the warehouse and generates the corresponding accounting documents. From the outbound delivery itself, we will also send an EDI message to a trading partner. The purpose of this is to notify the trading partner on the upcoming shipment and its details.

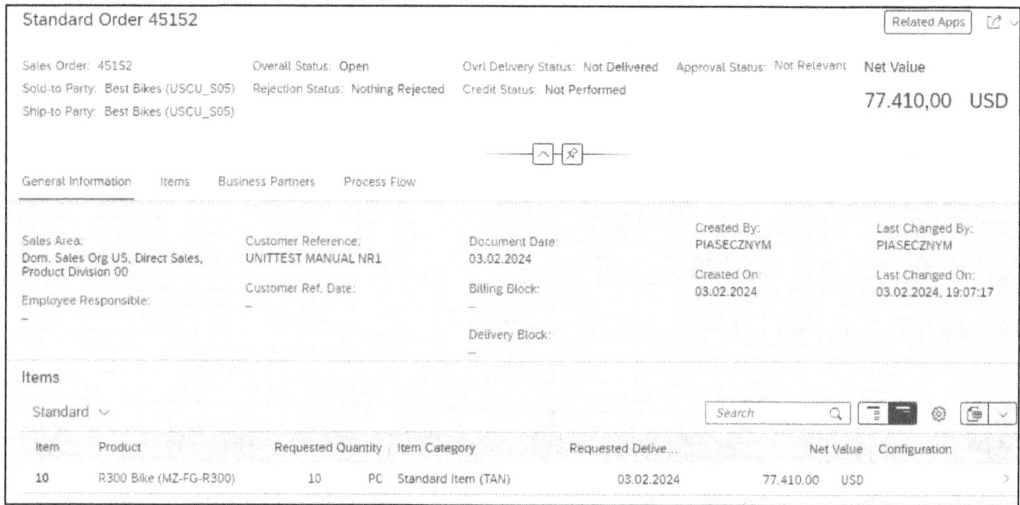

Figure 4.8 Sample Sales Order

A delivery document is grouped into two levels: header and item. The data is distributed across these levels as follows:

- **Delivery document header**
  The data located in the document header is valid for the entire document.
- **Delivery document item**
  Each item in the delivery document contains its own data. This includes data about the material, quantities, and weights, as well as stock information. Each delivery document can have several items.

Figure 4.9 shows the initial screen for creating an outbound delivery document with a reference to the sales order.

Figure 4.10 shows a sample outbound delivery document. Although the purpose of this document is different, you can find many analogies to the sales order document. There are again clearly visible header and item data.

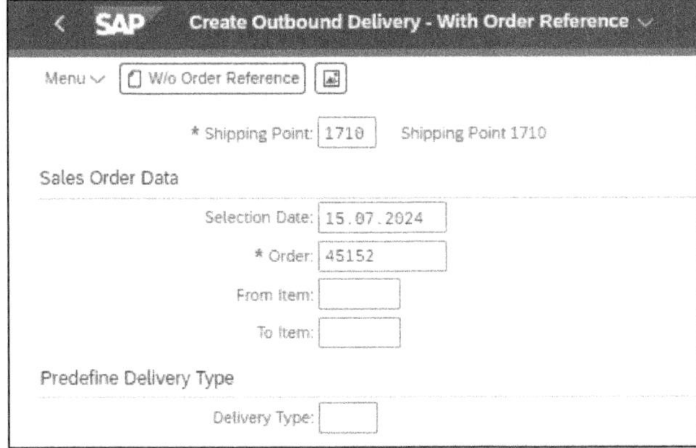

Figure 4.9 Creation of the Outbound Delivery Document with Reference

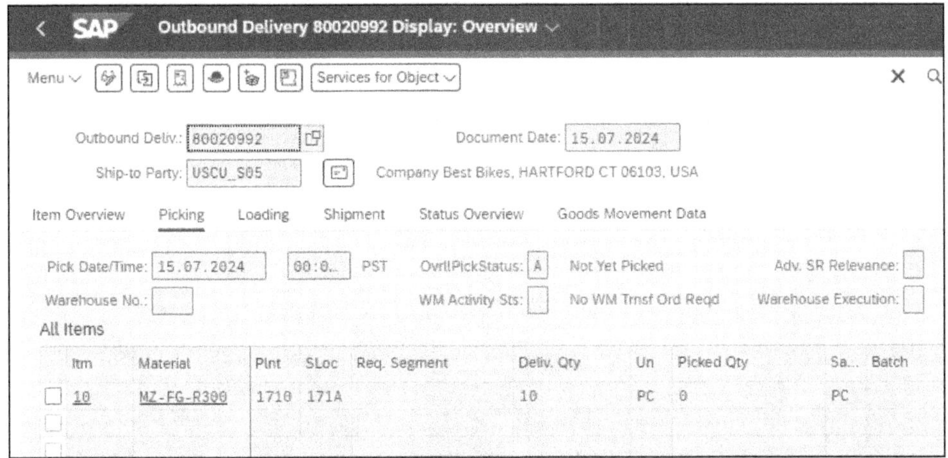

Figure 4.10 Sample Outbound Delivery Document

With the goods being delivered, we can go to the last step, which is billing. With a reference to the outbound delivery, we will create an invoice document. There is no need to maintain any additional data as most of the necessary information is pulled from subsequent sales order, the outbound delivery, and the finance-related Customizing and master data. From the invoice document, there will be also be an EDI message triggered for the trading partner. The structure of a billing document is similar to the previous sales order and outbound delivery documents. There is a header level and item level, and the header data again is applicable to the entire document, with the item data only specific to the given line item.

Figure 4.11 shows a sample invoice document with a clear distinction between header and item data.

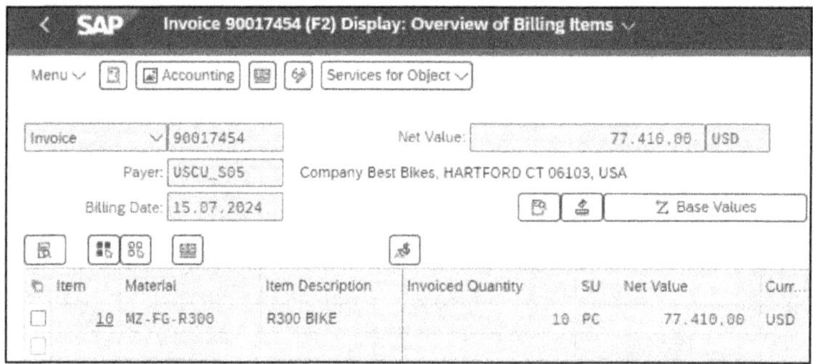

Figure 4.11 Sample Invoice Document

The last step mentioned in the diagram is accounting. This is critical for the company due to the legal requirements and because this is where payment for the goods happens, but it is out of scope of the EDI scenario presented here.

As you see, the process covers all steps starting from receiving the sales order from customer through the actual delivery of the goods and payment collection. It may not be so obvious from the short description here, but the process is highly integrated and customizable. There are many aspects that can be enhanced and automated, resulting in efficiency and superior customer service.

Document Flow

The documents within a sales and distribution process are linked to each other via the document flow. This enables you to access the history and current status of your sales and distribution processes at any time. You can display the document flow as a list with the linked documents, as shown in Figure 4.12. Depending on the document from which you call up the list, all the relevant preceding and subsequent documents are displayed. From this list, you can display the relevant documents or call up status overviews for the documents. This provides a quick overview of the progress of your sales processes at any time and can be used to answer customer questions quickly and reliably.

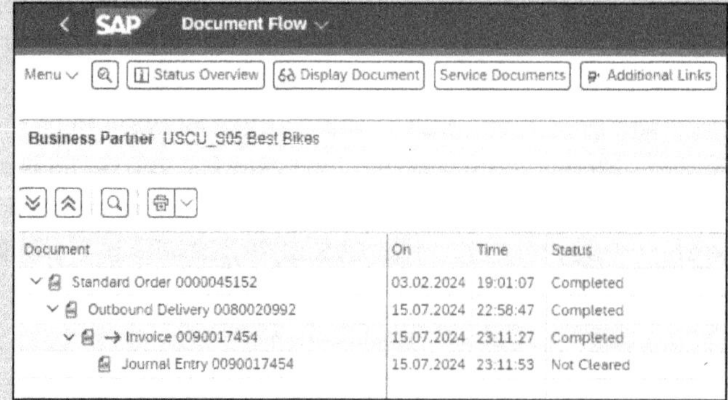

Figure 4.12 Sample Document Flow

4.1.5 Integration of Sales with External Buyers

In SAP S/4HANA, integrating sales with external buyers involves connecting and exchanging sales data with customers who use external systems. This can include online marketplaces, e-commerce platforms, or other external systems used by customers. Here are the key steps involved in integrating sales with external buyers in SAP S/4HANA:

1. **Creating a sales order**
   The first step in integrating sales with external buyers is to create a sales order in SAP S/4HANA. This sales order contains all the relevant information about the order, such as the products or services ordered, quantities, and pricing.
2. **Exchanging data with an external system**
   After the sales order is created, the system sends the data to the external system used by the customer. This can be done using various methods, such as EDI, API integration, or file transfer.
3. **Receiving order confirmation**
   Once the external system receives the sales order data, it can confirm the order and send back an order confirmation to SAP S/4HANA. This confirmation includes information about the order status, estimated delivery date, and other relevant information.
4. **Fulfilling the order**
   After receiving the order confirmation, SAP S/4HANA can begin the process of fulfilling the order. This includes picking, packing, and shipping the products or services ordered.
5. **Exchanging shipment information**
   Once the products or services are shipped, SAP S/4HANA can send shipment information to the external system used by the customer. This includes information about the shipment status, tracking number, and estimated delivery date.
6. **Receiving payment information**
   After the products or services are delivered, the external system can send payment information to SAP S/4HANA. This can include information about the payment status, payment method, and amount paid.

In summary, integrating sales with external buyers in SAP S/4HANA involves connecting and exchanging sales data with customers who use external systems. By automating these steps, organizations can streamline their sales processes, improve efficiency, and provide better customer service.

4.1.6 Implementation Context

In SAP S/4HANA, *implementation context* refers to the specific environment in which the implementation takes place. This includes factors such as the business requirements, organizational structure, existing systems and processes, and technical infrastructure. Here are some key aspects to consider when defining the implementation context for SAP S/4HANA:

- **Business requirements**
  The first step in defining the implementation context is to identify the business requirements. This includes understanding the goals of the implementation, the specific business processes that need to be supported, and any regulatory or compliance requirements.
- **Organizational structure**
  The implementation context also includes the organizational structure of the company. This includes understanding the various departments and roles involved in the implementation and how they will be impacted by the new system.
- **Existing systems and processes**
  Another key aspect to consider is the existing systems and processes in the organization. This includes understanding how data flows between systems, any existing integrations, and how data is currently managed and stored.
- **Technical infrastructure**
  The implementation context also includes the technical infrastructure that will be used to support the new system. This includes understanding the hardware and software requirements, the network infrastructure, and any other technical considerations.
- **Project scope**
  Finally, the implementation context includes defining the scope of the implementation project. This includes understanding the timeline, budget, and resources available for the project, as well as any risks or constraints that may impact the implementation.

By defining the implementation context in SAP S/4HANA, organizations can better understand the scope and requirements of the implementation project and ensure that the new system is aligned with the specific needs of the business.

4.2 Technical Solution Design

After reading Section 4.1, you should have elementary knowledge of the sales and distribution business process and its core transactions. You know exactly which business transactions to automate via your newly built EDI solution. But when defining the overall technical solution for the new EDI environment, the four transactions and the requirements of your first trading partner are not the only key factors. The core principle is that the overall solution should not be limited to a single standard: it should be flexible enough to handle multiple EDI standards, and to a certain extent nonstandard requirements like custom message formats as well.

4.2.1 Requirements

The first trading partner that will be part of our pilot implementation will act as a customer to our own company. We will be in the role of a supplier. The trading partner name is *Best Bikes*, and it comes with a requirement to enable four business transactions on EDI:

- Sales order—send from the trading partner to our own company
- Sales order processing—send from our own company to the trading partner
- Advance shipping notice—send from our own company to the trading partner
- Invoice—send from our own company to the trading partner

We already do business with this trading partner, so the process itself is not new, but at this point everything is performed manually, with communication relying on emails and phone calls. This is important as the EDI implementation for an existing business process is always simpler than one combined with the introduction of new trading partner. The latter is often connected with a need to clarify and agree on multiple functional aspects and then also maintain master data or even functional configuration. In this case, we can simply take the existing process flow and documents as a reference for EDI implementation, and all the required master data is already in place.

We already know the high-level functional requirements and transactions to be implemented, but let's also look at the additional technical requirements of the trading partner:

- Make use of the EDI message standard UN/EDIFACT D.96A S3. The exact message types are as follows:
  - UN/EDIFACT ORDERS D.96A S3
  - UN/EDIFACT ORDRSP D.96A S3
  - UN/EDIFACT DESADV D.96A S3
  - UN/EDIFACT INVOIC D.96A S3
- For each received interchange, we should send back the service report message UN/EDIFACT CONTRL D.96A S3.
- The transmission should be established via the Applicability Statement 2 (AS2) communication protocol:
  - Both parties should use the HTTPS protocol. Do not apply additional authentication; each payload should be encrypted as per the AS2 standard.
  - Both parties should use the message disposition notification (MDN) to confirm receipt of a message on AS2 transmission.
- We implement (near-)real-time processing with no specific time window for message processing. There will be no additional schedulers nor batch jobs.

Figure 4.13 shows a simplified view of the requirements that are starting point for our design.

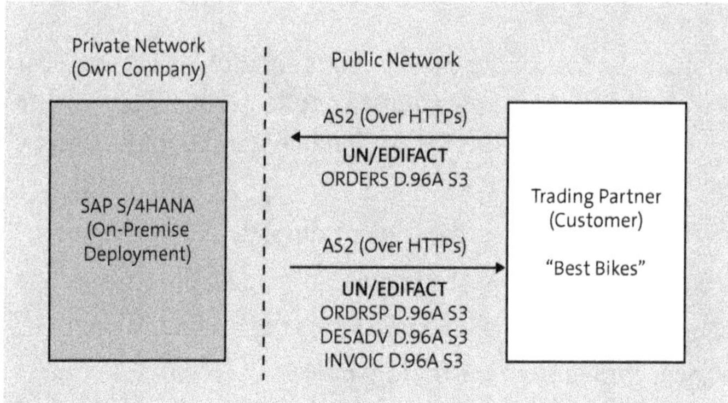

Figure 4.13 Requirements from Trading Partner's Point of View

These are only the key elements. In real-life cases, there are multiple other functional and technical details on which both parties need to agree when setting up EDI. For us, this is enough to begin. We can start defining the landscape with requirements for the initial flows in mind, but our decisions would be based on the fact that we need to support the full set of EDI standards. On top of that, we have our own requirements related to the final solution. Among these there are some operational aspects, which we will mention and explain in the following sections.

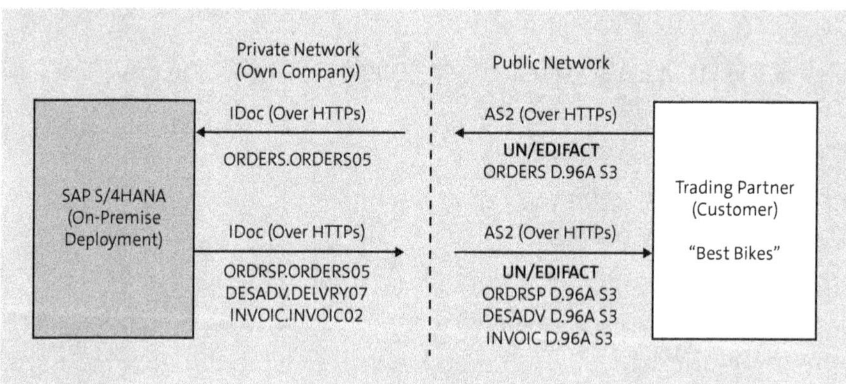

Figure 4.14 Requirements from Both Trading Partner and Your Company's Viewpoints

What we can reveal already, and it should not be a surprise to anybody, is that the SAP S/4HANA system itself can't handle UN/EDIFACT format or the AS2 connection. It does support the B2B EDI integration pattern in many ways but does not directly support these two global standards. On the SAP S/4HANA side, we will make use of different technologies: SAP IDocs for messaging and HTTPS for communication. Although AS2 protocol is HTTP(S)-based, it is not possible to connect a system communicating with "pure" HTTPS with another system that uses AS2 over HTTP(S). The conclusion is obvious: we need something in between to mediate the integration. But the purpose of the middleware system is not only to mediate on the technical aspects of the connected system and parties. For B2B EDI integration, the middleware has a much more important role: management of all trading partners and EDI transactions. This is a prerequisite for a transparent and efficient B2B EDI solution. Figure 4.14 shows the technology on SAP S/4HANA side and trading partner side.

Let's move to the next section, where we will detail the technical setup of the environment.

4.2.2 Diagram of the Landscape

When we look at the highest level, we have two main components:

- The ERP system of the company is on-premise SAP S/4HANA.
- SAP Integration Suite is the integration platform to support SAP and non-SAP integration, including B2B EDI.

The latter is a set of multiple components and services, and it needs to be detailed, but it is not different for SAP S/4HANA, which also offers multiple integration options. Then we also have different options for connecting these two main components. We need to detail this; we'll start with a diagram presenting the overall landscape, and from there we'll go step by step through each element.

Components Overview

Let's start with a high-level diagram that contains the on-premise SAP S/4HANA system and SAP Integration Suite running on SAP Business Technology Platform (SAP BTP; see Figure 4.15). All communication will make use of the HTTPS communication protocol, which gives basic security for exchanging data. To secure and fully control communication from SAP BTP to systems in the internal network, we will make use of the cloud connector. You will learn more about the cloud connector in Chapter 6, where we will walk you through the installation process and configuration for our use case.

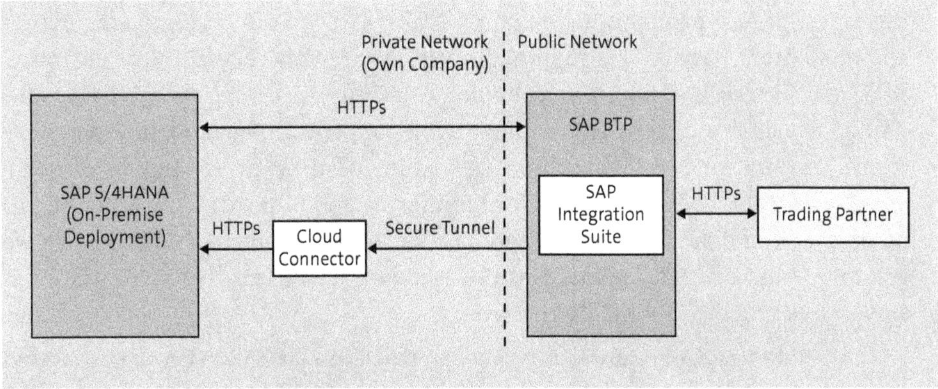

Figure 4.15 High-Level View of SAP S/4HANA Landscape for EDI Processing

As we mentioned earlier, each of the components visible in Figure 4.15 offers multiple options for integration. Let's detail the SAP S/4HANA part. Figure 4.16 shows the main building blocks and technology involved in the EDI processing on the SAP S/4HANA system.

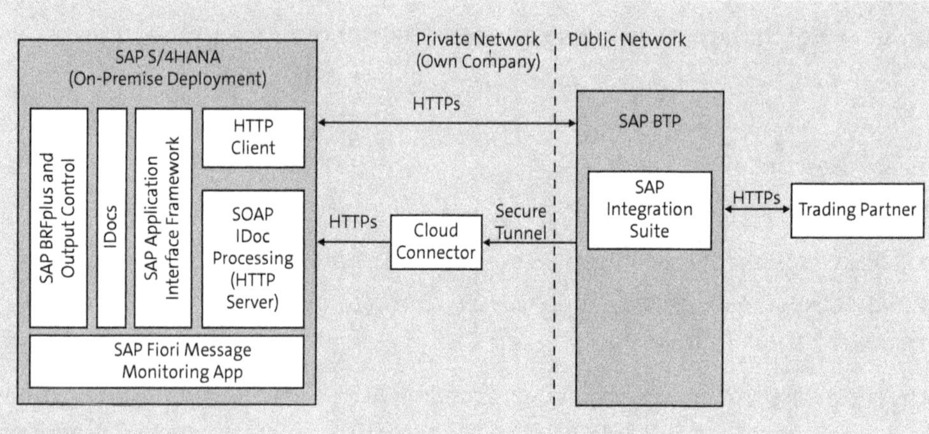

Figure 4.16 High-Level View of EDI Landscape with Detailed Components within SAP S/4HANA System

Let's walk through each of these components:

- **IDocs**
  On the SAP S/4HANA side, we will make use of IDocs. These are still available and supported by SAP products like on-premise SAP S/4HANA and SAP Integration Suite. From a technical point of view, the motivation behind this choice is that B2B EDI uses asynchronous message processing and IDocs are fully compliant and well proven for this integration pattern. The fact that IDocs are a well-known technology is yet another factor: almost each SAP consultant is familiar with the IDoc concept and there are numerous articles, blog posts, and forum discussions related to this topic. That's a significant factor that allows a quick start to the project and simplifies the aspect of finding appropriate team members to work on the implementation and in a latter phase on the support. This is also directly related to the costs and risks; making use of well-known technology makes the planning more reliable with limited possibility of something unexpected happening. Operational aspects are crucial as well, and the asynchronous nature of IDoc processing within SAP S/4HANA meets the requirements for monitoring and error handling according to established standards. You can learn more about IDocs in Chapter 7; at this stage, we will only mention that we make use of standard SAP IDocs. These are the ones already existing in SAP S/4HANA system:
  - The SAP S/4HANA system will receive an ORDERS.ORDERS05 IDoc for inbound sales orders.

- The SAP S/4HANA system will send an ORDRSP.ORDERS05 IDoc for outbound sales order responses.
- The SAP S/4HANA system will send a DESADV.DELVRY07 IDoc for outbound advance shipping notices.
- The SAP S/4HANA system will send an INVOIC.INVOIC02 IDoc for outbound invoices.

The advantage of choosing standard IDocs is that they come with predefined structures, and most of the configuration needed for processing these messages is already available in the system. The IDocs are also fully supported by SAP Integration Suite and its components, and we will take a closer look at this in the following sections.

Figure 4.17 shows the documents from the order-to-cash process and the corresponding IDoc types.

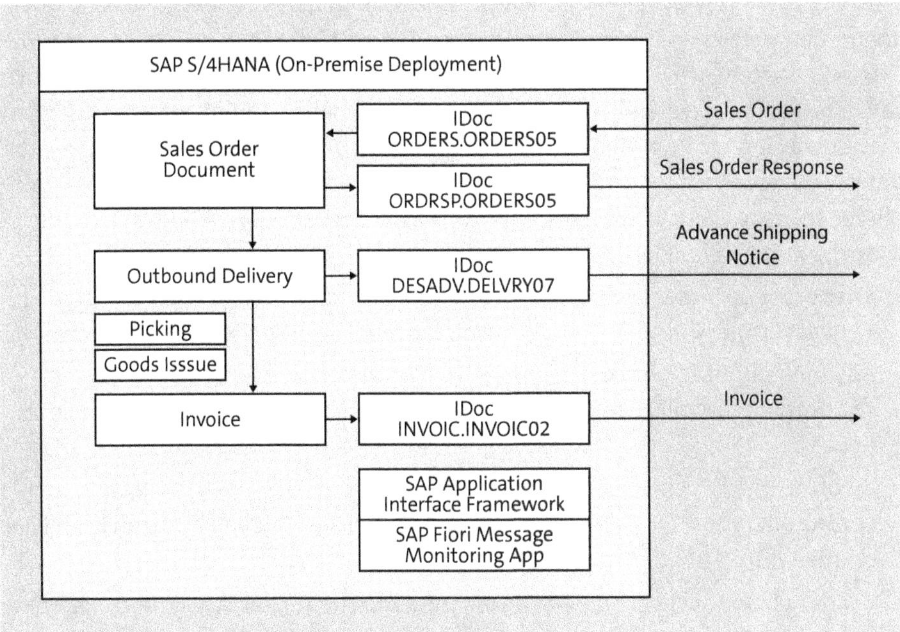

Figure 4.17 Documents in Order-to-Cash Process and Corresponding IDoc Types

- **HTTP client and SOAP IDoc processing (HTTP server)**
  The HTTP client and SOAP IDoc processing is something we'll elaborate on in Chapter 6, but let's give you a short introduction here. An IDoc is only a message structure, but it can be transmitted in different ways. The most common methods in the past have been the transactional remote function call (tRFC) and file-based transfer. In the latter, our IDoc is represented as a flat file on a file system and then it can be transferred like any other file. These two options are still available, but in our example, IDocs will be transmitted to and from SAP Integration Suite via the HTTPS pro-

tocol. The protocol used for IDoc transfer does not influence processing on the SAP S/4HANA side; the communication layer is fully independent from the technical and functional processing of the IDoc. The HTTP client and all the other components required for such functionality are included in the standard part of the SAP S/4HANA system. The transmission of inbound IDocs will also make use of HTTPS. On the SAP S/4HANA side there is a standard SOAP service capable of receiving and processing IDocs. Underneath that there is of course an HTTPS server, but we don't have to go into details of its functionality here.

All trading partners and all transactions will reuse the same connection. In more technical terms, there will be a single HTTPs port on the SAP S/4HANA side, and all outbound IDocs in the system will be sent to it. When we look at inbound IDoc to SAP S/4HANA system, there will be a single entry point (service) on the SAP S/4HANA side capable of receiving any IDoc type.

- **Business Rules Framework plus (BRFplus) and output control**
  This is closely related to the selection of IDocs. Whenever we have a business document, such as a sales order, and we need to send an EDI message from the SAP S/4HANA system, we need to have an appropriate output control mechanism that will define for which document we need output and what kind of output. This is just a very generic and simplified explanation, but the conditions for triggering the output can be really complex. There can be also multiple different output types. Limiting the scope to our IDoc-based scenarios, we need one or more solutions to trigger IDocs from the following:
  - **Sales order document**
    The output type will be an order confirmation and it will be sent via an IDoc channel.
  - **Advance shipping notice**
    The output type will be an advance shipping notice and it will be sent via EDI output.
  - **Invoice**
    The output type will be an advance shipping notice and it will be sent via an IDoc channel.

You can find more details on output configuration in Chapter 7, Section 7.3.3, where we will go through the configuration of all necessary outputs for our EDI scenario. As you may notice, there are two components mentioned and used: BRFplus, which is the newer and recommended approach for output control (and many more aspects); and the "old" output control functionality known from older SAP ERP systems. The reason behind this is that both output technologies are supported by SAP S/4HANA, but this comes with certain conditions and limitations. Looking at our EDI solution, the recommendation and the choice was to use BRF+, but due to limitation within SAP S/4HANA, it wasn't possible to use it for outbound delivery documents. We were forced to make use of the old output control for this specific document; the benefit is that you can see both solutions at work. Refer to Chapter 7, Section 7.3.3 for more details.

- **SAP Application Interface Framework**
  To facilitate the integration solutions on SAP S/4HANA side we decided to involve SAP Application Interface Framework. As this component is embedded within SAP S/4HANA system it comes with limited effort to start using it. Its capabilities are not limited to IDocs and EDI processing but that's one of the best use cases for it. The main motivations for choosing SAP Application Interface Framework are monitoring- and operations-related aspects. The key requirement for the overall solution is that there is full visibility into EDI processing within SAP S/4HANA. This transparent view also should allow for handling errors that may occur during the processing. Monitoring and error handling are key features offered by SAP Application Interface Framework. Alternatives to this could include usage of the SAP GUI–based transactions like Transaction WE02, but these are more technical transactions with a set of limitations, and they definitely lack transparency from the view of a usual business user. In addition, with an SAP S/4HANA implementation, it is quite often decided to make SAP Fiori a default and the only user interface for business users. In other words, business users may have no access to SAP GUI at all, and SAP GUI is only available for IT consultants who perform configuration or development on the system. (This is also the approach used in our case: while most of the configuration and development within SAP S/4HANA is performed in SAP GUI, the execution of the business process and validation of EDI transactions is fully performed in SAP Fiori.) The solution for error monitoring and full visibility of EDI processes is the Message Monitoring SAP Fiori applications.

In Chapter 8, you will go through all configuration steps required to enable an SAP Application Interface Framework monitoring scenario for each of the four interfaces. For now, let's give you a brief overview of the functionality we are going to enable there:

- We will create a dedicated AIF index table for each of the interfaces. This is a prerequisite for the AIF processing to work as the table stores details of each processed IDoc. Apart from the standard information related to the characteristic and the processing status, we will customize additional fields related to the content of the message. These so-called key fields can be used as search and filter attributes within the message monitoring transactions. In our case, these will be, for example, values related to the sales organization, plant, company code, customer name, purchase order number, delivery number, or invoice number.
- Searching and filtering within message monitoring is not the only functionality enabled by key fields. SAP Application Interface Framework has a set of features to support data protection and privacy, and the concept of *recipients*, in which the specific interfaces and data processed through them is visible only for a predefined set of users. This is managed via recipient groups: in a properly designed approach, we achieve a situation where IT support or business users get visibility and access only to interfaces for which they are responsible and only those that are applicable to

their role in the organization. In our case, we will prepare a foundation for such a concept by configuring the EDI interfaces in such a way that the recipient groups will be determined dynamically based on the organization unit from the message. For example, for a sales order interface, we will extract the sales organization value, and then we will use it to determine the recipient group that is specific to this sales organization.

- When we enable IDoc monitoring through SAP Application Interface Framework, we get the option to make use of SAP Fiori–based applications dedicated to (IDoc) message monitoring and error handling. There are two key apps: Message Monitoring Overview, and Message Monitoring. Both will be presented in Chapter 10 when we will verify the overall solution by processing all EDI transactions. Earlier, in Chapter 8, we will go through step-by-step instructions for enabling IDoc monitoring in SAP Application Interface Framework. Additional motivation for choosing SAP Application Interface Framework includes its features related to data security and access control. As you will see in Chapter 8, we will apply a concept in which visibility into and responsibility for interfaces is limited on an organizational level. This means that users will see only the interfaces to which they are assigned, and the visibility into business data also can be protected by additional authorization objects. On top of that we will enable notifications: there is no need for users to keep checking the SAP Fiori dashboards as they can be automatically notified by email of any unexpected situations (e.g., failing inbound sales order messages).

To summarize the design choices on SAP S/4HANA, all choices rely on functionality and features existing within the standard system. We will not introduce any changes to deviate from or overwrite standard SAP functionality. This doesn't come from the belief that we need to keep and use standard SAP functionality at all costs. There may be valid reasons and cases to introduce custom functionality or to extend the standard one. Building a future-proof EDI solution for an multinational enterprise operating on a global SAP S/4HANA system may require more functionality than offered within the standard. There is nothing wrong with using additional Customizing or custom code. More important is to keep a consistent approach and proper governance over all extensions that are being introduced. Especially in case of B2B EDI integration, where we may deal with hundreds or even thousands of different trading partners, it is recommended to define the solution as a framework with reusable building blocks rather than employ ad hoc handling of each unique deviation from the standard. With that said, our EDI scenario relies fully on standard SAP because we want to demonstrate the end-to-end solution in a basic use case. This is the starting point from which you can deviate, even to the point of using a custom solution in certain areas.

After this brief review of design choices related to SAP S/4HANA, let's now look at the next component: SAP Integration Suite.

SAP Integration Suite as an Integration Platform

Choosing SAP Integration Suite as a technical platform for B2B EDI processing is only part of the story. More important than the tool itself is its usage and the overall approach to this specific integration pattern. We could try to distinguish three different approaches for how to involve SAP Integration Suite with B2B EDI processing:

- Fully follow the recommended SAP solution for B2B EDI with all its components.
- Make use of SAP solutions for B2B EDI, but not necessarily with all components.
- Have a custom-built solution that avoids using an SAP solution for B2B EDI or uses only a small part of it.

All three scenarios still make use of SAP Integration Suite and Cloud Integration, but not necessarily of all the additional components like Integration Advisor or Trading Partner Management. The main difference is in the configuration and development part, where we can reuse packages delivered by SAP and deploy them as per guidelines, but we can also change them or even omit them and build custom integration flows from scratch. It is not that the latter makes sense, but you may have very specific requirements and valid reasons for doing this. Technically, it is an option.

Our EDI scenario will be fully configured and deployed on the recommended SAP solution for B2B EDI, so let's briefly introduce you to it.

We will make use of the *Cloud Integration—Trading Partner Management V2* SAP Integration Suite integration package. This package contains integration flows and scripts that define the pipeline for processing B2B EDI messages. The processing relies on and is controlled by configuration done in Trading Partner Management. Next to this there is also a need for runtime objects that define the messages and mapping. In our case, this will be defined and delivered by Integration Advisor; we will elaborate on this later.

The integration flows and scripts included in this package are available as read-only artifacts, allowing only basic but sufficient configuration options (via externalized parameters). The standard artifacts should not be changed; even if you need to add custom functionality, there are predefined steps in which this can be done. In our case, the standard artifacts delivered in this package are also enough: we will not be changing the standard artifacts nor developing any additional integration flows or scripts.

We will make use of the following SAP Integration Suite components:

- **Cloud Integration**
  As mentioned previously, Cloud Integration will be used as a runtime for the package mentioned earlier. We will not be developing any additional artifacts; that's the first benefit of using the standard SAP solution. Cloud Integration is also the component we will use to monitor our EDI transactions. We skip the standard message overview functionality as there is a more specialized B2B scenario with monitoring as an integral part of the SAP solution for B2B EDI.

- **Trading Partner Management**
  We could say that Trading Partner Management is the core of the overall B2B EDI solution. This is where we define our own company and all its trading partners. On top of these we define EDI transactions.
- **Integration Advisor**
  Integration Advisor is the component where we will define messages and mappings among them. Behind this simplified expression are message implementation guideline (MIG) and mapping guideline (MAG) objects, which we will study in detail in Chapter 10.

Figure 4.18 shows the different components of SAP Integration Suite and relations among them.

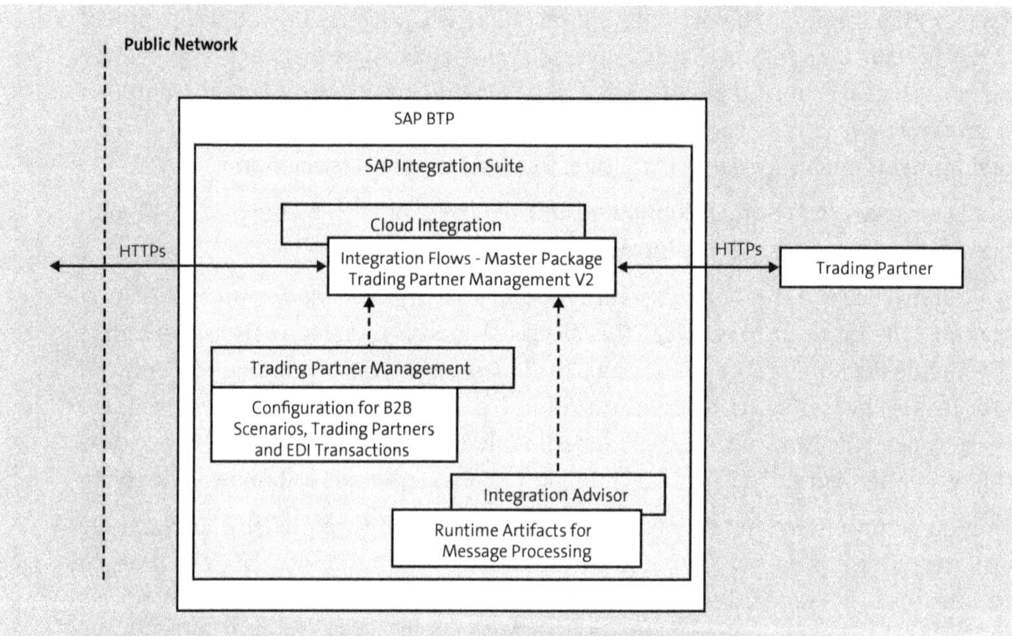

Figure 4.18 SAP Integration Suite Components Involved in B2B Processing

Let's take a closer look at the functionality of each component:

- **Cloud Integration**
  As mentioned earlier, we make use of the Trading Partner Management V2 integration package. It delivers artifacts that create the B2B EDI processing pipeline after deployment. This means that a set of integration flows (iFlows), supported by a script collection that process messages in a specific sequence. How the messages can be posted into this pipeline is also predefined, but it is not limited to one option. The output from the pipeline can also leave the pipeline in multiple ways. Both the input and output of the pipeline supports multiple connectivity protocols, starting from

AS2 and IDocs, which are required in our case, through support for ProcessDirect and SOAP. The pipeline solution has multiple steps in which we can link and trigger custom functionality, which gives us the option to implement additional connectivity options, like SFTP, AS4, and so on.

For proper operation and message processing, the pipeline solution requires a set of configuration parameters and a set of artifacts related to message definition and transformation (mappings). This is where the other two components from SAP Integration Suite play critical roles, discussed in the next two bullets.

- **Integration Advisor**
  As mentioned earlier, the Trading Partner Management V2 integration package delivers artifacts that create the B2B EDI processing pipeline after deployment. This means that a set of iFlows are executed in a specific sequence. How the messages can be posted into this pipeline is also predefined, but it is not limited to one option. There are different connectivity protocols supported by the overall solution.
- **Trading Partner Management**
  The artifacts require a set of parameters to control the processing and its conditions. These come from SAP Trading Partner Management configuration.
- **Partner directory**
  The partner directory is yet another key component of the overall B2B solution. The difference is that you do not use this component directly; we can even say that you do not use it at all. You can visualize it as a repository of parameters (string objects) and data (binary objects). The other components like Trading Partner Management and Integration Advisor make use of this repository to share all their configuration with Cloud Integration. In this way, the latter knows the details of your own company, trading partners, EDI agreements, and EDI transactions. In the partner directory, there are also runtime objects like MIGs and MAGs.

The overall concept of the B2B EDI processing pipeline is drafted in Figure 4.19. The idea here is not to show the technical details; we will study those in Chapter 9. The main purpose is to share the concept so that you can get a good understanding of the main building blocks and their functions. As stated a while ago, there are multiple connectivity options for the pipeline. For sender communication, it is implemented in such a way that there are multiple iFlows, where each contains a specific sender communication channel. You choose and deploy the one required by your scenarios. In our case, these are sender AS2 and sender IDoc communication iFlows. For receiver communication, there is a single iFlow that contains multiple receiver communication channels. The dynamic selection of an appropriate receiver channel is done within that single generic receiver communication iFlow based on the configuration parameters in the partner directory.

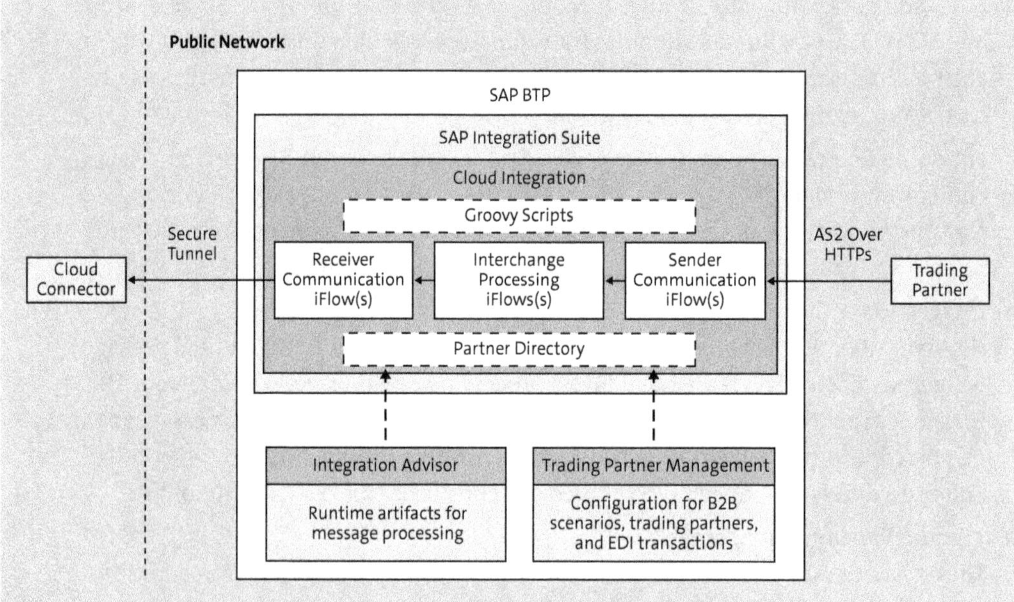

Figure 4.19 Illustration of B2B EDI Pipeline Concept

You should already understand how the pipeline processes B2B messages and how the messages are received and sent by this generic flow. In the center of the pipeline, there is also the interchange processing step, which takes care of message transformation. The obvious question at this point is how the pipeline recognizes the different messages and how it knows what to do with them. There can be thousands of different message types, and there can be as many different structures. There can be even custom structures, and for many of these we also need to define mappings. This is the exact place where Integration Advisor comes into play. Its role is to allow functional specification of the interfaces and then generate runtime artifacts out of these. The interface specification can be done without any coding (although some coding can be included if we need to handle very complex or specific requirements). There are two types of objects we create in Integration Advisor:

- **Message implementation guidelines**
  MIGs are the source or target messages in our B2B EDI flow. We need to create MIGs for each message we want to process in the pipeline. The specification involves steps like the selection of data sections and data elements that appear in our specific message; after that, we have the option to set multiple additional properties related to the message structure and its content. This step is simplified by the fact that within Integration Advisor, we can find a library of B2B standards that stores ready-to-use definitions of thousands of message types. Among many others, there are UN/EDIFACT and SAP IDocs messages, which we need.

For our EDI scenario, we need to define eight MIG objects: four related to SAP IDocs used by our company and four related to UN/EDIFACT messages used by the trading partner.

- **Mapping guidelines**
  Having definitions of source and target messages in MIG form is not enough to process messages. We need to define how the source message is transformed (mapped) into the target message. This is the role of MAG objects, which are also defined in Integration Advisor. The starting point of MAG definition is selection of the source MIG and target MAG. Like for MIGs, the specification can be done without any coding. For our EDI scenarios, we need to define four MAG objects, one for each EDI transaction in our flow.

For both MIG and MAG definition, it is possible to make use of proposals stored in a central integration knowledge base. We will not make use of that option when setting up our EDI scenario.

The MIG and MAG objects can be shared and reused across different trading partners and different EDI transactions.

Integration Advisor has also functionality to generate documentation based on MIG and MAG objects, which we will use in our sample implementation.

Figure 4.20 shows all the objects we will create in Integration Advisor and how they are positioned within the landscape. With the help of Trading Partner Management and the partner directory, the MIGs and MAGs will be available for Cloud Integration.

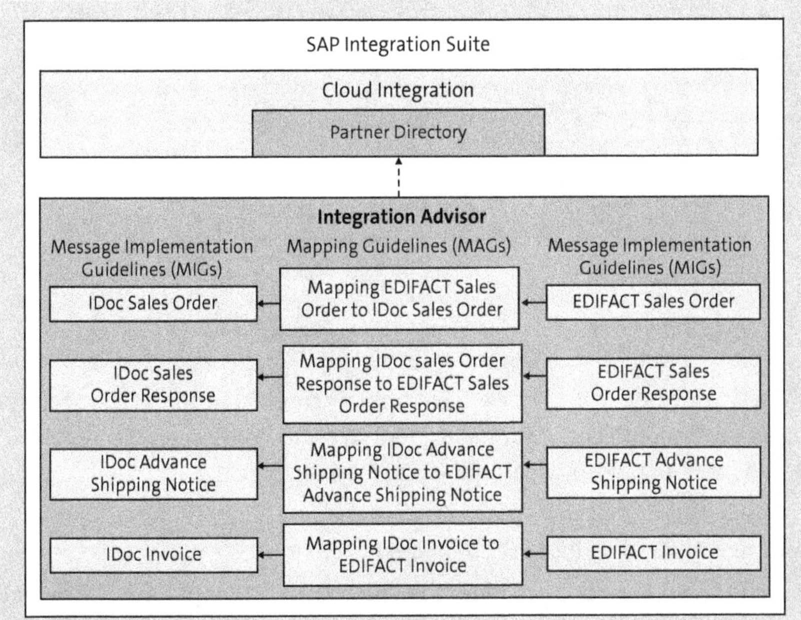

Figure 4.20 Key Configuration Items from Integration Advisor

To complete this introduction to the B2B concept in SAP Integration Suite, we can't skip Trading Partner Management. This component can be seen as the core one as it links the artifacts created in Integration Advisor with actual EDI transactions and then passed this configuration (via the partner directory) to Cloud Integration so that the message processing can be executed per our definitions (see Figure 4.21). The initial activity in Trading Partner Management is related to the definition of our own company and trading partners, which are in the scope of our B2B integration. The second part is the configuration of specific EDI scenarios (via an agreement object): this is where we link our own company, a trading partner, and specific EDI transactions and MIGs and MAGs.

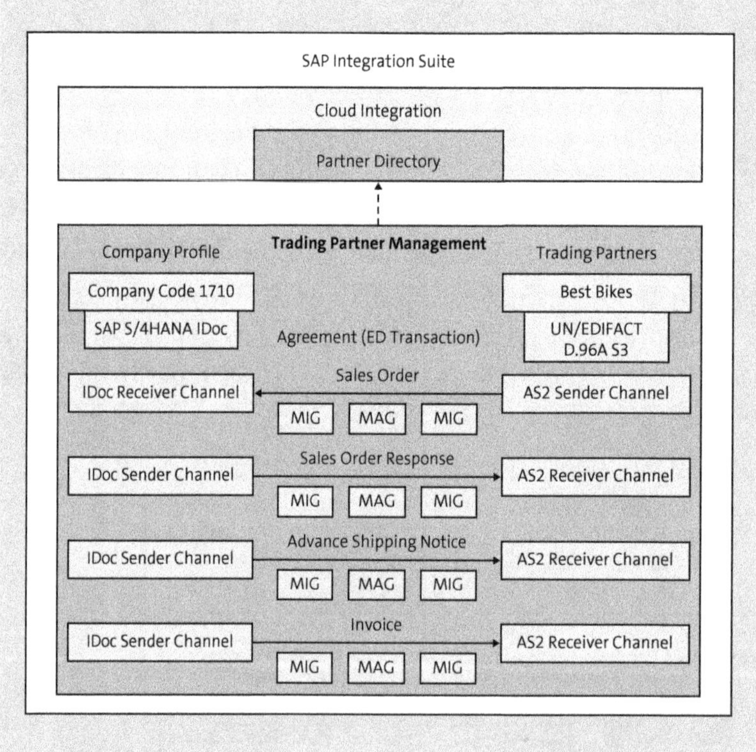

Figure 4.21 Key Configuration Items from Trading Partner Management

Last but not least, the whole solution we have implemented is in the scope of automated testing implemented via the Int4 Suite tool. We will go through the details of this solution, its functionality, and a sample scenario in Chapter 11.

Overall Landscape with Trading Partner Connected to It

In the earlier sections, you saw a few diagrams presenting different parts of the overall EDI solution in the SAP S/4HANA landscape. Figure 4.22 shows one more diagram, this one with the overall integration solution, including the trading partner connected to it.

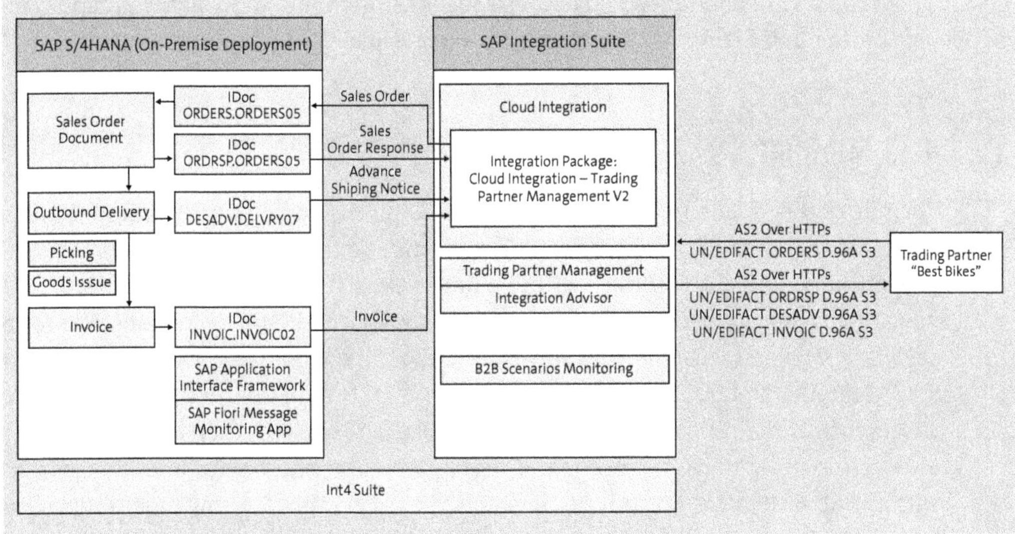

Figure 4.22 Overall Landscape with Trading Partner

We finish this chapter by leaving you with a complete technical solution for EDI processing within the SAP S/4HANA landscape. It is important to mention that the whole solution presented here is based on standard SAP components and recommendations. But that doesn't mean that there is only one solution for EDI processing within SAP S/4HANA that needs to exactly match the one presented here. You should instead look at this as a starting point, being aware that each component and the use of each component can be slightly or highly modified. Some of the components, technologies, and approaches may be modified or replaced by others. The given solution is flexible and scalable: you may have impression that by making use of standard SAP components we built a tightly coupled solution, but the opposite is true. Any additional message standards like ASC X12, TRADACOMS, GS1 XML, and many more can be supported out of the box. It isn't even necessary to make use of IDocs as the SOAP services from on-premise SAP S/4HANA are supported. The best example of the flexibility and openness of the solution is that even within SAP Integration Suite, we are not forced to make use of the whole set of predelivered components and features. Trading Partner Management is not limited to the use of MIGs and MAGs coming from Integration Advisor; you can even decide not to use mappings defined in Integration Advisor and instead make use of mappings developed in different ways, like with Groovy scripts or graphical mappings from Cloud Integration. This could be something worth considering for SAP S/4HANA or middleware platform migration projects, but that's another discussion. Last but not least, there are also many options to extend the given solution with almost any additional custom requirements related to business process or technical processing. Here of course, a standard note that any addition of custom functionality should be well planned applies. In the B2B EDI integration, the landscape may be very complex due to a high number of different standards, transactions, and trading partners. Quality, standardization, and harmonization are the keys to a reliable EDI solution.

4.3 Summary

The role of this chapter was to link Chapters 1–3 and the upcoming Chapters 5–10. The former chapters gave you a theoretical introduction to the SAP S/4HANA and EDI integration, but without any specific scenario or use case. The following chapters will give you step-by-step guidelines for how to configure your SAP S/4HANA landscape for a specific EDI scenario. This chapter aimed to fill the gap and allow a smooth transition to a very specific process, technology, and use case. In the first part of this chapter, we started with a definition and examples of business processes. Then we took a closer look at the order-to-cash process and its implementation in sales and distribution within SAP S/4HANA. To fully understand the SAP S/4HANA implementation, we examined the sales area, business partner, and material master record. With this basic information, we could then look at sales-related documents, which will be part of our EDI scenario: the sales order, outbound delivery, and invoice.

The second part of this chapter described the technical requirements and solution design. It started with a look at the EDI standard being used in our EDI scenario: UN/EDIFACT version D.96A S3. To complement this part, we also stated that IDocs will be used as the technology on SAP S/4HANA side and HTTPS will be the communication protocol, which also is used for the communication with the trading partner via an HTTPS-based AS2 protocol. We listed the components used within SAP S/4HANA and discussed the purpose of SAP Application Interface Framework. After that, we looked at SAP Integration Suite and the role of the selected components: Cloud Integration, Trading Partner Management, and Integration Advisor. The different technical choices were enriched with graphic diagrams, and at the end of the chapter we presented a diagram of the overall landscape with all its components. All this should have prepared you well for the following chapters, where you will have a chance to configure each aspect of the EDI solution. Next, Chapter 5 will give you some information on the prerequisites related to the identifiers used within EDI processing. You will also become familiar with the identifiers and master data objects used in our specific scenario.

Chapter 5
Prerequisites

There is a set of prerequisites and decisions that you should be familiar with before going into system configuration. In this chapter, you will learn about identifying the different parties involved in the EDI process. We also will summarize the approach to and choices in our model EDI scenario.

In the first part of this book, we went through the theory and gave a short introduction to EDI in SAP S/4HANA. At the end of that first part, we explained the basics of EDI, the procure-to-pay process, and the key elements of the enterprise structure defined within SAP S/4HAHA. We should already have an idea of how these are linked together, but before we start with configuration of our dummy EDI scenario on all the related components, we need to go into details and be more specific about the scenarios. The starting point is the agreement between the two parties involved in the EDI process. What are the exact business transactions for which we build the interfaces? What is the format of the EDI message, what is the communication protocol, what are the other nonfunctional requirements? We'll ask these and many other questions. We also need to look at the identifiers and master data that will be used by the other party and by us. We will go through all of this in this chapter. The intention is to make a smooth transition from the theoretical part to the practice. It can't be a direct switch from one to another.

In other words, we will be listing all the details necessary to set up a fictional EDI scenario. This is between our fictional company that resides on an on-premise SAP S/4HANA system and a fictional third-party trading partner acting as a customer and residing in its own remote IT system.

5.1 Prerequisites Checklist

We are implementing EDI processing on a sales order process in which our fictional SAP S/4HANA company acts as a supplier. The other fictional third party is a customer; we will call it a *trading partner*. The set of transactions we are enabling for EDI is as follows:

- **Purchase order**
  The trading partner sends a purchase order. From our view, it is an order.

- **Purchase order response**
  The trading partner gets a purchase order response. From our view, it is an order response.
- **Advance shipping notification (ASN)**
  The trading partner gets shipment advice. From our view, it is a notification for the goods we are going to deliver.
- **Invoice**
  The trading partner gets an invoice. This is a document based on which we get payment for the goods.

Now we will go into the technical details. Based on the mutual agreement between the supplier and customer, we agreed on the following:

- We use UN/EDIFACT standard, version D.96A S3 and the following message types:
  - ORDERS, purchase order message
  - ORDRSP, purchase order response message
  - DESADV, dispatch advice message
  - INVOIC, invoice message
- We use AS2 as the communication protocol with all its security features (payload encryption, signature) and MDN confirmation. All details are already in place, including the trading partner's certificate.
- We agreed on identifiers to be used for the sender and receiver in the EDI message headers.
- We agreed on identifiers to be used for the vendor, sold-to and ship-to parties, and payer within each EDI message.
- The trading partner shared test messages in UN/EDIFACT, which can be used for the configuration and development process on our side.

We will review all of this in the following sections.

UN/EDIFACT specification

The official and complete specification of the UN/EDIFACT format can be found at *http://s-prs.co/v573601.*

5.1.1 Agreed-Upon Identifiers

Let's assume that we already have a connection with a trading partner or, even better, with hundreds of them. Now the question is how to know what we are supposed to do with each of the thousands of messages we receive every day. Of course, we need to identify who the message is from; that would be the so-called message sender. Second, we need to identify who the message was sent to; that would be the message receiver. But this is a very simplified definition: in practice, the process should be extended a bit, and such verification should be done in multiple steps. Figure 5.1 shows the different processing steps for a case in which we receive an EDI message from a trading partner. This is a technology-agnostic view; it's applicable for an SAP S/4HANA landscape but not limited to it. Let's elaborate on these steps:

1. **Network layer**
   A trading partner sends an EDI message. For the AS2 protocol, it is an HTTP(S) call. And this step is where we identify this inbound connection (HTTP call) to the infrastructure. The destination is the AS2 entry point on which the trading partner posts the message. It is highly recommended to have the AS2 entry point protected; do not have it publicly and fully exposed. One common approach is to have a check on the IP address of the caller. Only the known IP addresses are allowed to post a message. Any unknown IP addresses are rejected and not allowed to post a message. This is a technic called *IP whitelisting*. Our first verification step would be to check the sender's network/IP address. Note that there is nothing here about checking the message type or content, but just who's trying to deliver something.
2. **AS2 adapter**
   After we've allowed the sender to make a connection and post a message, we can check what's in the first envelope. Basically, the AS2 message consists of the AS2 header and AS2 payload. The latter should be encrypted, and before we even start working on it we should verify the AS2 header. As per the AS2 standard, there are two mandatory header values:
   - AS2-From
   - AS2-To

   These are for the identification of the sending system and receiving system. That's exactly what we should do. First, we check if the AS2-From value belongs to a trading partner with whom we have an EDI agreement and EDI communication. Second, we check if AS2-To contains our own AS2 identifier.

   As per the formal specification, these values can have up to 128 characters and are case-sensitive. The exact values are up to the agreement between the trading partners, but the recommend approach is to use meaningful and unique values instead of imaginary ones.

   The values could be company-specific, using something like Data Universal Numbering System (DUNS) or Global Location Number (GLN), but they could be also any strings agreed upon between EDI parties.

   If there is a match, then we can continue with encrypting the AS2 payload and passing the message to the integration platform.

AS2 Headers

The official definition and usage of AS2 headers is specified in a publicly available document at *http://s-prs.co/v573602*.

GLN and DUNS Numbers

Powered by GS1 standards, GLN provides businesses the ability to know who is involved in transactions and where things are located throughout the supply chain. By uniquely identifying parties and locations, GLN helps with tracking products, optimizing processes, and providing greater visibility to transactions taking place around the world. GLN does for parties and locations what barcodes do for products.

DUNS numbers are unique nine-digit identification numbers provided by Dun & Bradstreet. A DUNS number is site-specific. Therefore, each distinct physical location of an entity (a branch, division, headquarters, etc.) may be assigned a DUNS number.

3. **Integration platform**
   In the earlier steps, we verified that the network connection (HTTPS call) was made from a known IP address and that the AS2 communication has been addressed to us (see the AS-To header parameter). Now we can decrypt the AS2 payload and start processing the content. The first step should not be to blindly forward the message to our ERP business system but rather to perform a thorough verification. As you will recall from the earlier introduction to EDI formats, most of the EDI standards have a dedicated section within each message to identify and specify the interchange. If we take the UN/EDIFACT as a reference, there is a UNB segment that contains the interchange sender and receiver details. These are the values that should be used in the middleware platform to define the processing steps and destination (ERP) system for our EDI message. We also use the UNH segment, which tells us the message type and version, but the sender and receiver are the ones to build the routing on.

   For a simple analogy, imagine that you find a letter in your mailbox. You read the envelope carefully to check the sender and receiver; you don't just open the envelope because it was in your mailbox—or at least, you shouldn't. But if it was in your mailbox and it was your name on the envelope, you may open it—and that's the next step.

4. **ERP business system**
   We can trigger the processing of messages and related business transactions. In most cases, the initial step will be related to identification of the parties involved in the transaction. If we again take the sales order as an example, we need to know who the customer, supplier, and maybe also ship-to party and payer are. All this depends on the characteristics of our ERP system, but only when we know all the mandatory actors can we create the sales order. If we move one step back and look at the received UN/EDIFACT standard, there we can see NAD segments, which hold information on the parties involved in this specific transaction and their functions. There are also additional address or reference data, but in most cases the main key is the identifier of each partner, like sold-to party, ship-to party, or supplier. In most scenarios, the party identifiers are extracted from the original EDI message and not recognized by the ERP business system, which has its own master data and thus also its

own business partner definitions. We need to link these external identifiers from the EDI message to internal ERP partner numbers. As you may know, SAP ERP has built-in mechanics for handling external partner references, and luckily these still exist in SAP S/4HANA. On the following pages, we will give you an example, and in Chapter 7 we will go through the step-by-step configuration.

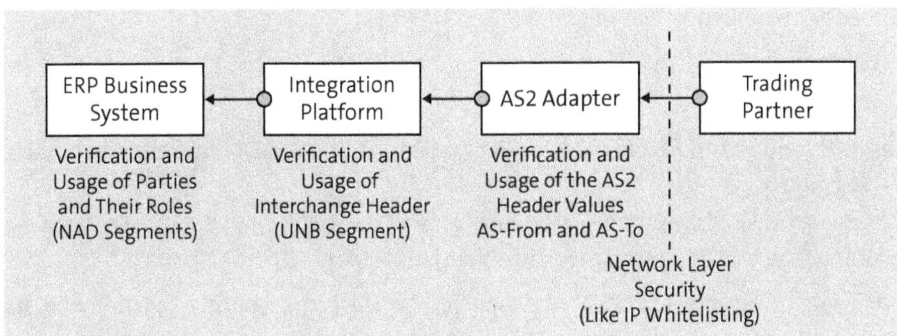

Figure 5.1 Simplified View of EDI Message Flow when Received from Trading Partner

You may be tempted on reflection to think that you don't need to validate and handle all these qualifiers on all these steps. You may say, "I only have a few trading partners and only one ERP business system, so there is no way they're going to a different system." Yes, you may be right in some cases, at the very beginning of your EDI landscape. But what we are looking for here is to apply a future-proof approach; consistent and transparent rules are always better than leaving it to chance. Hopefully your organization will grow, along with the number of EDI trading partners. Having hundreds of EDI partners and thousands of messages per day will prove that having a proper design and strict rules is critical to keep everything under control. Also, in both business and IT, we see a constant need for change. There are multiple harmonization initiatives, acquisitions, and carve-outs, and situations in which you only have a single ERP business system and a single integration platform are rare.

Let's summarize what we need and the exact values we'll use for our sample implementation:

- **AS2 system identifiers used within the AS2 transmission**

  As our own AS2 identifier, we will use `AS2_COMPANY_1710`. The trading partner will use AS2 identifier `AS2_BEST_BIKES`.

  In both cases, these values are not related to any standard (not GLN nor DUNS); they are just dummy values mutually agreed upon between both parties involved in AS2 communication. It is a good practice and strong recommendation to use more formal and standardized values like GLN or DUNS values. This is a more reliable and transparent approach. Also, the nonproduction and production environments should be distinguished by usage of different identifiers. We recommend having at least two sets of AS2 identifiers: one nonproduction and one production.

For AS2 communication where a trading partner sends a sales order, we would see the following in the AS2 header:

- AS2-From: AS2_BEST_BIKES
- AS2-To: AS2_COMPANY_1710

For AS2 communication where we send a sales order response to our partner, the AS2 header would look like this:

- AS2-From: AS2_COMPANY_1710
- AS2-To: AS2_BEST_BIKES

■ **Message sender and receiver identifiers within UN/EDIFACT interchange header (UNB segment)**

As our own UNB interchange identifier, we will use EDI_COMPANY_1710. The trading partner will use UNB interchange identifier EDI_BEST_BIKES.

Looking at a sample EDIFACT message received from a trading partner and its header, in the UNB segment, we can see content like that shown in Figure 5.2.

```
1  UNB+UNOE:2+EDI_BEST_BIKES:ZZZ+EDI_COMPANY_1710:ZZZ+240227:1830+16862
```

Figure 5.2 UNB Segment in EDIFACT Message Received from Trading Partner

Let's study each element in detail:

- **UNB**
  This indicates the start of the UNB (Interchange Header) segment.
- **UNOE:2**
  This is the syntax identifier:
  - UNOE refers to the service string advice code. It indicates the syntax level or standards agreed upon by the trading partners. The UNO part typically refers to the UN/ECE syntax rules.
  - The 2 indicates the version number of the syntax.
- **EDI_BEST_BIKES:ZZZ**
  This is the sender's identification.
  - EDI_BEST_BIKES is an identifier for the sender of the interchange. It could be a company name, a division within a company, or an application.
  - ZZZ is a qualifier that specifies the type of ID. In EDIFACT, ZZZ is often used to indicate a mutually defined identifier.
- **EDI_COMPANY_1710:ZZZ**
  This is the recipient's identification:
  - EDI_COMPANY_1710 is an identifier for the receiver of the interchange.
  - ZZZ again is a qualifier that specifies the type of ID, with the same meaning as before.

- **240227:1830**
  This is the date and time of preparation:
  - 240227 stands for the date (in YYMMDD format); here, this corresponds to February 27, 2024.
  - 1830 is the time in 24-hour clock format (HHMM); here, this corresponds to 6:30 p.m.
- **16862**
  This is the *interchange control reference*, a unique number assigned by the sender to identify the interchange. Here, the unique reference number is 16862.
  - In S002 (interchange sender header), the data element 0004 (interchange sender identification) has the value EDI_BEST_BIKES.
  - In S003 (interchange receiver header), the data element 0010 (interchange receiver identification) has the value EDI_COMPANY_1710.
  - In both segments, S002 (interchange sender header) and S003 (interchange receiver header), the data element 0007 (identification code qualifier) is filled with the ZZZ value. Per the UN/EDIFACT standard, that's an indication that the values are mutually defined between trading partners. It does not come from or follow any standard.

- **Trading partner identifiers within the UN/EDIFACT message (NAD segment)**
  The purpose of NAD segment is to specify the name and address details for the parties involved in specific business transaction. These could be the sold-to party, ship-to party, payer, supplier, and so on.

  As our own external identifier, we will use COMPANY_1710_VN. This will be our external representation of the sales organization within SAP S/4HANA. For a trading partner, this will be the supplier to be used when they release a purchase order.

  The trading partner will use BEST_BIKES_HQ as an external identifier for its purchasing organization and BEST_BIKES_WMS to represent its warehouse location—that is, the ship-to location.

  A sample NAD segment in the EDIFACT ORDERS message received from a trading partner may look like Figure 5.3.

```
NAD+VN+COMPANY_1710_VN::ZZZ'
NAD+BY+BEST_BIKES_HQ::ZZZ'
NAD+DP+BEST_BIKES_WHS::ZZZ'
```

Figure 5.3 NAD Segments in EDIFACT Message Received from Trading Partner

In EDIFACT, the NAD segment is used to specify the name and address of the parties involved in the transaction. *NAD* stands for *name and address*. Each NAD segment provides different party information based on the qualifier at the beginning of the segment. Here's what each of these NAD segments represents:

- NAD+SU+COMPANY_1710_VN::ZZZ'
  - NAD
    Indicates the start of the name and address segment.
  - SU
    This is the party qualifier that stands for *supplier*. It identifies the party being referred to as the supplier in this transaction.
  - COMPANY_1710_VN
    This is the identification of the supplier. The VN portion might indicate that this is the identification for the supplier's headquarters.
  - ZZZ
    This indicates that the identification code is mutually defined. It's not a standard EDIFACT code but one that the trading partners have agreed upon.
- NAD+BY+BEST_BIKES_HQ::ZZZ'
  - NAD
    As before, this marks the name and address segment.
  - BY
    This qualifier indicates the *buyer*. It identifies the party being referred to as the buyer or the purchasing organization in the transaction.
  - BEST_BIKES_HQ
    This is the identification of the buyer, again possibly referring to the headquarters.
  - ZZZ
    As before, this indicates a private, mutually agreed-upon identification code.
- NAD+DP+BEST_BIKES_WMS::ZZZ'
  - NAD
    Indicates the name and address segment.
  - DP
    This qualifier stands for *delivery party*. It identifies the party to which the goods should be delivered.
  - BEST_BIKES_WMS
    This is the identification for the delivery party, which in this context appears to be a warehouse management system (WMS) for Best Bikes.
  - ZZZ
    This is a mutually defined code, indicating that BEST_BIKES_WMS is not a standard EDIFACT party identification code.

In the NAD segment, field 3035 is filled with BY; that's the party function. Other functions include the following:

- BY
  The buyer—or in SAP terminology, the sold-to party

- **DP**
  The delivery recipient—or in SAP terminology, the ship-to party
- **SU**
  The supplier—or in SAP, the sales organization

Then we have section C082, and field 3039 has the external ID values. Here, 3055 has the ZZZ value, again indicating that it is mutually defined between trading partners.

Let's summarize the values of the identifiers used in the dummy scenario. Figure 5.4 shows the flow in which we receive UN/EDIFACT ORDERS D01B from a trading partner. Then in Figure 5.5, you can see the process in the opposite direction: identifiers used when the sales order response is sent from our business system to the trading partner.

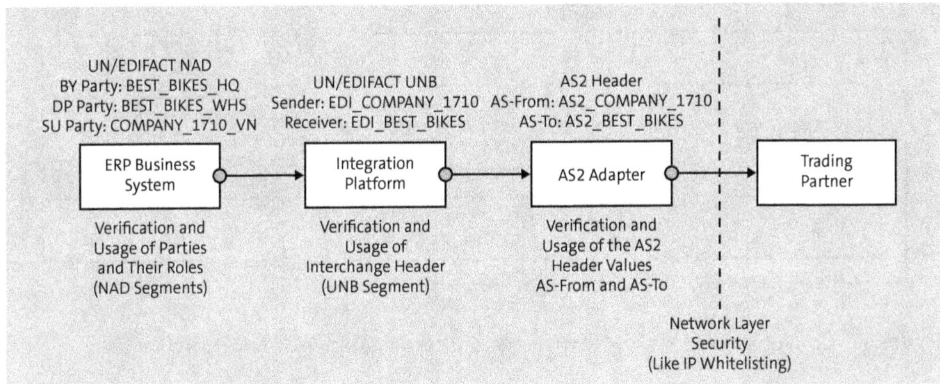

Figure 5.4 Identifiers Defined for Inbound Sales Order Flow

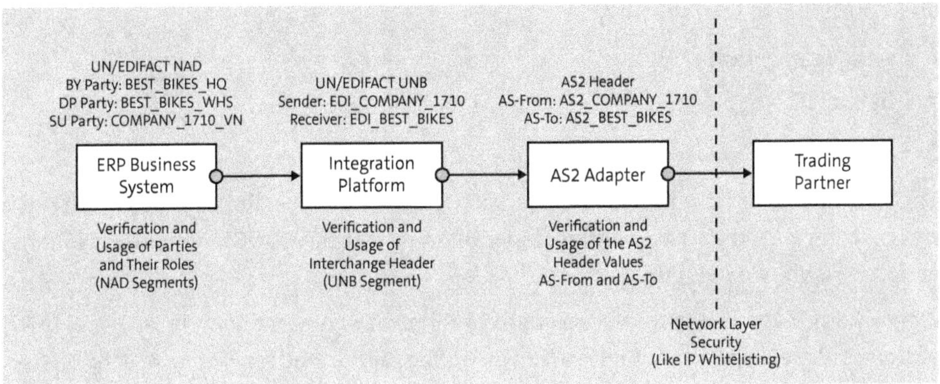

Figure 5.5 Identifiers Defined for Outbound Flows

5.1.2 SAP S/4HANA Sales Organization and Master Data

One topic still not fully clarified is what the process is within an ERP business system. We said that the sales order message arrives to the business system with external identifiers for the customer and supplier. We need to convert these identifiers to internal numbers known by the business system in use. The different parties/trading partners involved in the EDI process have their own approach and number ranges for handling master data. This is also the case for SAP S/4HANA, which does not know which partner is behind the BEST_BIKES_HQ sold-to party value used in our scenario. The external numbers need to be translated to internal business partner numbers. As we already discussed when going over the sales organization concept within SAP S/4HANA, we need a sales area to create a sales order. Determination of the sales area is also standard SAP functionality, which is triggered based on the internal numbers. A simplified view of the concept is shown in Figure 5.6. The exact configuration steps required within the SAP S/4HANA system will be explained in Chapter 7.

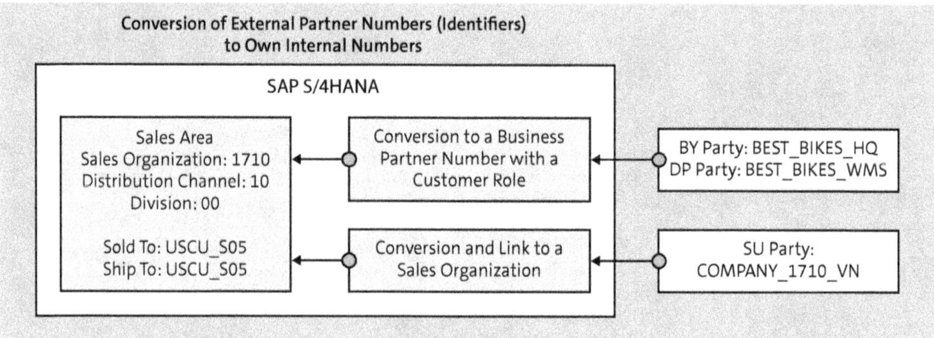

Figure 5.6 Conversion of External Identifiers to Internal SAP S/4HANA Entities

To summarize all the descriptions and diagrams, in the EDI scenario deployed on the SAP S/4HANA landscape, we aim to create and process a sales order within the sales area consisting of the following:

- Sales Organization 1710
- Channel 10
- Division 00

The trading partner used in our EDI scenario is defined within SAP S/4HANA as a business partner with the customer role under BP number USCU_S005. A preview of Transaction BP is shown in Figure 5.7.

Figure 5.8 shows a header of the standard order created manually in the SAP S/4HANA system. This sales document is created within the mentioned sales area, which consists of sales organization 1710, channel 10, and division 00. The sold-to party is identified by the business partner ID mentioned earlier, USCU_S05, for the customer named Best Bikes. The intention is that our EDI transaction for sales order will result in the creation of a sales orders as in this example.

Figure 5.7 Business Partner (Customer) Used in EDI Scenario

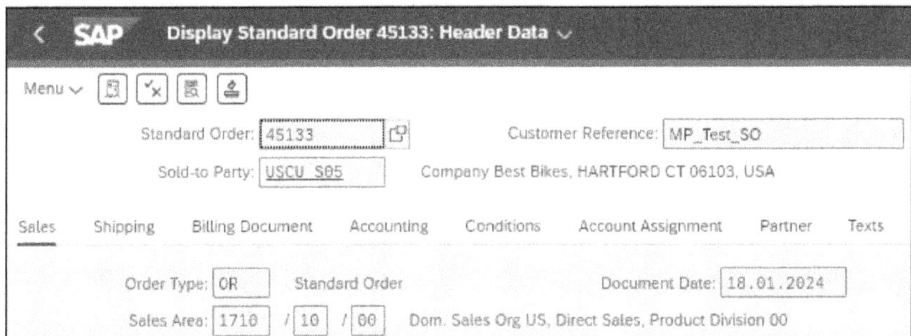

Figure 5.8 Sample Manually Created Sales Order in Selected Sales Area

5.1.3 Material Master Data

Master data is key to successful EDI processing. We keep our scenario simple in terms of functional requirements and master data. The trading partner knows our product portfolio and their IT system can handle our material number. All the different EDI messages that we will exchange with our trading partner contain material numbers as defined within our SAP S/4HANA. There is no need for additional material determination in our scenario. But it's good to mention that it is a standard functionality in SAP ERP as well as in SAP S/4HANA. It is a master data object where the customer's material information can be maintained: that data includes the customer's part number and a product description, but also shipping information and much more. The most important part is that we define a link between the customer's material number and our SAP S/4HANA material number. After doing this, it is possible to use the customer's number as an input for the sales order transaction. The customer material info record will automatically translate this external material number into appropriate material from the SAP S/4HANA master data.

Figure 5.9 shows a sample material we will use in our EDI transaction. This is a view from Display Material (Transaction MM03), showing sales organization–specific data.

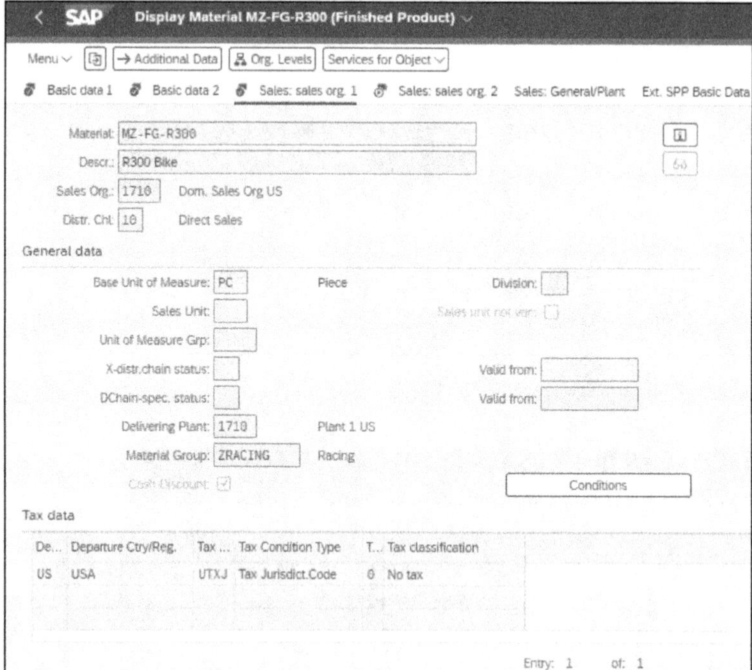

Figure 5.9 Sample Material Active in Sales Organization 1710, Distribution Channel 10

5.2 Summary

In this chapter, we explained how handling of the different identifiers should be done within a properly designed EDI solution. That does not necessarily mean that this is the only correct approach, but it is definitely a good starting point that you can adjust depending on your specific requirements, dependencies, or limitations. While explaining the concept, we also introduced very specific but still fictional values that will be used during the following chapters of this book when we go through the actual configuration of our dummy EDI scenarios.

Before we start using the identifiers, we need to work through the next chapter, in which we perform initial setup of the environment. This will involve enabling and connecting the various components that form our EDI landscape.

Chapter 6
Initial Setup of the Environment

In this chapter, we start configuring the EDI environment. After making sure we have met all the prerequisites, we will enable bidirectional communication between SAP S/4HANA and Cloud Integration. We will also explain the role of the cloud connector and lead you through its deployment.

Before you start with the configuration directly related to EDI messaging, you need to make sure that the different technical components from your landscape can communicate with each other. The most important is the connectivity between on-premise SAP S/4HANA and SAP Business Technology Platform (SAP BTP) integration services. In this chapter, we will explain the overall concept and detail what the inbound and outbound connections look like. A significant part of this chapter is filled with topics related to the cloud connector, which plays a key role in the environment by linking the on-premise and cloud components. When you first begin to setup your environment, you should go through the list of prerequisites and assumptions; these are items that must be taken care of. Most of these do not have a direct relation to the EDI setup and are linked to the infrastructure itself. They are listed in the first section but excluded from the scope of this EDI guideline. In the later part of this chapter, you can find step-by-step instructions to configure connectivity for data flowing from SAP S/4HANA to Cloud Integration. Right after this, you can find similar step-by-step instructions for setting up connectivity for data flowing in the opposite direction.

6.1 Readiness of the Technical Components

The key topic of this section is the complete end-to-end setup of the EDI scenario in an SAP S/4HANA environment. There are multiple aspects that need to be taken care of before such configuration can happen. There are certain technical and nontechnical areas related to the overall setup that cannot be covered within this book; items not directly linked to the EDI area are excluded. To make the starting point clear, you can find all the key prerequisites listed in the following sections.

6.1.1 Prerequisites for the Components

The prerequisites for technical components include the following:

- There is a running SAP S/4HANA on-premise system.
- There is an activated SAP BTP global account and the SAP Integration Suite service is on it.
- The subaccount with SAP Integration Suite runs on the Cloud Foundry environment.
- You have all the required access and authorizations within your SAP S/4HANA system, SAP BTP, and all the other components.

6.1.2 Assumptions

The following items are excluded from scope of this guide. Some of them may be briefly mentioned, but we will not elaborate on topics related to the following:

- Licenses, internal change procedures, (country-specific) legal requirements, and so on.
- Infrastructure and operational aspects like setup of the internal network, traffic rules, and firewalls.
- Authorization—although you can find some instructions and advice related to authentication and authorization in the context of EDI communications.
- Security. This broad topic definitely goes outside the scope of this book, though you can still find a few recommendations related to security.

Note

Two words often create confusion—authentication versus authorization:

- *Authentication* is a process of verifying identity. There are several authentication methods; in the most common basic authentication, a user confirms their identity via a password.
- *Authorization* defines what privileges a specific identity has and what actions they can perform. For example, there can be a user who is known as an employee but who has a job role in which no activities in the IT system are required. He can be authenticated (recognized by the IT system), but he may have no privileges to perform any actions within the system.

6.2 Deployment of the Trading Partner Management Package

Starting from this chapter and running through Chapter 9, you will be configuring multiple SAP components. In the center will be the Cloud Integration runtime executing integration flows (iFlows) predelivered by SAP in a package called Cloud Integration—Trading Partner Management V2. It will still be a while before we execute any EDI transactions, but you can already copy and deploy the necessary iFlows. Later in this chapter, we will need to obtain some of information from the runtime itself.

Follow these steps to copy and deploy the necessary artifacts:

1. Using the left-hand menu in SAP Integration Suite, navigate to **Discover • Integration**.
2. Type "Trading Partner Management" in the **Search** box at the top, as shown in Figure 6.1.

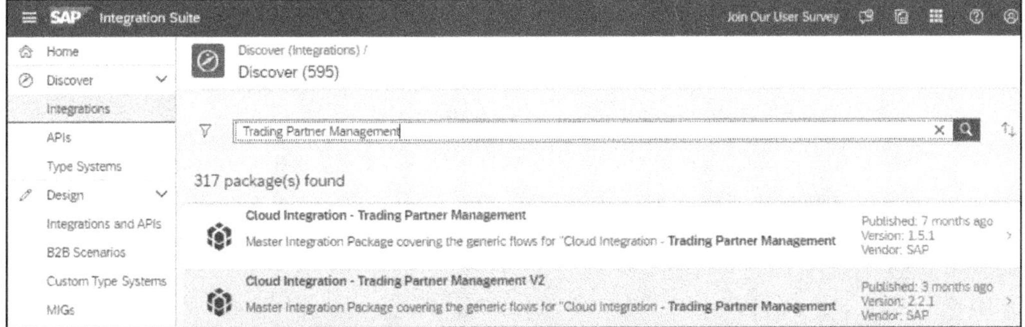

Figure 6.1 Searching for "Trading Partner Management" in Predelivered Integrations and Packages

3. **Click the Cloud Integration—Trading Partner Management V2** result to open it.
4. Click **Copy** in the top-right corner of the screen, as shown in Figure 6.2. This will copy the package to your tenant.

Figure 6.2 Header of Predelivered Cloud Integration—Trading Partner Management V2 Package

5. Navigate to **Design • Integrations and APIs**. You should see the package copied to your own tenant, similar to what's shown in Figure 6.3.

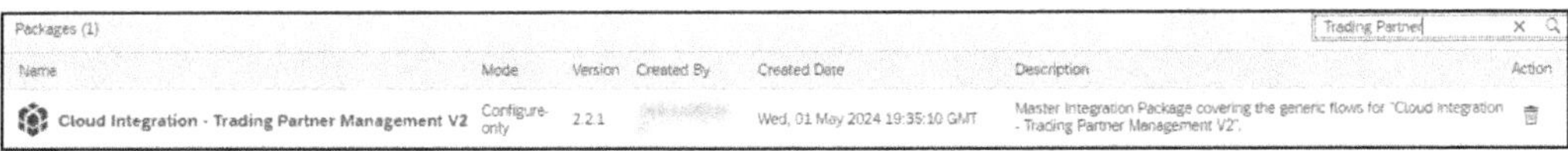

Figure 6.3 Details of Trading Partner Management V2 Package Copied to Cloud Integration Tenant

6. As you can see, the package is marked **Configure-only**. And in fact, we won't even modify the default parameters; it will work for us as is. Click the package and navigate to the list of **Artifacts**, as shown in Figure 6.4. We will deploy the following artifacts, which will be used in our B2B scenario:
   - **Reusable Groovy Script V2**
   - **Step 1—Sender AS2 Communication Flow V2**
   - **Step 1—Sender AS2 MDN Flow V2**
   - **Step 1—Sender IDOC Communication Flow V2**
   - **Step 1b—Write Message to Message queue**
   - **Step 2—Interchange Processing Flow V2**
   - **Step 3—Receiver Communication Flow V2**

   We can skip the iFlows related to Process Direct and SOAP communication. These will not be used in our scenario.

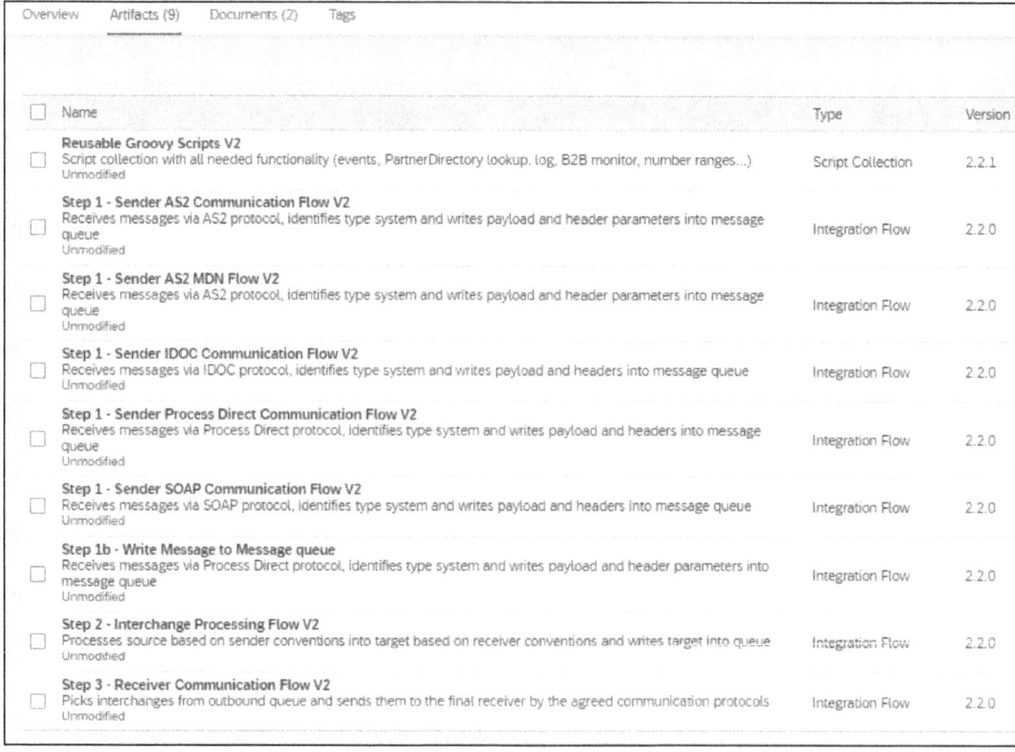

Name	Type	Version
Reusable Groovy Scripts V2 Script collection with all needed functionality (events, PartnerDirectory lookup, log, B2B monitor, number ranges...) Unmodified	Script Collection	2.2.1
Step 1 - Sender AS2 Communication Flow V2 Receives messages via AS2 protocol, identifies type system and writes payload and header parameters into message queue Unmodified	Integration Flow	2.2.0
Step 1 - Sender AS2 MDN Flow V2 Receives messages via AS2 protocol, identifies type system and writes payload and header parameters into message queue Unmodified	Integration Flow	2.2.0
Step 1 - Sender IDOC Communication Flow V2 Receives messages via IDOC protocol, identifies type system and writes payload and headers into message queue Unmodified	Integration Flow	2.2.0
Step 1 - Sender Process Direct Communication Flow V2 Receives messages via Process Direct protocol, identifies type system and writes payload and headers into message queue Unmodified	Integration Flow	2.2.0
Step 1 - Sender SOAP Communication Flow V2 Receives messages via SOAP protocol, identifies type system and writes payload and headers into message queue Unmodified	Integration Flow	2.2.0
Step 1b - Write Message to Message queue Receives messages via Process Direct protocol, identifies type system and writes payload and header parameters into message queue Unmodified	Integration Flow	2.2.0
Step 2 - Interchange Processing Flow V2 Processes source based on sender conventions into target based on receiver conventions and writes target into queue Unmodified	Integration Flow	2.2.0
Step 3 - Receiver Communication Flow V2 Picks interchanges from outbound queue and sends them to the final receiver by the agreed communication protocols Unmodified	Integration Flow	2.2.0

Figure 6.4 List of Artifacts in SAP Trading Partner Management V2 package

7. To deploy an artifact, click the **Actions** icon located on the right side of the screen (next to the **Version** column) and choose **Deploy**, as shown in Figure 6.5. Perform this action for each artifact from the earlier list.

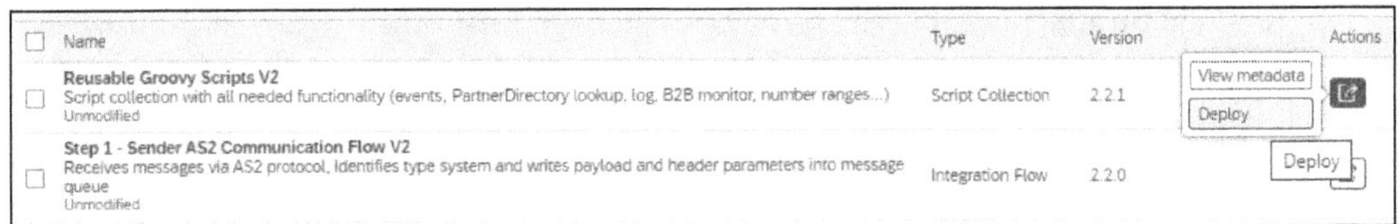

Figure 6.5 Trading Partner Management V2 Package: Deployment of Artifacts

8. When you deploy an artifact, you may be asked on which runtime profile the artifact should be deployed, as shown in Figure 6.6. By default, there is only one profile, so clicking **Yes** should be the most common choice. If you have more runtime profiles, choose the appropriate one.

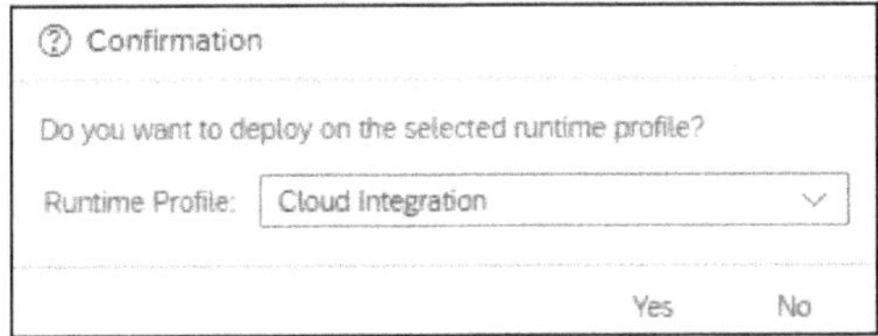

Figure 6.6 Prompt for Confirmation on Runtime Profile

9. When you are done with the deployment of all artifacts, navigate to **Monitor • Integrations and APIs**, then in the **Manage Integration Content** section, choose the tile to display **All** artifacts. This should display all artifacts deployed on the runtime and their status. As shown in Figure 6.7, all seven artifacts are successfully deployed.

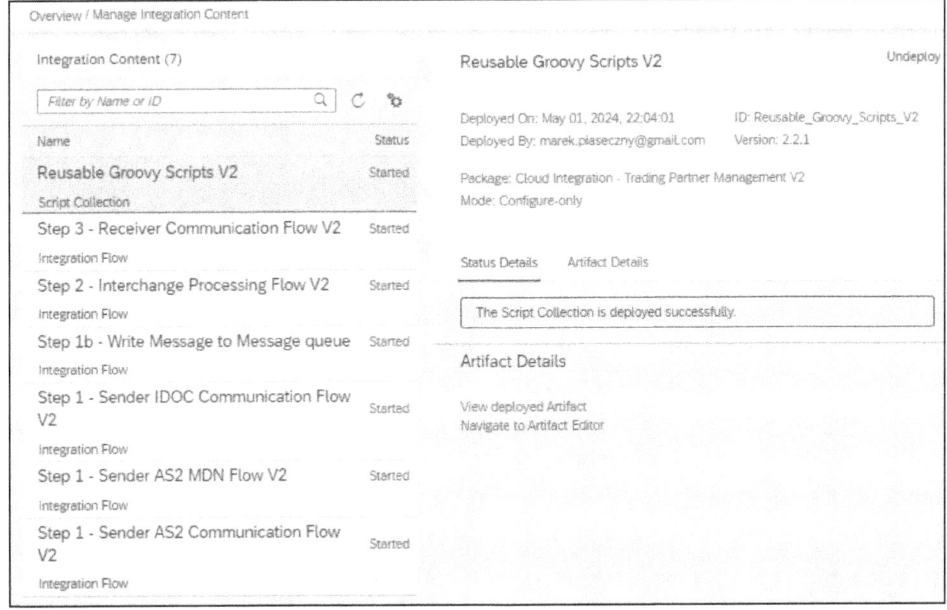

Figure 6.7 Overview of Deployed Artifacts

10. While we are reviewing the overview of deployed artifacts, we can look at the iFlow responsible for receiving IDocs: **Step 1—Sender IDOC Communication Flow V2**. In our scenario, the SAP S/4HANA system will be sending IDocs to this iFlow. It will be necessary to know its endpoint (URL) when setting up connectivity from SAP S/4HANA to Cloud Integration. We can copy the endpoint now: simply choose the iFlow in the overview list and navigate to the **Endpoint** tab, as shown in Figure 6.8. The entry point can be copied easily via the **Copy** icon located at the end of the line.

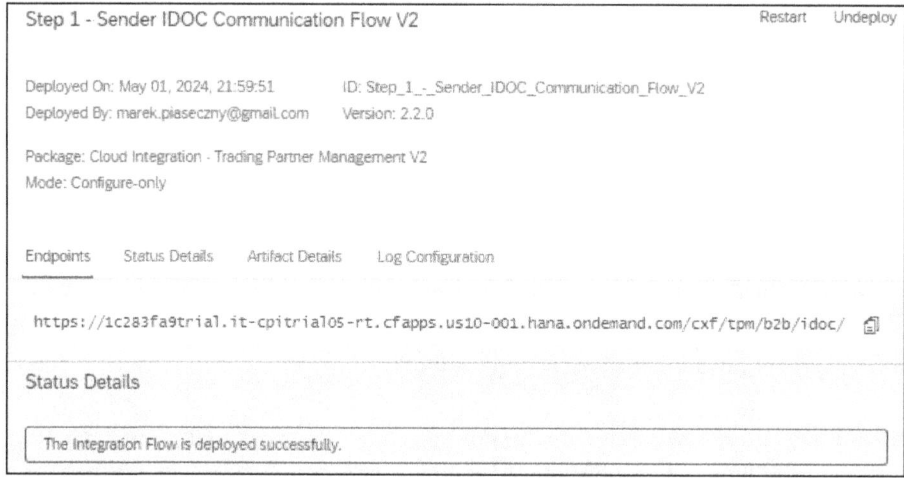

Figure 6.8 View of Sender IDOC Communication Flow V2 iFlow and Its Endpoint

In our example, the endpoint to which SAP S/4HANA will be sending IDocs is *https://1c283fa9trial.it-cpitrial05-rt.cfapps.us10-001.hana.ondemand.com/cxf/tpm/b2b/idoc/*.

After deployment of the necessary artifacts related to Trading Partner Management, we can move to the next section and set up the connectivity between SAP S/4HANA and SAP BTP.

6.3 Connecting SAP S/4HANA and SAP BTP

In this section, we will handle the bidirectional communication between SAP S/4HANA and SAP BTP. We will start with an explanation of how the different components communicate with each other. In the next step, we go through the configuration required for sending data from SAP S/4HANA to a specific iFlow running on Cloud Integration. In this step, we will also implement a secure authentication method with the use of a client certificate.

The second part of this section will guide you through the configuration steps to enable data flow from Cloud Integration to the SAP S/4HANA system. You will quickly notice that this is a more complex task than the previous one because for the SAP S/4HANA inbound flows, we make use of the cloud connector for security reasons. You will also find instructions related to installation and initial configuration of the cloud connector. We will also take a close look at the activation of the SOAP IDoc processing service on the SAP S/4HANA system.

6.3.1 Overview

As explained earlier in Chapter 4, there are three key parts to how SAP S/4HANA connects to SAP BTP:

- SAP S/4HANA connects directly to Cloud Integration.
- SAP BTP does not connect to SAP S/4HANA directly.
- There is an additional component in between called the cloud connector.

Let's add some more details to explain the overall concept related to the connectivity between SAP S/4HANA and SAP BTP. Figure 6.9 shows additional details to illustrate the connectivity between SAP S/4HANA and SAP BTP, including a dotted line to separate the internal network from the public internet where SAP BTP services are located.

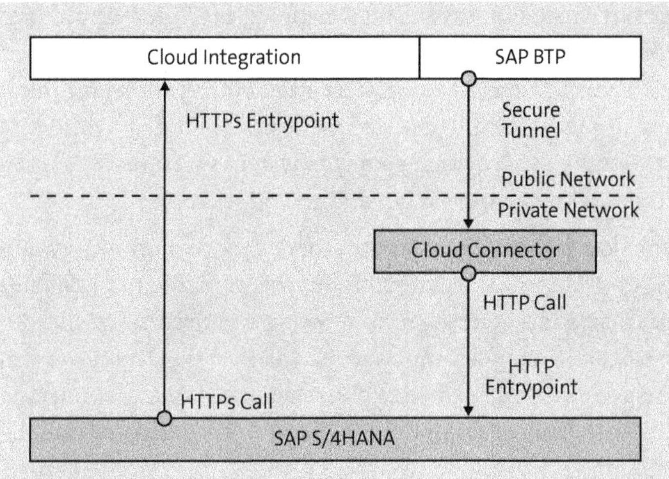

Figure 6.9 How Different Components of SAP S/4HANA and SAP BTP Are Connected

In most cases, we have some network security components: forward/reverse proxies, firewalls, and so on. We don't address that part of network architecture here; the assumption is that SAP S/4HANA and the cloud connector, which are located in an internal network, are able to reach the remote components.

Let's start with an outbound flow where SAP S/4HANA sends data to Cloud Integration. This is represented by the arrow on the left side of the diagram. The small circle indicates a place where the connection is triggered—in this case, in SAP S/4HANA, which acts as the HTTP client. It will make a call to the HTTP endpoint on Cloud Integration, which acts as an HTTP server. This may sound complex, but the HTTP endpoint is created automatically after deploying an iFlow with an IDoc sender channel. Note that SAP S/4HANA connects directly to the Cloud Integration runtime component.

HTTP versus HTTPS

For the sake of formality, let's clarify the difference between HTTP and HTTPS:

- HTTP
  HTTP stands for *Hypertext Transfer Protocol*. It is a communication protocol—in other words, a set of rules for how to exchange data between a client (sender) and a server (recipient).
- HTTPS
  HTTPS stands for *Hypertext Transfer Protocol Secure*. The one big disadvantage of HTTP is the fact that there is no mechanism to protect the data being transferred (i.e., no channel encryption). HTTPS is not a different protocol than HTTP; instead, it is a variant via which an entire transmission can be encrypted via the TLS/SSL protocol.

SAP S/4HANA, as well as earlier versions like SAP ERP, has the capability to communicate via HTTP(S), acting as both client and server. This functionality is offered by a component called Internet Communication Manager (ICM). At this stage, being aware of this component is enough; there is no need to know its architecture nor its technical details. Although it may be useful to look at some of the features offered by the ICM transactions, especially the ones related to tracing; they may help a lot for any troubleshooting. We will look at this later when we test our end-to-end solution.

The right side of the diagram looks a bit more complex: there are two more arrows and an additional box for the cloud connector. You read about this component in Chapter 2, so you already know that it acts as a reverse proxy allowing secure connection from a cloud to an on-premise system. The most important fact is that the data flow from SAP BTP to the on-premise system is enabled and controlled by this small connector. From a network perspective, the cloud environment does not have direct access to the on-premise systems: there is no need to allow any network traffic from the cloud to the on-premise environment as the data flows within a tunnel established by the cloud connector. The communication protocol used in this case is also HTTP(S).

6.3.2 SAP S/4HANA Outbound Flows

There are two key actions within SAP S/4HANA to configure the outbound communication:

1. Create an HTTP connection to an external system. You do that in Transaction SM59 (RFC Destinations), in the same way as you define ABAP or TCP/IP connections.
2. Create a port. When sending IDocs, we don't push them directly to a destination; we do that via a port. Such an approach gives us some additional flexibility, and the port

may also contain some additional parameters that define how the data should be processed for the given destination.

Transaction SM59, RFC Destinations

The name of this transaction may be confusing. To maintain HTTP, we need to go to a transaction related to RFC destinations? But RFC is a different communication method than HTTP! While the name may be misleading, from a functional point of view, Transaction SM59 is the correct transaction to maintain not only RFC destinations but also destinations related to other communication protocols.

We'll walk through these steps in more detail in the following sections.

Creating an HTTPS Connection to the External System

To configure our HTTP destination, there are two key items that need to be obtained from the called system (Cloud Integration) and then maintained within the calling system (SAP S/4HANA):

- The target URL to which we connect
- The logon and security configuration (for the credentials to authorize posting IDocs on the remote service)

We already collected the URL in the previous section when we deployed the artifacts from the Trading Partner Management integration package. This is the URL we will maintain as our HTTPS destination: *https://1c283fa9trial.it-cpitrial05-rt.cfapps.us10-001.hana.ondemand.com/cxf/tpm/b2b/idoc/*.

When you have the destination address for the outbound IDocs, you can start the configuration within SAP S/4HANA. You start with Transaction SM59, where initially you will see a tree in which the different connection types are grouped, as shown in Figure 6.10. **HTTP Connections to External Server** is the type we are interested in.

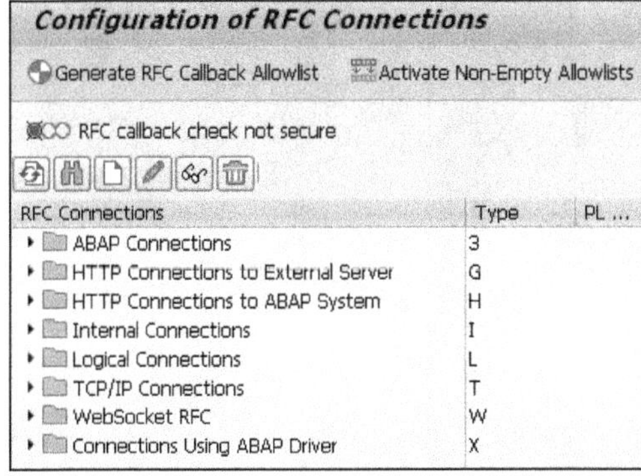

Figure 6.10 Initial Screen of Transaction SM59 to Maintain RFC Destinations

Choose [ ] to create a new destination. A new popup appears as shown in Figure 6.11. Here you need to type the name of the new **Destination** and choose its **Connection Type** from a predefined list. Both elements are mandatory.

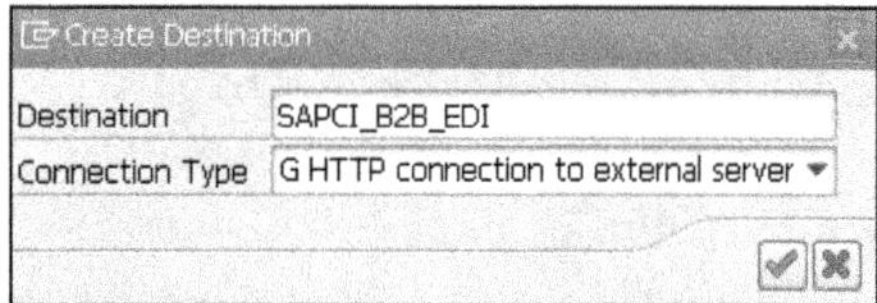

Figure 6.11 Creating New Destination in Transaction SM59

After confirming the small popup, there is a new window where you define all the key properties of the new connection. There are four tabs: the first one, called **Administration**, is filled automatically with the creation and last update dates. The next tab, called **Technical Settings**, is shown in Figure 6.12, and it holds the details of the target system. That's where you maintain the target URL.

The destination address of the SAP Integration Suite service will be maintained in three different fields on the **Technical Settings** tab:

- **Host**
  This is the first part of the URL. Here you skip the initial part indicating the protocol (i.e., *https://*), so maintain the following value: *1c283fa9trial.it-cpitrial05-rt.cfapps.us10-001.hana.ondemand.com*.
- **Port**
  The default port for HTTPS is 443, and that's also the value to use and maintain here.
- **Path Prefix**
  Here, maintain the following value: */cxf/tpm/b2b/idoc/*.

Depending on the architecture of the network in your organization, you may need to use an HTTP proxy. You can maintain parameters related to the HTTP proxy in the bottom section, **HTTP Proxy Options**.

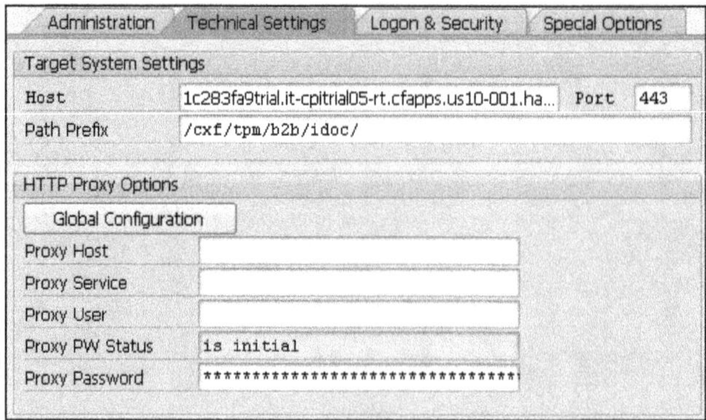

Figure 6.12 Technical Settings of External Connection

After maintaining the host and path of the target system, you go to the third tab, **Logon & Security**. Here you will maintain the necessary credentials and enable the SSL protocol. When you send a message to Cloud Integration, it is necessary to authenticate. There is no option for configuring the iFlow entry point for public or anonymous access—but that's also not recommended approach. You want to have the IDoc sender adapter protected. The key available authentication methods are as follows:

- Basic authentication
- Client certificate authentication
- OAuth2 authentication

Basic authentication is the simplest option to activate, but it is not recommended. Its usage should be limited to nonproduction environments and temporary situations. The other two authentication options, client certificate and OAuth2, are technically different but equally secure. Your choice may depend on security policies within your company or other details. For our scenario, we will use the client certificate option, which is easier.

The are two variants for client certificate authentication available within SAP BTP:

- We can generate a new certificate (key pair) within SAP BTP and then load the appropriate certificate (private key) in the calling system, SAP S/4HANA.
- We can export an existing client certificate from SAP S/4HANA and then load it into SAP BTP as an external certificate. This is the approach we will follow here.

As a first step, you export the client certificate from SAP S/4HANA. Go to Transaction STRUST and click **SSL Client (Standard)**. You will see your own SSL client certificate, which needs to be exported (see Figure 6.13).

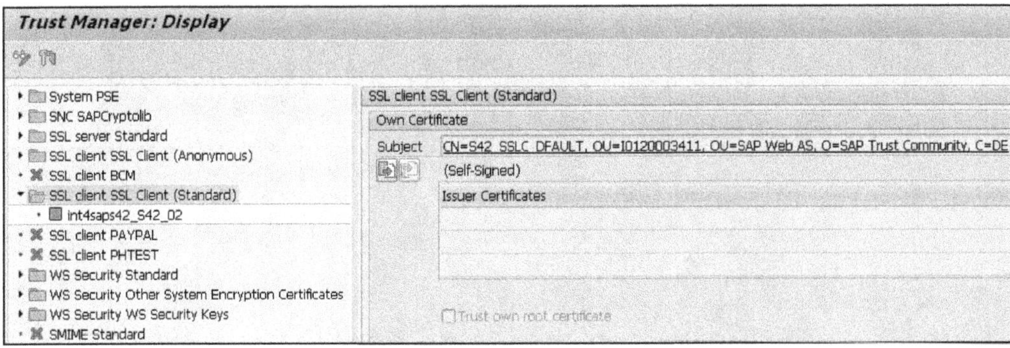

Figure 6.13 Trust Manager: Own SSL Client Certificate

At the bottom of this screen, you can see the details of the own certificate and make use of the **Export Certificate** option (see Figure 6.14).

Export the certificate in **Base64** format as shown in Figure 6.15.

Certificate	
Subject	CN=S42_SSLC_DFAULT, OU=I0120003411, OU=SAP Web AS, O=SAP Trust Community, C=DE
Subject (Alt.)	
Issuer	CN=S42_SSLC_DFAULT, OU=I0120003411, OU=SAP Web AS, O=SAP Trust Community, C=DE
Serial Number (Hex.)	0A:20:17:06:25:11:27:01
Serial Number (Dec.)	729608454793144065
Valid From	25.06.2017 11:27:01 to 01.01.2038 00:00:01
Algorithm	RSA
Key Strength	1024
Signature Algorithm	RSA+SHA256
Check Sum (MD5)	71:A7:16:99:31:B7:AA:0E:FA:40:2E:B4:BA:ED:7B:4A
Checksum (SHA1)	A7:E0:D8:9E:E3:E1:61:85:1A:28:3C:C6:75:13:CB:1A:FF:FE:16:23

Add to Certificate List

Export certificate

Figure 6.14 Trust Manager: Own SSL Client Certificate: Export Certificate

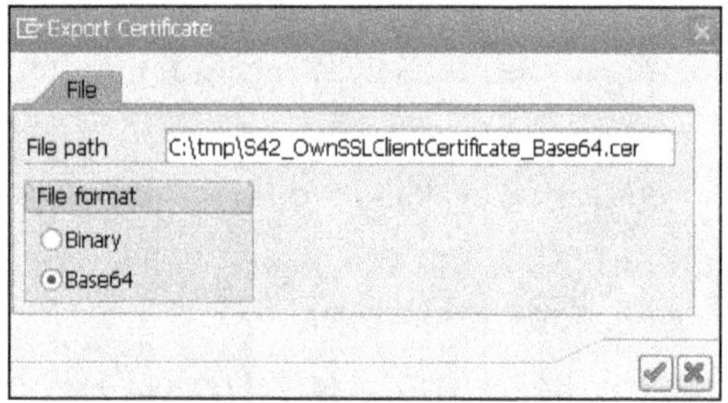

Figure 6.15 Trust Manager: Exporting Own SSL Client Certificate to Base64 File Format

When you have the certificate exported, you can move to the next step, which is creating a service key in SAP BTP. The steps are as follows:

1. Go to the SAP BTP cockpit and then to the subaccount on which you have subscribed to the SAP Integration Suite application. Find an instance of **Process Integration Runtime** in the **Service** column with **integration-flow** in the **Plan** column, then select **Create Service Key** from the actions menu, as shown in Figure 6.16.

Instances (4)

Service instances created in: Cloud Foundry | Kyma/Kubernetes | Other environments

Instance	Service	Plan	Runtime Environment	Scope	Credentials	Status
dev-api-instance	Process Integration Runtime	api	Cloud Foundry	dev	3 keys	Created
dev-apiportal-apiaccess-instance	API Management, API portal	apiportal-apiaccess	Cloud Foundry	dev	1 key	Created
dev-pir-instance	Process Integration Runtime	integration-flow	Cloud Foundry	dev	6 keys	Created
new-test-only	Process Integration Runtime	integration-flow	Cloud Foundry	dev	1 key	Created

View Parameters
Update
Create Binding
Create Service Key
Add Labels
Delete

Environments (1)

Environment instances created in this subaccount.

Figure 6.16 SAP BTP Cockpit: Create Service Key

2. Maintain the necessary information as follows:
   - **Service Key Name: B2B_S42_ClientCertificate**
   - **Key Type: External Certificate**
   - **External Certificate** (only applicable for key type **External Certificate**): Paste the certificate previously exported from SAP S/4HANA via Transaction STRUST. This is simply the content of the Base64 file.
   - Confirm by clicking **Create** (see Figure 6.17).

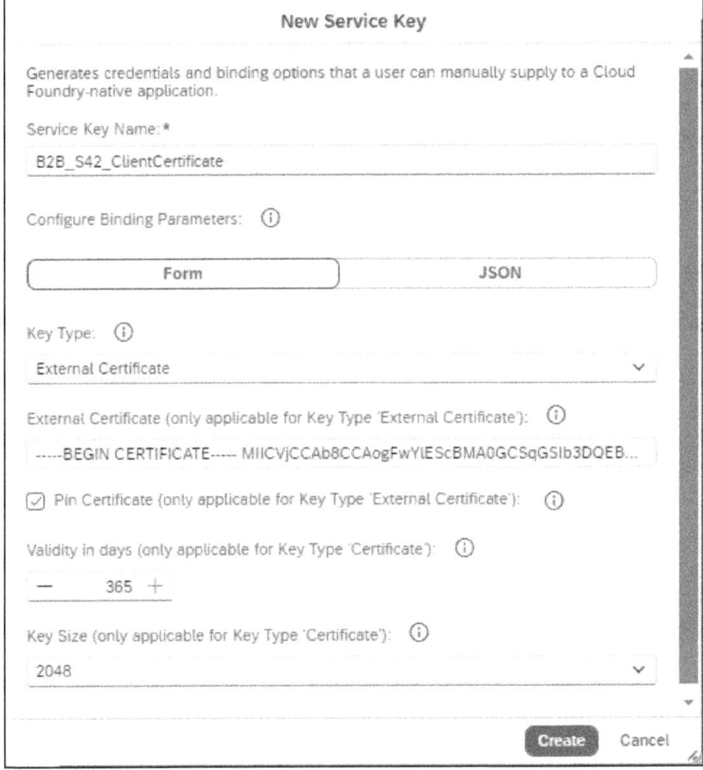

Figure 6.17 SAP BTP Cockpit: Create Service Key with External Certificate Key Type

After a few seconds, you should see your newly created service key in the list (see Figure 6.18).

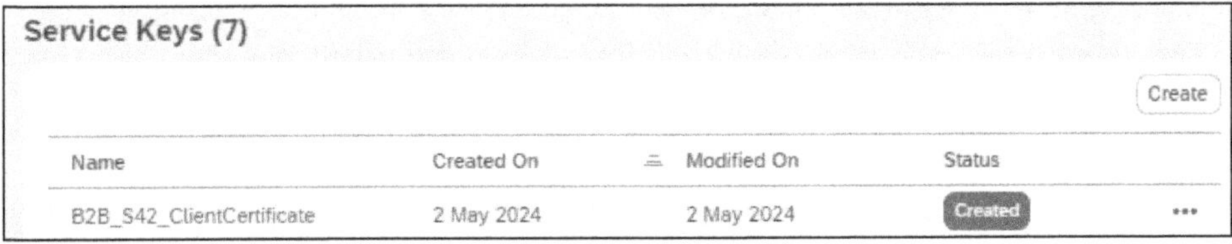

Figure 6.18 Newly Created Service Key of External Certificate Type

Now you can go back to SAP S/4HANA and finalize the configuration of the HTTP destination. Make sure that the following options are maintained on the **Logon & Security** tab:

- Choose **Do Not Use a User** in the **Logon with User** section.
- Choose **Active** for the **SSL** parameter in **Security Options • Status of Secure Protocol**.
- Choose **DFAULT SSL Client (Standard)** for **SSL Client PSE ID** in **Security Options • Status of Secure Protocol**.

Figure 6.19 shows the final configuration.

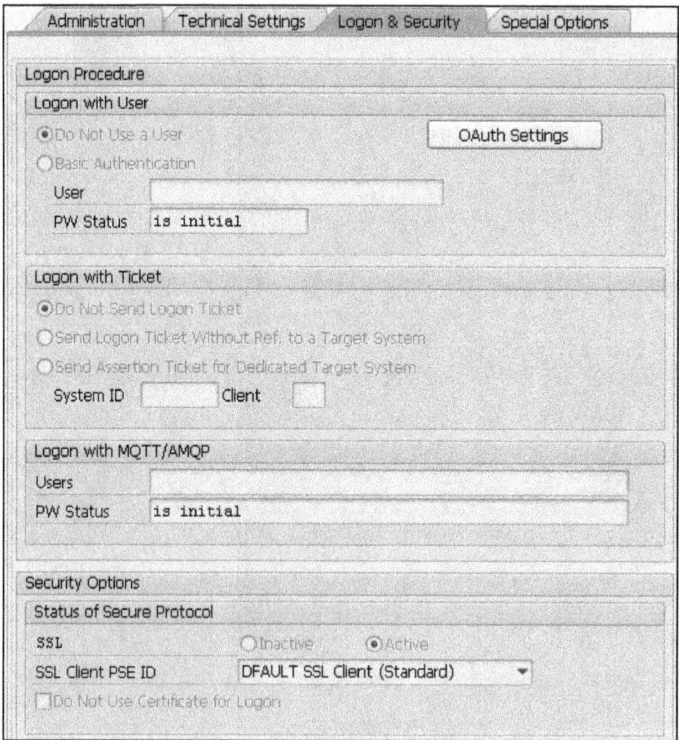

Figure 6.19 HTTP Destination: Logon & Security Tab

Connection Test

As you may have noticed, a feature called a *connection test* is available when you open any destination in Transaction SM59. For the new HTTP connection to Cloud Integration, this test may help check some aspects of the connection, but it will not give a full picture of the status.

It uses the GET method on the remote destination, which in our case is not supported by the remote service on Cloud Integration. We get an error related to a bad request, which is a good sign as it proves that we can reach the destination and were authenticated successfully. But it does not confirm that the remote service is able to receive and process IDocs.

At this stage, you already have the HTTP(S) connection configured within the backend system, so you can refer to it by the RFC destination with a specific name. But when sending IDocs to SAP S/4HANA, we do not use an RFC destination directly. Instead, there is a port of a specific type and with specific parameters, and this port is linked to the RFC destination. All outbound IDocs are sent to the port.

HTTPS Port

To create a port, use Transaction WE21. Follow these steps:

1. From the different port types shown after starting the transaction, choose **XML HTTP** and click the first icon, **Create** (or press [F7]; see Figure 6.20).

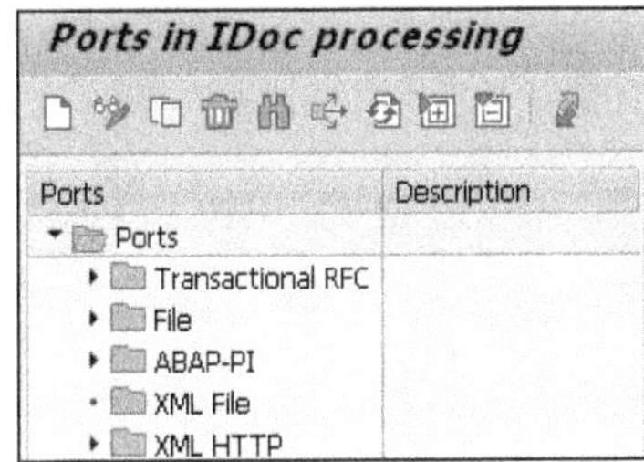

Figure 6.20 Logon and Security Parameters Related to External Connection

2. Fill in the required fields as follows, as shown in Figure 6.21:
   - **Port**
     Give the new port a name as per the naming conventions in your organization—or just any meaningful and unique name. In our example, we enter "SAPCI_B2B".
   - **Description**
     It's a good practice to give a self-explanatory description here.
   - **RFC destination**
     This is where you link the previously created external connection to the port. From the selection screen, choose **SAPCI_B2B_EDI**, which is the name used earlier in Transaction SM59.
   - **Content Type**
     Keep the selection set to **application/x-sap.idoc**.
   - Keep all the other parameters set to their default values.

Let's summarize the activities performed to this point in the SAP S/4HANA system: First, you configured an HTTP connection pointing to an entry point on Cloud Integration. Second, you created an XML HTTP port and assigned the external HTTP destination to it. From now on, we can use the port to pass IDocs for the purpose of processing them via a Cloud Integration iFlow responsible for B2B processing. You also completed the configuration related to the outbound connectivity.

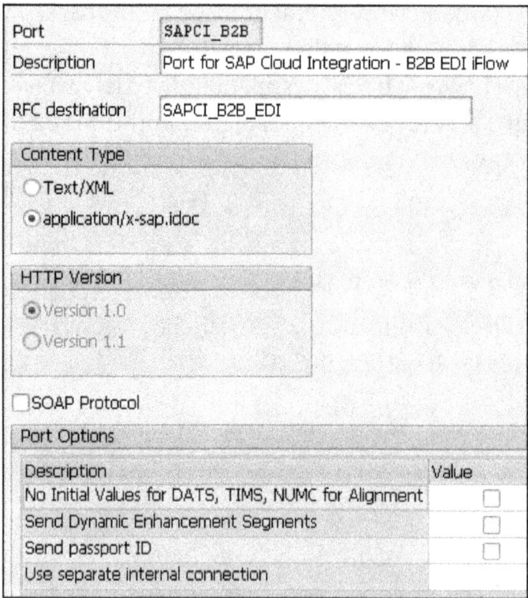

Figure 6.21 Details Related to HTTP XML Port

In the next section, you'll go through the steps required to set up inbound connectivity to SAP S/4HANA.

6.3.3 SAP S/4HANA Inbound Flows

As shown in Figure 6.9, the inbound flow toward SAP S/4HANA seems more complex. Comparing it to the outbound flow, it may seem to be unnecessarily complex. Why do we need to put an additional box between SAP S/4HANA and Cloud Integration? The answer is simple: security. SAP S/4HANA runs on premise in a private network. Because it is a key system, in most cases it runs in a specific segment of a private network called a *high-security zone*, where even the internal network traffic is strictly controlled. We definitely don't want to allow traffic from a public network to come directly into the high security of our private network. This problem is solved by the cloud connector, which acts as a reverse invoke proxy.

The most important part is that we don't need to open our firewalls for any inbound traffic from a public network. The cloud connector is a connectivity agent deployed in our private network, most often in the demilitarized zone (DMZ). It triggers the connection to SAP BTP and establishes a secure tunnel between the on-premise environment and the SAP BTP subaccount. All services and applications of that specific subaccount may use this tunnel to send data to on-premise destinations. Each such connection still needs to be explicitly defined and allowed on the cloud connector. That covers the secure communication from the public to the private network. In the following sections, we will deploy and configure the cloud connector and demonstrate how it works.

A second key aspect of the SAP S/4HANA inbound flow is that it uses HTTPS communication and an HTTPS endpoint. For many, that may be a surprise: there is no RFC communication here, which was always seen as an inseparable couple with IDocs. But this was wrong: we don't need RFC to transfer IDocs. Here we will use HTTP(S) to transfer IDocs from Cloud Integration, which means that we need an HTTP(S) server on SAP S/4HANA; more precisely, we need an HTTP(S) service that will be able to receive IDocs. We will look at this in the following section.

Enabling HTTP(S) IDoc Service on SAP S/4HANA

To clarify, this is not a new feature of SAP S/4HANA; this service has been in SAP ERP for many years and continues in SAP S/4HANA.

There are two standard services capable of receiving IDocs: one is plain HTTP, and the second is based on SOAP. We will use the latter as it has one big advantage: it returns the IDoc number in the response (if the IDoc was successfully created). The HTTP service only confirms if the IDoc was accepted or not; it does not return the number of the newly created IDoc.

Let's enable the SOAP service for inbound IDoc processing. There are two transactions you need to use:

- Transaction SICF, where you maintain services for HTTP communication
- Transaction SRTIDOC, where you register the IDoc service on the SOAP runtime

Start in Transaction SICF to find the service and check its properties. After running the transaction, you will see a selection screen as shown in Figure 6.22. You can go to the next step without any filters; just click the **Execute** icon or press F8.

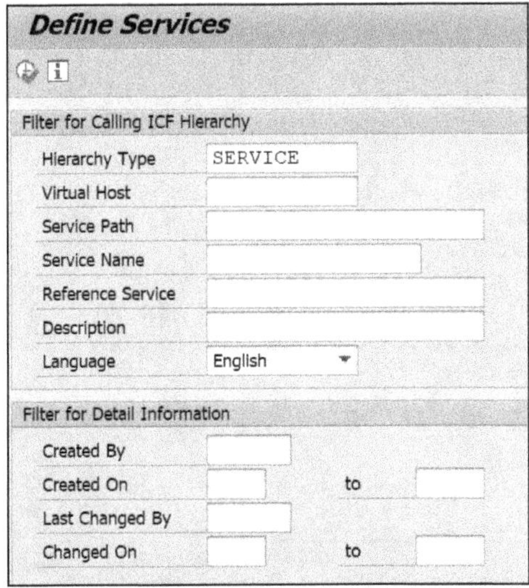

Figure 6.22 Transaction SICF: Define Services

The next screen looks like the one shown in Figure 6.23. In the top part, you once again have filtering options that help you find the service. In the bottom part, you have a hierarchical list of all services. You need to expand and browse through the list to find the service. The SOAP service related to processing inbound IDocs is in path */sap/bc/srt/idoc*.

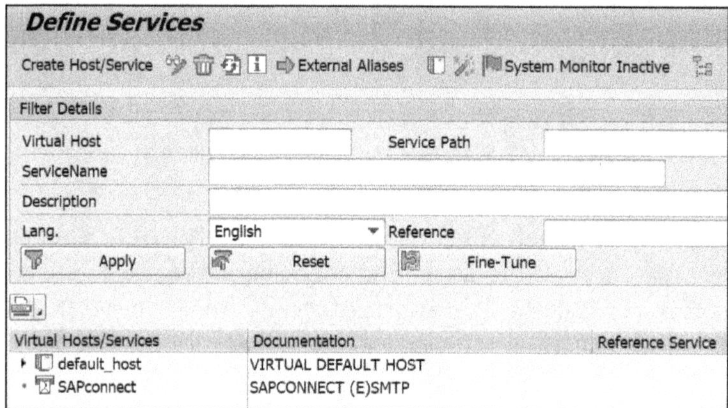

Figure 6.23 Transaction SICF: Define Services

In Figure 6.24, you can see that the **Inbound SOAP for IDoc** service is active.

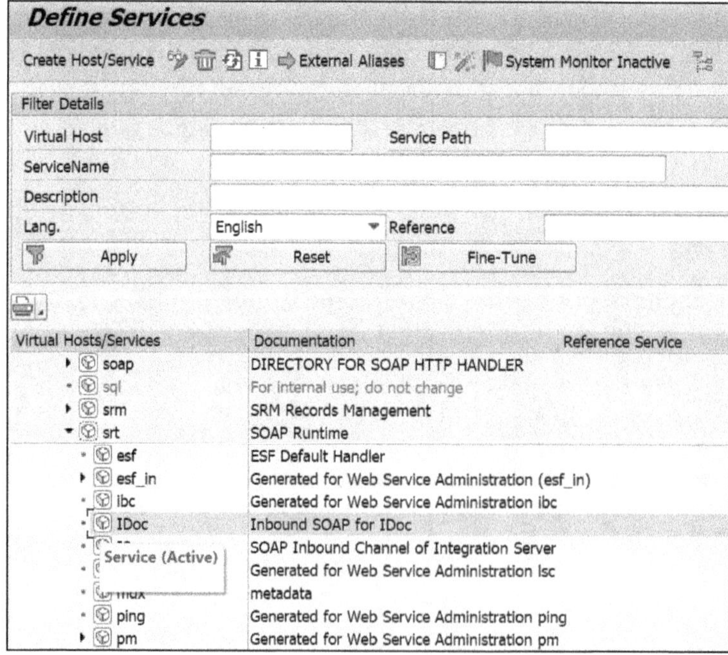

Figure 6.24 Transaction SICF: Inbound SOAP for IDoc

After making sure that the IDoc SOAP service is active, you can register it for the SOAP runtime. You do this using Transaction SRTIDOC. As shown in Figure 6.25, the required parameters are already prefilled on the selection screen. It is not necessary to change any of the default values here. Trigger the registration by choosing **Execute** or pressing [F8].

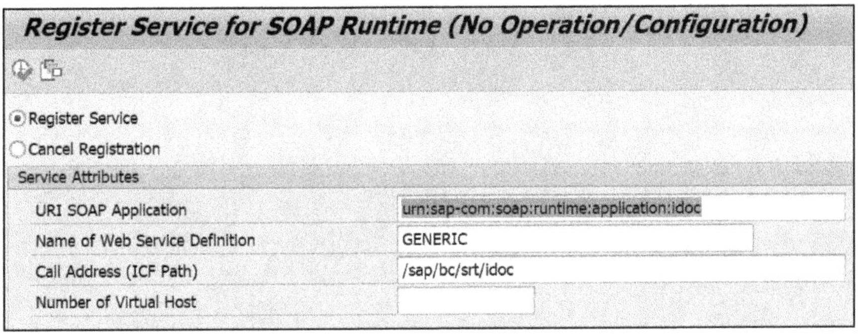

Register Service for SOAP Runtime (No Operation/Configuration)

Register Service

Cancel Registration

Service Attributes

URI SOAP Application: urn:sap-com:soap:runtime:application:idoc

Name of Web Service Definition: GENERIC

Call Address (ICF Path): /sap/bc/srt/idoc

Number of Virtual Host:

Figure 6.25 Transaction SRTIDOC with Default Properties for Service Registration

At this stage, you have an active SOAP IDoc service on the SAP S/4HANA system. It is ready to receive an IDoc via an HTTPS call. The only open question may be what the URL of this service is and how to authenticate it.

To find the URL of the SOAP IDoc service, use the test service function from Transaction SICF. After finding the service in the navigation tree, open the context menu as shown in Figure 6.26 and select **Test Service**. By executing this function, you will open the HTTP service via the default web browser in your system.

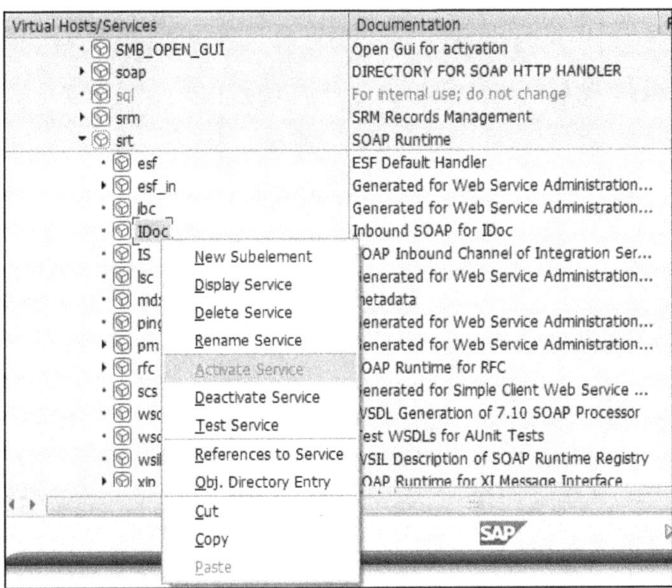

Figure 6.26 Context Menu in Transaction SICF

Using the test service function opens your default web browser, which makes a call to the HTTP service, as shown in Figure 6.27.

Figure 6.27 Results of Test Service Function in Transaction SICF

You can copy the URL from the web browser as a starting point; note that it uses HTTP. The HTTPS version runs on a different port. You can find the port used by HTTPS protocol via Transaction SMICM (see Figure 6.28), under menu path **Goto • Services**. After calling this function, you will be redirected to an overview of all active services running on the given SAP S/4HANA system.

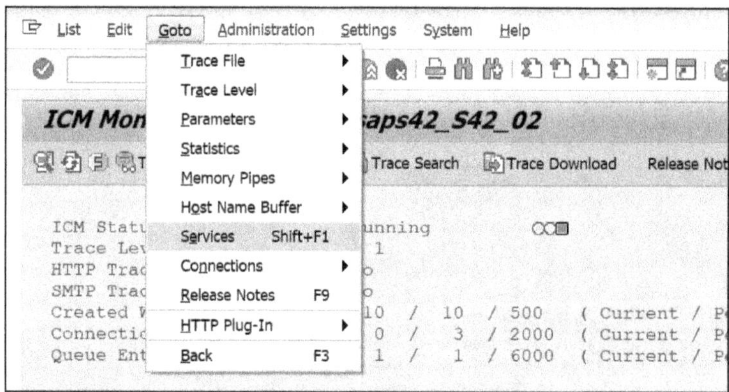

Figure 6.28 Transaction SMICM: Menu Goto • Services

In Figure 6.29, you can see that there are two active services: HTTP on port 8002 and HTTPS on port 44301.

After all these steps, you finally have a URL under which the SOAP IDoc service is running. But it is a service on SAP S/4HANA located in on-premise environment. Now we need to involve the cloud connector to expose that service to our services on SAP BTP located in the public network.

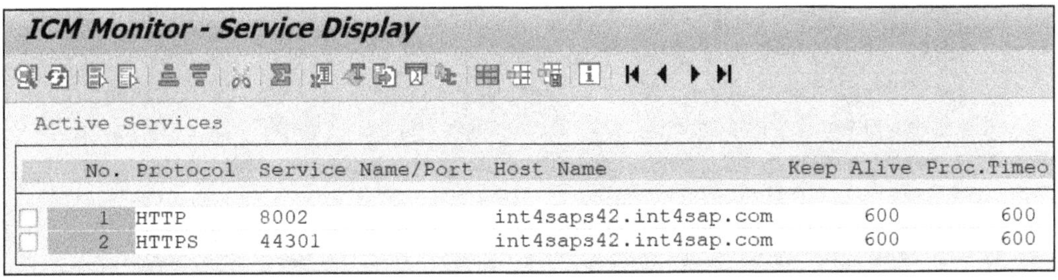

ICM Monitor - Service Display

Active Services

No.	Protocol	Service Name/Port	Host Name	Keep Alive	Proc.Timeo
1	HTTP	8002	int4saps42.int4sap.com	600	600
2	HTTPS	44301	int4saps42.int4sap.com	600	600

Figure 6.29 Transaction SMICM: Service Display

Deployment of Cloud Connector

The cloud connector is a connectivity agent that can be downloaded from the SAP website and deployed on a machine within your network. There are multiple operating systems supported. There are two prerequisites for the network layer: the cloud connector must be allowed to make an outbound connection to the appropriate SAP BTP subaccount, and it must be allowed to make an outbound connection to an SAP S/4HANA server and the services running on it. In our scenario, both connections will be via HTTP. There is no need for HTTPS as the cloud connector has its own mechanism to encrypt data and to create a secure communication channel even within HTTP transmission.

Cloud Connector Version

We recommend that you always use the latest version of the cloud connector. This guarantees that you have the most stable and secure environment. In addition, with every new release you can expect a set of new or improved features. For setting up our scenario, we used version 2.1.15.

There are two aspects to be considered when selecting a machine on which to deploy the cloud connector:

- **System requirements**
  The best approach is to check the latest hardware and software prerequisites in the documentation directly available in SAP Help Portal.
- **Network zone**
  As mentioned earlier, there are two requirements from the network point of view. The machine must have an internet connection at least to the SAP BTP subaccount with which we are opening the communication channel. At the same time, the agent must be allowed to connect to the internal, on-premise system we are exposing to the SAP BTP subaccount. In most cases, and this is the recommended approach, the cloud connector is installed in a DMZ network.

> **Demilitarized Zone**
>
> A *demilitarized zone* is an additional physical or logical subnetwork in the organization's network. It's purpose is to introduce an additional security layer between the trusted, high-security internal zone and an untrusted public network like the internet.

The deployment of the cloud connector is a straightforward process. First you must download the appropriate installation package from a public SAP website at *https://tools.eu1.hana.ondemand.com/#cloud* (see Figure 6.30).

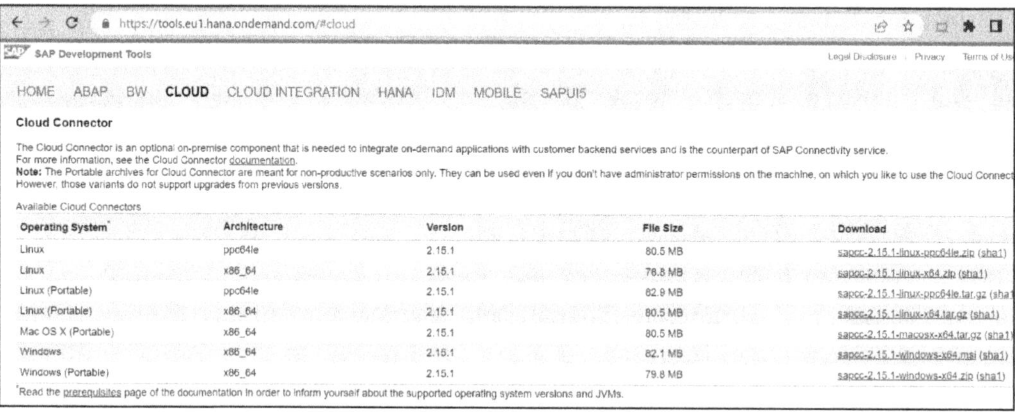

Figure 6.30 SAP Website with Cloud Connector Installation Packages

There are multiple versions depending on the target operating system. There are also some portable packages, which could be a good choice when your aim is to get familiar with the product on a local, nonproduction environment. The portable versions are not recommended for production environments.

The installation process for the cloud connector is not much different than installation of any other software. We will not describe this part in detail. After all the steps, you can check if the process was successful by opening the administration console via a web browser. Enter "https://<hostname>:<port>", where <hostname> is the machine on which cloud connector was installed and <port> is the port configured during installation. The default port is 8443 in most cases. With default installation settings, opening *https://localhost:8443* should get you to the console.

The first time you log in, you use the default credentials that can be found in SAP documentation. The initial login screen looks as shown in Figure 6.32.

In the next step, it is mandatory to change the password and choose the installation type (see Figure 6.31). There are two options under **Choose Installation Type**:

- **Master (Primary Installation)**
  Choose this for your primary installation, the one actively handling all connections.

- **Shadow (Backup Installation)**
  Choose this for a backup installation. This is part of a high availability setup and initially stays in a standby mode. You can find additional information on high availability ahead.

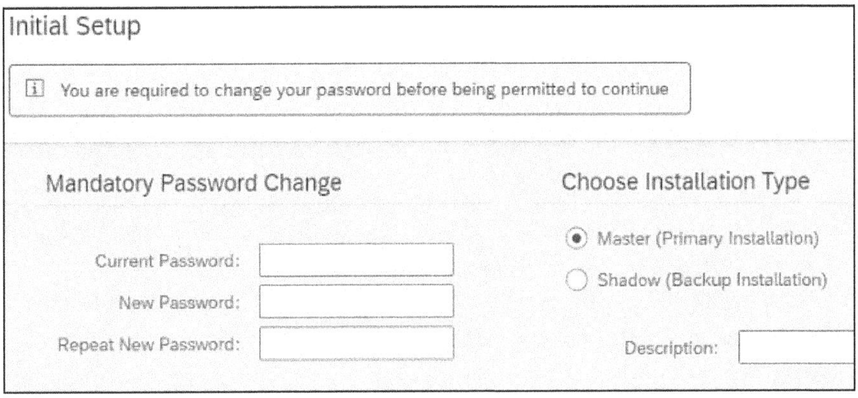

Initial Setup

You are required to change your password before being permitted to continue

Mandatory Password Change

Current Password:
New Password:
Repeat New Password:

Choose Installation Type

Master (Primary Installation)
Shadow (Backup Installation)
Description:

Figure 6.31 Mandatory Changes to Initial Password and Selection of Installation Type

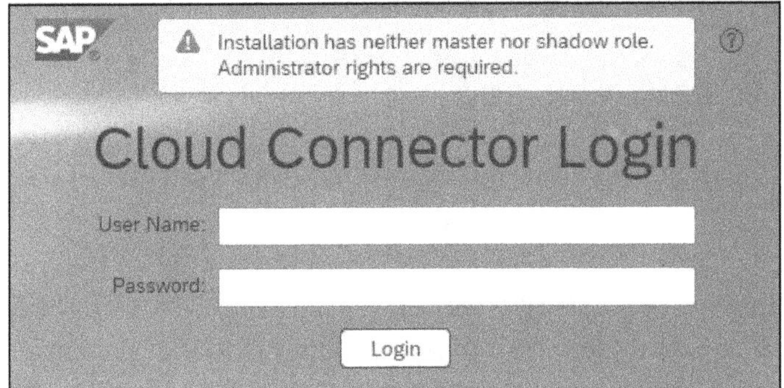

Figure 6.32 Login Screen During First Logon to Cloud Connector

Tomcat

The cloud connector runs on Apache Tomcat, a well-known HTTP server. That's why we connect to the administration console via a web browser.

Initial Configuration of Cloud Connector

After successful deployment, you may start the process of initial configuration, in which you make sure that the service runs in a secure and reliable way.

There are three key items that should be taken care of:

- **Authentication**
  The cloud connector supports basic authentication to the administration console. Just after the installation, you may log on using the predelivered user named Administrator; the password is publicly shared and can be found in the SAP documentation. The default user has unlimited privileges and can perform any activity. You must change the initial password during the first logon.

  But the problem starts when you want to create more accounts and define different privileges on them. There is no built-in feature to manage local users: you may only have one account called Administrator, and this account will always have administrator privileges.

  An alternative is to use Lightweight Directory Access Protocol (LDAP) to configure user authentication on the cloud connector. This is the recommended approach. It allows you to tighten the security of your cloud connector instance and to create additional user accounts limited to specific roles. The predefined authorization roles are as follows:

  - sccadmin (or admin)
  - sccsubadmin
  - sccdisplay
  - sccsupport
  - sccmonitoring

  Refer to the SAP documentation to find more information on enabling LDAP authentication.

- **High availability**
  A second key item that should be managed on a freshly installed instance is the high availability setup. The idea is that there are two separate installations of cloud connector, and they should be running on different servers—maybe even in different data centers. That depends on your corporate high availability strategy. One of the instances is defined as the master (primary installation) and the second as a shadow (backup installation). You define the installation type during the first logon. Only the master instance is involved in routing the connections. The shadow instance stays in standby mode listening for the status of master instance. Refer back to Figure 6.31 to see the selection options. In the second configuration step, you define a link between these two instances and the time after which the master should be considered unavailable. When the master instance is down for the predefined amount of time (e.g., 60 seconds) the shadow instance takes over the master role.

- **Security**
  Apart from the mentioned authentication element, there are a few more security aspects that should be taken into account and managed before productive usage of the cloud connector. However, these fall far outside the scope of this book. Refer to the SAP Help Portal for more information on cloud connector security.

Enabling a Secure Connection from the Cloud to On-Premise System

In the previous sections, we deployed and went through the initial configuration of the cloud connector. The component is now ready to be involved in our main task, which is connecting an on-premise system to the SAP BTP subaccount. The cloud connector will be mediating and controlling the communication; from this, we can see that there are two independent activities to be performed: connecting the cloud connector to the SAP BTP subaccount, and connecting the cloud connector to SAP S/4HANA. We'll cover these in the following sections.

Connecting the Cloud Connector to the SAP BTP Subaccount

To configure this connection, follow these steps:

1. **Obtain the details of your SAP BTP subaccount that you want connect via the cloud connector.**
   The mandatory information is the region and the subaccount ID. One way to find this information is to open the chosen subaccount on the global account view in the SAP BTP cockpit. You can find all the required information on the overview screen, as shown in Figure 6.33.

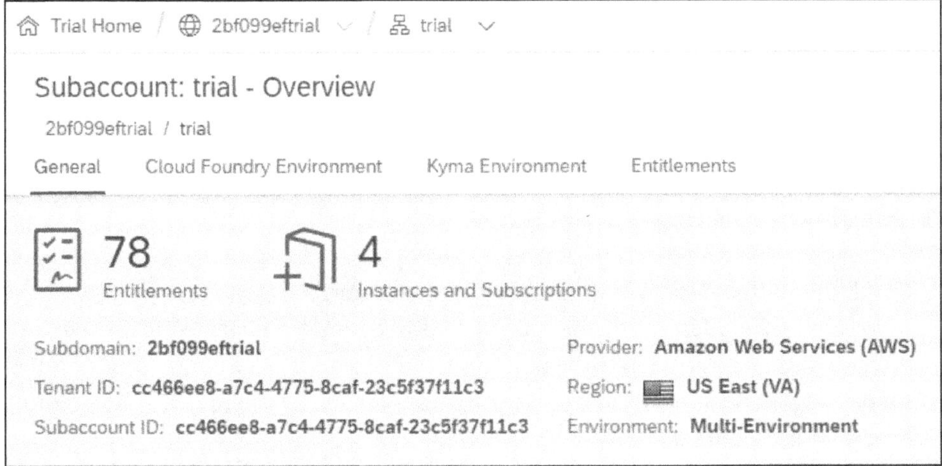

Figure 6.33 Overview of SAP BTP Subaccount

2. **Add the new subaccount in the cloud connector.**
   After you are logged into the cloud connector, click **Add Subaccount**, which is located in the top-right corner of the main page (see Figure 6.34). A new popup window titled **Add Subaccount** appears where you can maintain all the necessary parameters of your subaccount. As mentioned before, there are only three mandatory items (the ones with red asterisks), but it is a good practice to give a meaningful display name and description. See Figure 6.35 for the configuration properties used in our case. Click **Save** when you are done.

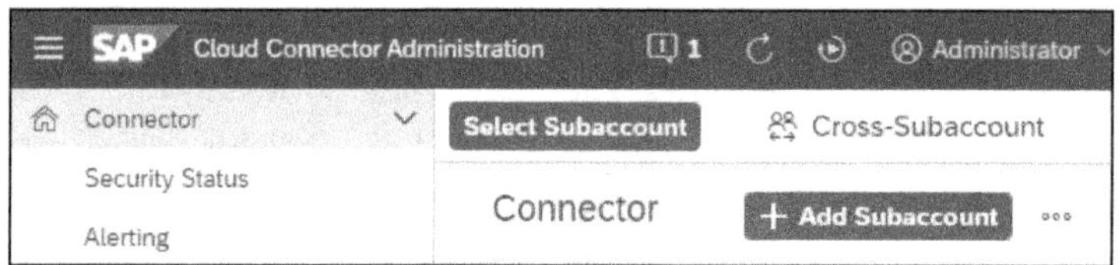

Figure 6.34 Top Part of Cloud Connector Home Page

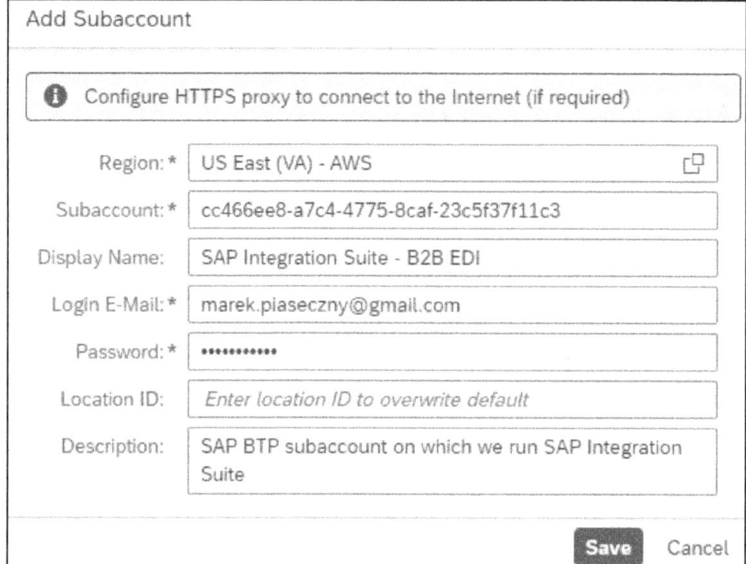

Figure 6.35 Adding Subaccount in Cloud Connector

3. **Check the status of the newly connected subaccount.**
   After successful addition of the SAP BTP subaccount, you can see it on the subaccount dashboard located in the bottom part of the main page of the cloud connector (see Figure 6.36). You can also display more details by opening the subaccount overview (see Figure 6.37).

As you can see, there is a successfully established tunnel between the cloud connector and the SAP BTP subaccount. You can see the details of the subaccount and who initiated the connection, and you have the option to disconnect the tunnel at any time.

You can also export and import the configuration at any time. This is a useful feature for any migration-related activities, and it can be also used as a backup option.

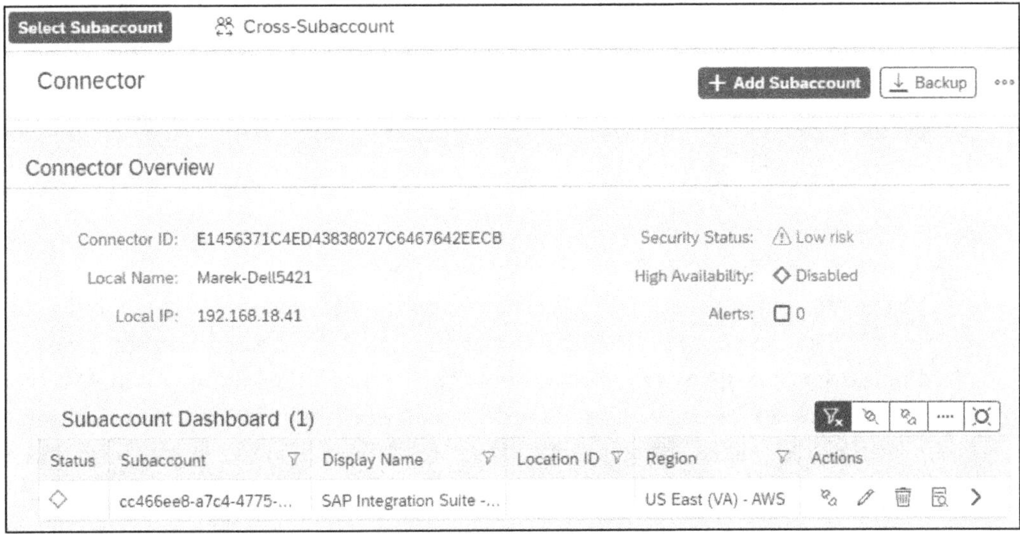

Figure 6.36 Cloud Connector View: Subaccount Dashboard

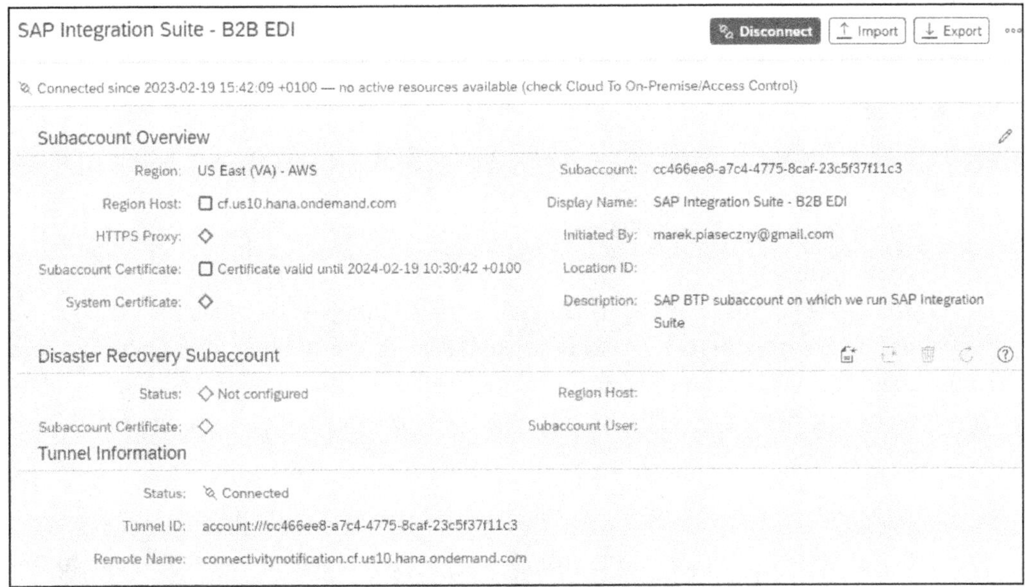

Figure 6.37 Cloud Connector View: Subaccount Details

At this stage, you have a secure tunnel between the cloud connector located in your network and your SAP BTP subaccount. This tunnel can be used by any service on SAP BTP to connect to on-premise resources. As mentioned earlier and proven in this chapter, it is not required to allow any connection coming from an untrusted network to the internal network where the cloud connector runs.

As you may have noticed in Figure 6.36, there is an orange icon in the **Status** column indicating that there is a warning. It says **No active resources available**, which means that the subaccount can indeed make a connection to cloud connector, but there is nothing exposed from the internal network. There are no resources that can actually be called and used by the subaccount.

Next, you will expose the HTTP service of SAP S/4HANA.

Connecting the Cloud Connector to On-Premise SAP S/4HANA

You need to configure each internal server or service that you want expose for use by the SAP BTP subaccount explicitly. You may think that this creates unnecessary overcomplexity and costs extra effort without much added value. But if you look at this from the perspective of security and the need to have all the data flows transparent and under control, it makes perfect sense. This part of the configuration is done within a specific subaccount already added to the cloud connector. It happens on the subaccount level, meaning that whatever we expose from our internal network is available only for a subaccount for which we did the whole configuration.

Before you start, make sure that the details related to the on-premise resource are known. There are three key elements that are must-haves for the cloud connector configuration:

- **Host**
  This is the host on which SAP S/4HANA is running.
- **Path**
  This is the path of IDoc HTTP service (ICF path). The default is */sap/bc/srt/idoc*.
- **Port number**
  This is the port used by the HTTP service on the SAP S/4HANA system. Note that we use HTTP here; there is no need to use HTTPS as the cloud connector has its own mechanism to encrypt the data in transfer. At the beginning of this chapter, while activating the IDoc HTTP service, you learned that for our test system we use port 8002.

To expose the on-premise resource via the cloud connector, follow these steps:

1. In the cloud connector, use the list on left side of the screen to go the **Cloud to On-Premise** configuration section for the subaccount previously added (see Figure 6.38).

Cloud to On-Premise

The **Cloud to On-Premise** section is named for the data flow direction. In this case, it is about setting up a communication channel where data flows from the cloud to an on-premise system.

2. Trigger the process of adding a new on-premise system by clicking **Add** +, located on the right side of the screen.

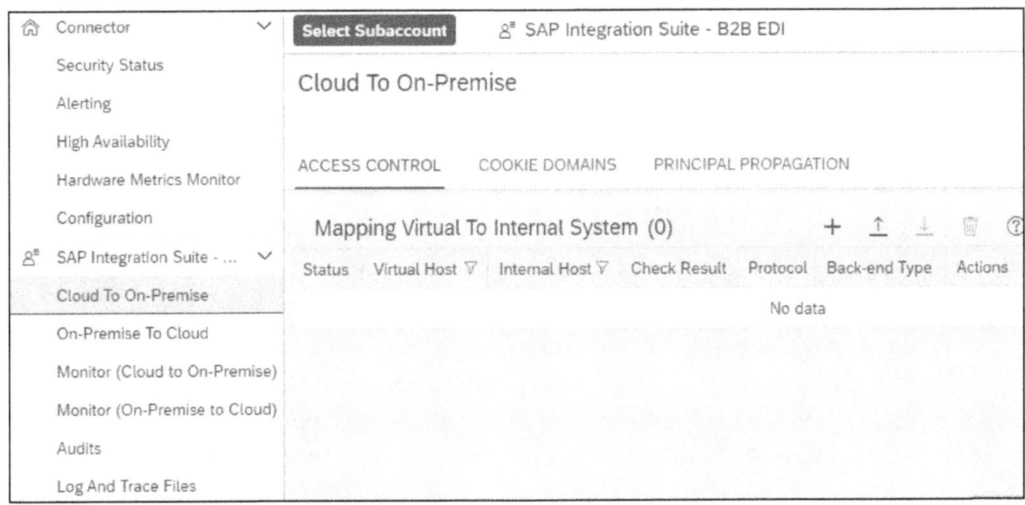

Figure 6.38 Cloud to On-Premise Section

3. A new window called **Add System Mapping** appears. From the dropdown list, choose **ABAP System** for **Backend Type**, and confirm by clicking **Next** (see Figure 6.39).

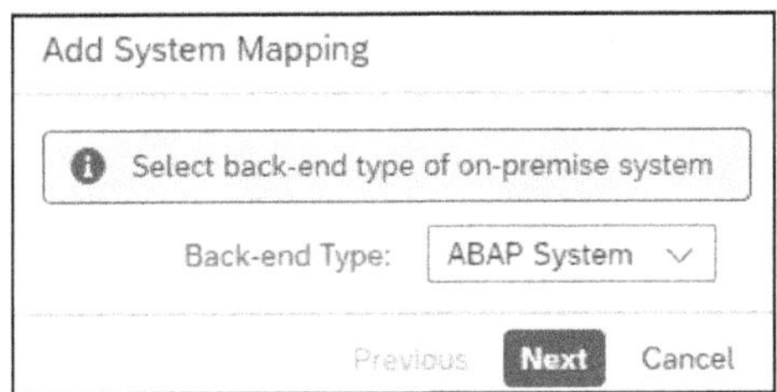

Figure 6.39 Selection of Backend Type

4. Select **HTTPS** for **Protocol** (see Figure 6.40).

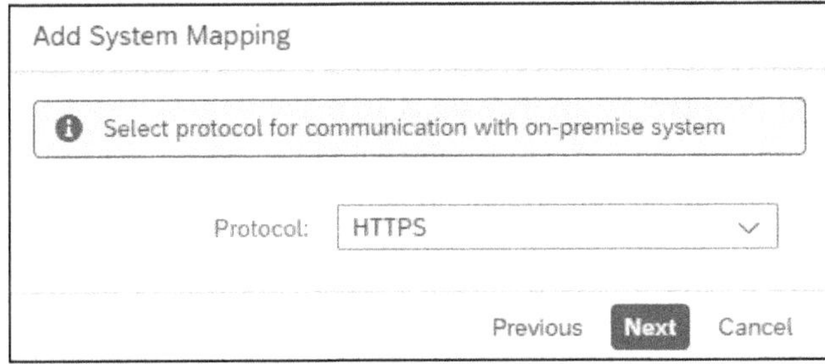

Figure 6.40 Add System Mapping: Select Protocol

5. Maintain the **Internal Host** and **Internal Port** (see Figure 6.41).

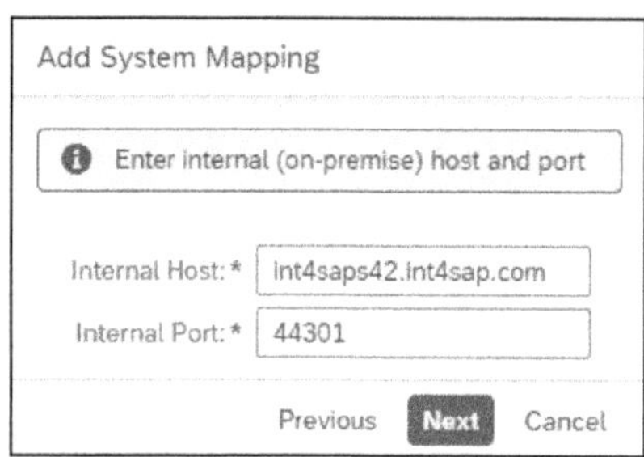

Figure 6.41 System Mapping: Maintenance of Virtual Host

6. Confirm the **Virtual Host** and **Virtual Port** settings, which by default are filled by the values of **Internal Host** and **Internal Port** (see Figure 6.42). For simplicity here, we keep the virtual host and port the same as the internal host and port. But in the real world, it is recommended to apply different names to the virtual host and port.

Virtual Host

The purpose of the virtual host is to hide the actual address of your on-premise system. Within SAP BTP, only the virtual host and virtual port are visible. These values do not need to contain any technical characteristic of the actual technical host; they can be any descriptive name.

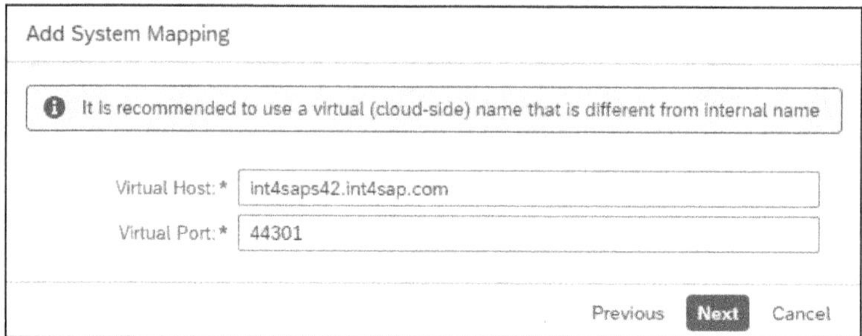

Figure 6.42 System Mapping: Maintenance of Virtual Host

7. Make sure that **Allow Principal Propagation** is disabled. See Figure 6.43.

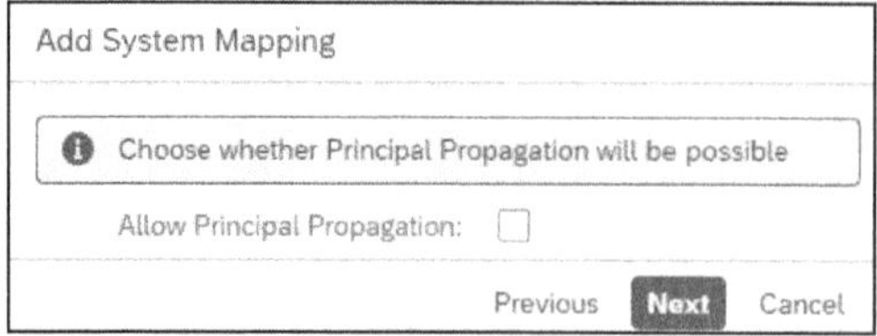

Figure 6.43 Deactivate Allow Principal Propagation

8. Confirm usage of a virtual host by selecting **Use Virtual Host** on the next screen (see Figure 6.44).

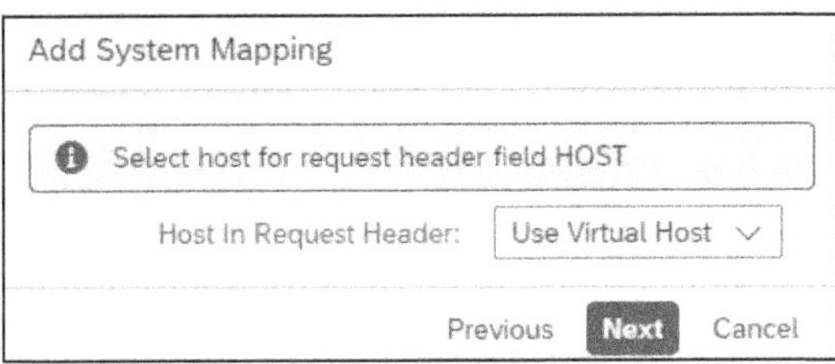

Figure 6.44 Use Virtual Host

9. Maintain a description (see Figure 6.45). This is optional, but it can definitely help with transparency as your landscape grows.

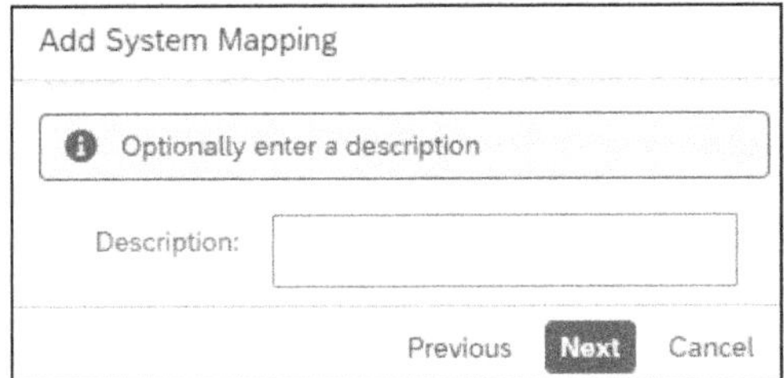

Figure 6.45 Maintain Optional Description

10. Check the data on the summary screen and confirm by clicking **Finish** (see Figure 6.46).

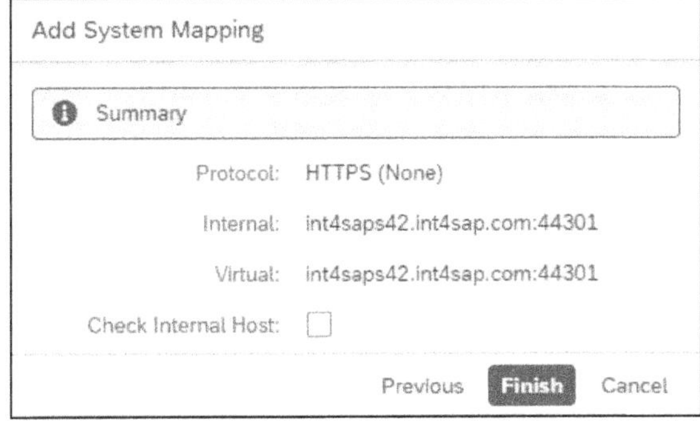

Figure 6.46 Summary for Mapped System

11. In Figure 6.47, you can see that the internal system was added. But you still need to specify which internal resources you will expose. For HTTP, this is done by adding the URL path of the on-premise resource. Click **Add** + to maintain it.

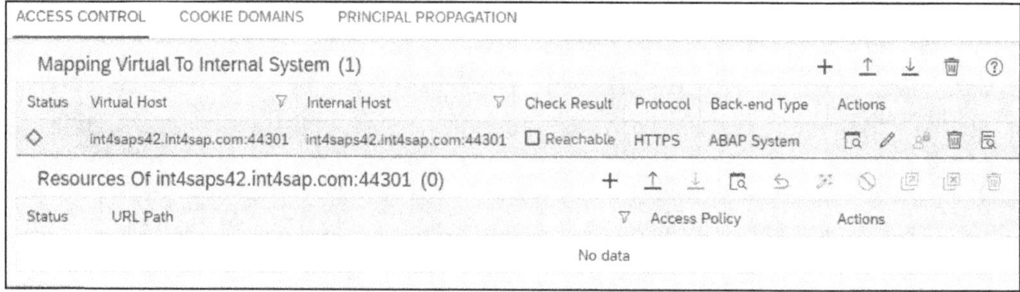

Figure 6.47 Successfully Mapped Internal System without Added Resources

12. In the configuration window shown in Figure 6.48, you maintain the details of the internal resource. In this example, **URL Path** contains **/sap/bs/srt/idoc**, which is the path pointing to the IDoc HTTPS SOAP service. By selecting **Path Only (Sub-Paths Are Excluded)** for **Access Policy**, you indicate that this is the only service that can be accessed via the cloud connector.

Figure 6.48 Adding Resource of On-Premise System to Be Reachable from Cloud Connector

13. After executing all these steps, you can see the on-premise ABAP system mapped via HTTPS and with successfully exposed resources pointing to the SOAP IDoc service (see Figure 6.49).

14. There is also an option to see the status from a perspective in the SAP BTP cockpit. After logging onto the subaccount, navigate to the **Connectivity** section and choose **Cloud Connectors**. On the overview page, you can find information about the connected cloud connector instance. At the bottom of the screen, there is information about the exposed backend systems. The system you just added is listed there (see Figure 6.50).

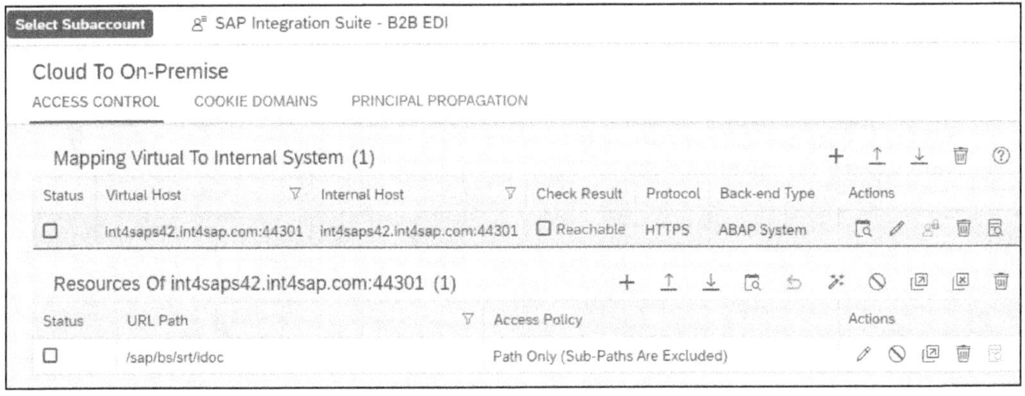

Figure 6.49 Successfully Mapped Internal System with Added Resources

Exposed Back-End Systems

Host	Protocol	Back-End Type	Resources
int4saps42.int4sap.com:44301	HTTP	ABAP System	Available

Figure 6.50 SAP BTP Cockpit and Backend Systems Exposed by Cloud Connector

6.4 Summary

In this chapter, we went through all the steps to establish connectivity between an SAP BTP subaccount and an on-premise SAP S/4HANA system. While some of the steps we executed may be handled by different teams like basis or even infrastructure, we included them with all the details to give as broad a view of the solution as possible.

The activities started with the deployment of necessary artifacts from the master package predelivered by SAP: Cloud Integration—Trading Partner Management V2. In the following chapters, we will go into the details of SAP Integration Suite, but the deployment was necessary first to collect endpoint-related parameters. In the next steps, we configured the HTTPS destination and port, which will be used by SAP S/4HANA to send IDocs to the iFlow deployed on Cloud Integration. In this part, we also ensured the proper security by enabling client certificate authentication.

The last activity was the most time-consuming and most complex as we had to install and then configure the cloud connector, which involves enabling the SOAP IDoc service on SAP S/4HANA and finding the appropriate system parameters.

At this stage, you have the connectivity channels in place. Data can be transferred from SAP S/4HANA to Cloud Integration and the other way around. What you're still missing is the data to be transferred. In the next chapter, you will work on the SAP S/4HANA configuration that enables IDocs messages.

Chapter 7
Configuration and Development in SAP S/4HANA

The core concept of IDocs is still available in the SAP S/4HANA system. After a short introduction, we will go through all the configuration steps to enable inbound and outbound IDoc processing for our EDI scenario. At the end of this chapter, we will look at Business Rules Framework plus for output processing.

Finally, we're at the stage where we can focus on configuring the SAP S/4HANA system itself. In this chapter, we'll focus mainly on IDocs. There have already been a few books and hundreds of articles describing all different aspects and views of IDocs, so we'll keep the introduction part short but concise here. We'll take a closer look at our sample implementation and then will go through the full configuration process to enable processing of the selected IDocs in the selected steps of the order-to-cash process. Our approach is to use as much standard SAP functionality as possible because most of the background processing does not differ from that of previous SAP ERP systems.

7.1 IDoc Fundamentals

IDocs are structures or data containers within SAP. IDoc is a native SAP standard to exchange data between SAP and non-SAP systems. There are hundreds of predefined IDocs embedded in each SAP business system. The number comes from the fact that each business object—sales order, delivery, material, customer, and so on—has its own IDoc definition, its own structure, which can be filled by the object's specific data.

IDocs are key elements of SAP integration. They are also broadly used within the any IT landscape within an SAP system and play a critical part in the SAP integration area. Even with the introduction of SAP S/4HANA (on-premise) and the move toward other technologies like OData APIs and Event Mesh, IDocs can't be fully replaced and they will still play a significant role in the future SAP S/4HANA landscape.

On top of that, we have and will have custom requirements, business scenarios, and use cases that can't be fully covered by the predelivered, standard IDoc definitions. With that comes the topic of IDoc extensions.

There are three key elements related to the IDoc definition:

- **Message type**
  This specifies the data type being distributed by the IDoc and the function of the IDoc. The message type itself does not define the structure of the IDoc. It is a characteristic used to configure and to distinguish different flows and use cases. Message types can be freely defined, which is the key element enabling IDoc reusability. Predelivered message types include ORDERS, ORDRSP, and ORDCHG, among others.
- **Basic type**
  This is the actual definition of the (technical) IDoc structure. It defines what the data container looks like and what kind of data it can store. Predelivered basic types include ORDERS05, INVOIC02, and MATMAS05, among others.
- **Extension (optional)**
  If none of the standard IDoc structures meet your needs, you can create your own IDoc structure. This custom structure can be defined from scratch, but the most common approach is to take a basic type and add segments/fields. In this way, you create an IDoc extension to the basic type. Such custom, extended IDocs are defined by the message type plus the basic type plus the extension. In our scenario, we will not use extensions; our flows will make use of the standard IDocs. Therefore, we will not go into the details of IDoc extensions.

There are a few mandatory configuration elements for exchanging data via IDocs. There could be some additional elements related to IDoc types outside of our EDI scenario, but these are the basic minimum requirements:

- IDoc definition must be set by means of all the artifacts like the message type and basic type (and of course related segments, fields, data elements, etc.), and the configuration around them.
- The port is a prerequisite for sending any IDoc out of the SAP S/4HANA system. It's the destination for our message. We created a port in Chapter 6, pointing to an iFlow on Cloud Integration. The port is only applicable for outbound traffic.
- The process to consume the inbound IDoc is most often a function module or BAPI.
- The process to generate the outbound IDoc is most often a function module, a BAPI, or a standalone program.
- There is an ALE/IDoc processing layer that links all these elements and takes care of handling the IDocs, but the exact condition that we process and for which partner must be defined in a partner profile.
- A partner profile is mandatory; it's maintained in Transaction WE20. Here we define with whom we want to enable IDoc-based communication. The starting point is a partner type: this could be a customer, vendor, bank, logical system, and so on. In our example, it is simply a customer. In a second step, we determine what outbound and inbound messages to enable for that partner/customer.

An IDoc instance contains the following:

- **Control record**
  This is a technical header of the IDoc message and includes metadata about the IDoc, such as the sender, receiver, IDoc type, and message type. It contains mandatory and critical information for routing and processing of the document. It does not contain any business data, but when you send an IDoc to an SAP business system, you still must take care of filling in this section properly.
- **Data records**
  These are the segments that carry the actual data being transferred. Each segment contains a specific set of fields related to a business transaction or entity, such as customer information, order details, or invoice data. The structure and content of these segments depend on the IDoc type and the specific business context.
- **Status records**
  These records track the processing status of the IDoc at various stages within the sender and receiver systems. They are used for monitoring, error handling, and ensuring the successful transmission and processing of the IDoc. The status record is not part of any IDoc transmission; it is generated within SAP S/4HANA based on the different processing steps. Usually, each IDoc within the system has multiple status records, but the most recent one indicates the current status.

7.2 IDoc Selections

The previous section gave you a very brief description of IDocs. The intention of this section is now to look at our specific use case and choose IDoc messages appropriate to support it. This should clarify many doubts you may still have about IDoc technology, and at the same time we will move ahead with our EDI implementation. All the choices will be explained and enriched with a description of the IDoc structure. At the end, we will take a close look at the sample ORDERS.ORDERS05 IDoc. We will look at its structure and sample data, then consider Transaction WE60, where you can not only find IDoc documentation but also export it out of the system.

7.2.1 Message Types and Basic Types

We'll now select the exact IDocs for our EDI scenario. The good news is that within SAP S/4HANA, we can already find hundreds of predefined IDocs with which we can cover most business processes. To give you some idea, there are around 2,400 predefined message types and 2,500 IDoc types. In our case, using standard IDocs is enough; we will not be customizing any extensions. We'll go directly to the selected IDocs, as explained in the following sections.

Sales Order

The sales order has the following characteristics:

- Message type: ORDERS
- IDoc type: ORDERS05

Note that the common notation is MessageType.IDocType. Following this approach, ORDER.ORDERS05 is how we will be referring to the IDoc used in the sales order flow.

For message type ORDERS and basic type ORDERS05, there is no need to customize anything additional to enable this IDoc. This is the most common scenario, and all the necessary artifacts and configuration elements are already within the system. Although we don't need to modify anything, we will walk you through the key elements so that you can get a full understanding.

The starting point is a definition of a logical message, as can be seen in Transaction WE81 (Message Types), shown in Figure 7.1. ORDERS is already there with the **Short Text** column showing **Purchase order/order**. This indicates that we have a message defined for a process purchase order/order.

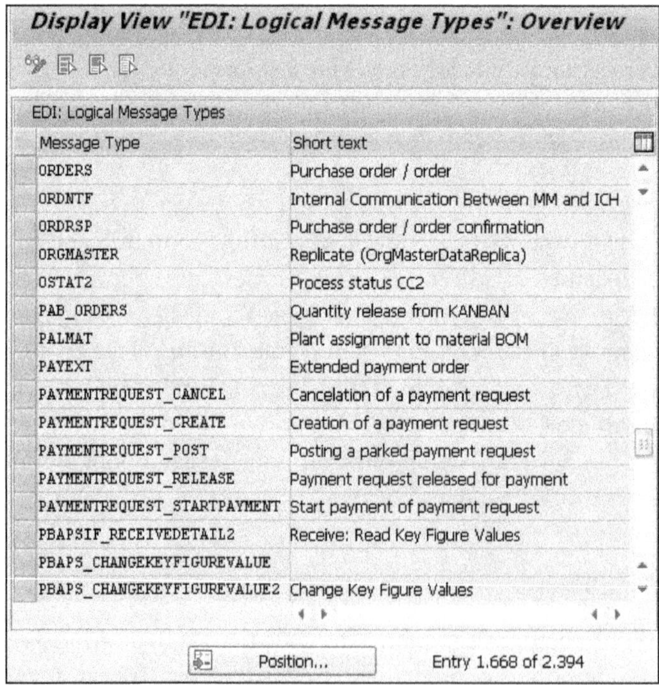

Figure 7.1 Message Type Definitions Seen in Transaction WE81

As you already know, we also need a structure that can be used together with the IDoc message type, and our selection is ORDERS05. How the structure looks is shown in Transaction WE30 (IDoc Types). Figure 7.2 shows the selection screen of Transaction WE30.

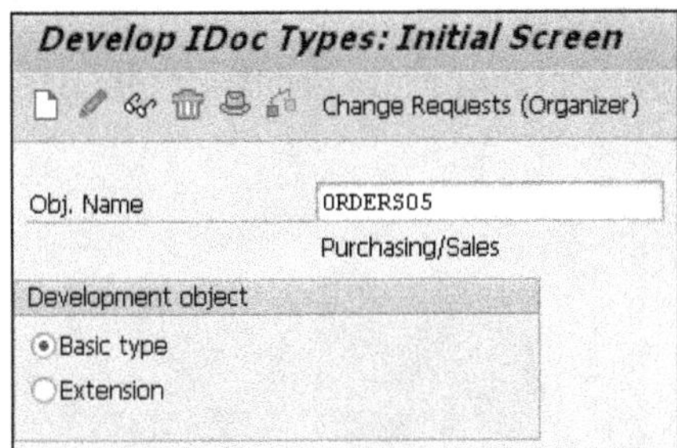

Figure 7.2 Selection Screen for Transaction WE30: IDoc Types

After maintaining the object name (**Obj. Name**) and object type (in the **Development Object** area)—in this case, ORDERS05 and **Basic Type**—we can display the IDoc structure. The display view is seen in Figure 7.3.

Display basic type: ORDERS05

```
ORDERS05      Purchasing/Sales
   E1EDK01    IDoc: Document header general data
   E1EDK14    IDoc: Document Header Organizational Data
   E1EDK03    IDoc: Document header date segment
   E1EDK04    IDoc: Document header taxes
   E1EDK05    IDoc: Document header conditions
   E1EDKA1    IDoc: Document Header Partner Information
   E1EDK02    IDoc: Document header reference data
   E1EDK17    IDoc: Document Header Terms of Delivery
   E1EDK18    IDoc: Document Header Terms of Payment
   E1EDK35    IDoc: Document Header Additional Data
   E1EDK36    IDOC: Doc.header payment cards
   E1EDKT1    IDoc: Document Header Text Identification
   E1EDP01    IDoc: Document Item General Data
   E1CUCFG    CU: Configuration data
   E1EDL37    Handling unit header
   E1EDS01    IDoc: Summary segment general
```

Figure 7.3 Transaction WE30: Display View of Basic Type ORDERS05 Used in Sales Order Creation and Sales Order Response

The IDoc basic type and message type must be linked together to create a fully functional IDoc assignment of the message type to IDoc type. This is done in Transaction WE82 (IDoc Type/Messages). Figure 7.4 shows what this look like in the system. Among many other entries, you can see ORDERS being linked to ORDERS05. That's the pair we are going to use for our sales order flow.

For inbound IDocs, you must also assign a function module to the IDoc type. This assignment defines which function module can be used to process specific IDocs. There can be multiple entries for the same function module, which one is used for the actual processing in a specific scenario is defined in the partner profile. The entry for ORDERS.ORDERS05 is already part of the standard configuration; it is shown in Figure 7.5. From this screen, you can already conclude that this type of IDoc can be processed by function module IDOC_INPUT_ORDERS and that this function is also an object already existing in the system. That's correct. This transaction is only applicable for inbound IDoc processing. We do not need such entries for outbound IDoc processing.

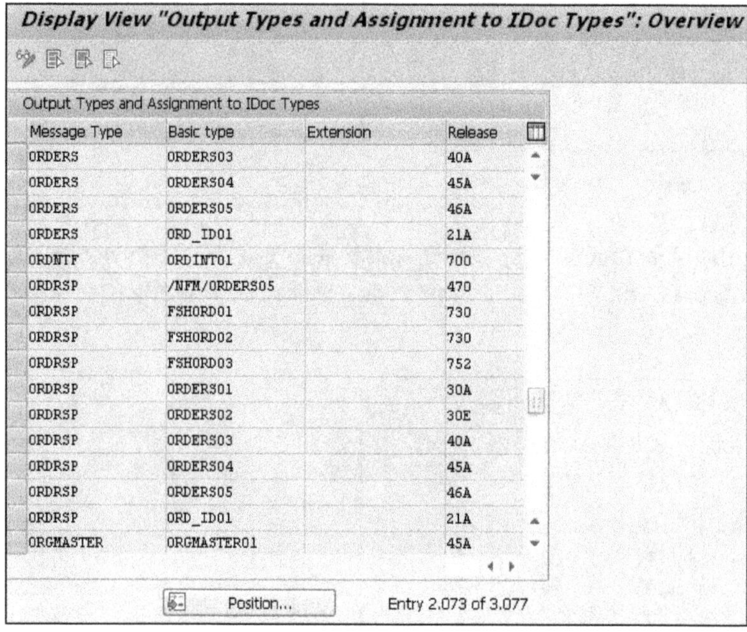

Display View "Output Types and Assignment to IDoc Types": Overview

Output Types and Assignment to IDoc Types

Message Type	Basic type	Extension	Release
ORDERS	ORDERS03		40A
ORDERS	ORDERS04		45A
ORDERS	ORDERS05		46A
ORDERS	ORD_ID01		21A
ORDNTF	ORDINT01		700
ORDRSP	/NFM/ORDERS05		470
ORDRSP	FSHORD01		730
ORDRSP	FSHORD02		730
ORDRSP	FSHORD03		752
ORDRSP	ORDERS01		30A
ORDRSP	ORDERS02		30E
ORDRSP	ORDERS03		40A
ORDRSP	ORDERS04		45A
ORDRSP	ORDERS05		46A
ORDRSP	ORD_ID01		21A
ORGMASTER	ORGMASTER01		45A

Position... Entry 2.073 of 3.077

Figure 7.4 Transaction WE82: Message Type Assignment to Basic Type

Display View "IDoc: Assignment of FM to Log. Message and..": Details

Function module	IDOC_INPUT_ORDERS
Function type	Function module
Basic type	ORDERS05
Extension	
Message Type	ORDERS
Message Variant	
Mess. function	
Object Type	BUS2032

IDoc: Assignment of FM to Log. Message and IDoc Type

Direction	Inbound
Description	Purchase order / order
Name	Sales Order

Figure 7.5 Transaction WE57: Assignment of Functional Module to Message Type

Sales Order Response

The sales order response has the following characteristics:

- Message type: ORDRSP
- IDoc type: ORDERS05

As you're already familiar with the sales order and ORDERS05 structure, this one will be simple. We already said that the basic type can be reused in multiple scenarios and processes. The sales order response is exactly such situation. Note that we still operate on the same business object (sales order) and we use the same structure like in the sales order flow. To distinguish the sales order response from the sales order creation and to have the possibility to apply the necessary configuration, we will use a different message type: ORDRSP. If you look back at Figure 7.1, you can see that the ORDRSP message type is already there.

Shipping Notification

The shipping notification has the following characteristics:

- Message type: DESADV
- IDoc type: DELVRY07

In the shipping notification flow, we will make use of DESADV.DELVRY07, as shown in Figure 7.6, where you can see the structure. Looking at the content and data types, this differs a lot from ORDERS05. Both are defined for different business objects and different purposes. The customizing required for using this IDoc is already in place. In Transaction WE81, you can see that message type DESADV is already defined. In Transaction WE82, you can find an existing relationship between the DESADV message type and the DELVRY07 IDoc type.

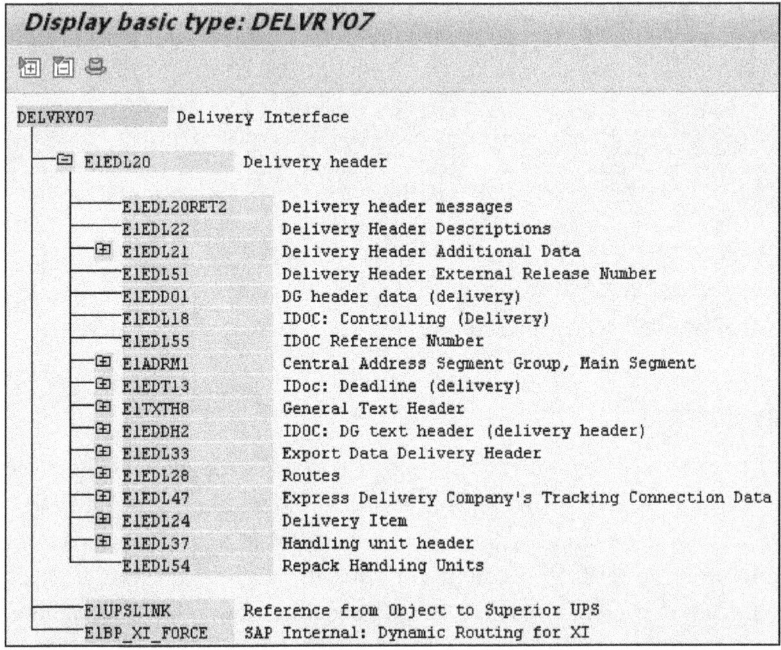

Figure 7.6 Transaction WE30: Display View of Basic Type DELVRY07 Used in Shipping Notification Flow

Invoice

The invoice has the following characteristics:

- Message type: INVOIC
- IDoc type: INVOIC02

For an invoice, we will make use of INVOIC.INVOIC02, as shown in Figure 7.7, where you can see the structure. Looking at the content and data types, it differs from DELVYR07, but looks close to the ORDERS05 structure. The customizing required for using this IDoc is already in place. In Transaction WE81, you can see that message type INVOIC is already defined. In Transaction WE82, you can find an existing relationship between the INVOIC message type and the INVOIC02 IDoc type.

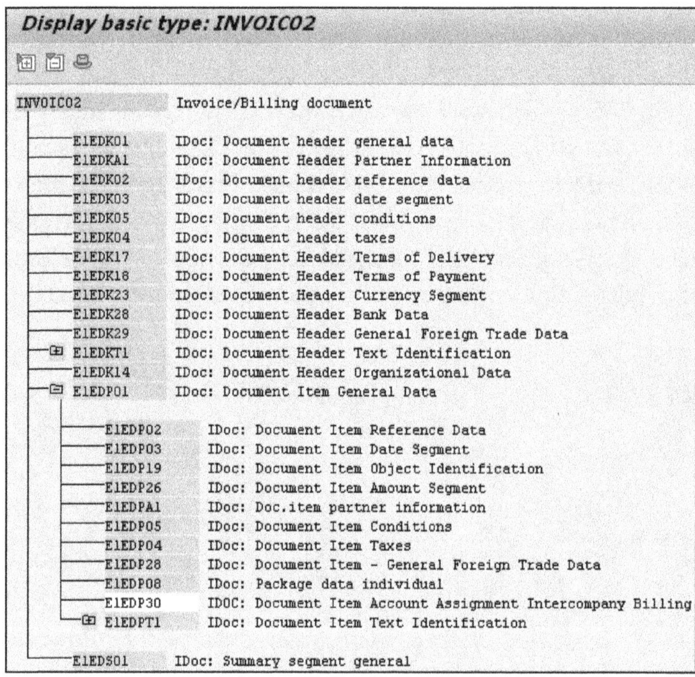

Figure 7.7 Transaction WE30: Display View oF Basic Type INVOIC Used in Invoice Flow

7.2.2 Segments and Fields

Next, we'll take a brief look at segments and fields. If we look once again at the basic ORDERS05 type, we can see that it is built from multiple blocks like E1EDK01, E1EDK14, E1EDK03, and so on. These are segments. Segments can be linked to and reused across multiple different IDoc types. Each segment contains from a few to several fields related to a specific part of the business object; it is a specific, often related set of data. If we choose one such segment in Transaction WE30, we can see its attributes as shown in Figure 7.8. In this example, we have segment E1EDKA1, whose role is to store partner-related information in the document header. For a sales order, this segment may occur multiple times: there can be one with sold-to data, one with ship-to data, one with supplier data, and so on.

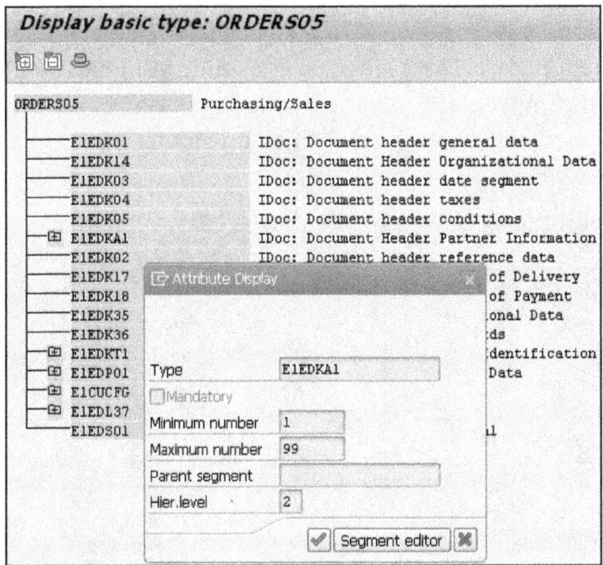

Figure 7.8 Transaction WE30: Display Attributes of Segment E1EDKA1, Part of Basic Type ORDERS05

The two building blocks which define the IDoc structure are as follows:

- **IDoc segments**
  Segments are the constituents of an IDoc that contain actual data related to a specific aspect of a business transaction. Each segment represents a logical grouping of related information, such as header details, item information, or address data. Segments are organized hierarchically within an IDoc, allowing for the representation of complex documents in a structured manner. They are defined by segment types, which determine the structure and meaning of the data contained within.

- **IDoc fields**
  Within each segment, fields are the individual data elements that store the values of the information being transferred. Each field has a specific data type and length, defined according to the requirements of the information it holds, such as a numeric order quantity, a date, or a text description. Fields are where the actual transaction data, like customer names, addresses, product IDs, and quantities, are entered and transmitted. The arrangement and content of fields are crucial for the correct interpretation and processing of the IDoc by the receiving system.

We can maintain segments using Transaction WE31. But again, there is a huge set of standard (predelivered by SAP) segments that are already linked to the standard IDoc types. We should not modify these; instead, we should create our own segments. But that goes into the topic of IDoc extensions, and we will not do so in detail here as it is not part of our EDI scenario. One important aspect of which you should be aware is that there is also a versioning mechanism and release concept applicable for IDocs and their elements, like segments. In the end, these are also ABAP dictionary objects. In Figure 7.9, you can see header information for segment E1EDKA1; note that it has multiple definitions, each linked with a different release. These are updates coming from SAP itself: all new definitions are backward-compatible. That means that the newest definition still contains all the fields from previous releases, and then on top of that it may contain some additional fields.

In Figure 7.10, you can see the definition of segment E1EDKA1 in its latest version.

Development segments: Initial screen

Segment type: E1EDKA1

IDoc: Document Header Partner Information

Definitions

Version	Definition	Release	Release	Appl....	No. o...	Lgth	Last Chan...	Last Ch...
003	E2EDKA1003	☑	45A		44	995	30.04.1998	09:18:31
002	E2EDKA1002	☑	40A		43	980	11.08.1997	13:20:41
001	E2EDKA1001	☑	30F		41	943	10.11.1996	10:33:31
000	E2EDKA1	☑	30A		39	843	03.04.1995	13:57:33

Figure 7.9 Header Information for Segment E1EDKA1

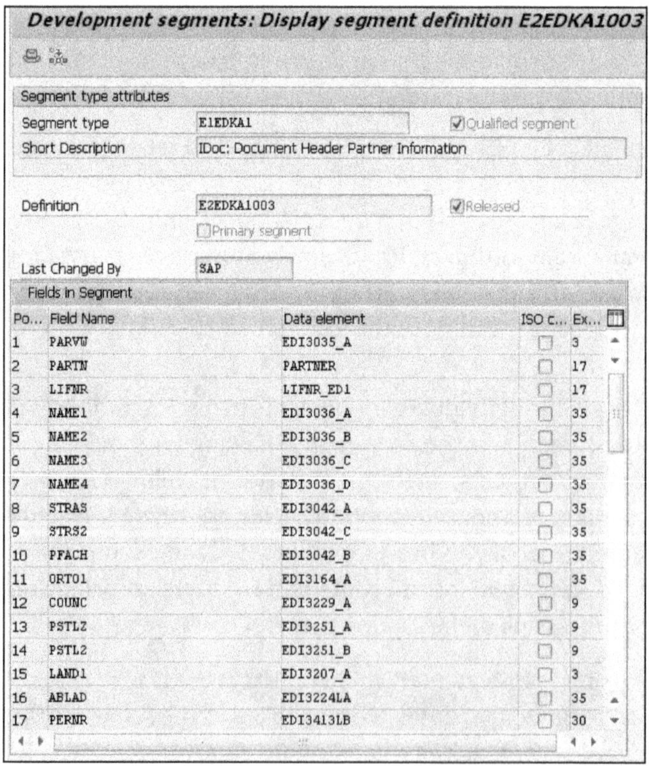

Development segments: Display segment definition E2EDKA1003

Segment type attributes

Segment type: E1EDKA1 ☑ Qualified segment

Short Description: IDoc: Document Header Partner Information

Definition: E2EDKA1003 ☑ Released

☐ Primary segment

Last Changed By: SAP

Fields in Segment

Po...	Field Name	Data element	ISO c...	Ex...
1	PARVW	EDI3035_A	☐	3
2	PARTN	PARTNER	☐	17
3	LIFNR	LIFNR_ED1	☐	17
4	NAME1	EDI3036_A	☐	35
5	NAME2	EDI3036_B	☐	35
6	NAME3	EDI3036_C	☐	35
7	NAME4	EDI3036_D	☐	35
8	STRAS	EDI3042_A	☐	35
9	STRS2	EDI3042_C	☐	35
10	PFACH	EDI3042_B	☐	35
11	ORT01	EDI3164_A	☐	35
12	COUNC	EDI3229_A	☐	9
13	PSTLZ	EDI3251_A	☐	9
14	PSTL2	EDI3251_B	☐	9
15	LAND1	EDI3207_A	☐	3
16	ABLAD	EDI3224LA	☐	35
17	PERNR	EDI3413LB	☐	30

Figure 7.10 Transaction WE31: Display Field in Segment E1EDKA1

7.2.3 Sample IDocs

You already have some basic knowledge of IDocs and you know where you can find some additional details related to their content. Now let's see what a sample IDoc actually looks like. By *sample*, we mean an instance of an IDoc object that is created in line with the appropriate structure definition and that contains business data that allows the function processing. In short, this will be an IDoc instance that is correct from technical and functional points of view.

There is also one important clarification: when we look at SAP S/4HANA, the IDocs are stored in their database, and in most cases we look at them via the predelivered transaction for IDoc display—for example, Transaction WE02. Alternatively, you can display the IDoc by looking at the database tables. We'll start this section by covering Transaction WE60, the documentation for IDoc types. Then, we'll discuss viewing sample IDocs in Transaction WE02 and database tables.

Transaction WE60

Before you look at sample IDocs, it's good to get familiar with Transaction WE60 (Documentation for IDoc Types). If you want to get details on a specific IDoc structure—that is, a specific IDoc basic type or extension—then this is the transaction to go to. The alternative to this approach is to open the technical definition of the IDoc in the previously mentioned transactions like WE30 and WE31 and navigate to the necessary elements and their characteristics.

When you start Transaction WE60, you will see the selection screen shown in Figure 7.11. Enter "ORDERS05" in the **Basic Type** field and confirm the selection by clicking the first icon on the left-hand side at the top, **HTML Format.** You will see documentation of the IDoc structure in HTML format, asshown in Figure 7.12. From this list you can jump into the details of the selected segment by clicking the **Structure** hyperlink. Figure 7.13 shows the details of the E1EDK01 segment.

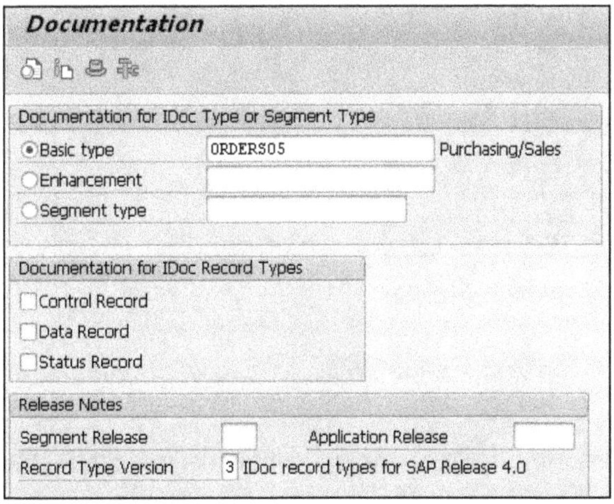

Figure 7.11 Transaction WE60: Documentation for IDoc Types

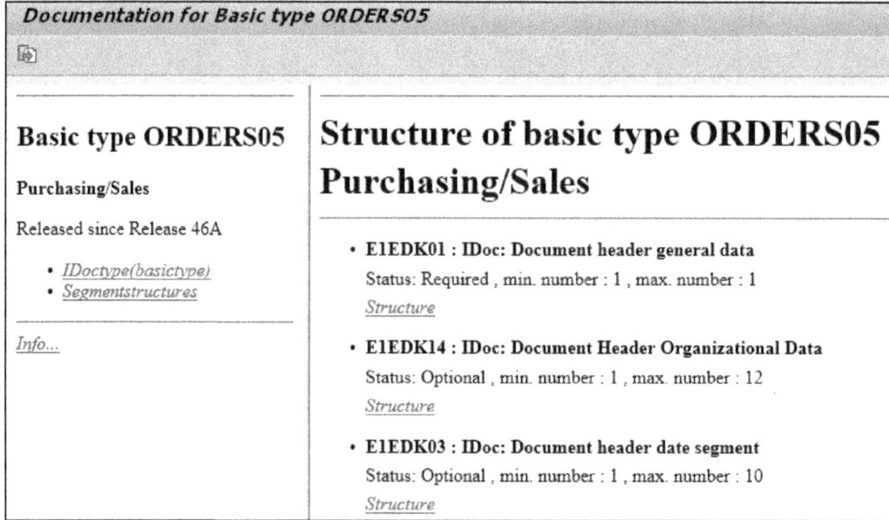

Figure 7.12 Transaction WE60: ORDERS05 Structure

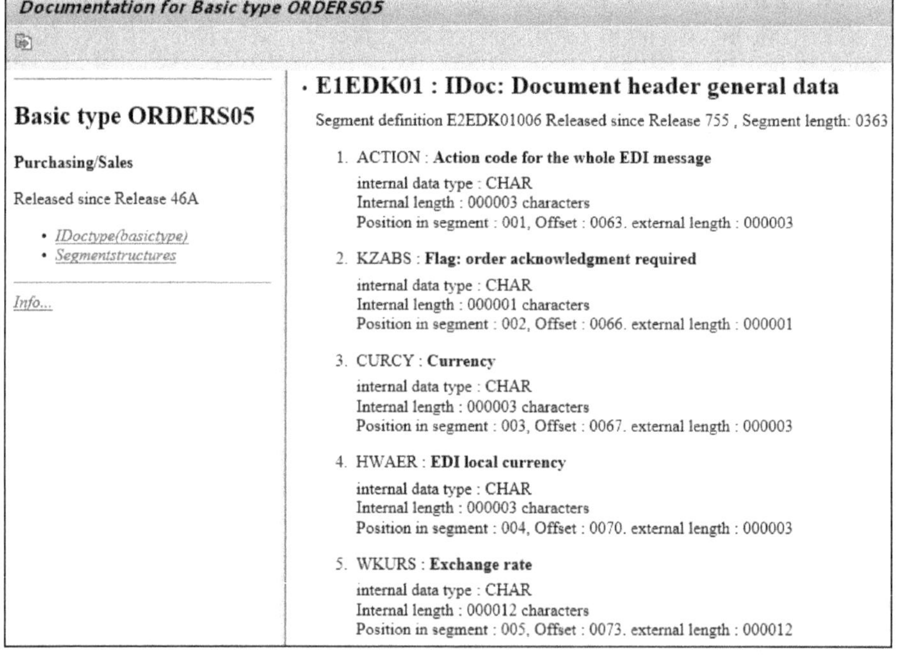

Figure 7.13 Transaction WE60: E1EDK01 Segment

There are two additional features within Transaction WE60 that may be required or useful:

- You can export the IDoc documentation in HTML format. When viewing the HTML documentation, navigate to the **Documentation-Export** menu function.

- You can export the IDoc definition in XSD format. From the main selection screen, navigate to **Documentation-XML Schema**, as shown in Figure 7.14, and complete these steps:
  - Confirm the default action, **Create documentation for Unicode file?**, by clicking **Yes**.
  - You will see the XML schema as shown in Figure 7.15. Navigate to menu function **XML-Download**, which allows downloading the XML schema to a local XSD/XML file.

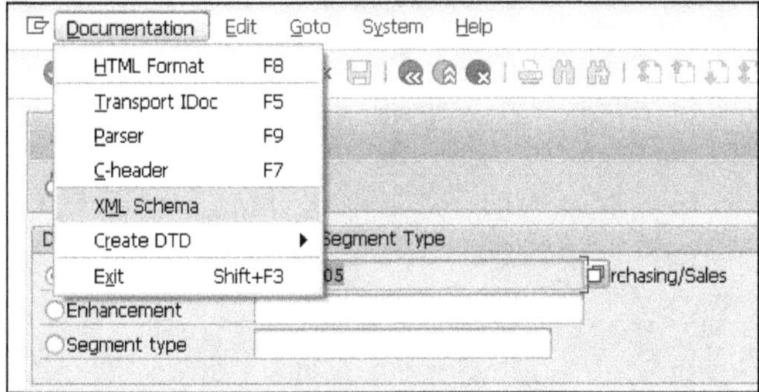

Figure 7.14 Transaction WE60: Creation of XML Schema

Documentation

```
  <?xml version="1.0" encoding="UTF-8"?>
- <xsd:schema version="1.0" xmlns:xsd="http://www.w3.org/2001/XMLSchema">
  - <xsd:element name="ORDERS05">
    - <xsd:annotation>
        <xsd:documentation>Purchasing/Sales</xsd:documentation>
      </xsd:annotation>
    - <xsd:complexType>
      - <xsd:sequence>
        - <xsd:element name="IDOC">
          - <xsd:complexType>
            - <xsd:sequence>
              - <xsd:element name="EDI_DC40">
                - <xsd:annotation>
                    <xsd:documentation>IDoc control record for the interface to the external
                      system</xsd:documentation>
                  </xsd:annotation>
                - <xsd:complexType>
                  - <xsd:sequence>
                    - <xsd:element name="TABNAM" fixed="EDI_DC40" type="xsd:string">
                      - <xsd:annotation>
                          <xsd:documentation>Name of Table Structure</xsd:documentation>
                        </xsd:annotation>
                      </xsd:element>
                    - <xsd:element name="MANDT" minOccurs="0">
                      - <xsd:annotation>
                          <xsd:documentation>Client</xsd:documentation>
                        </xsd:annotation>
                      - <xsd:simpleType>
                        - <xsd:restriction base="xsd:string">
                            <xsd:maxLength value="3"/>
                          </xsd:restriction>
                        </xsd:simpleType>
                      </xsd:element>
```

Figure 7.15 Transaction WE60: XSD Structure Generated for IDoc Basic Type ORDERS05

Viewing IDoc in Transaction WE02

In our demo system, we already have a sample IDoc 392002, an inbound IDoc for sales order creation. It has message type ORDERS and basic type ORDERS05, and it is already a valid IDoc for our EDI scenario. We'll go over how to configure it in Section 7.3.1, but let's take a quick look at it first. When we start Transaction WE02, we are welcomed by a selection screen as shown in Figure 7.16. We already have the IDoc number, and that's our input; we also adjusted the timeframe as we know that the IDoc was created only a few days ago.

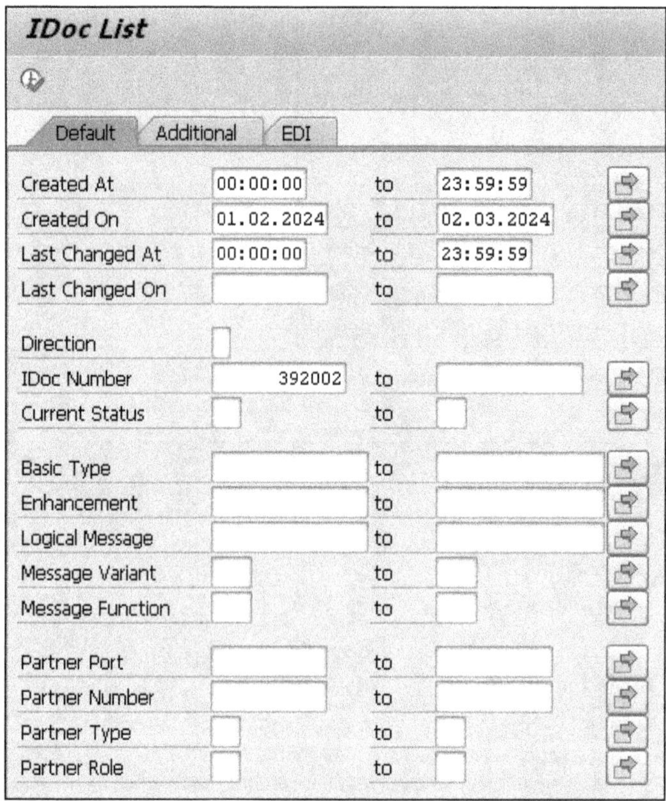

Figure 7.16 Selection Screen for Transaction WE02: IDoc List

As a result, we will get only a single IDoc, and it will be directly displayed as shown in Figure 7.17. On the left-hand side we can navigate to the different data segments. On the right-hand side, we see **Short Technical Information** (coming from the IDoc control record) and the **Content of Selected Segment** fields. In our example, we are looking at the E1EDKA1 segment with field **PARVW** and a partner function with value **LF**, indicating that the data is related to supplier. The second field is **PARTN**, partner number.

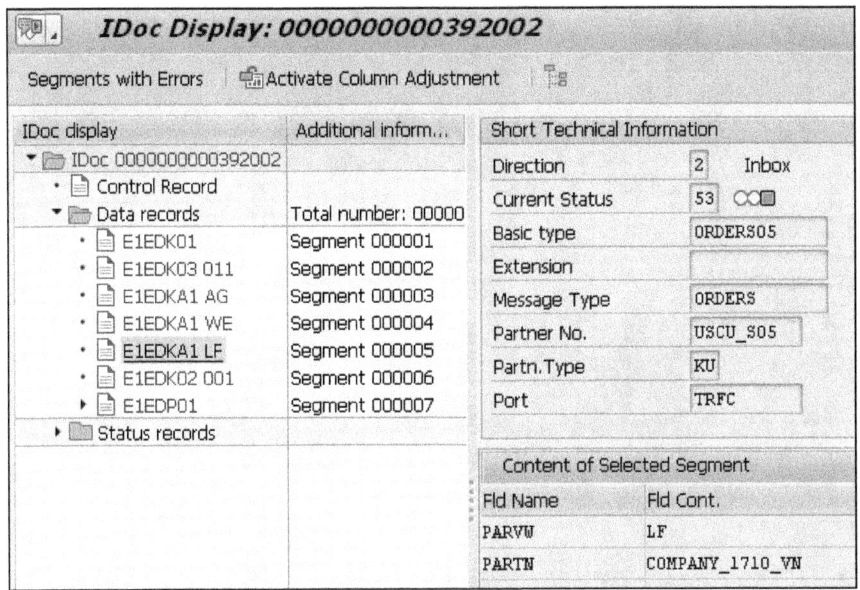

Figure 7.17 Displaying Sample IDOC via Transaction WE02: IDoc List

There is an option to see the complete data of the selected IDoc. Click **Data Display**, the third icon at the top of the screen. In the newly opened window, click **Expand Subtree** and you will see a view like that shown in Figure 7.18. It is a bit of a spoiler, but this is exactly the ORDER IDoc that we will create and process in our EDI scenario. It is a very basic message, but it contains enough information to create a sales order in the SAP S/4HANA system:

- **E1EDK02 011** contains the IDoc creation date.
- **E1EDKA1 AG** contains information on the sold-to party—in this case, external identifier BEST_BIKES_HQ.
- **E1EDKA1 WE** contains information on the ship-to party—in this case, external identifier BEST_BIKES_WHS.
- **E1EDKA1 LF** contains information on the supplier party—in this case, identifier COMPANY_1710_VN.
- **E1EDK02 001** contains the purchase order number.
- **E1EDP01** contains item-related data. Field MENGE is the quantity (10), and field WERKS is a plant—but eventually it will be removed from the IDoc (it is not necessary in our case).
- **E1EDP19 003** contains the EAN material number. MZ_FG_R300 is an existing and active material in our SAP S/4HANA system.

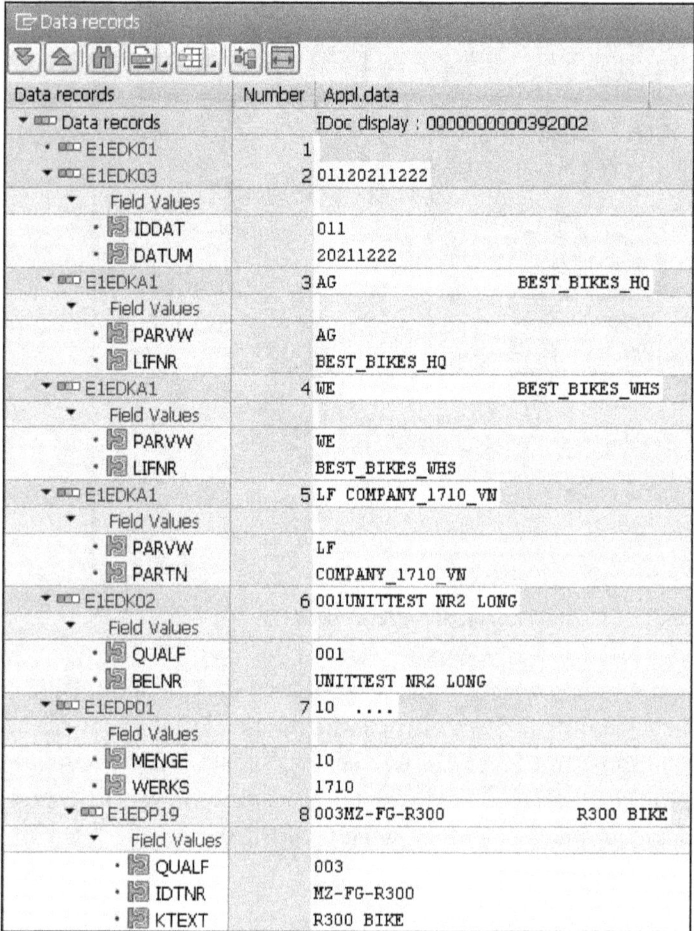

Data records	Number	Appl.data	
Data records		IDoc display : 0000000000392002	
E1EDK01	1		
E1EDK03	2	01120211222	
Field Values			
IDDAT		011	
DATUM		20211222	
E1EDKA1	3	AG	BEST_BIKES_HQ
Field Values			
PARVW		AG	
LIFNR		BEST_BIKES_HQ	
E1EDKA1	4	WE	BEST_BIKES_WHS
Field Values			
PARVW		WE	
LIFNR		BEST_BIKES_WHS	
E1EDKA1	5	LF COMPANY_1710_VN	
Field Values			
PARVW		LF	
PARTN		COMPANY_1710_VN	
E1EDK02	6	001UNITTEST NR2 LONG	
Field Values			
QUALF		001	
BELNR		UNITTEST NR2 LONG	
E1EDP01	7	10	
Field Values			
MENGE		10	
WERKS		1710	
E1EDP19	8	003MZ-FG-R300	R300 BIKE
Field Values			
QUALF		003	
IDTNR		MZ-FG-R300	
KTEXT		R300 BIKE	

Figure 7.18 Displaying Data Record via Transaction WE02: Complete ORDERS IDoc

Viewing IDocs in Database Tables

It was mentioned that an IDoc consists of three elements: the control record, the data records, and the status records. All these elements are stored in different database tables: tables EDIDC, EDID4, and EDIDS. It is not necessary on a daily basis, but if you are involved in EDI configuration, then definitely you will look at these tables from time to time. Our sample IDoc is shown in these tables in Figure 7.19, Figure 7.20, and Figure 7.21.

If you have worked with older SAP ERP systems, you may already know both of these options. They are still available within SAP S/4HANA, but none of these options is good when we think about end users. Using Transaction WE02 is the common approach, but apart from the complexity there are also multiple limitations that do not make life for the end user and IT personnel easy. So, through the years, we could see multiple custom solutions built for IDoc handling. For our EDI scenario, we will go a different way: the IDoc monitoring and error handling will be handled by SAP Application Interface

Framework. On top of that, we also have SAP Fiori applications that offer a new user experience in this area. We will go through this topic in the following sections.

Search in Table: EDID4 — IDoc Data Records from 4.0 onwards
Number of Hits: 8
Runtime: 0 — Maximum No. of Hits: 500
Insert Column

Details

IDoc no.	Counter	Number	Segment	Number	Hier.level	Data Aging	Length	Application data
392002		1	E1EDK01		1		1.000	
392002		2	E1EDK03		2		1.000	01120211222
392002		3	E1EDKA1		2		1.000	AG BEST_BIKES_HQ
392002		4	E1EDKA1		2		1.000	WE BEST_BIKES_WHS
392002		5	E1EDKA1		2		1.000	LF COMPANY_1710_VN
392002		6	E1EDK02		2		1.000	001UNITTEST NR2 LONG
392002		7	E1EDP01		2		1.000	10
392002		8	E1EDP19	7	3		1.000	003MZ-FG-R300 R300 BIKE

Figure 7.19 Table EDID4 and Sample ORDERS05 IDoc

EDIDC: Display of Entries Found

Search in Table: EDIDC — Control record (IDoc)
Number of Hits: 1
Runtime: 0 — Maximum No. of Hits: 500
Insert Column

Details

IDoc no.	Release	Status	IDoc Type	Directn	Port	Part. Type	Partn.no.	Address	Client	SADRintRec	SADR comm.type
392002	757	53		2	SAPS42	LS	S42CLNT400				

Figure 7.20 Table EDIDC and Sample ORDERS05 IDoc

Search in Table: EDIDS — Status Record (IDoc)
Number of Hits: 5
Runtime: 0 — Maximum No. of Hits: 500
Insert Column

Details

IDoc no.	Date stat.error	Time	Statcount.	Date created	Time	Status	User Name	Program	Function module	Code	Status text
392002	27.02.2024	22:31:07		27.02.2024	22:31:07	74	PIASECZNYM				
392002	27.02.2024	22:31:07	1	27.02.2024	22:31:07	50	PIASECZNYM				
392002	27.02.2024	22:31:07	2	27.02.2024	22:31:07	64	PIASECZNYM	SAPLEDIR	IDOC_STATUS_64OR65_SET		& &, &, &.
392002	27.02.2024	22:31:07	3	27.02.2024	22:31:12	62	PIASECZNYM	SAPLBD20			Direct call started
392002	27.02.2024	22:31:12	4	27.02.2024	22:31:12	53	PIASECZNYM				&1 &2 has been saved.

Figure 7.21 Table EDIDS and Sample ORDERS05 IDoc

7.3 IDoc Processing

We have thoroughly covered the theoretical foundations and acquired the essential information necessary to advance with the configuration of inbound processing for our IDocs. But what we are still missing in the system is a detailed definition of what kinds of IDocs we receive and from whom and what kinds of IDocs we generate and for whom. Last but not least, for each inbound and outbound IDoc that already has a receiver or sender, we have to define and link the processing function. All this and more will be explained in the next sections. We will start with inbound processing.

7.3.1 Inbound Processing

Figure 7.22 shows a simplified diagram with the key elements involved in the inbound IDoc processing. The starting point is that we have an instance of the IDoc in the system. It is irrelevant how it was transmitted to and created in the system, be it through RFC, a SOAP service, or even a flat file. The end results are always the same: we have one occurrence of an IDoc with the correct structure and syntax. In the next step, the system goes through the control record; if all the header information is correct and complete, we should have a link to a partner profile that determines how this specific IDoc should be processed.

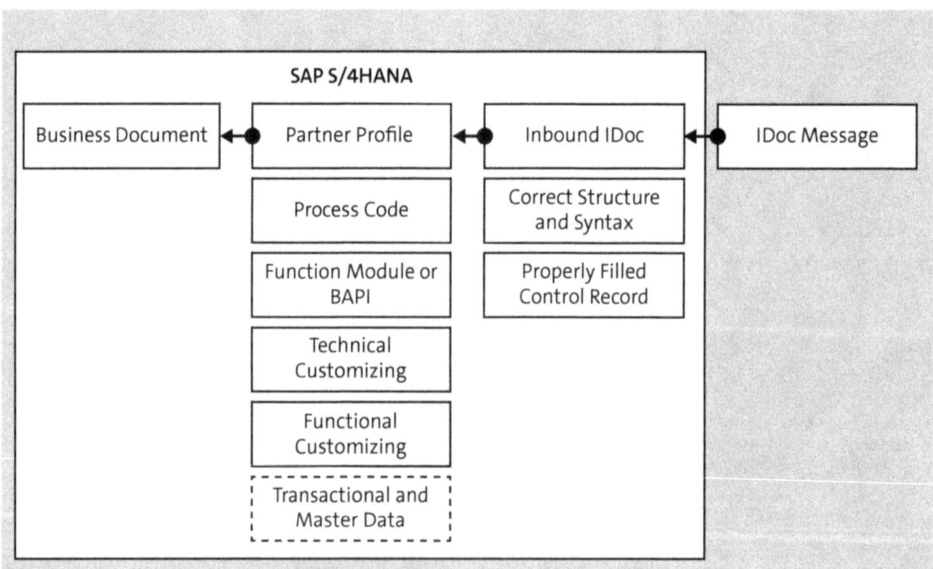

Figure 7.22 Key Elements within Inbound IDoc Processing

Let's explain in a few simple words what we will configure here in this section: there will be an inbound IDoc from customer USCU_S05, and it will be of message type ORDERS and basic type ORDERS05. That already tells us what we receive and from whom: Now, how do we process this message? In our case, we will be processing the IDoc immediately after it appears in the system, making use of process code ORDE. This is a customizing element defining how the IDoc should be processed. There are several standard process codes already predelivered by SAP, including ORDE. Behind this process code there is a standard function module, `IDOC_INPUT_ORDERS`. We'll see this in more detail later in Figure 7.25 and Figure 7.26.

Let's go through the different elements:

- **Correct structure and syntax of the IDoc**
  This will be taken care of within our integration platform. We make sure that we use proper basic types and message types.
- **Properly filled IDoc control record**
  This step is also addressed to the sender of the message, as before.
- **Process code, function module or BAPI**
  This is already part of the technical configuration within SAP S/4HANA. This and the overall technical configuration will be clarified in the following sections of this chapter.
- **Functional customizing, transactional, and master data**
  This is out of the scope of this book, but some of the key elements will be briefly mentioned. The base for our configuration is that we already have a (demo) system with at least basic organization and a set of data that allows execution of a purchasing business process.

In case of an inbound process, the starting point for SAP S/4HANA is that we have an IDoc with the correct structure syntax and proper data in the control record and data records.

A common point where we get stuck is when we already know the business transaction and related IDoc type but are not sure about the process code to be used for processing the inbound IDoc or to generate the outbound IDoc. One solution for that would be to make use of Transaction WE64 (Process Codes in IDoc Outbound and IDoc Inbound), as shown in Figure 7.23. In this transaction, you can search and display all process codes. You can navigate using the structure on left-hand side, where the process codes are grouped per direction (outbound and Inbound messages) and then, on the lower level, by IDoc message type. The mentioned figure shows process code SD10, which is related to outbound message ORDRSP; this is the one we will use in our case.

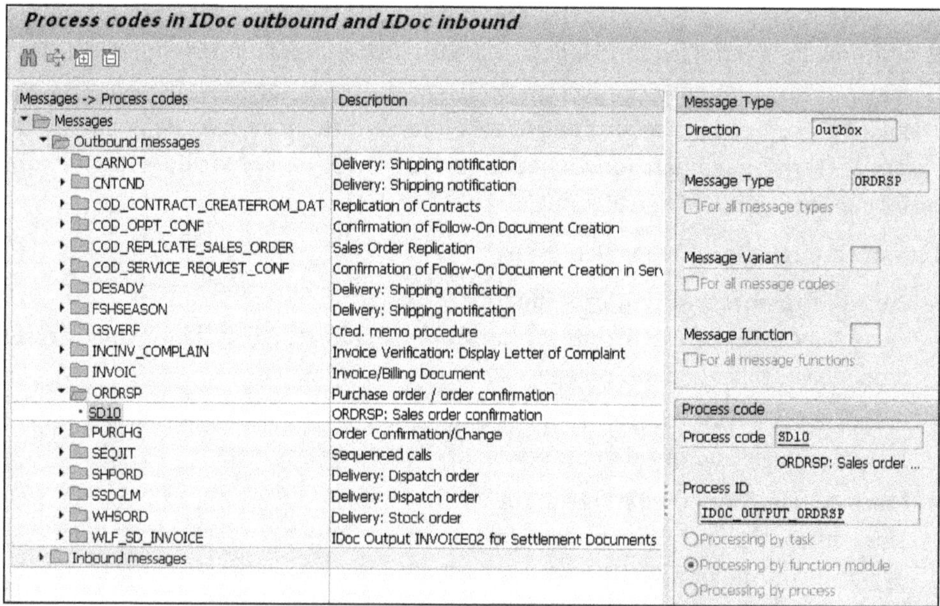

Figure 7.23 Transaction WE64: Process Code for Outbound Message ORDRSP

Figure 7.24 shows process code ORDE related to inbound message ORDERS.

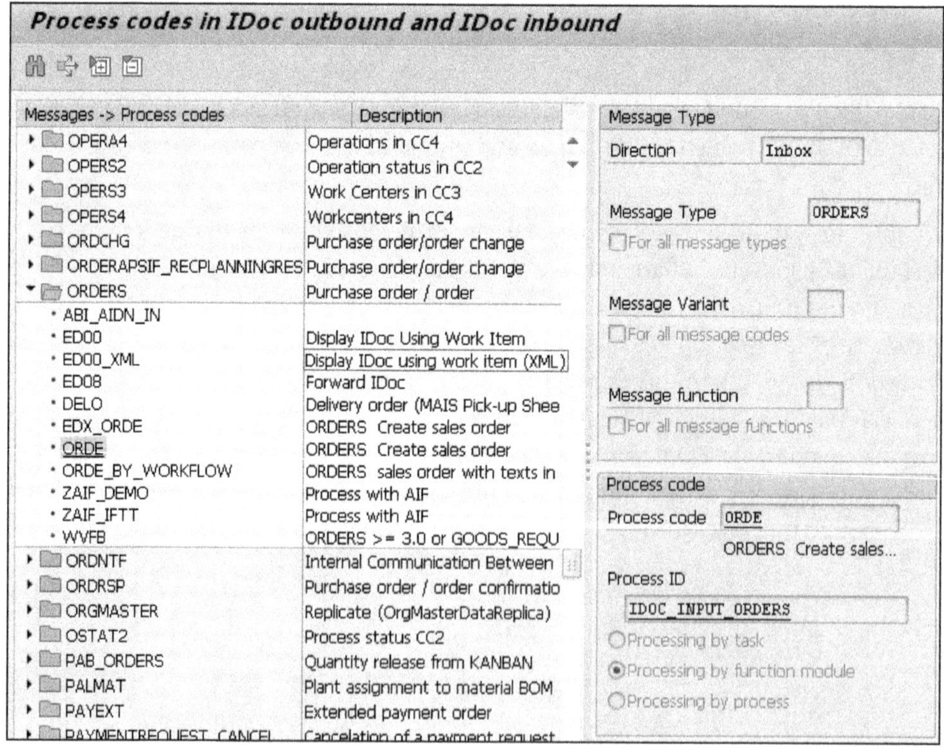

Figure 7.24 Transaction WE64: Process Codes in Outbound and Inbound IDocs

The mentioned transaction allows only for display of the process codes from the perspective of direction and IDoc message type. There are two additional transactions related to process code maintenance: Transaction WE42 for inbound process codes and Transaction WE41 for outbound process codes. Figure 7.25 shows the display view of inbound process codes from Transaction WE42. It is possible to navigate to the details of each inbound process code as shown in Figure 7.26. In both of these transactions, you can switch to edit mode and maintain the necessary entries, including creation of your own, custom process codes.

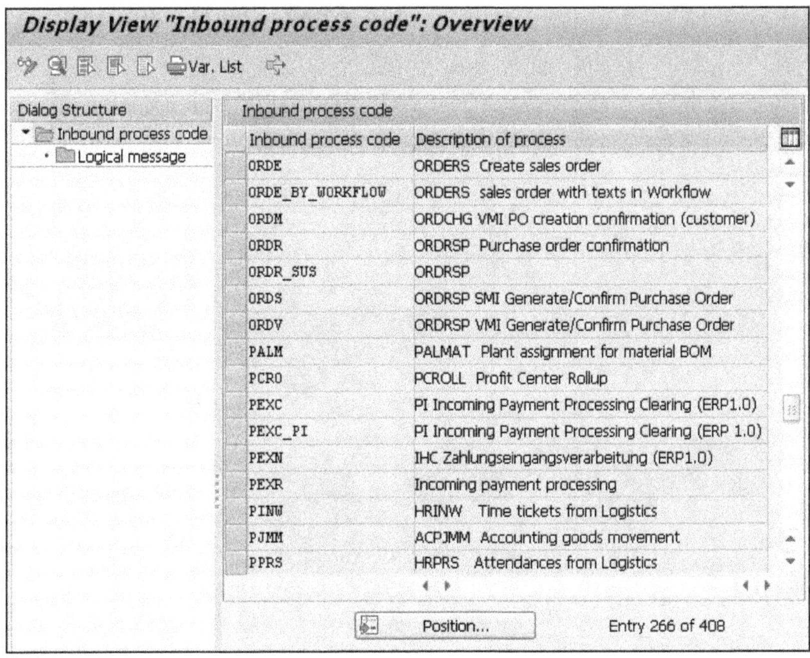

Figure 7.25 Inbound Process Codes in Transaction WE42

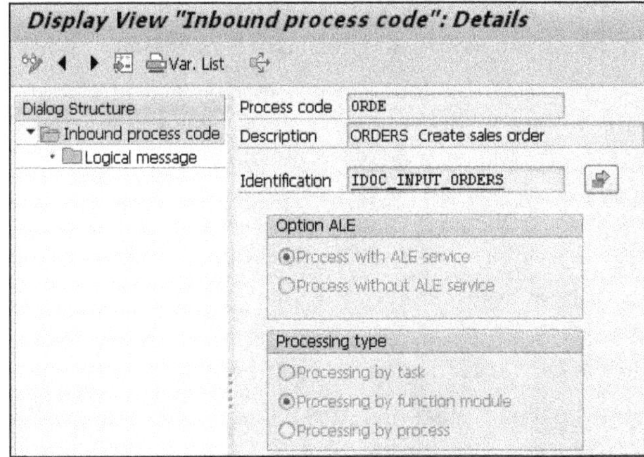

Figure 7.26 Details of ORDE Process Code for Inbound Message ORDERS

After getting familiar with these additional details, we can start with the configuration of the EDI partner profile. After starting Transaction WE20, on the left-hand side we have a list with the different partner types. The EDI scenarios are configured directly with the specific trading partner. The partner type in our case will be customer KU. On the right-hand side, we will view or manage details of the selected partner profile entry. Figure 7.27 shows what the transaction screen looks like.

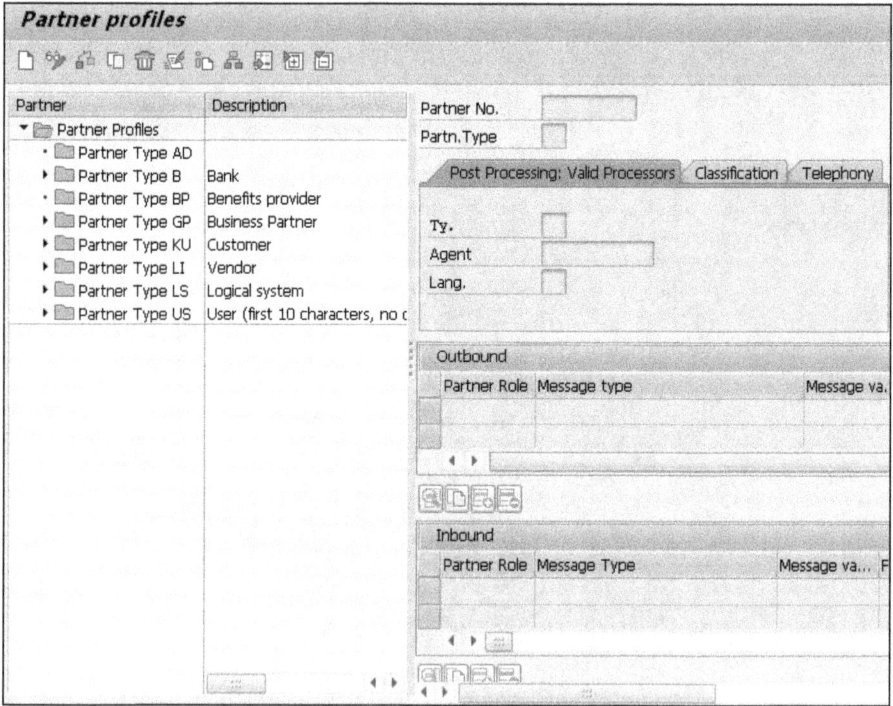

Figure 7.27 Transaction WE20: Starting Point for Creation of Partner Profile

Follow these steps to create the partner profile for our scenario:

1. Click the **Create** icon, the first icon at the top of the screen (see Figure 7.28).

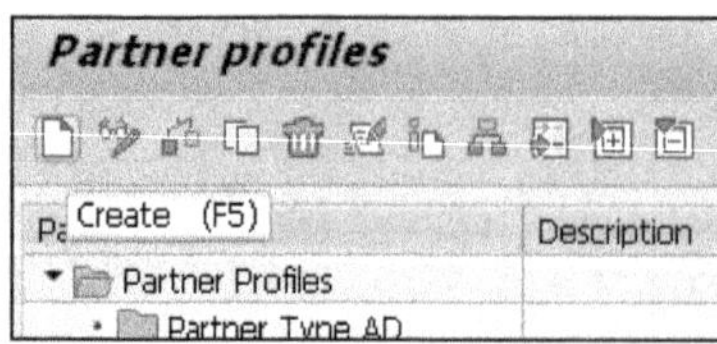

Figure 7.28 Transaction WE20: Create New Partner Profile

2. Maintain the two key fields in the header of the partner profile entry:
   - **Partner No.** is for the (business partner) customer number in your SAP S/4HANA system. In our case, this is **USCU_S05**.

- **Partn.Type** is the partner type. In this case, this is **KU**.

  It is also mandatory to maintain values in the **Post Processing: Valid Processors** tab. This is related to workflow notifications, and even if we don't use those, we still have to add some values to be able to continue. We have maintained our own user there, but it is irrelevant as we will be using a different approach for IDoc error handling.

What we just created is a main partner profile entry. It doesn't yet define the exact inbound nor outbound IDocs that we want to process with our trading partner. Neither does it define how we want to process these IDocs. All these details need to be configured in the **Outbound** and **Inbound** sections, which you can see in Figure 7.29.

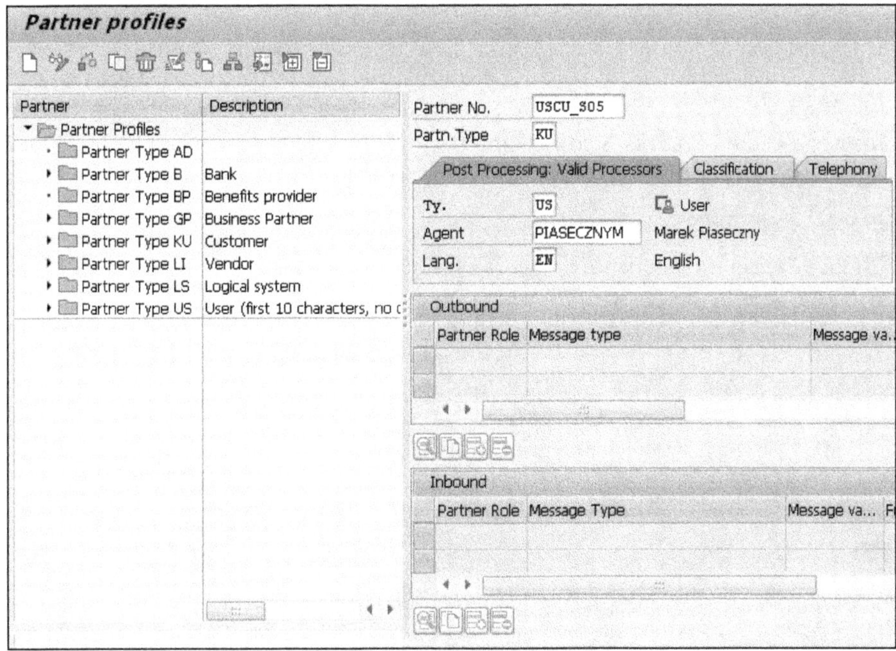

Figure 7.29 Creating Partner Profile

Follow these steps to create inbound parameters in the **KU USCU_S05** partner profile:

1. Make sure you choose the appropriate KU profile and that you are in edit mode. Switching between display and edit mode is controlled via the second icon located at the top of the list (see Figure 7.30).

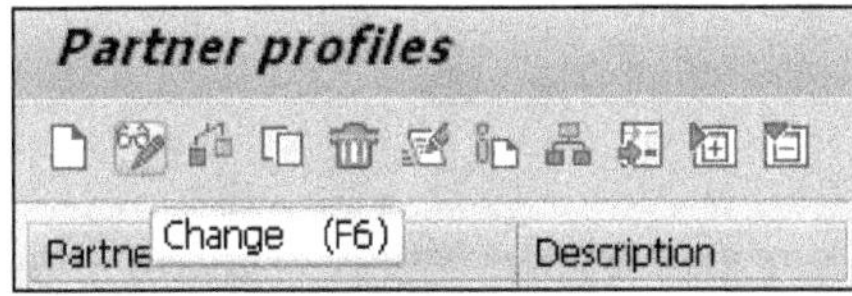

Figure 7.30 Switching to Change Mode in Partner Profile Transaction (WE20)

2. Go to the **Inbound** section at the bottom of the right-hand side section. Click **Create Inbound Parameter** (see Figure 7.31).

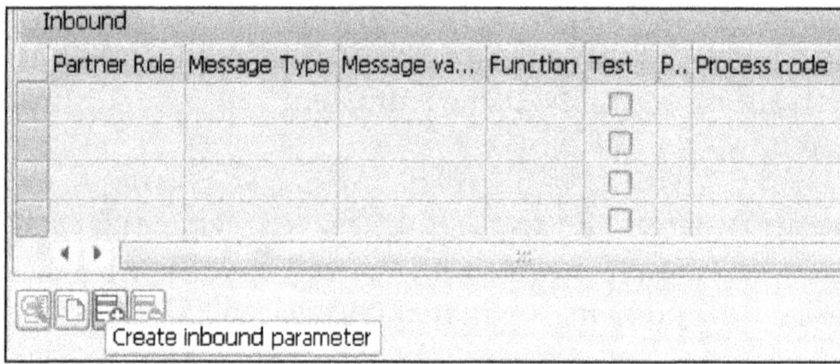

Figure 7.31 Icon for Creating Inbound Parameter

3. Maintain the following parameters (these are also shown in Figure 7.32):
   - **Message Type**: ORDERS.
   - **Message code**: B2B. this is something that we introduce to distinguishing IDocs related to B2B EDI scenarios from all the other IDocs.
   - **Process code**: ORDE.
   - Choose **Cancel Processing after Syntax Error**.
   - Choose **Trigger Immediately**.
   - We can skip the other two tabs: **Post Processing: Valid Processors** and **Telephony**.

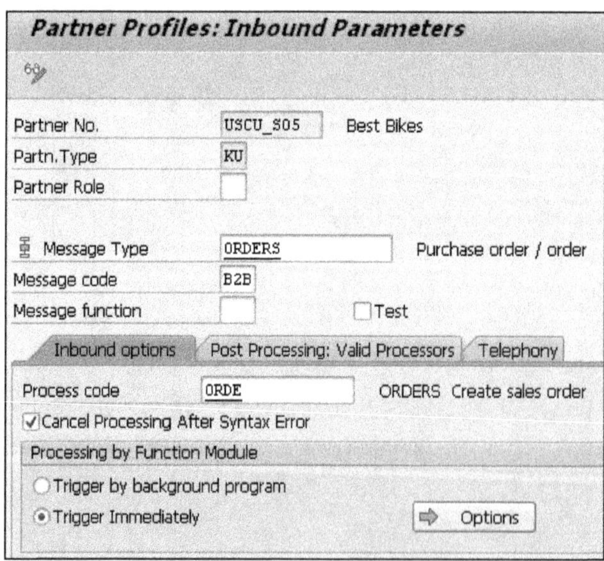

Figure 7.32 Definition of Inbound Parameters

The partner profile entry with inbound parameters allows us to process the inbound ORDER.ORDERS05 IDoc received from partner KU USCU_S05 with process code ORDE.

We already know that the whole setup for the ORDER message type and ORDERS05 IDoc type is in place and that behind ORDE we have a IDOC_INPUT_ORDERS function module, which is the standard function for processing sales order messages.

7.3.2 Outbound Processing

In a nutshell, this is what we will configure in this section: There will be outbound (IDoc) message types ORDRSP, DESADV, and INVOIC that we want to create and send to our customer USCU_S05. For each of these message types, we define process codes to be used for IDoc creation. What's also very important is that for each message type we define a port, a remote destination where the IDoc will be sent. For this, Cloud Integration will take care of the proper transformation from IDoc to EDIFACT format and transmission to the trading partner system.

Follow these steps to create entries for outbound parameters in our **KU USCU_S05** partner profile, as shown in Figure 7.33:

1. Again, make sure you choose the appropriate KU profile and that you are in edit mode.
2. Go to the **Outbound** section at the bottom of the right-hand side. Click **Create Outbound Parameter.**

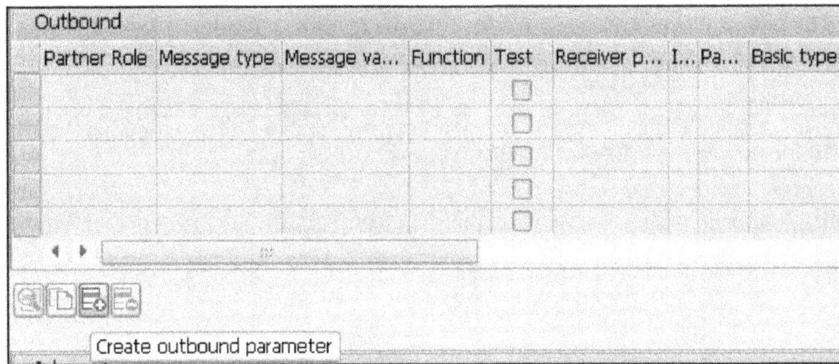

Figure 7.33 Transaction WE20: Create Outbound Parameter

3. In total, we need three entries, one for each message type. We start with the configuration related to the sales order response message. Maintain the parameters in the **Outbound Options** tab as shown in Figure 7.34:
   - **Message Type**: ORDRSP.
   - **Message code**: B2B. This is something that we introduce to distinguish IDocs related to B2B EDI scenarios from all the other IDocs.
   - **Basic Type:** ORDERS05.
   - Choose **Cancel Processing after Syntax Error.**

– **We don't have to choose any specific release** in **Seg. Release in IDoc Type** or **Segment App.Rel.**, but by default recent releases will be used.
– We can skip three other tabs, **Post Processing: Valid Processors**, **Telephony**, and **EDI Standard**, but we must fill **Message Control**.

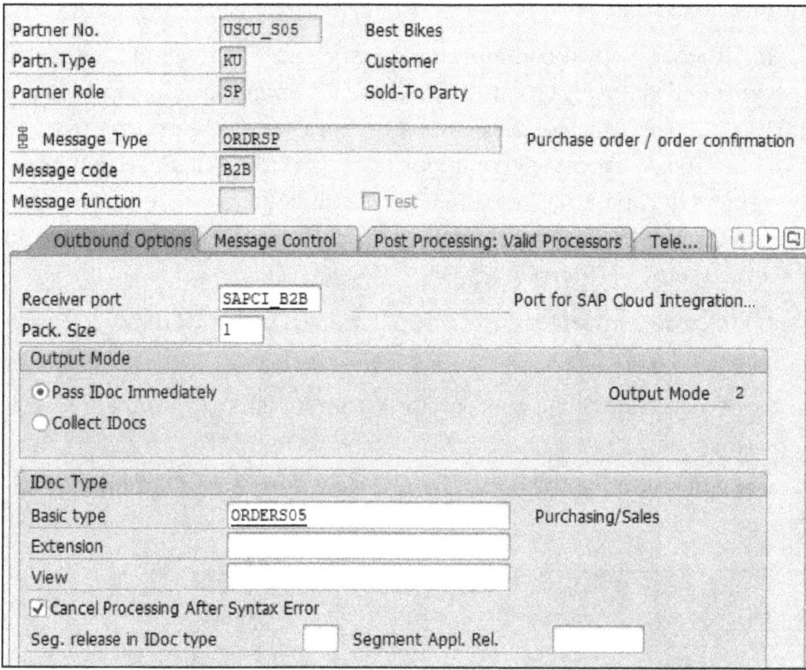

Figure 7.34 Outbound Parameters: Outbound Options for ORDRSP

4. Maintain two entries on the **Message Control** tab, as shown in Figure 7.35:
   – First entry:
     - **Application: V1**
     - **Message Type: BA01**
     - **Process Code: SD10**
     - **Change Message:** not set
   – Second entry:
     - **Application: V1**
     - **Message Type: BA01**
     - **Process Code: SD10**
     - **Change Message:** set

   Message control is the core setting in the outbound partner profile. As shown in the initial diagram, this is the link between the EDI output added to the business document and the IDoc processing itself. There are three key elements:

- **Application V1** defines that we operate within the sales application area. This is the SAP approach to cluster and distinguish configuration artifacts from the different functional areas. As an example, V2 is shipping and V3 is billing area.
- **Message Type** is the output type, which we trigger from the respective business documents from the sales area. In this case, we have **BA01**, which is the predelivered output for EDI order response.
- **Process Code** plays a similar role to that we know from the inbound processing. Behind the process code we have a function module or BAPI that can generate specific IDoc types. If you look at Transaction WE41, which allows maintenance of outbound process codes, you can find **SD10**. While checking details of this process code, you can see that it is configured to use the `IDOC_OUTPUT_ORDRSP` function module and to operate on message type ORDRSP.

For a sales order response, we make use of the **Change** checkbox available next to each message control entry. This setting allows us to trigger the outbound IDoc for a newly created sales order but then also to retrigger the same output for each change in the existing sales order document.

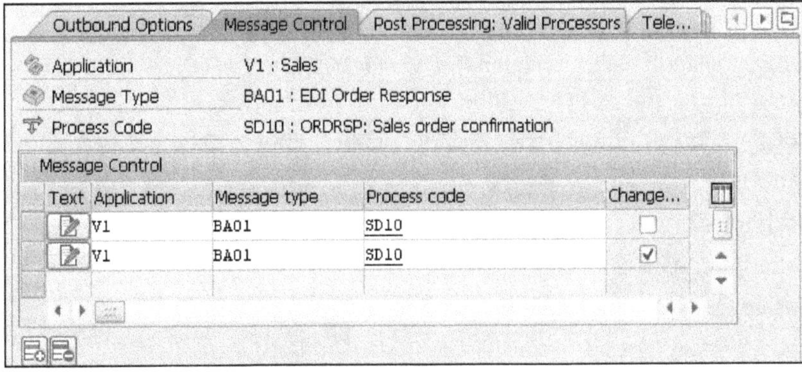

Figure 7.35 Message Control Tab for ORDRSP Message

5. Next in the sequence is a partner profile configuration for shipment advice. Maintain the parameters on the **Outbound Options** tab as shown in Figure 7.36:
   - **Message Type: DESADV.**
   - **Message code: B2B.**
   - **Basic Type: DELVRY07.**
   - Choose **Cancel Processing after Syntax Error.**
   - **We don't have to choose any specific release** in **Seg. Release in IDoc Type** or **Segment App.Rel.**, but by default, the recent releases will be used.
   - We can skip three other tabs, **Post Processing: Valid Processors**, **Telephony**, and **EDI Standard**, but we must fill **Message Control**.

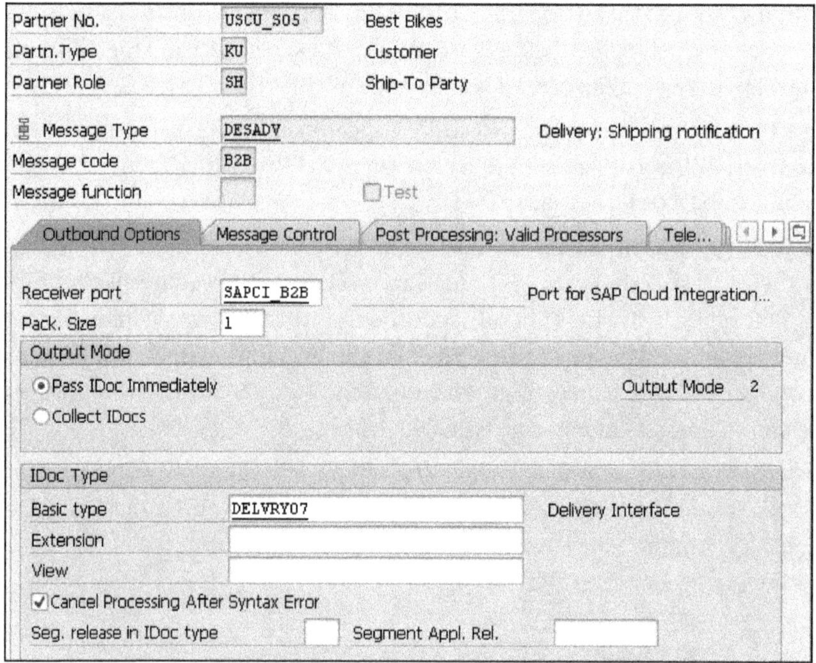

Figure 7.36 Outbound Parameters: Outbound Options for DESADV

6. Maintain the parameters on the **Message Control** tab as shown in Figure 7.37:
   - First entry:
   - **Application: V2**
   - **Message Type: ASNO**
   - **Process Code: DELV** (this one is linked with IDOC_OUTPUT_DELVRY)
   - **Change Message:** not set

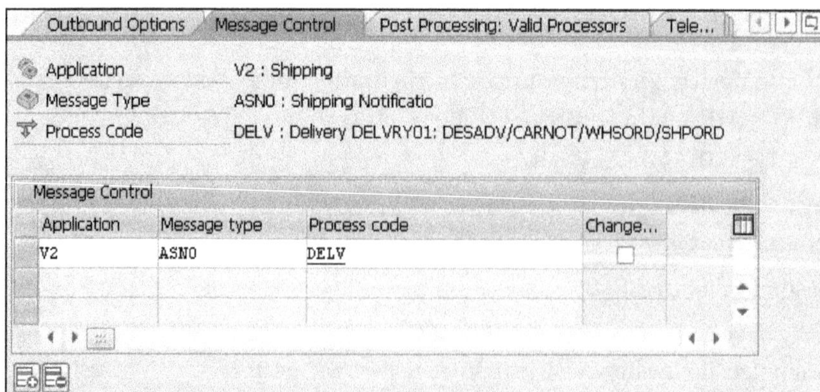

Figure 7.37 Message Control Tab for DESADV Message

7. Last in the sequence is a partner profile configuration for invoice. Maintain the parameters on the **Outbound Options** tab as shown in Figure 7.38:
   - **Message Type: INVOIC**
   - **Message Code: B2B**
   - **Basic Type: INVOIC02**
   - Choose **Cancel Processing after Syntax Error**.
   - We don't have to choose any specific release in **Seg. Release in IDoc Type** or **Segment App.Rel.**, but by default, the recent releases will be used.
   - We can skip three other tabs, **Post Processing: Valid Processors**, **Telephony**, and **EDI Standard**, but we must fill **Message Control**.

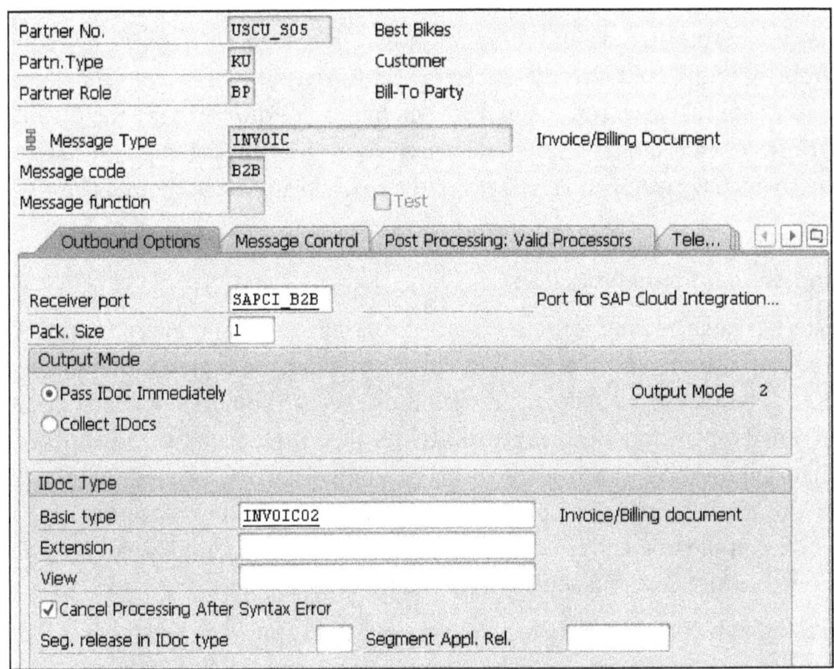

Figure 7.38 Outbound Parameters: Outbound Options for INVOIC

8. Maintain the parameters on the **Message Control** tab as shown in Figure 7.39:
   - **Application: V3**
   - **Message Type: RD00**
   - **Process Code: SD09** (this one is linked with `IDOC_OUTPUT_INVOIC`)
   - **Change Message:** not set

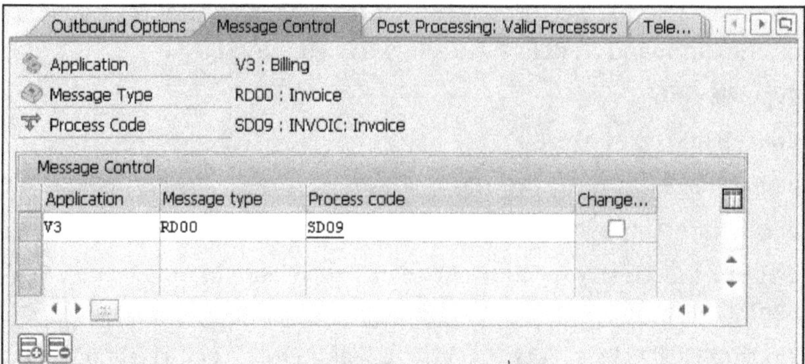

Figure 7.39 Message Control Tab for INVOIC Message

Now we're done with partner profile configuration. At this stage, we have the key configuration in place and would already succeed with technical processing of the given IDocs. But there are still two elements that we will handle before verifying the overall SAP S/4HANA setup. In the next section, we briefly look at Business Rules Framework plus (BRFplus), which we will involve in automatic triggering of the outbound IDocs.

7.3.3 Business Rules Framework Plus

While BRFplus has been around for many years, its usage for output determination is something new, unique to SAP S/4HANA. BRFplus is the default engine for managing outputs in SAP S/4HANA in the sales and distribution area. Although there is an option to switch off the new functionality and go back to the old output control, we do not recommend that. The new approach gives more flexibility and it is something worth investing in. Unfortunately, at this point in time the scope of the new output management is not yet comparable with the scope of the old output control (based on table NAST); it simply does not cover all applications and functional areas.

We start directly with checking if the new output control is properly activated in our system. All the necessary functions can be found in Transaction SPRO via menu path **Cross Application Components • Output Control • Define Output Control • Define Output Types**, as shown in Figure 7.40.

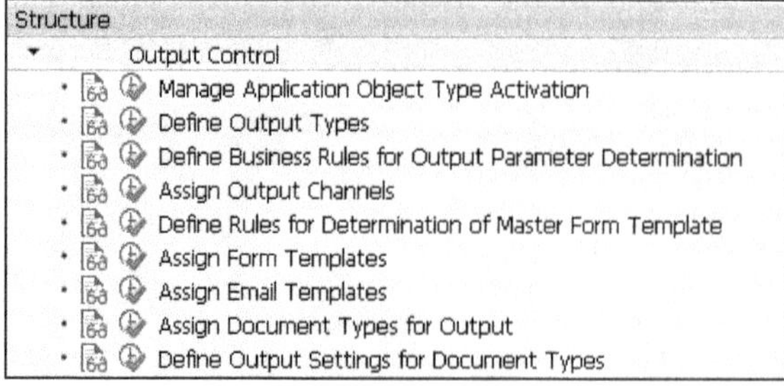

Figure 7.40 Customizing Transactions Related to Output Control

The first task is to check if all our application object types are in the active status. These are the object types (see Figure 7.41)

- SALES_DOCUMENT
- OUTBOUND_DELIVERY
- BILLING_DOCUMENT

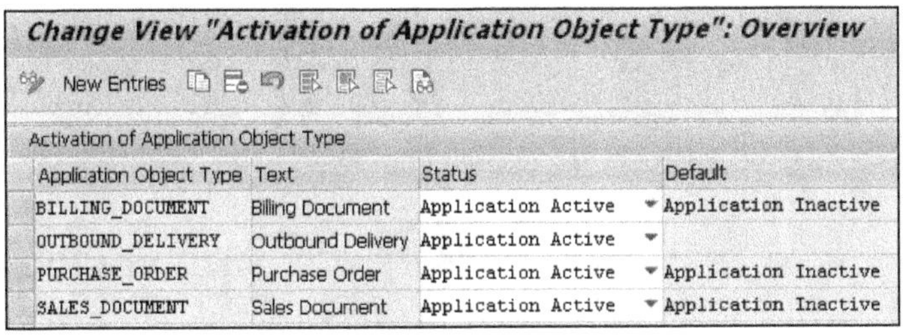

Change View "Activation of Application Object Type": Overview

New Entries

Activation of Application Object Type

Application Object Type	Text	Status	Default
BILLING_DOCUMENT	Billing Document	Application Active	Application Inactive
OUTBOUND_DELIVERY	Outbound Delivery	Application Active	
PURCHASE_ORDER	Purchase Order	Application Active	Application Inactive
SALES_DOCUMENT	Sales Document	Application Active	Application Inactive

Figure 7.41 Activation of Application Object Type

In the next step, for each application object type, we need to assign an appropriate output type. The concept behind this is that for a single object type like, for example, the sales document, we may have multiple different outputs. This will actually happen in our scenario, where the sales order document will trigger two different output types. First we will have the order confirmation output triggered during the sales order creation, then changes in the existing sales order will trigger the order confirmation change output. We need to make sure the two output types are linked to the sales document. For the other two documents, outbound delivery and billing document, we will only use one order type. The complete configuration we need is as follows:

- SALES_DOCUMENT output type ORDER_CONFIRMATION
- SALES_DOCUMENT output type ORDER_CONFIRMATION_CHANGE
- OUTBOUND_DELIVERY output type DELIVERY_NOTE
- BILLING_DOCUMENT output type BILLING_DOCUMENT

Each of these output types needs to have the IDoc channel assigned in the **Assign Output Channels** step (see Figure 7.42).

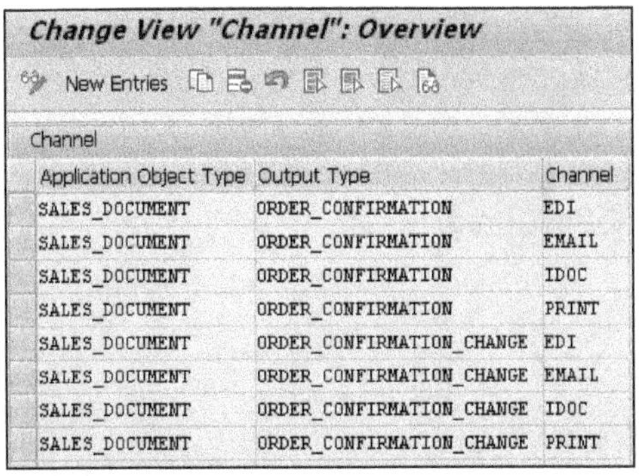

Change View "Channel": Overview

New Entries

Channel

Application Object Type	Output Type	Channel
SALES_DOCUMENT	ORDER_CONFIRMATION	EDI
SALES_DOCUMENT	ORDER_CONFIRMATION	EMAIL
SALES_DOCUMENT	ORDER_CONFIRMATION	IDOC
SALES_DOCUMENT	ORDER_CONFIRMATION	PRINT
SALES_DOCUMENT	ORDER_CONFIRMATION_CHANGE	EDI
SALES_DOCUMENT	ORDER_CONFIRMATION_CHANGE	EMAIL
SALES_DOCUMENT	ORDER_CONFIRMATION_CHANGE	IDOC
SALES_DOCUMENT	ORDER_CONFIRMATION_CHANGE	PRINT

Figure 7.42 Assignment of Channels to Output Types

Output for Delivery Not Available

At this moment, we have the new output management activated for all object types that are part of our scenario. This is done for the IDoc channel. Before we continue with the next step, which is about defining business rules for triggering the specific outputs on a specific condition, we need to stop for a moment. At the time of writing this book, there is still a functionality gap in on-premise SAP S/4HANA, the new output management framework is not yet supported for delivery documents. This means that for outbound delivery documents we still need to work with the "old" output control framework (table NAST). The positive about this situation is that we may get familiar with both options.

There are two SAP Notes you can read to learn more about the current gap related to output management for deliveries, which you can find at the following URLs:

- *http://s-prs.co/v573603*
- *http://s-prs.co/v573604*

We recommend that you check the status at the time of reading this book in case the gap closes in the future.

7.3.4 Adding Output Determination Rules via Business Rules Framework Plus

In this section, you will configure a trigger for a sales order response message for each new sales order with sold-to USCU_S05. The sales order response message also will be triggered on changes within such sales order documents.

From a BRFplus perspective, our scope is as follows:

- Application object type: SALES DOCUMENT
- Output type: ORDER_CONFIRMATION
- Output type: ORDER_CONFIRMATION_CHANGE
- Channel: IDoc
- Customer number, sold-to: USCU_S05

The configuration itself is done via Transaction ODP (Output Parameter Determination). After starting this transaction, you are redirected to the web browser and SAP Fiori–based application, as shown in Figure 7.43.

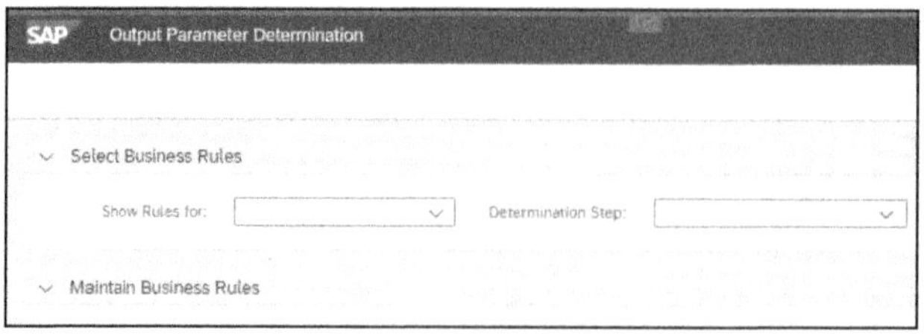

Figure 7.43 Output Parameter Determination

The steps to configure a trigger for the sales order response are as follows:

1. Within the section at the top of the screen called **Select Business Rules**, choose the appropriate application object type and determination step. In the bottom part of the screen, in the **Maintain Business Rules** section, configure details related to the area selected.
2. Choose **Sales Document** from the combo list under **Show Rules For**.
3. **Choose Determination Step: Output Type**:
   - Here you define the output types you want to trigger. Add two entries as you want two output types triggered—the first when a new sales order is created and the second when an existing sales order is changed. For **Output Type**, select **ORDER_CONFIRMATION** and **ORDER_CONFIRMATION_CHANGE.**
   - There are multiple different sales documents and types, but in this case you need the output just for the standard order. That means selecting **C** for **Document Category** and **TA** for **Sales Doc. Type**.
   - You want the output triggered immediately, so the value to choose for **Dispatch Time** is **1 (Immediately)**.
   - Figure 7.44 shows the complete configuration for this case.

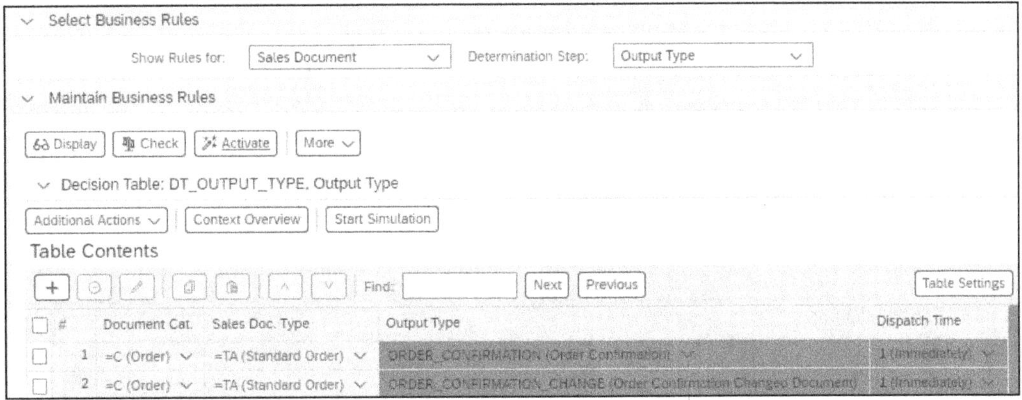

Figure 7.44 Definition of Output Type for Sales Document

4. **Choose Determination Step: Receiver**:
   - In this step, you define for which partner role the output should be generated. In this case it is the sold-to party. Maintain AG for Role for both **ORDER_CONFIRMATION** and **ORDER_CONFIRMATION_CHANGE**. (If you're unsure where this sold-to party is coming from, review how we configured the EDI partner profile in Transaction WE20 [Section 7.3.2]).
   - Set Exclusive Indicator to false in order to not trigger other channels. Now, whenever you trigger an IDoc output, you don't want to trigger printing or email.
   - Figure 7.45 shows the complete configuration for this case.

Figure 7.45 Definition of Receiver for Sales Document

5. **Choose Determination Step: Channel:**
   - In this step, you define via which channel you want to send the message, but as you do not want to send the response IDoc from all sales order documents or for all customers, you need to include an additional condition at this step. There are multiple options within BRFplus to achieve similar results, but the one to apply in this case makes use of the fact that all the EDI customers are in group Z1, and so the message on the IDoc channel will be only triggered for customers in that group. At this moment there is only one customer, but whenever you add the next one you don't have to maintain BRFplus output conditions for it; it will be enough to assign any new EDI customers to customer group Z1.
   - Figure 7.46 shows the complete configuration for this case.

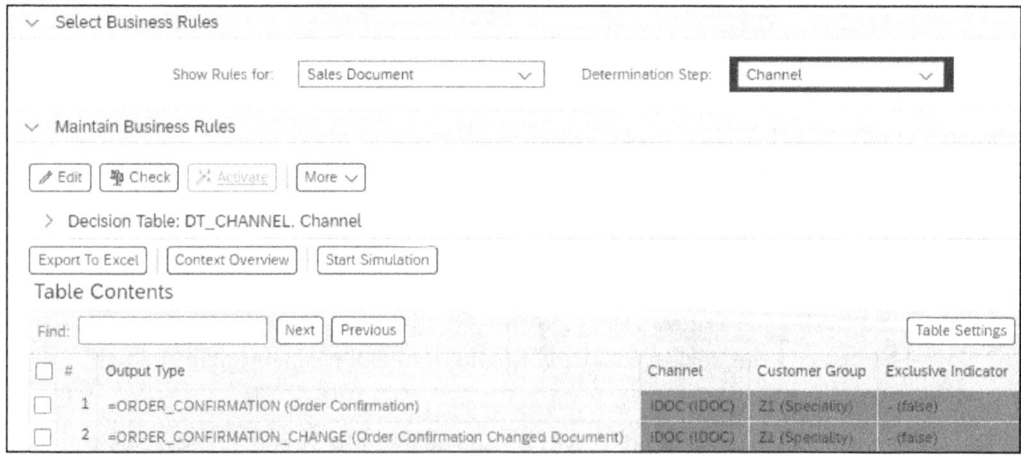

Figure 7.46 Setting Up Conditions for IDoc Channel: Z1 Customer Group

We are done with the BRFplus configuration for triggering a sales order response via an IDoc channel. You must be aware that the tool has broad functionality, including options for additional enhancements. It can handle very simple and very complex scenarios. In our example, we had the comfort of not having any additional dependencies coming from already existing rules. To configure invoice-related EDI output, we follow similar steps to those for a sales document. Choose the billing document and then maintain the three key determination steps:

1. **Output type**
   Configure output type **BILLING_DOCUMENT** for billing type **=L2**, which is the standard document type applicable for this case. As you don't want any delay, set **Dispatch Time** to **1 (Immediately)**. Figure 7.47 shows the appropriate entry.

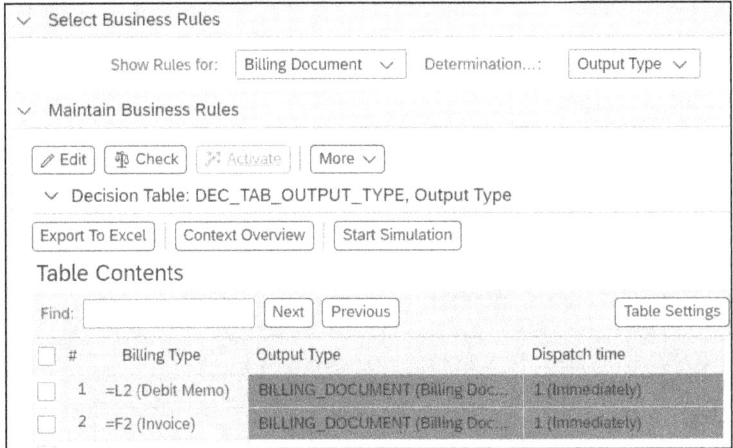

Figure 7.47 Setting Up Output Type for Billing Document

2. **Receiver**
   The main difference compared to the sales document setup is that for the billing document, you make use of partner role RE (the bill-to party configured in the partner profile).

3. **Channel**
   For an IDoc channel, once again make use of customer group Z1 to trigger the IDoc only for selected customers. Figure 7.48 shows the appropriate entry.

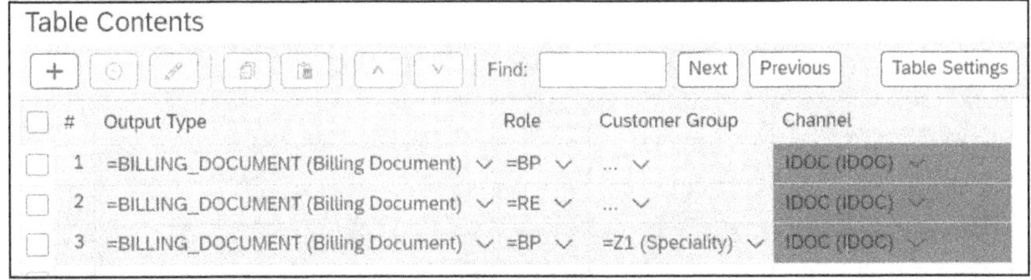

Figure 7.48 Setting Up IDoc Channel for Billing Document

As mentioned earlier, at the time of writing this book, the BRFplus-based output control for delivery documents is not available yet. We need to configure triggering the ASNO delivery output in the old way using output control. It is not ideal to have to apply different solutions, but there is an alternative solution. We need to remember that although BRFplus is the approach recommended by SAP, it is not mandatory; there is still an option to switch back and make use of the classic output control. It happens for a variety of reasons, but it is still the case that many organizations decide to continue with the old output control for new SAP S/4HANA implementations.

Let's summarize our requirements and what we want to configure in this section:

- We want to trigger ASNO output for any outbound delivery document with sales organization 1710 and customer USCU_S05.
- The output should be triggered immediately via an EDI medium and on partner USCU_S05 for the ship-to function.

The steps to configure as are follows:

1. Start Transaction VV21 to create condition records.
2. Enter "ASNO" in the **Output Type** field, as shown in Figure 7.49, and confirm the selection.

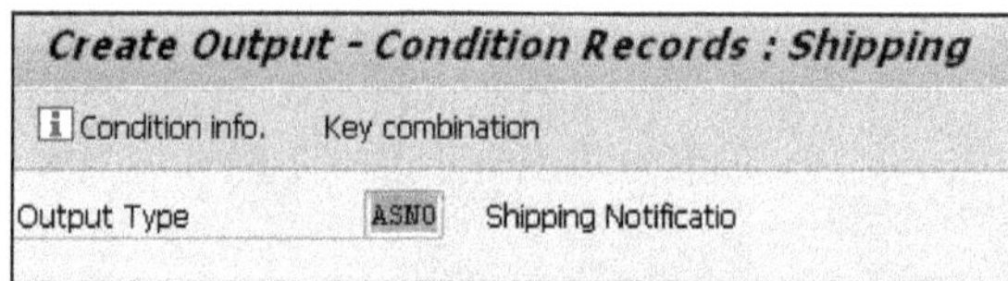

Figure 7.49 Selection Screen for Transaction VV21: Create Output Transaction

3. If you click **Key Combination**, you can see the possible conditions for triggering this output. In this standard output, there is only one combination: **Sales Organization/ Customer Number**. It is possible to change or add combinations to the output, but it requires additional customizing. Luckily in our case, the default option is exactly what we need (see Figure 7.50).

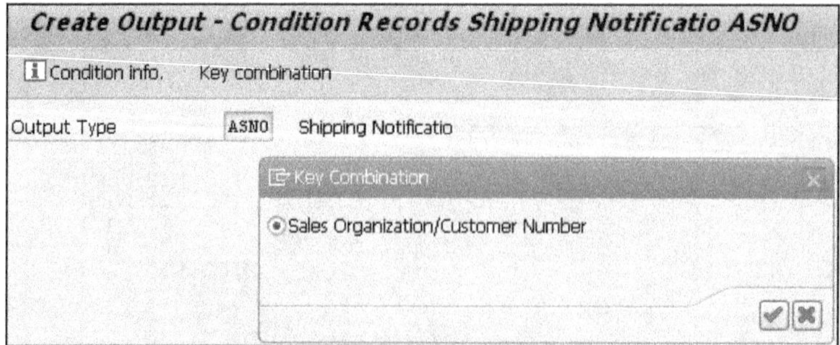

Figure 7.50 Key Combinations for ASNO Shipping Notification Output

4. Create the condition record by adding the following values (see Figure 7.51):
   - **Sales Organization: 1710**
   - **Customer: USCU_S05**
   - **Function: SH**
   - **Partner: USCU_S05**
   - **Medium: 6 EDI**
   - **Date: 4 (Immediate Trigger)**
   - **Language: EN**
5. Save the entry.

Create Condition Records (Shipping Notificatio): Fast Entry

Communication

Sales Organization 1710 Dom. Sales Org US

Condition Recs.

Customer	Name	Funct	Partner	M...	Da...	Lan...
USCU_S05		SH	USCU_S05	6	4	EN

Figure 7.51 Creation of Condition Records for EDI Customer USCU_S05 and Organization 1710

7.4 Handling External Identifiers during IDoc Processing

With all the configuration done so far, we can already technically process inbound ORDER IDocs from our EDI customer, but there are some master data/functional aspects we need to take care of.

When we receive a sales order EDI message from our trading partner, we will transform it to IDoc format and post it to SAP S/4HANA. The identifiers of trading partners involved in the sales transaction will be moved 1:1 from an EDIFACT message to the IDoc, but as explained in Chapter 5, these will be the *external identifiers*. We will need to convert them to values recognizable by SAP S/4HANA. We already know that there will be three E1EDKA1 segments with trading party–related information:

- **Segment E1EDKA for sold-to party**
  Field PARVW='AG' and field PARTN='BEST_BIKES_HQ'
- **Segment E1EDKA for ship-to**
  Field PARVW='WE' and field PARTN='BEST_BIKES_WHS'
- **Segment E1EDKA for supplier**
  Field PARVW='LF' and field PARTN='COMPANY_1710_VN'

Within the standard SAP S/4HANA functionality, there are already some tables that can be involved in the conversion of external to internal numbers. The one we'll involve in our inbound process is table EDPAR. Let's go through a short description and then see the exact entries.

Table EDPAR is managed via Transaction VOE4. We will make use of it for the sales order, sales order response, and invoice. Function modules related to the mentioned processes already have appropriate logic to transform external numbers to the related internal numbers and vice versa. All that's left is proper maintenance of the table. The view for adding new entries looks as shown in Figure 7.52.

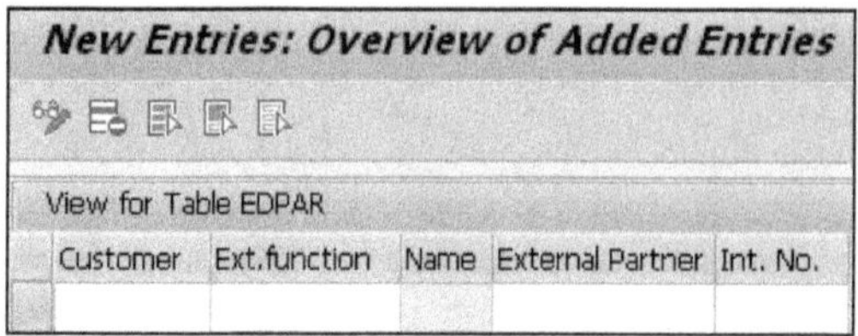

Figure 7.52 Adding EDPAR Entries via Transaction VOE4

There are four fields to be filled in as follows:

- **Customer**
  This is number of the customer used in partner profile configuration—here, **USCU_S05**.
- **Ext.function**
  This is the partner function, like sold-to party, ship-to party, or supplier. The selection can be made from the dropdown list.
- **Name**
  This is not editable. Tt will be automatically filled with the name associated with the partner function.
- **External Partner**
  This is the external identifier that you receive/send in an EDI message. This reference number is not known to the SAP business system.
- **Int. No.**
  This is the internal SAP account number that corresponds to the earlier external partner.

To process the sales order IDoc with three segments, we need to add two entries in table EDPAR. Maintain the first entry for converting the external sold-to party number to its internal number:

- **Customer: USCU_S05**
- **Ext.function: SP**
- **Name: will be determined automatically**
- **External Partner: BEST_BIKES_HQ**
- **Into. No.: USCU_S05**

Maintain the second entry for converting the external ship-to party number to its internal number:

- **Customer: USCU_S05**
- **Ext.function: SH**
- **Name: will be determined automatically**
- **External Partner: BEST_BIKES_WHS**
- **Into. No.: USCU_S05**

Figure 7.53 shows both entries.

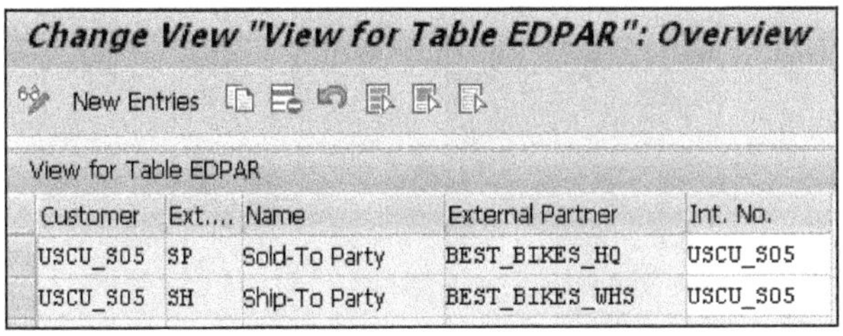

Change View "View for Table EDPAR": Overview

New Entries

View for Table EDPAR

Customer	Ext....	Name	External Partner	Int. No.
USCU_S05	SP	Sold-To Party	BEST_BIKES_HQ	USCU_S05
USCU_S05	SH	Ship-To Party	BEST_BIKES_WHS	USCU_S05

Figure 7.53 Content of Table EDPAR with Sold-To and Ship-To Party Identifiers

At this point, you may ask why we did not add an entry for the vendor which is in segment E1EDKA1 LF with value COMPANY_1710_VN. It is definitely not an actual business partner in our SAP S/4HANA system—neither a proper company code nor a sales organization. The reason for not adding a conversion rule for the LF party in table EDPAR is that there is another table that supports determination of the sales organization. This is table EDSDC, and it can be maintained via Transaction VOE2. Apart from the sales area details, it also allows you to define the sales order type. Let's maintain the entries required for processing our inbound sales order IDoc.

Start Transaction VOE2 and create a new entry with values as follows:

- **Customer: USCU_S05**
- **Vendor number: COMPANY_1710_VN**
- **Sorg.: 1710**
- **DChl: 10**
- **Dv: 00**
- **SaTy: OR**

You can see the complete entry in Figure 7.54. Note that the key values for each entry are the customer and vendor number. This is the pair for which we define the sales organization, distribution channel, division, and sales order type. (**Name** is not an editable field; it is just a description for the division.) You can see that for the customer number, we used an internal value, determined from the table EDPAR conversion. For the vendor number, we maintained the external identifier as we have no conversion entry in table EDPAR.

Our approach and the way we maintained tables EDPAR and EDSDC is enough for our use case and to demonstrate how the basic functionality works, though you should be aware that it is not the only option. Especially for table EDSDC, it is possible to maintain more generic entries, which may force a specific sales area or order type on the vendor level, and then you don't need an entry for each vendor. Also, table EDPAR is not used only during sales order processing. We will prove that later when setting up an invoice. Anyway, the key message is that it is recommended to review the overall scope and all requirements and then, based on this, define the approach for using tables EDPAR, EDSDC, and others for conversion of external/internal numbers. The most important element here is transparency and consistency.

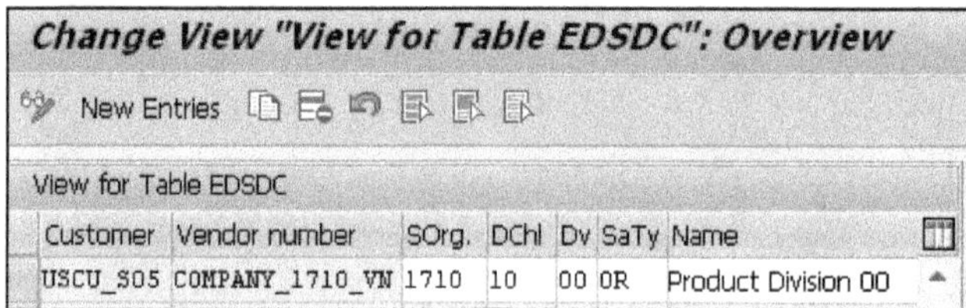

Figure 7.54 Table EDSDC Entry to Determine Sales Area from Customer and Vendor Number

With this configuration, we can successfully process the inbound IDoc. Now let's look at the requirements related to outbound messages:

- For the sales order response, we have no requirement to send partner-related information. There will be no NAD segment in the EDIFACT message delivered to the EDI trading partner.
- For shipping notifications, we have a requirement to include address details for the sold-to and ship-to parties, including the external reference number.
- For invoice messages, we have a requirement to send details related to the invoicing party; that's the address data and the external identifier of company code 1710.

Let's start with the shipping notification. Table EDPAR is not involved in this process, but there is a table that works in similar way: table PUMA. There is no standard transaction for its maintenance, but at least there is a maintenance view. We can make use of the standard transaction for view maintenance, Transaction SM30. The view generated for table PUMA is called V_PUMA.

Table PUMA is only used within the output processing from the outbound delivery. Function modules related to outbound delivery IDocs already contain the necessary logic, so adding appropriate entries to table PUMA is the only task for us. To maintain the necessary entries, follow these steps:

1. Start Transaction SM30, enter "V_PUMA" in the **Table/View** field, and click **Edit**, as shown in Figure 7.55.

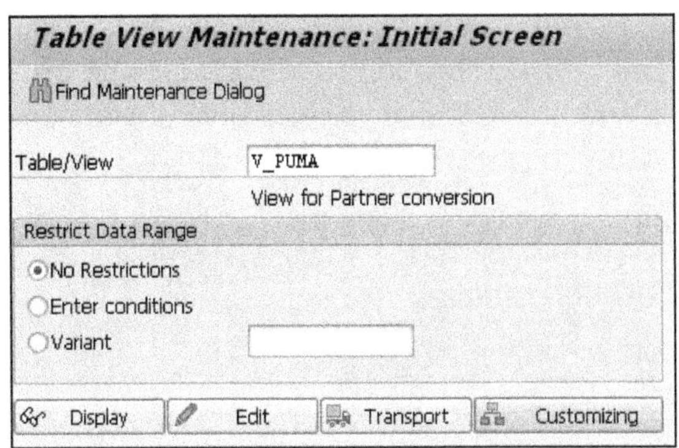

Figure 7.55 Selection Screen for Transaction SM30: View Maintenance for V_PUMA

2. Click **New Entries** and add two entries with the following details, as shown in Figure 7.56:
   - The first entry is for converting internal sold-to number USCU_S05 to external partner number BEST_BIKES_HQ. We maintain this entry on ship-to USCU_S05 as this is our receiver partner number as maintained in the partner profile:
     - **Rec.func.:SH**
     - **RecPartNo:USCU_S05**
     - **Conv.func.: SP**
     - **PartNoConv: USCU_S05**
     - **External partner no.: BEST_BIKES_HQ**
   - The second entry is for converting internal sold-to number USCU_S05 to external partner number BEST_BIKES_HQ. We maintain this entry on ship-to USCU_S05 as this is our receiver partner number as maintained in the partner profile:
     - **Rec.func.:SH**
     - **RecPartNo:USCU_S05**
     - **Conv.func.: SH**
     - **PartNoConv: USCU_S05**
     - **External partner no.: BEST_BIKES_WHS**

Rec.func.	RecPartNo	Conv.func.	PartNoConv	External partner no.
SH	USCU_S05	SP	USCU_S05	BEST_BIKES_HQ
SH	USCU_S05	SH	USCU_S05	BEST_BIKES_WHS

Figure 7.56 Table PUMA Entries for Converting Internal Sold-To and Ship-To Pary Numbers

The last one is an invoice, where we need to include the external reference of the invoicing party. As mentioned earlier, table EDPAR is not only used in inbound processing. Among many others, the outbound processing of an invoice also makes use of this table.

Once again, start Transaction VOE4 and add an entry as follows:

- **Customer: USCU_S05**
- **Ext.function: PI (Invoicing Party)**
- **External Partner: COMPANY_1710_VN**
- **Into. No.: keep empty**

You can see the complete entry in Figure 7.57. Note that we kept the internal number empty. The standard logic that exists in the function module requires exactly such an entry to populate the external partner in the E1EDKA segment related to the invoicing party (the RS qualifier). We will verify the results by running some tests later on.

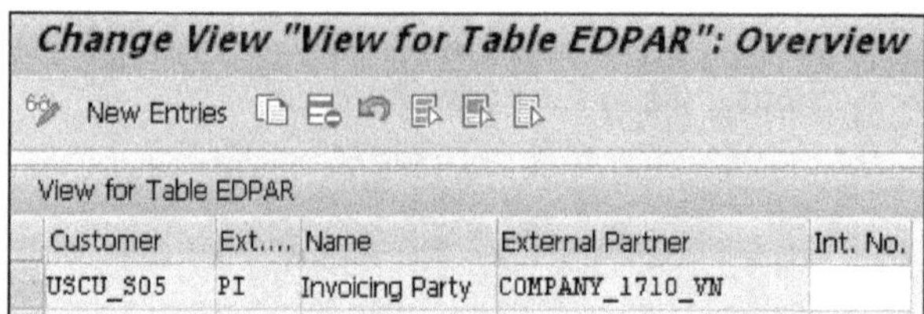

Figure 7.57 Table EDPAR Entry for Populating External Partner for Invoicing Party

7.5 Summary

This chapter started with a quick reminder of the basics of IDocs. The key is to understand the concepts of IDoc types, message types, and extensions. For the basic EDI scenario we introduce in this book, it may seem not so important, but when the EDI landscape keeps growing and we need to manage several different transactions in a few variants and then multiply this by a few functional areas, then it is necessary to make proper choices for IDoc structures and their logical definition. The second important point is that each IDoc consists of three parts: the control record, data records, and status records. Using the first one, we can define and influence the technical processing. Data records are the keys for successful execution of the business transaction. Last but not least is the status record. Its main purpose is to give feedback on IDoc processing, and we have limited to no options to influence its content. This and the short mention of IDoc segments and fields from the initial part of this chapter will pay off in the next chapter, where we will work on the definition of IDoc message mapping.

The second part of this chapter covered the step-by-step configuration related to both inbound and outbound IDoc processing. As you probably noticed, the inbound processing requires a bit more effort, while the setup required for creation of an outbound

IDoc may seem a bit simpler. We also briefly looked at BRFplus output management. The intention here was to give you an idea of this tool but not necessarily present the full scope of its features. In practice, BRFplus is usually not the sole responsibility of the integration specialist or the integration team; it should be a common effort of the technical and functional teams to define an appropriate, consistent approach and implement it. We need to be aware of the risk when the different rules may overlap or overwrite each other, and the EDI-/IDoc-based outputs are usually in the minority; there can be a high number of very complex rules related to printing, emails, and other outputs coexisting in the system. Due to a gap in the standard SAP S/4HANA functionality, we had a chance to make use of the old (table NAST-based) output control for the delivery document. That gives us the opportunity to compare both solutions.

The final part of this chapter was related to conversion of the external identifiers to internal numbers and vice versa. Again, we only briefly went through this topic, demonstrating the use case of the standard tables and functionality.

This chapter should give you a good idea of what's required to process IDocs within SAP S/4HANA and how the inbound and outbound processing works at a high level. All the configuration maintained in this chapter results in a fully functioning solution. In the next chapter, we will configure and activate SAP Application Interface Framework on top of all our IDocs. This will simplify the monitoring and error handling of our EDI interfaces and will move us to the SAP Fiori–based applications for performing these activities.

Chapter 8
SAP Application Interface Framework

In this chapter, we will go through configuration of SAP Application Interface Framework. Not only does this component streamline interface implementation, but its message monitoring applications using SAP Fiori also are a game changer on the UI front.

In this chapter, we will continue with configuration within SAP S/4HANA. While we already have the basic configuration related to IDocs and their processing in place, we can enhance the overall setup by enabling SAP Application Interface Framework monitoring. This will be done for all interfaces in scope of our B2B EDI scenario. This will significantly improve the overall monitoring and error handling capability within the business system, empowering key users and business users to handle functional errors. While SAP Application Interface Framework itself is not a new functionality, its overall functionality within SAP S/4HANA has been significantly improved by several added features. Another great feature is that SAP Fiori offers dedicated applications for message monitoring. Finally, business and key users can say goodbye to the technical transactions known from SAP ERP.

8.1 Enabling SAP Application Interface Framework

Our approach is that we create an SAP Application Interface Framework interface for each unique message flow. It will be an IDoc-based interface enabled for the monitoring scenario. We will not enable full SAP Application Interface Framework processing as at this stage as we do not have any additional requirements that could require structure mapping, checks, value mapping, and so on. The main requirement is that we want to have a transparent and real-time view of interface processing within the SAP S/4HANA system.

There are four main phases we can distinguish in this process:

1. Collect details on all the interfaces we want to activate and define the functionality.
2. Define and create the necessary ABAP/workbench objects.
3. Customize the interfaces.
4. Enable or customize any additional features.

Starting from the first point, the overall scope for us would be a full set of all our B2B interfaces that are active for our sales process. We will group all our interfaces and related configuration using the *SAP Application Interface Framework namespace*. This is an element for logical structuring of objects and configuration within SAP Application Interface Framework. It can contain multiple interfaces, checks, value mapping, and actions, and the recommended approach is that we use it to collect objects relevant to a specific project, functional area, country, and so on. In our case, we use *B2B* as a namespace. This will be where we will develop and govern B2B-related integrations.

Additional required details are listed in Table 8.1. The IDoc types do not require explanation as these were discussed in Chapter 7. The description is any text that describes the interface and its function in a meaningful way. The last one is the interface name, where we have a choice of how to use the 12-character length given by SAP. We keep it simple here: we concatenate the IDoc message type with a suffix, either _IN or _OUT, indicating the direction from an SAP S/4HANA point of view.

IDoc Type	Namespace	Interface Name	Description
ORDERS.ORDERS05	B2B	ORDERS_IN	Sales order
ORDRSP.ORDERS05	B2B	ORDRSP_OUT	Sales order response
DESADV.DELVRY07	B2B	DESADV_OUT	Delivery
INVOIC.INVOIC02	B2B	INVOIC_OUT	Invoice

Table 8.1 Overview of SAP Application Interface Framework Interfaces Defined per IDoc Type

At this stage, we have a clear list with technical interfaces to activate. But there are still a few more details to define to make the solution more efficient.

The next important aspects we need to choose before defining the technical part are the key fields. As the name suggests, these are the most important data elements within the message. They can be used to identify the transaction, the parties involved, or other (most often) unique characteristics or values among all the messages being exchanged. A good example of a key field for a sales order document could be a purchase order number or sold-to party. Key fields are used in multiple places, including content-based message search, authorizations, and dynamic recipient determination. We will configure and use all these.

While selecting the key fields, we also define the (ABAP) data elements behind them. Table 8.2, Table 8.3, Table 8.4, and Table 8.5 show the selections made for our set of interfaces.

Key Field	Field	Data Element
Sender	SNDPRN	EDI_SNDPRN
Customer name	NAME1	NAME1
Sales organization	VKORG	VKORG
Purchase order number	PO_NUMBER	BSTKD

Table 8.2 Key Fields Defined for Sales Orders Interface in SAP Application Interface Framework

Key Field	Field	Data Element
Receiver	RCVPRN	EDI_RCVPRN
Sales organization	VKORG	VKORG
Purchase order number	PO_NUMBER	BSTKD

Table 8.3 Key Fields Defined for Sales Order Response Interface in SAP Application Interface Framework

Key Field	Field	Data Element
Receiver	RCVPRN	EDI_RCVPRN
Plant	WERKS	EMPFW
Delivery number	VBELN	VBELN_VL

Table 8.4 Key Fields Defined for Outbound Delivery Interface in SAP Application Interface Framework

Key Field	Field	Data Element
Receiver	RCVPRN	EDI_RCVPRN
Company code	COMP_CODE	BUKRS
Invoice number	VBELN	VBELN_VF

Table 8.5 Key Fields Defined for Invoice Interface in SAP Application Interface Framework

8.2 Customizing IDoc-Based Interfaces on SAP Application Interface Framework

The previous chapter covered configuring IDocs in detail, but there is still a set of ABAP/workbench objects we need to create before we customize the SAP Application Interface Framework interfaces. That's because these objects need to be maintained with SAP Application Interface Framework customizing, so they must be known and must exist at the time of configuration.

8.2.1 Creating Index Tables and IDoc Structures

The purpose of index tables is to track processing by storing detailed information on each unique message. Next to the standard information being captured in the index table, we have an option to include data from custom key fields. We'll discuss index tables further in this section, in addition to three additional tables for dynamic recipient determination and the SAP Application Interface Framework IDoc structure.

Index Tables

We will make use of a so-called single index table here. While there will be one index table per each interface, there will be a 1:1 relation between the index table and the SAP Application Interface Framework interface. It is possible to reuse the same index table for multiple interfaces (especially when they have the same key fields), but it is not recommended for performance reasons. In total, we will create four tables, as shown in Table 8.6.

IDoc Type	Index Table	Description
ORDERS.ORDERS05	ZAIF_TBL_INDEX01	Index table for B2B ORDERS_IN
ORDRSP.ORDERS05	ZAIF_TBL_INDEX02	Index table for B2B ORDRSP_OUT
DESADV.DELVRY07	ZAIF_TBL_INDEX03	Index table for B2B DESADV_OUT
INVOIC.INVOIC02	ZAIF_TBL_INDEX04	Index table for B2B INVOIC_OUT

Table 8.6 List of Index Tables Used for IDoc-Based Interfaces

The base for our index tables will be the template table predelivered by SAP. We start by making a copy of the table /AIF/STD_IDX_TBL structure, and after that we can make the necessary adjustments to the newly created object. Adding our custom key fields as defined above and adjusting the description of the table are the key activities.

Here are the step-by-step instructions for creating the first index table related to the sales order interface:

1. Start Transaction SE11.
2. In the **Database Table** field, enter the name of the predelivered template index table, "/AIF/STD_IDX_TBL", as shown in Figure 8.1.

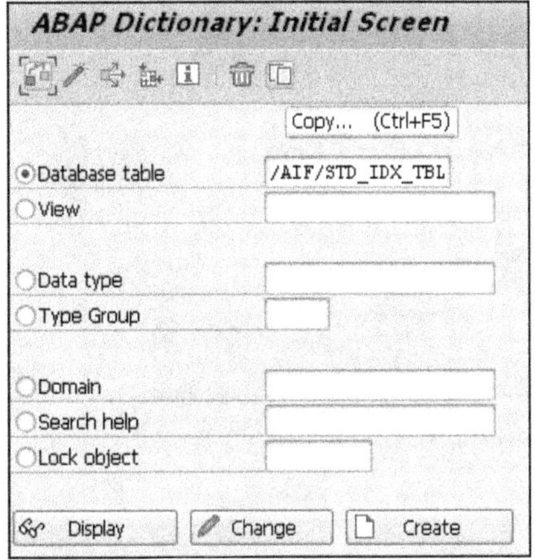

Figure 8.1 Copying Template Index Table /AIF/STD_IDX_TBL

3. Click the last icon in the top row, **Copy Object**.
4. Give a new name for the newly created table; make use of the overview table here. For the sales order interface, enter "ZAIF_TBL_INDEX01".
5. Confirm creation of the table. After a successful operation, you should see a message like this: **Table was copied from /AIF/STD_IDX_TBL to ZAIF_TBL_INDEX01.**
6. You have the table, so now open it in Transaction SE11 to make the necessary adjustments.
7. Adjust the description of the table by entering "Index table for B2B ORDERS_IN" in the **Short Description** field at the top.
8. To add key fields, you need to enter the field name and data element of each of the additional key fields in the empty fields starting from the bottom of the existing structure. The end result should look like Figure 8.2.
9. Save and activate the object. The latter can be done while you are on the selection screen of Transaction SE11 via menu path **Dictionary Object • Activate**.

You can repeat the same steps for the other three interfaces, taking the appropriate key field details from the tables in Section 8.1. At the end, we should have four index tables created, one for each interface.

If you display one of the index tables, you will see that there are a lot of fields coming from the template; these are related to message tracing details. We do not modify any of these; they are all used within the SAP Application Interface Framework engine. Figure 8.3 shows a table related to the sales order response.

Transparent Table ZAIF_TBL_INDEX01 Active(Revised)
Short Description Index table for B2B ORDERS_IN
Attributes | Delivery and Maintenance | Fields | Input Help/Check | Currency/Quantity Fields | Indexes
Srch Help | Built-In Type 1 / 10

Field	Key	Ini...	Data element	Data Type	Length	Deci...	Coordinate	Short Description
MANDT	✓	✓	MANDT	CLNT	3	0	0	Client
MSGGUID	✓	✓	GUID_32	CHAR	32	0	0	GUID in 'CHAR' Format in Uppercase
.INCLUDE			/AIF/IFKEYS	STRU	0	0	0	Keys
.INCLUDE			/AIF/ADMIN	STRU	0	0	0	Additional information structure (be appent to tran. table)
PID			SXMSPID	CHAR	40	0	0	Integration Engine: Pipeline ID
SNDPRN			EDI_SNDPRN	CHAR	10	0	0	Partner Number of Sender
NAME1			NAME1	CHAR	30	0	0	Name
VKORG			VKORG	CHAR	4	0	0	Sales Organization
PO_NUMBER			BSTKD	CHAR	35	0	0	Customer Reference

Figure 8.2 View of First Index Table for Sales Order Interface

Transparent Table ZAIF_TBL_INDEX02 Active
Short Description Index table for B2B ORDRSP_OUT
Attributes | Delivery and Maintenance | Fields | Input Help/Check | Currency/Quantity Fields | Indexes
Srch Help | Built-In Type 1 / 44

Field	Key	Ini...	Data element	Data Type	Length	Deci...	Coordinate	Short Description
MANDT	✓	✓	MANDT	CLNT	3	0	0	Client
MSGGUID	✓	✓	GUID_32	CHAR	32	0	0	GUID in 'CHAR' Format in Uppe
.INCLUDE			/AIF/IFKEYS	STRU	0	0	0	Keys
NS			/AIF/NS	CHAR	6	0	0	Namespace
IFNAME			/AIF/IFNAME	CHAR	10	0	0	Interface Name
IFVER			/AIF/IFVERSION	CHAR	5	0	0	Interface Version
.INCLUDE			/AIF/ADMIN	STRU	0	0	0	Additional information structur
.INCLUDE			/AIF/ADDITIONAL...	STRU	0	0	0	Additional information structur
NUMBEROFABORTS			/AIF/NUMBEROFAB...	INT4	10	0	0	Abort Number
NUMBEROFERRORS			/AIF/NUMBEROFER...	INT4	10	0	0	Error Number
NUMBEROFWARNINGS			/AIF/NUMBEROFWA...	INT4	10	0	0	Warning Number
NUMBEROFSUCCESS			/AIF/NUMBEROFSU...	INT4	10	0	0	Success Number
NUMBEROFINFOS			/AIF/NUMBEROFIN...	INT4	10	0	0	Information Number
.INCLUDE			/AIF/ADDITIONAL...	STRU	0	0	0	Additional information structur
NSRECIP			/AIF/NS	CHAR	6	0	0	Namespace
RECIPIENT			/AIF/ALRT_REC	CHAR	30	0	0	Alert Management Recipient

Figure 8.3 Display View of Index Table Related to Outbound Sales Order Response

Table for Dynamic Recipient Determination

There are three additional tables which we create for dynamic recipient determination. The purpose of these tables is to make a link between the message characteristics and the recipient group. To be precise, the message characteristics are our key fields. The approach is similar to the one used for index table. Here we also have a template table predelivered by SAP: /AIF/T_ALRT_DEF. Using Transaction SE11, we again make a copy of the template to our own table. Then we adjust the table, adding a description and one additional field in each subsequent table. In total, we need three tables created from the template table:

- Table ZAIF_ALRT_VKORG with the description "AIF Alerting B2B recipients—sales organization"
- Table ZAIF_ALRT_BUKRS with the description "AIF Alerting B2B recipients—company code"
- Table ZAIF_ALRT_WERKS with the description "AIF Alerting B2B recipients—plant"

In the first table from the top, we add the sales organization (VKORG) key field and we adjust the description to the text shown in the previous list. At the end, we expect the results shown in Figure 8.4.

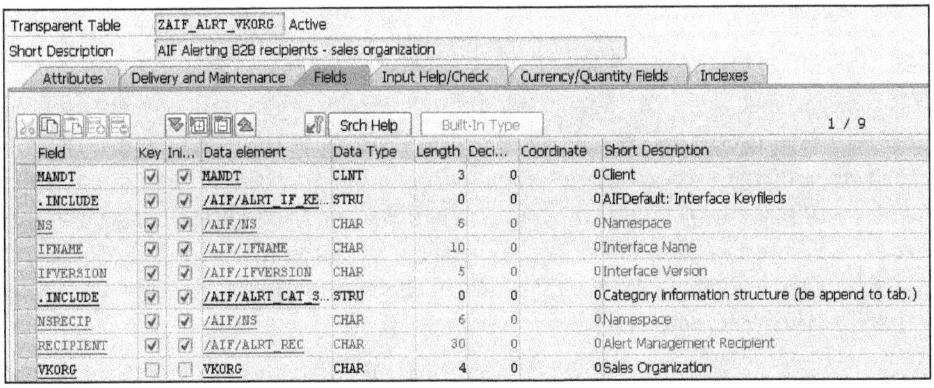

Transparent Table ZAIF_ALRT_VKORG Active
Short Description AIF Alerting B2B recipients - sales organization
Attributes | Delivery and Maintenance | Fields | Input Help/Check | Currency/Quantity Fields | Indexes
Srch Help | Built-In Type 1 / 9

Field	Key	Ini...	Data element	Data Type	Length	Deci...	Coordinate	Short Description
MANDT	☑	☑	MANDT	CLNT	3	0	0	Client
.INCLUDE	☑	☑	/AIF/ALRT_IF_KE...	STRU	0	0	0	AIFDefault: Interface Keyfileds
NS	☑	☑	/AIF/NS	CHAR	6	0	0	Namespace
IFNAME	☑	☑	/AIF/IFNAME	CHAR	10	0	0	Interface Name
IFVERSION	☑	☑	/AIF/IFVERSION	CHAR	5	0	0	Interface Version
.INCLUDE	☑	☑	/AIF/ALRT_CAT_S...	STRU	0	0	0	Category information structure (be append to tab.)
NSRECIP	☑	☑	/AIF/NS	CHAR	6	0	0	Namespace
RECIPIENT	☑	☑	/AIF/ALRT_REC	CHAR	30	0	0	Alert Management Recipient
VKORG	☐	☐	VKORG	CHAR	4	0	0	Sales Organization

Figure 8.4 Table for Dynamic Recipient Determination Based on Sales Organization

In a similar way, we create the other two tables. The second contains the company code key field COMP_CODE as shown in Figure 8.5. To the third table we add the WERKS field as shown in Figure 8.6.

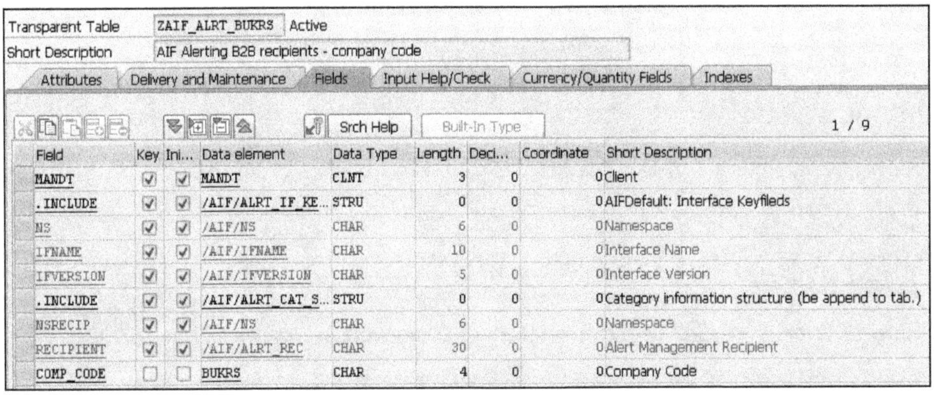

Transparent Table ZAIF_ALRT_BUKRS Active
Short Description AIF Alerting B2B recipients - company code
Attributes | Delivery and Maintenance | Fields | Input Help/Check | Currency/Quantity Fields | Indexes
Srch Help | Built-In Type 1 / 9

Field	Key	Ini...	Data element	Data Type	Length	Deci...	Coordinate	Short Description
MANDT	☑	☑	MANDT	CLNT	3	0	0	Client
.INCLUDE	☑	☑	/AIF/ALRT_IF_KE...	STRU	0	0	0	AIFDefault: Interface Keyfileds
NS	☑	☑	/AIF/NS	CHAR	6	0	0	Namespace
IFNAME	☑	☑	/AIF/IFNAME	CHAR	10	0	0	Interface Name
IFVERSION	☑	☑	/AIF/IFVERSION	CHAR	5	0	0	Interface Version
.INCLUDE	☑	☑	/AIF/ALRT_CAT_S...	STRU	0	0	0	Category information structure (be append to tab.)
NSRECIP	☑	☑	/AIF/NS	CHAR	6	0	0	Namespace
RECIPIENT	☑	☑	/AIF/ALRT_REC	CHAR	30	0	0	Alert Management Recipient
COMP_CODE	☐	☐	BUKRS	CHAR	4	0	0	Company Code

Figure 8.5 Table for Dynamic Recipient Determination Based on Company Code

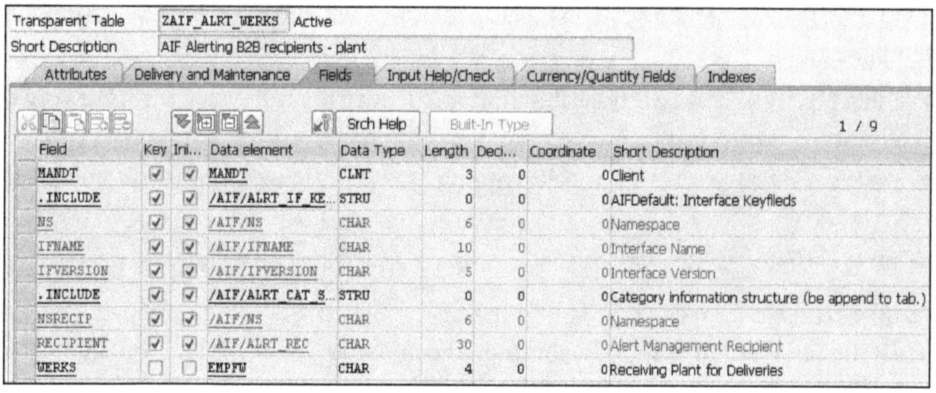

Transparent Table	ZAIF_ALRT_WERKS Active
Short Description	AIF Alerting B2B recipients - plant

Attributes | Delivery and Maintenance | Fields | Input Help/Check | Currency/Quantity Fields | Indexes

Srch Help | Built-In Type | 1 / 9

Field	Key	Ini...	Data element	Data Type	Length	Deci...	Coordinate	Short Description
MANDT	☑	☑	MANDT	CLNT	3	0	0	Client
.INCLUDE	☑	☑	/AIF/ALRT_IF_KE...	STRU	0	0	0	AIFDefault: Interface Keyfileds
NS	☑	☑	/AIF/NS	CHAR	6	0	0	Namespace
IFNAME	☑	☑	/AIF/IFNAME	CHAR	10	0	0	Interface Name
IFVERSION	☑	☑	/AIF/IFVERSION	CHAR	5	0	0	Interface Version
.INCLUDE	☑	☑	/AIF/ALRT_CAT_S...	STRU	0	0	0	Category information structure (be append to tab.)
NSRECIP	☑	☑	/AIF/NS	CHAR	6	0	0	Namespace
RECIPIENT	☑	☑	/AIF/ALRT_REC	CHAR	30	0	0	Alert Management Recipient
WERKS	☐	☐	EMPFW	CHAR	4	0	0	Receiving Plant for Deliveries

Figure 8.6 Table for Dynamic Recipient Determination Based on Plant

SAP Application Interface Framework IDoc Structure

It is required that every SAP Application Interface Framework interface have a source and destination structure. These SAP Application Interface Framework structures are called the SAP and RAW structures. Depending on the direction of the interface, the RAW or SAP structure is used as the source or destination. To process messages, the SAP Application Interface Framework engine needs to know the structure of the RAW and SAP messages. For IDocs, we can make use of the existing report /AIF/IDOC_GEN to generate the necessary structure.

We follow specific naming conventions for the structure name, keeping in mind that the IDoc basic type is the leading element determining the technical structure. If this is the case, we can reuse the same structure for the sales order and sales order response for the SAP Application Interface Framework interface. It's the same case as it is for IDoc processing, where the IDoc message type is only the logical name to indicate the different process and allow the application to process it differently.

Table 8.7 shows three structures we want to generate.

IDoc Type	Structure
ORDERS.ORDERS05	ZAIF_STR_ORDERS05
ORDRSP.ORDERS05	ZAIF_STR_ORDERS05
DESADV.DELVRY07	ZAIF_STR_DELVRY05
INVOIC.INVOIC02	ZAIF_STR_INVOIC02

Table 8.7 Overview of SAP Application Interface Framework Structures Defined per IDoc Type

To create an SAP Application Interface Framework structure for IDocs, we use Transaction /AIF/IDOC_GEN. Follow these steps to create the SAP Application Interface Framework IDoc structure for the ORDERS.ORDERS05 IDoc:

1. After starting Transaction /AIF/IDOC_GEN, you will see an empty screen as shown in Figure 8.7.

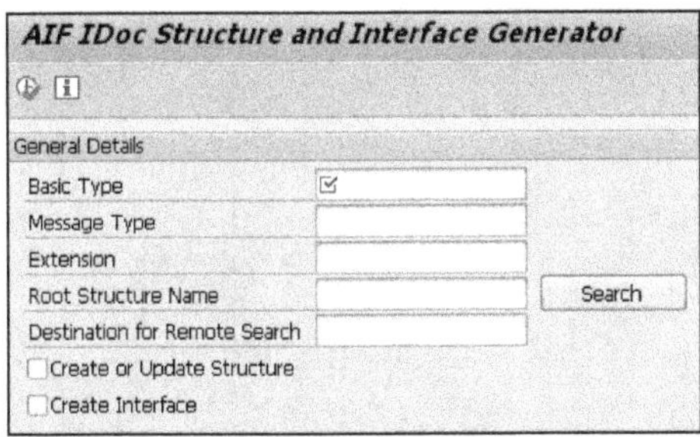

Figure 8.7 Selection Screen with Parameters Related to IDoc Structure Generation

2. Maintain the necessary fields as shown in Figure 8.8:
   - In the **General Details** section:
     - **Basic Type: ORDERS05**
     - **Root Structure Name: ZAIF_STR_ORDERS05**
     - **Create or Update Structure:** Check this option
   - In the **Structure Details** section:
     - **Structure Prefix: ZAIF_**
     - **Package: ZAIF** (depending on your system)
     - **Workbench Request**: Appropriate transport ID (depending on your system)
3. Execute the report with these parameters. If everything goes well, you will see log messages, as shown in Figure 8.9.
4. You can see the created structure in Transaction SE11, as shown in Figure 8.10.

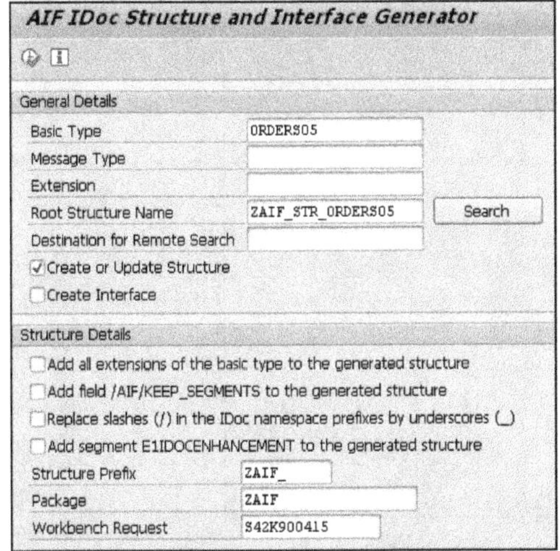

Figure 8.8 Prefilled Selection Screen to Generate SAP Application Interface Framework Structure

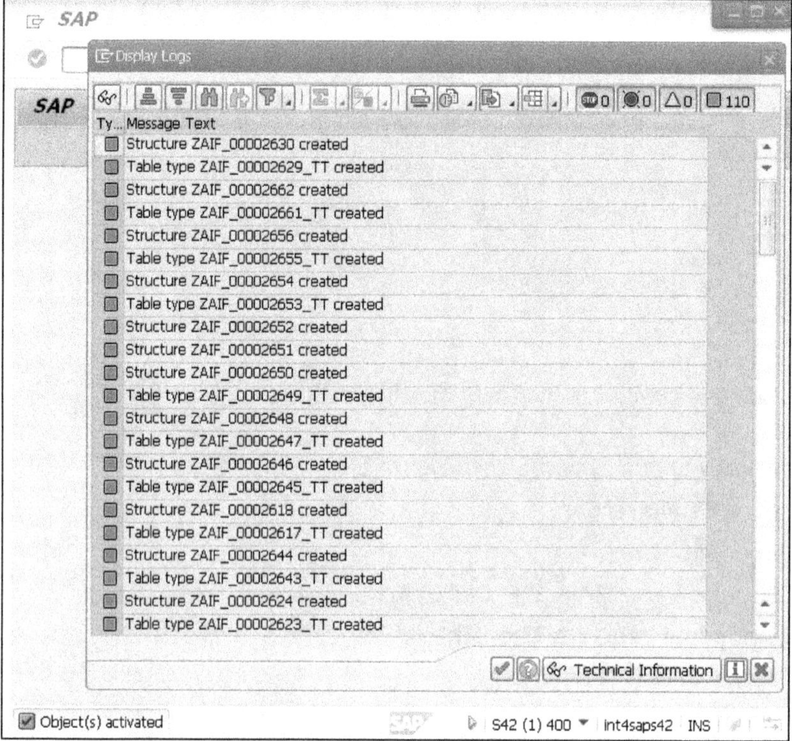

Figure 8.9 Log Messages after Successful Generation of IDoc Structure

Dictionary: Display Structure

Hierarchy Display Append Structure... Proxy Editor

Structure: ZAIF_STR_ORDERS05 Active
Short Description: Purchasing/Sales

Attributes | Components | Input Help/Check | Currency/quantity fields

Built-In Type 1 / 17

Component	Typing Method	Component Type	Data Type	Length	Deci...	Coordinate	Short Descr
EDIDC	Types	EDIDC		0	0	0	Control recc
E1EDK01	Types	ZAIF_00002581		0	0	0	IDoc: Docur
E1EDK14	Types	ZAIF_00002582_TT		0	0	0	IDoc: Docur
E1EDK03	Types	ZAIF_00002584_TT		0	0	0	IDoc: Docur
E1EDK04	Types	ZAIF_00002586_TT		0	0	0	IDoc: Docur
E1EDK05	Types	ZAIF_00002588_TT		0	0	0	IDoc: Docur
E1EDKA1	Types	ZAIF_00002592_TT		0	0	0	IDoc: Docur
E1EDK02	Types	ZAIF_00002594_TT		0	0	0	IDoc: Docur
E1EDK17	Types	ZAIF_00002596_TT		0	0	0	IDoc: Docur
E1EDK18	Types	ZAIF_00002598_TT		0	0	0	IDoc: Docur
E1EDK35	Types	ZAIF_00002600_TT		0	0	0	IDoc: Docur
E1EDK36	Types	ZAIF_00002602_TT		0	0	0	IDOC: Doc.h

Figure 8.10 View of SAP Application Interface Framework IDoc Structure for Sales Order and Sales Order Response Interface

In the same way, we run report /AIF/IDOC_GEN for the other two IDoc types. There are two, not three, as the sales order and sales order response share the same IDoc basic type and thus also the same structure. There is also no need to have two different SAP Application Interface Framework structures for the two sales order–related interfaces. Finally, we should have two more structures, one for basic type DELVRY07, as shown in Figure 8.11, and a second for INVOIC02, as shown in Figure 8.12.

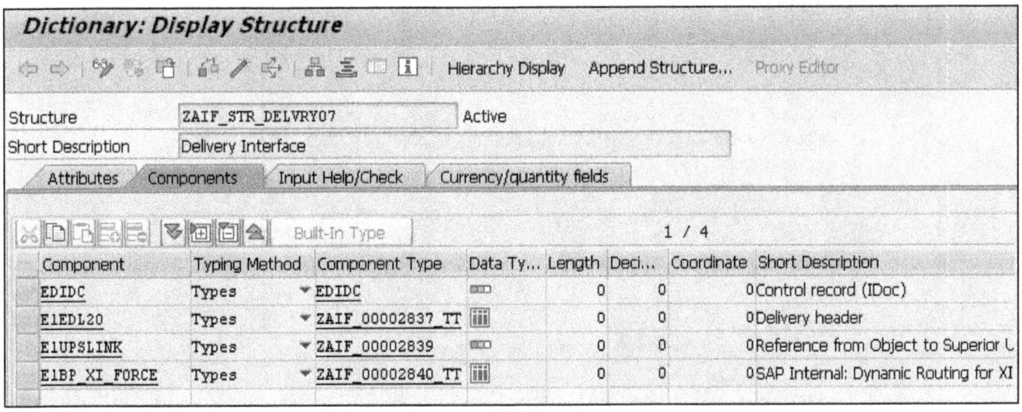

Dictionary: Display Structure

Hierarchy Display   Append Structure...   Proxy Editor

Structure: ZAIF_STR_DELVRY07   Active

Short Description: Delivery Interface

Attributes | Components | Input Help/Check | Currency/quantity fields

Built-In Type   1 / 4

Component	Typing Method	Component Type	Data Ty...	Length	Deci...	Coordinate	Short Description
EDIDC	Types	EDIDC		0	0	0	Control record (IDoc)
E1EDL20	Types	ZAIF_00002837_TT		0	0	0	Delivery header
E1UPSLINK	Types	ZAIF_00002839		0	0	0	Reference from Object to Superior U
E1BP_XI_FORCE	Types	ZAIF_00002840_TT		0	0	0	SAP Internal: Dynamic Routing for XI

Figure 8.11 View of SAP Application Interface Framework IDoc Structure for Delivery Interface

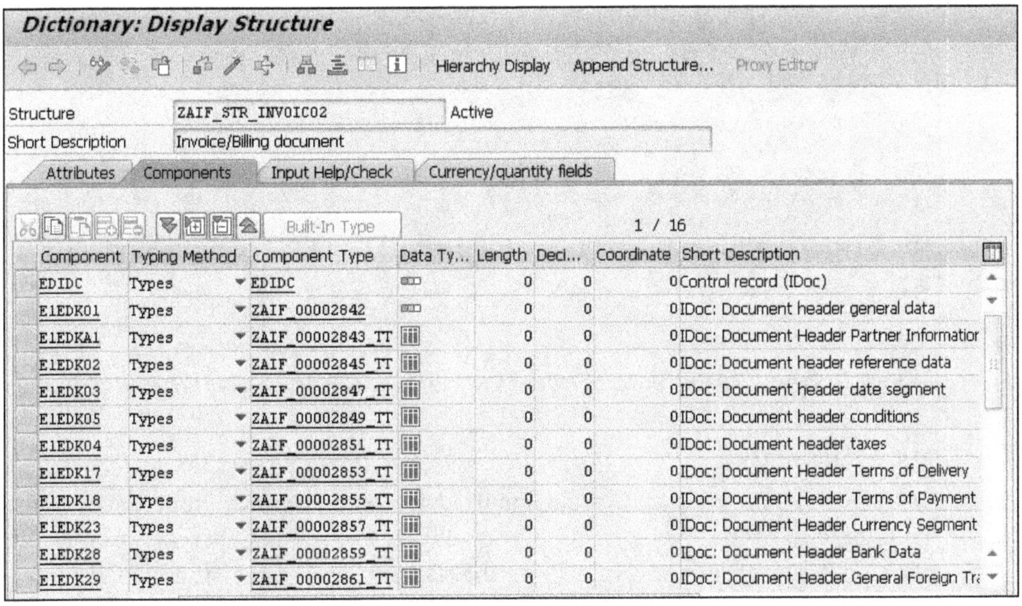

Dictionary: Display Structure

Hierarchy Display   Append Structure...   Proxy Editor

Structure: ZAIF_STR_INVOIC02   Active

Short Description: Invoice/Billing document

Attributes | Components | Input Help/Check | Currency/quantity fields

Built-In Type   1 / 16

Component	Typing Method	Component Type	Data Ty...	Length	Deci...	Coordinate	Short Description
EDIDC	Types	EDIDC		0	0	0	Control record (IDoc)
E1EDK01	Types	ZAIF_00002842		0	0	0	IDoc: Document header general data
E1EDKA1	Types	ZAIF_00002843_TT		0	0	0	IDoc: Document Header Partner Informatior
E1EDK02	Types	ZAIF_00002845_TT		0	0	0	IDoc: Document header reference data
E1EDK03	Types	ZAIF_00002847_TT		0	0	0	IDoc: Document header date segment
E1EDK05	Types	ZAIF_00002849_TT		0	0	0	IDoc: Document header conditions
E1EDK04	Types	ZAIF_00002851_TT		0	0	0	IDoc: Document header taxes
E1EDK17	Types	ZAIF_00002853_TT		0	0	0	IDoc: Document Header Terms of Delivery
E1EDK18	Types	ZAIF_00002855_TT		0	0	0	IDoc: Document Header Terms of Payment
E1EDK23	Types	ZAIF_00002857_TT		0	0	0	IDoc: Document Header Currency Segment
E1EDK28	Types	ZAIF_00002859_TT		0	0	0	IDoc: Document Header Bank Data
E1EDK29	Types	ZAIF_00002861_TT		0	0	0	IDoc: Document Header General Foreign Tr

Figure 8.12 View of SAP Application Interface Framework IDoc Structure for Invoice Interface

At this stage we have index tables, recipient determination tables, and SAP Application Interface Framework IDoc structures related to all our IDoc flows. Now we can move to the next step, where we make use of these during the configuration process.

8.2.2 Interface Development

The first step is to open the SAP Application Interface Framework customizing menu by starting Transaction /AIF/CUST, as shown in Figure 8.13. We will go from the top to the bottom through all required areas and do the necessary customizing in the following sections. First, you need to define a namespace.

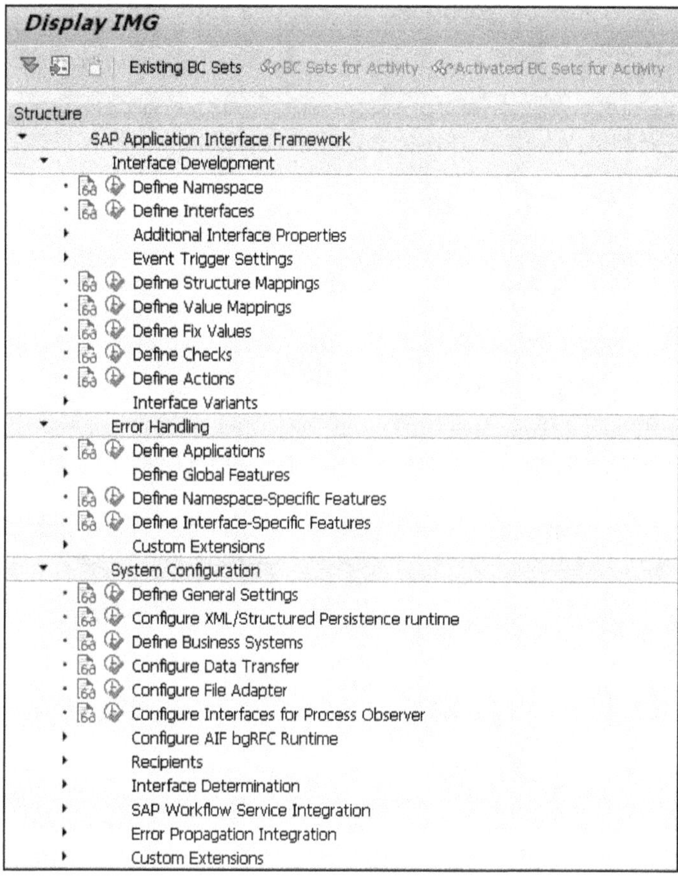

Figure 8.13 View of All Customizing Options Available after Starting Transaction /AIF/CUST

Define Namespace

Follow menu path **Interface Development • Additional Interface Properties • Define Namespace** to define the namespace. You will see a screen like that shown in Figure 8.14. Click **New Entries** and add "B2B" in the **NS** (namespace) field and "B2B EDI interfaces" in the **Namespace Description** field, as shown in Figure 8.15.

After creating a namespace, we can start defining the unique interfaces and configuration artifacts within it. From now on, we will start using the information collected and the workbench (ABAP) objects that we generated earlier in this chapter.

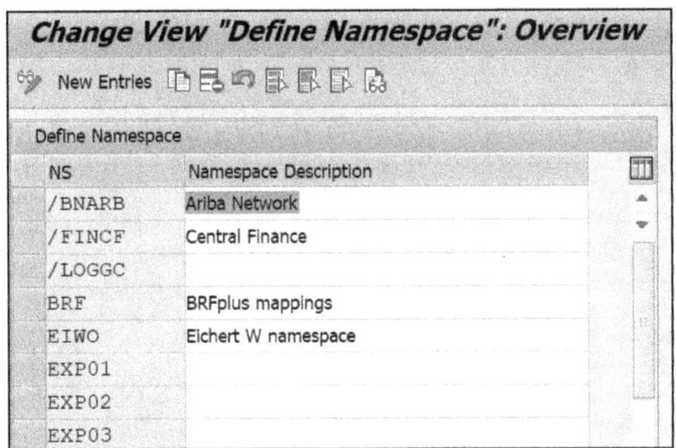

Figure 8.14 Change View for Defining SAP Application Interface Framework Namespaces

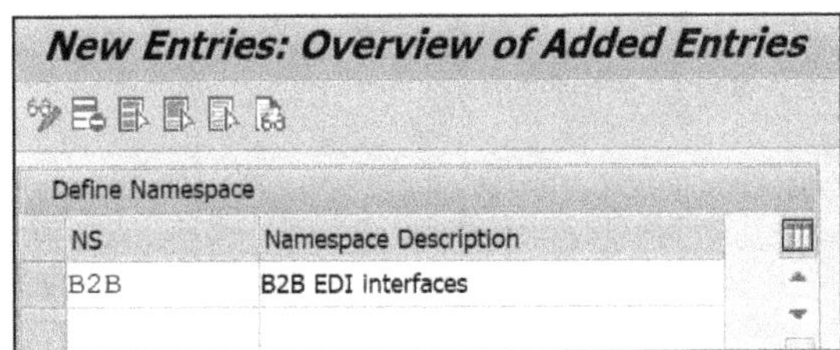

Figure 8.15 Definition of B2B Namespace

Define Interfaces

From now on, we will define and customize specific interfaces. This requires the same steps for each interface. To not make this section too long, we will limit the step-by-step instructions to one inbound and one outbound interface. We use an inbound sales order and an outbound sales order response for reference. If there is something specific to know in these steps related to the delivery and invoice interfaces, that will be also mentioned. To define the first sales order interface, follow menu path **Interface Development • Additional Interface Properties • Define Interface** and then follow these steps:

1. On the empty list, click **New Entries**.
2. Maintain the following fields (see Figure 8.16):
   - **Interface Name: ORDERS_IN**

     This is the most important element. It will be reused in the following configuration steps and will be the main identification for the interface. You can use any text but are limited to 10 characters. ORDERS_IN is the value to use for this example.
   - **Description: Sales orders**

     The importance of this element comes from the fact that this is the value on which most end users rely when working with the monitoring transaction. Simply enter "Sales order" here; combined with an interface name, it is fully meaningful.

- **SAP Data Structure: ZAIF_STR_ORDERS05**
- **RAW Data Structure: ZAIF_STR_ORDERS05**
- **Move Corresponding Structures**: This selection box needs to be checked
- **Interface Direction: Inbound**
- **Pre-Processing**: This selection box needs to be checked

3. Save the entry.

New Entries: Details of Added Entries

Namespace	B2B
Interface Name	ORDERS_IN
Interface Version	1
Define Interfaces	
Description	Sales orders
SAP Data Structure	ZAIF_STR_ORDERS05
Raw Data Structure	ZAIF_STR_ORDERS05
Record Type in Raw Structure	
☑ Move Corresponding Structures	
Check Function Module	
Init Function Before Mapping	
Init Function Before Processing	
Separate Commit	No Separate Commit
Lifetime of Application Log	
☐ Test Mode	
Proxy Class Inbound	
Proxy Class Outbound	
Proxy Method	
Field for the Sending System	
Status Handling	
☑ Pre-Processing	
☐ Proxy XML Transformation	
Interface Direction	Inbound

Figure 8.16 Interface Definition for Sales Order Interface

You should repeat these steps for the other three interfaces using the parameters shown in Table 8.8 and setting the proper interface direction; all the other parameters stay the same.

Interface Name	Description	SAP Data Structure	Raw Data Structure
ORDRSP_OUT	Sales order response	ZAIF_STR_ORDERS05	ZAIF_STR_ORDERS05
DESADV_OUT	Outbound delivery	ZAIF_STR_DELVRY07	ZAIF_STR_DELVRY07
INVOIC_OUT	Invoice	ZAIF_STR_INVOIC02	ZAIF_STR_INVOIC02

Table 8.8 Overview of SAP Data Structures per Interface Name

At the end, you should have four interfaces defined as shown in Figure 8.17.

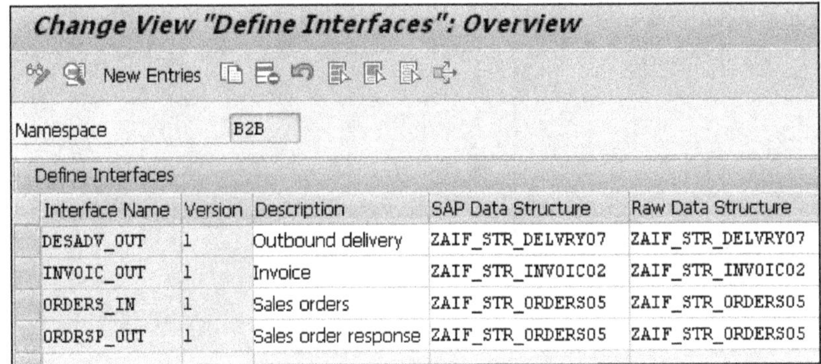

Figure 8.17 Overview of All SAP Application Interface Framework Interfaces for Example

Specify Interface Engines

In this step, we will define engines to be used for handling messages in our IDoc-based interfaces. Follow path **Interface Development • Additional Interface Properties • Specify Interface Engines** in the familiar Transaction /AIF/CUST.

After starting the transaction, you may see a somewhat confusing list where all the IDoc interfaces are showing some default values, like **Proxy** under **ApplEngine**. Figure 8.18 shows as sample of what you may see. Don't worry about it: this information comes from the default SAP Application Interface Framework settings, which can't know what exactly you are going to activate. Open each entry by double-clicking it and adjust the engine properties with the following values:

- **Application Engine: IDoc**
- **Persistance Engine: IDoc**
- **Selection Engine: AIF Index Tables**
- **Logging Engine: AIF Application LOG**

You can see an example in Figure 8.19.

Define Interfaces (Engine Fields)

Interface	Version	Description	ApplEngine	NS	Custome...	PersEngine
DESADV_OUT	1	Outbound delivery	Proxy			Proxy
INVOIC_OUT	1	Invoice	Proxy			Proxy
ORDERS_IN	1	Sales orders	Proxy			Proxy
ORDRSP_OUT	1	Sales order response	Proxy			Proxy

Figure 8.18 Overview with Default Values for Application and Persistence Engines

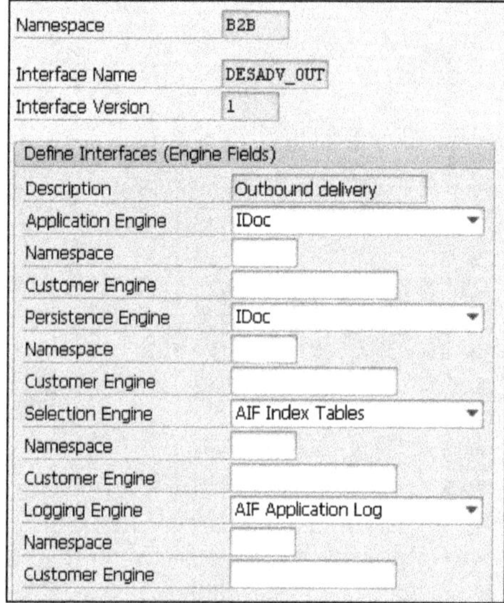

Namespace	B2B
Interface Name	DESADV_OUT
Interface Version	1

Define Interfaces (Engine Fields)

Description	Outbound delivery
Application Engine	IDoc
Namespace	
Customer Engine	
Persistence Engine	IDoc
Namespace	
Customer Engine	
Selection Engine	AIF Index Tables
Namespace	
Customer Engine	
Logging Engine	AIF Application Log
Namespace	
Customer Engine	

Figure 8.19 Proper Definition of Engines for IDoc-Based Interface

Assign IDoc Types

As mentioned earlier, we will not enable the full SAP Application Interface Framework processing here but only the simplified monitoring scenario. To be able to see the related IDocs in the monitoring tool, we need to define which IDocs belong to which interface.

Follow menu path **Interface Development • Additional Interface Properties • Assign IDoc Types** to define IDoc types. After starting the transaction, you can see a view like that in Figure 8.20. All our interfaces are already on the list, but without any specific IDoc type assigned.

Namespace: B2B

Define Interfaces (IDoc fields)

Interface	Version	Message Type	Basic type	Extension	Msg. Var.	Msg.funct
DESADV_OUT	1					
INVOIC_OUT	1					
ORDERS_IN	1					
ORDRSP_OUT	1					

Figure 8.20 Initial Screen for Assigning IDoc Types with No Default Values

Like in other transactions, when you double-click each entry, you can enter the required values in the following fields, as shown in Figure 8.21:

- **Namespace: B2B**
- **Interface Name: ORDERS_IN**
- **Interface Version: 1**
- **Description**: This value is provided automatically
- **Message Type: ORDERS**
- **Basic Type: ORDERS05**

There is no need to maintain the other values nor to set any of the additional options at the bottom.

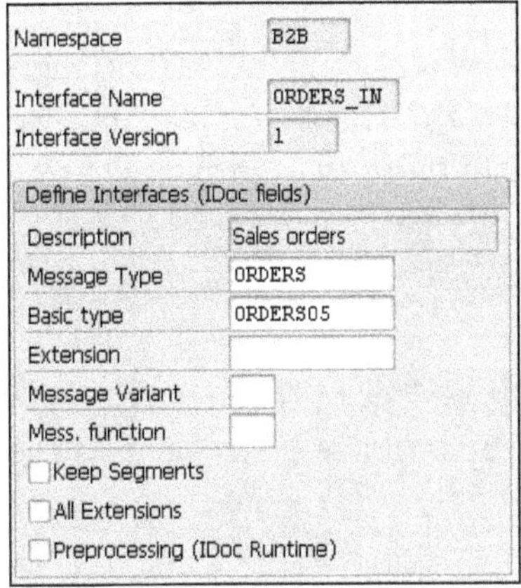

Figure 8.21 Assigned IDoc Type for Sales Order Interface

In a similar way, we maintain the values for the other interfaces using the appropriate message type and basic type values. The end results should appear as shown in Figure 8.22.

Namespace B2B

Define Interfaces (IDoc fields)

Interface	Version	Message Type	Basic type	Extension	Msg. Var.	Msg.funct.	Keep Seg.	All Ext.	Preprocessing (IDoc Runtime)
DESADV_OUT	1	DESADV	DELVRY07				☐	☐	☐
INVOIC_OUT	1	INVOIC	INVOIC02				☐	☐	☐
ORDERS_IN	1	ORDERS	ORDERS05				☐	☐	☐
ORDRSP_OUT	1	ORDRSP	ORDERS05				☐	☐	☐

Figure 8.22 Overall View of Assigned IDoc Types

After assigning the IDoc types, we can leave the **Additional Interface Properties** customizing section. We are done with the key configuration, but there are still a few more steps required in the **Error Handling** section.

Full SAP Application Interface Framework Monitoring

We skip all the options in the middle part: **Define Structure Mappings, Define Value Mappings, Define Fix Values, Define Checks, Define Actions,** and so on. These are related to SAP Application Interface Framework processing in full processing mode. In this powerful mode, SAP Application Interface Framework is capable of performing almost unlimited sets of operations on structures and their content. This is one of the most important features of SAP Application Interface Framework itself and where we can see the full potential of the tool. But as we only need basic interface monitoring here, we do not need to enable structure mapping and all the following features.

8.2.3 Error Handling

We already configured the interface itself, but we now have to configure what the interface and its data look like in the monitoring transaction and which error handling features are enabled for the end user. We'll walk through those steps in the following sections.

Define Namespace-Specific Features

Start by following menu path **Error Handling • Namespace-Specific Features.** Select the view from the left-hand list and then maintain the appropriate parameters on the right-hand side. There are three areas to configure:

- **Define Interface-Specific Features**
  This is where we make a link between the specific index table and the interface. We also configure parameters related to how the interface should be displayed in the monitoring transaction.
- **Configure Alerts**
  In this section, we configure details related to alerting and notifications: when they should be triggered, what the exact content of the notification is, and who is the receiver.
- **Define Recipients**
  In this section, we define the appropriate receiver groups, which need to be created before configuration of the Configure Alerts step.

You can see the initial screen in Figure 8.23.

Follow these steps to configure the first section, starting with the sales order interface:

1. Select **Define Interface-Specific Features** in the left-hand menu.
2. Click the **New Entries** icon at the top of the screen.

3. We are already maintaining objects within the B2B namespace as this was our selection when we started this transaction. The fields we need to fill in are as follows:
   - **Interface Name: ORDERS_IN**
   - **Interface Version: 1**
   - **Message Idx Table: ZAIF_TBL_INDEX01**
   - Check **Display Root Field**, **Display Description**, **Hide Empty Comp.**, and **Disp Source Button**
   - **Interface Disp. Name:** Enter "Sales orders"

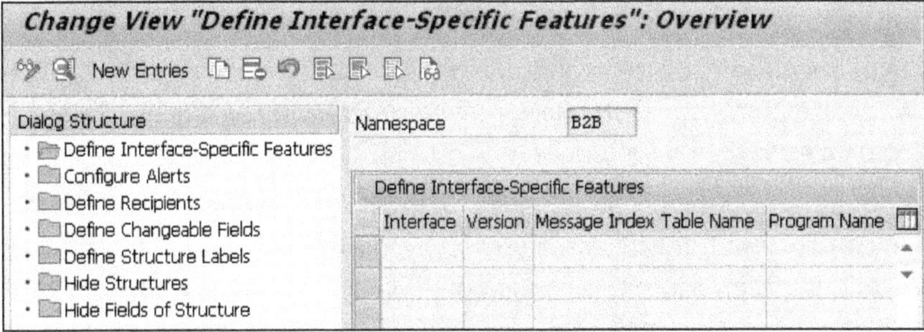

Figure 8.23 Starting Point for Configuration

At the end, the screen should look as shown in Figure 8.24. Save the entry.

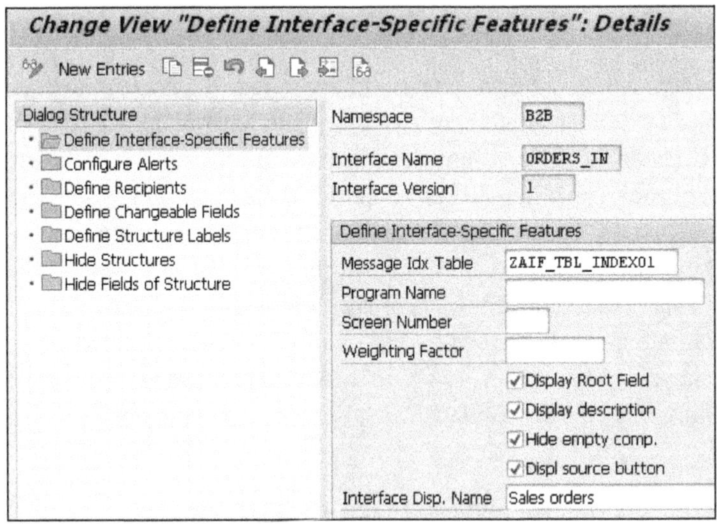

Figure 8.24 Define Interface-Specific Features for Sales Order Interface

Repeat the process for the other three interfaces using the appropriate index table and descriptions. The expected end results are as shown in Figure 8.25.

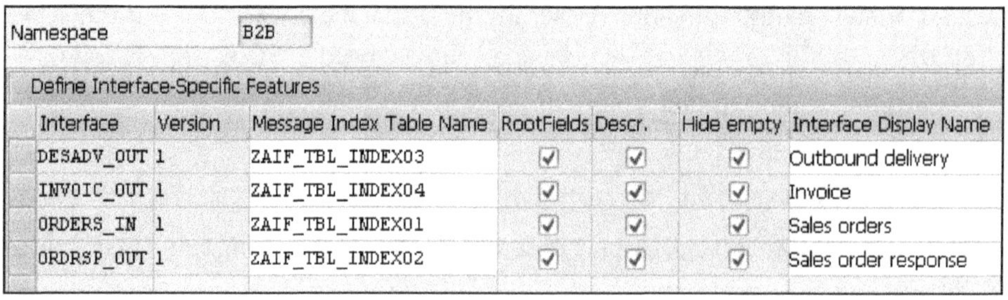

Namespace: B2B

Define Interface-Specific Features

Interface	Version	Message Index Table Name	RootFields	Descr.	Hide empty	Interface Display Name
DESADV_OUT	1	ZAIF_TBL_INDEX03	☑	☑	☑	Outbound delivery
INVOIC_OUT	1	ZAIF_TBL_INDEX04	☑	☑	☑	Invoice
ORDERS_IN	1	ZAIF_TBL_INDEX01	☑	☑	☑	Sales orders
ORDRSP_OUT	1	ZAIF_TBL_INDEX02	☑	☑	☑	Sales order response

Figure 8.25 Overview of Interface-Specific Features

Define Recipients

Although **Configure Alerts** is the next option in the left-hand menu, we cannot maintain it yet. We need to first jump to **Define Recipients**; only after maintaining the recipient groups can we go back to the alerts.

We will make things a bit complex here, but it's for the sake of having a flexible framework that can support multiple different organizations and different organizational entity types, and that can distribute alerts and notifications to the proper recipients. We will define four recipient groups:

- SUPPORT_IT
- BUSINESS_VKORG_1710
- BUSINESS_WERKS_1710
- BUSINESS_BUKRS_1710

The concept we apply here is that we have one group dedicated to global IT support, which would receive notifications related to any interface and any organization. Then we have three groups related to business entities, separate groups for the company code, sales organization, and plant. These are to support our scenario, which operates within company code 1710, sales organization 1710, and plant 1710. (The naming conventions may be a bit confusing, but we'll keep it as is within SAP S/4HANA). The whole approach is flexible in such a way that additional organizations can be added and supported easily. As an example, if we need to add a sales organization—say, 1810—then we can create an additional BUSINESS_VKORG_1810 group and, with the usage of dynamic recipient determination, enable the appropriate distribution of alerts and notifications.

Definition of recipients goes as follows:

1. Select the **Define Recipients** view in the menu on the left-hand side.
2. Click the **New Entries** icon at the top of the screen.
3. There are only two values we need to maintain:
   - **Recipient for Alert: SUPPORT_IT**
   - **Alert Recipient Description: Users from IT support team**

   The end results should be as shown in Figure 8.26.

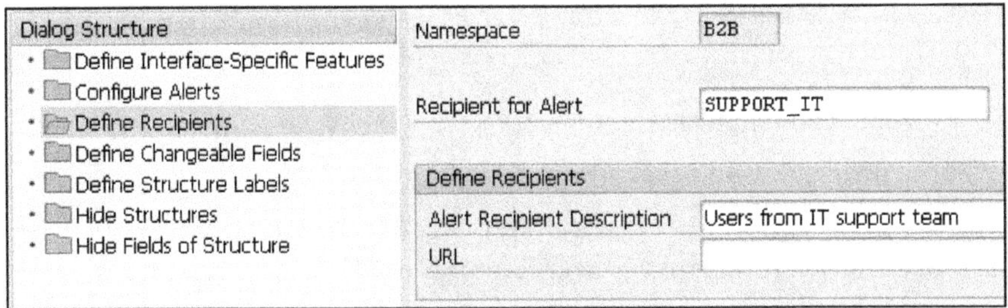

Figure 8.26 Recipients Group for IT Support Users

4. In a similar way, we create the following groups, starting with BUSINESS_VKORG_1710:
   - **Recipient for Alert: BUSINESS_VKORG_1710**
   - **Alert Recipient Description: Business users sales org. 1710**

BUSINESS_BUKRS_1710 should have the following values:

- **Recipient for Alert: BUSINESS_BUKRS_1710**
- **Alert Recipient Description: Business users company code 1710**

BUSINESS_WERKS_1710 should have the following values:

- **Recipient for Alert: BUSINESS_WERKS_1710**
- **Alert Recipient Description: Business users plant 1710**

All four recipient groups are created on the B2B namespace level, and they should look as shown in Figure 8.27. In the following section, we will explain exactly how they are used and how the final solution works.

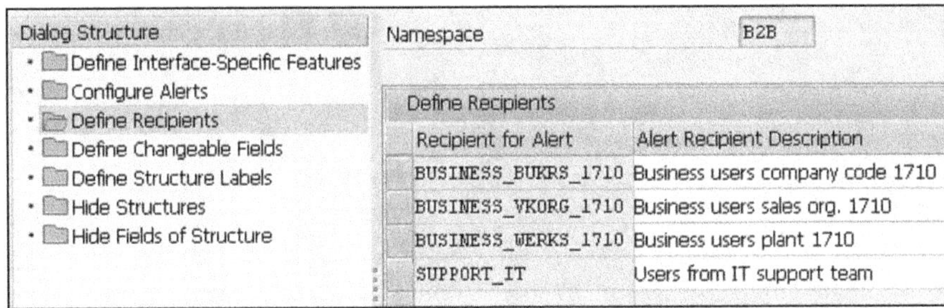

Figure 8.27 Overview with All Recipient Groups Created for EDI Scenario

Configure Alerts and Alert Categories

One of the key features of our monitoring scenario will be alerting, which we partly configured in the previous section and will continue in the next sections.

Before doing this, we make a check based on the alert category. This is the definition of the layout and content of the notification message that is triggered based on a predefined event, like an IDoc error.

To check the existing definition, follow these steps:

1. Start Transaction ALRTCATDEF.
2. Choose **SAP Application Interface Framework** from the left-hand menu.
3. In the right-hand list, find and click **/AIF/ALERT_CAT_DEF**.
4. Navigate to the **Long and Short Text** tab at the bottom of the screen; that's where the definition of the alert message can be seen. This is the default alert category. It shows the format and the content of the notification that users receive from SAP Application Interface Framework. This is the one that we will use in our EDI scenario. Figure 8.28 shows how it looks in our demo system.

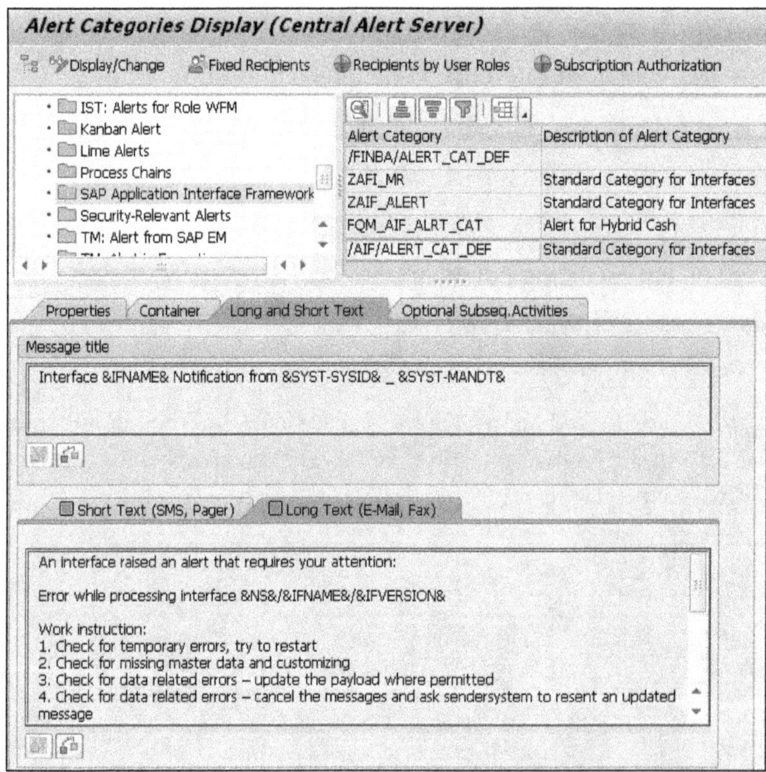

Figure 8.28 Standard Category Alert for SAP Application Interface Framework

We can now go back to the configuration of the alert in the interface-specific features. For every IDoc that ends in an error, warning, or abort message, we will generate an alert and send an email-based notification. Instead of using a static link between an interface and recipient group, we are defining a solution where appropriate recipient groups will be determined based on the message content (organization entity). This is yet another place where we make use of the key fields defined earlier. They already store the organization-related data like sales organization, company code, and plant.

Before we go into the alert configuration, let's summarize per interface which key fields we are going to use for the dynamic recipient determination and what ABAP objects would be involved in this:

- For the sales order and sales order response, we will make use of a sales organization (VKORG) to determine the appropriate recipient groups. We already have this value defined as a key field, we already have recipient group BUSINESS_VKORG_1710, and we already have a table that can link all the variables, table ZAIF_ALRT_VKORG.
- For a delivery, we will make use of a plant (WERKS) to determine the appropriate recipient groups. We already have this value defined as a key field, we already have recipient group BUSINESS_WERKS_1710, and we already have a table that can link all the variables, table ZAIF_ALRT_WERKS.
- For an invoice, we will make use of a company code (BUKRS) to determine the appropriate recipient groups. We already have this value defined as a key field, we already have recipient group BUSINESS_BUKRS_1710, and we already have a table that can link all the variables, table ZAIF_ALRT_BUKRS.

8

To configure alerts for a sales order interface, follow these steps:

1. Select **Configure Alerts** in the menu on the left-hand side.
2. Click the **New Entries** icon at the top of the screen.
3. Enter the required values as shown in Figure 8.29:
   - **Interface Name: ORDERS_IN**
   - **Interface Version: 1**
   - **Rec Assgn Table: ZAIF_ALRT_VKORG**
   - **Alert Category: /AIF/ALERT_CAT_DEF**
   - **Namespace: B2B**

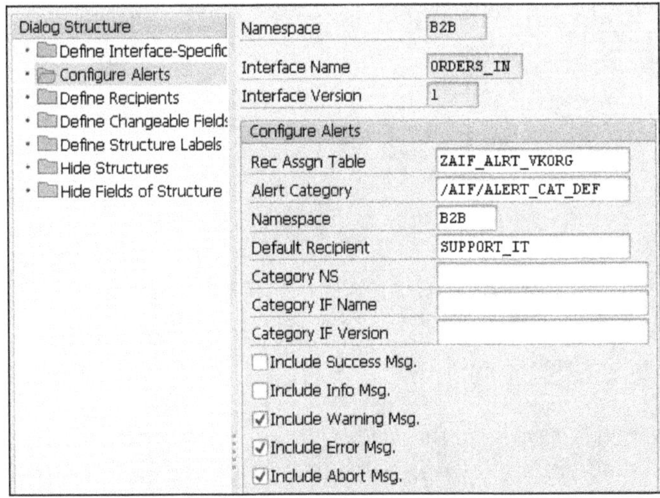

Figure 8.29 Alert Configuration for Sales Order Interface

- **Default Recipient: SUPPORT_IT**
- Check the following options: **Include Warning Msg.**, **Include Error Msg.**, and **Include Abort Msg.**

In a similar way, configure alerts for the other three interfaces. Pay attention to **Rec Assgn Table** as this is the value that changes; all the other parameters stay the same. You can see the end results in Figure 8.30.

Namespace: B2B

Configure Alerts

Interface	Version	Recipient Assignment Table	Alert Category	NS	Default Recipient
DESADV_OUT	1	ZAIF_ALRT_WERKS	/AIF/ALERT_CAT_DEF	B2B	SUPPORT_IT
INVOIC_OUT	1	ZAIF_ALRT_BUKRS	/AIF/ALERT_CAT_DEF	B2B	SUPPORT_IT
ORDERS_IN	1	ZAIF_ALRT_VKORG	/AIF/ALERT_CAT_DEF	B2B	SUPPORT_IT
ORDRSP_OUT	1	ZAIF_ALRT_VKORG	/AIF/ALERT_CAT_DEF	B2B	SUPPORT_IT

Figure 8.30 Overview of Alert Configuration for All Four Interfaces

Define Interface-Specific Features

In this step, we define namespace-specific features for error handling.

Start Transaction /AIF/CUST and navigate to **Error Handling • Namespace-Specific Features**. That's the step in which we make use of the key fields. We will link them to a given interface and define how their values should be determined. Note that when you start Transaction /AIF/CUST and then run **Error Handling • Interface-Specific Features**, you will be asked to define the work area as in Figure 8.31. The reason behind this is that, as the name says, we maintain interface-specific features, so before doing this we need to precisely define for which interface we do so. All the configuration will be performed in the preselected work area.

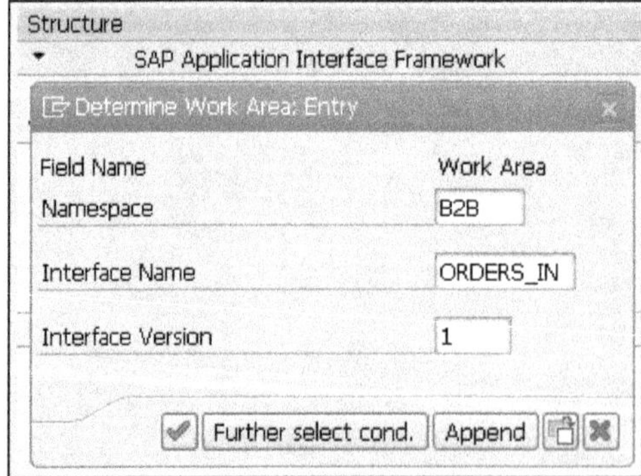

Figure 8.31 Selection Screen for Determination of Work Area

At the beginning of this chapter, we selected the fields from our messages that act as key fields. Then, for each of the interfaces, we created index tables, and each includes the related key fields.

In the "Define Namespace-Specific Features" section earlier in this chapter, we linked each index table to the appropriate interfaces. What is still missing is the assignment of the key fields to the interface fields. In other words, we need to know what data from the content of our message should be taken and used for each key field.

Select the correct view in the menu on the left-hand side for the action you want to execute. You will see a screen like the one shown in Figure 8.32.

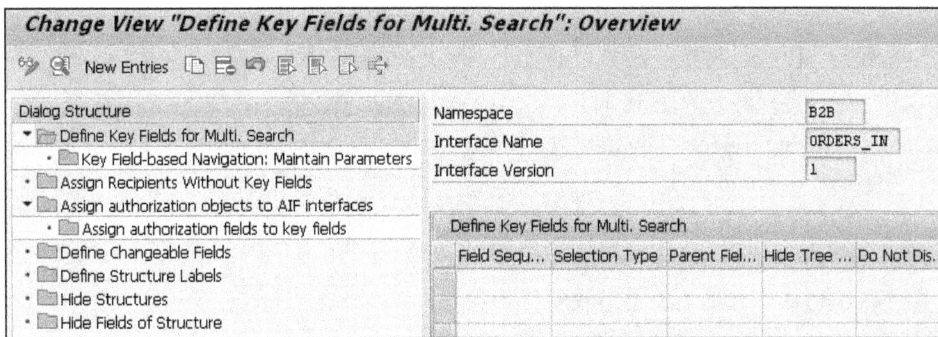

Figure 8.32 Interface-Specific Features

Let's go through all steps required for the sales order interface. As a starting point, let's discuss the table from the beginning of this chapter, with the key fields to configure for the sales order interface. Let's revise the table to include the details of the source of the data (see Table 8.9). As you can see, for the first field we can easily find value of the key field in the IDoc: it is enough to read the SNDPRDN field from the IDoc control records, in the EDIDC segment. The other three key fields are trickier; we cannot read the related values by simply giving the name of the segment and field. To find values for the customer name and sales organization, we need to perform a more complex lookup, including selection of data from the SAP S/4HANA tables. It is part of the standard functionality of SAP Application Interface Framework to use function modules for finding and filling in values of the key fields. Table 8.9 shows references to the sample ABAP code we'll show later in this chapter.

Key Field	Field	Data Element	Source of the Data
Sender	SNDPRN	EDI_SNDPRN	IDoc control record, field SNDPRN.
Customer name	NAME1	NAME1	Customer name from KNA1 table. Lookup performed by function module ZAIF_KEYFIELD_ORDERS_DESCR, as shown later in Listing 8.3.
Sales organization	VKORG	VKORG	Sales organization from table EDSDC. Lookup performed by function module ZAIF_KEYFIELD_ORDERS_SORG, as shown later in Listing 8.2.

Table 8.9 Sales Orders ORDERS.ORDERS05

Key Field	Field	Data Element	Source of the Data
Purchase order number	PO_NUMBER	BSTKD	IDoc segment E1EDK02 segment with qualifier 001, field BELNR. Lookup performed by function module ZAIF_KEYFIELD_ORDERS_PO, as shown later in Listing 8.1.

Table 8.9 Sales Orders ORDERS.ORDERS05 (Cont.)

To maintain the key fields for the sales order interface, follow these steps:

1. Click the New Entries icon at the top of the screen.
2. Start defining the four key fields by filling in the fields as follows:
   - SNDPRD key field
     - **Field Sequence No.: 10**
     - **Key Field Name: SNDPRN**
     - **Data element: EDI_SNDPRN**
     - **Field Name: EDIDC-SNDPRN**
     - **Raw or SAP structure: Source structure (raw for inbound, SAP for outbound)**
     - **Multiple selection type: Single selection**
     - In the **SAP Fiori only Features** section, set the option **Display in Message Monitoring by Default**

   Figure 8.33 shows how this should look.

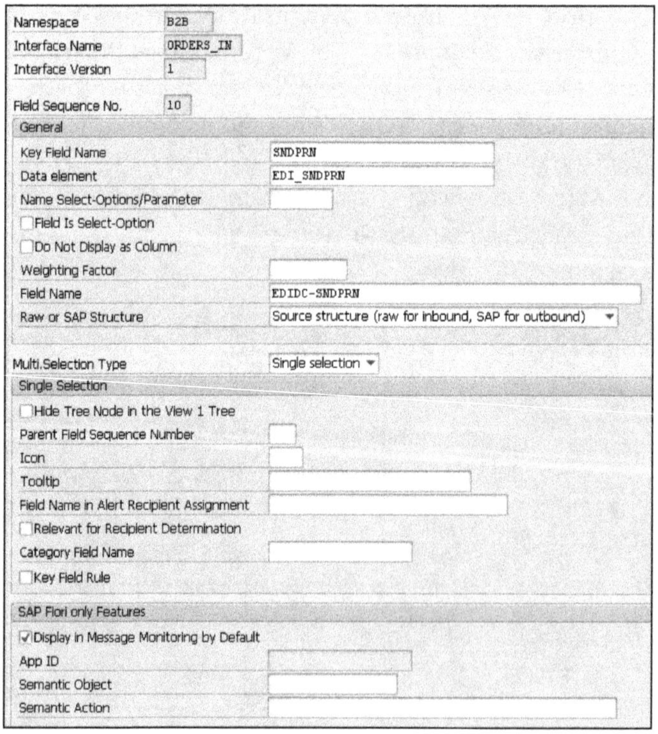

Namespace B2B
Interface Name ORDERS_IN
Interface Version 1
Field Sequence No. 10
General
Key Field Name SNDPRN
Data element EDI_SNDPRN
Name Select-Options/Parameter
Field Is Select-Option
Do Not Display as Column
Weighting Factor
Field Name EDIDC-SNDPRN
Raw or SAP Structure Source structure (raw for inbound, SAP for outbound)
Multi.Selection Type Single selection
Single Selection
Hide Tree Node in the View 1 Tree
Parent Field Sequence Number
Icon
Tooltip
Field Name in Alert Recipient Assignment
Relevant for Recipient Determination
Category Field Name
Key Field Rule
SAP Fiori only Features
Display in Message Monitoring by Default
App ID
Semantic Object
Semantic Action

Figure 8.33 Definition of SNDPRN Key Field for Sales Order Interface

- NAME1 key field
  - **Field Sequence No.: 15**
  - **Key Field Name: NAME1**
  - **Data element: NAME1**
  - **Field Name:** leave empty (we are not pulling this data directly from IDoc field)
  - **Raw or SAP structure: Source structure (raw for inbound, SAP for outbound)**
  - **Multiple selection type: Single selection**
  - Set the **Key Field Rule** option and enter the **Function Module** as **ZAIF_KEYFIELD_ORDERS_DESCR**
  - In the **SAP Fiori only Features** section, set the option **Display in Message Monitoring by Default**

  Figure 8.34 shows how this should look.

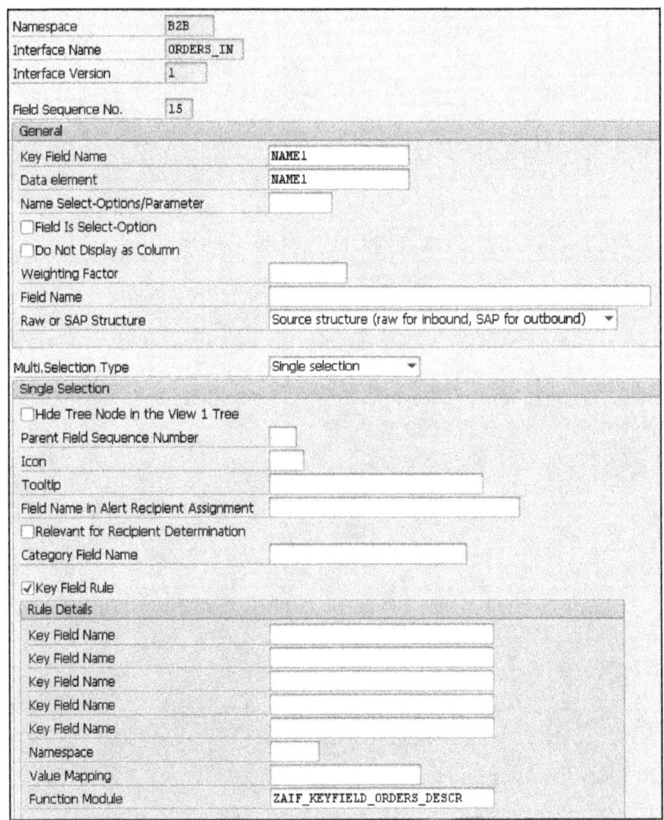

Figure 8.34 Definition of NAME1 Key Field for Sales Order Interface

- VKORG key field
  - **Field Sequence No.: 20**
  - **Key Field Name: VKORG**
  - **Data element: VKORG**

- **Field Name**: Leave empty (we are not pulling this data directly from an IDoc field)
- **Raw or SAP structure: Source structure (raw for inbound, SAP for outbound)**
- **Multiple selection type: Single selection**
- **Field Name in Alert Recipient Determination: VKORG**
- Set the option **Relevant for Recipient Determination**
- Set the option **Key Field Rule** with the **Function Module** set to **ZAIF_KEYFIELD_ORDERS_SORG**
- In the **SAP Fiori only Features** section, set the option **Display in Message Monitoring by Default**

Figure 8.35 shows how this should look.

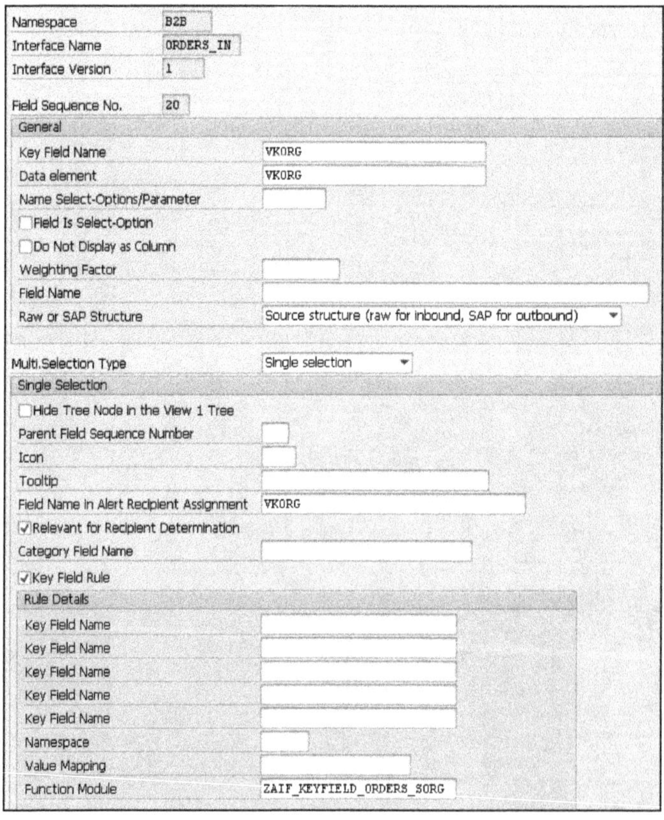

Figure 8.35 Definition of VKORG Key Field for Sales Order Interface

- PO_NUMBER key field
  - **Field Sequence No.: 30**
  - **Key Field Name: PO_NUMBER**
  - **Data element: BSTKD**
  - **Field Name:** Leave empty (we are not pulling this data directly from an IDoc field)

- **Raw or SAP structure: Source structure (raw for inbound, SAP for outbound)**
- **Multiple selection type: Single selection**
- Set the option **Key Field Rule** with **Function Module** set to **ZAIF_KEYFIELD_ORDERS_PO**
- In the **SAP Fiori only Features** section, set the option **Display in Message Monitoring by Default**

Figure 8.36 shows how this should look.

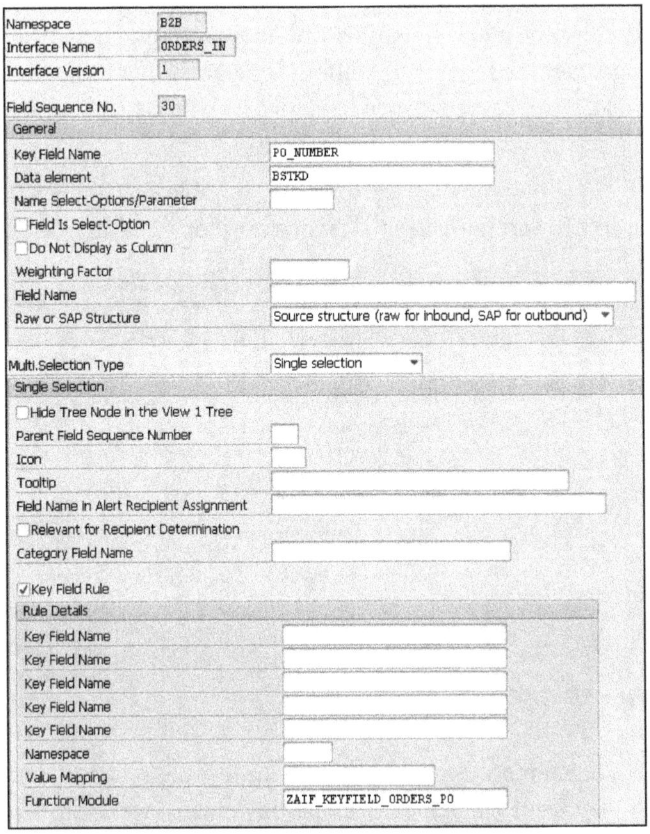

Figure 8.36 Definition of PO_NUMBER Key Field for Sales Order Interface

There are a few aspects we need to clarify about these configurations:

- At the bottom of the screen, there is a **Display in Message Monitoring by Default** selection box. It is set for all key fields.
- The value for the sender key field (SNDPRN) is taken from the IDoc control records (EDIDC-SNDPRN).
- The values for key fields customer name (NAME1), sales organization (VKORG), and purchase order number (PO_NUMBER) cannot be taken directly from the IDoc by a simple reference; we must make use of function modules. Using these, we can have a more

precise lookup on the IDoc segment (purchase order number) or even a lookup on table DB (customer name and sales organization). The ABAP listings can be found on the following pages.

- The sales organization key field (VKORG) will be used for recipient determination and that's something we need to configure. We keep the same field name and set the **Relevant for Recipient Determination** option.

We've finished key field configuration for the sales order interface, but there are three more interfaces to go. Table 8.10, Table 8.11, and Table 8.12 list the key fields and the source of the data. There are still some additional function modules to pull data from segments where we need to consider the segment qualifier. Using a qualifier is not possible within the **Field Name** field, so whenever we need to pull data with a specific qualifier, we need to make use of a key field rule and function module.

Taking the sales order interface as a reference and the details from Table 8.10, Table 8.11, and Table 8.12, you should be able to configure key fields for the other three interfaces.

Key Field	Field	Data Element	Source of the Data
Receiver	RCVPRN	EDI_RCVPRN	IDoc control record, field RCVPRN.
Sales organization	VKORG	VKORG	IDoc segment E1EDK14 with qualifier 008, field ORGID. Lookup performed by function module ZAIF_KEYFIELD_ORDRSP_VKOR, as shown later in Listing 8.4.
Purchase order number	PO_NUMBER	BSTKD	IDoc segment E1EDK02 with qualifier 001, field BELNR. Lookup performed by function module ZAIF_KEYFIELD_ORDERS_PO, as shown later in Listing 8.1.

Table 8.10 Sales Order Response ORDRSP.ORDERS05

Note that for Table 8.11, there is one remark necessary: in our dummy organization and use case, we have a 1:1 relation between the plant and shipping point, so we can safely take the shipping point value as a plant. And we do that as it is a simpler approach—and in our case, nothing significant for the overall EDI scenario and concept.

Key Field	Field	Data Element	Source of the Data
Receiver	RCVPRN	EDI_RCVPRN	IDoc control record, field RCVPRN
Plant	WERKS	EMPFW	IDoc segment E1EDL20 field VSTEL (*shipping point)
Delivery number	VBELN	VBELN_VL	IDoc segment E1EDL20 field VBELN

Table 8.11 Outbound Delivery

Key Field	Field	Data Element	Source of the Data
Receiver	RCVPRN	EDI_RCVPRN	IDoc control record, field RCVPRN.
Company code	COMP_CODE	BUKRS	IDoc segment E1EDK14 with qualifier 003, field ORGID. Lookup performed by function module ZAIF_KEYFIELD_INVOIC_C, as shown later in Listing 8.5.
Invoice number	VBELN	VBELN_VF	IDoc segment E1EDK01 field BELNR.

Table 8.12 Invoice

Function Modules for Key Fields Determination

In Listing 8.1, Listing 8.2, Listing 8.3, Listing 8.4, and Listing 8.5, you can find the ABAP code related to the function modules used in earlier sections.

```
FUNCTION ZAIF_KEYFIELD_ORDERS_PO .
*"----------------------------------------------------------------------
*"*"Local Interface:
*"  IMPORTING
*"     REFERENCE(IS_RAW_STRUCTURE) TYPE  ANY OPTIONAL
*"     REFERENCE(IS_SAP_STRUCTURE) TYPE  ANY OPTIONAL
*"     REFERENCE(IS_KFLDS) TYPE  /AIF/T_INF_KFLDS
*"  CHANGING
*"     REFERENCE(CS_SIDX_ENTRY) TYPE  ANY
*"----------------------------------------------------------------------

constants: segment_name    type edilsegtyp value 'E1EDK02',
           qualifier_value    type edi_qualfo     value '001'.

  data: source_e1edk02 type e1edk02.

  field-symbols: <fs_ifname>        type /aif/ifname,
                 <fs_idoc_data>     type edidd_tt,
                 <fs_idoc_segment> type edidd,
                 <fs_kfld>          type any.

  assign component 'IFNAME' of structure cs_sidx_entry to <fs_ifname>.
  if <fs_ifname> is not assigned.
    return.
  endif.

  assign component 'IDOC_DATA' of structure is_raw_structure to <fs_idoc_data>.
  if <fs_idoc_data> is not assigned.
```

```
    return.
  endif.

  loop at <fs_idoc_data> assigning <fs_idoc_segment> where segnam = segment_
name.
    move <fs_idoc_segment>-sdata to source_e1edk02.
    if source_e1edk02-qualf = qualifier_value.
      assign component 'PO_NUMBER' of structure cs_sidx_entry to <fs_kfld> .
      <fs_kfld>  = source_e1edk02-belnr.
      condense <fs_kfld>.
    endif.
  endloop.

ENDFUNCTION.
```

Listing 8.1 ABAP Function Module ZAIF_KEYFIELD_ORDERS_PO

```
FUNCTION ZAIF_KEYFIELD_ORDERS_SORG .
*"----------------------------------------------------------------------
*"*"Local Interface:
*"  IMPORTING
*"     REFERENCE(IS_RAW_STRUCTURE) TYPE  ANY OPTIONAL
*"     REFERENCE(IS_SAP_STRUCTURE) TYPE  ANY OPTIONAL
*"     REFERENCE(IS_KFLDS) TYPE  /AIF/T_INF_KFLDS
*"  CHANGING
*"     REFERENCE(CS_SIDX_ENTRY) TYPE  ANY
*"----------------------------------------------------------------------
TABLES: EDSDC.              "ermitteln VKORG, VTWEG, SPART

constants: segment_name    type edilsegtyp value 'E1EDKA1',
           parvw_value    type edi_qualfo     value 'LF'.

  data: source_e1edka1 type e1edka1.

  field-symbols: <fs_ifname>        type /aif/ifname,
                 <fs_idoc_data>     type edidd_tt,
                 <fs_idoc_segment> type edidd,
                 <fs_kfld>          type any.

  assign component 'IFNAME' of structure cs_sidx_entry to <fs_ifname>.
  if <fs_ifname> is not assigned.
    return.
  endif.

  assign component 'IDOC_DATA' of structure is_raw_structure to <fs_idoc_data>.
```

```
  if <fs_idoc_data> is not assigned.
    return.
  endif.

  loop at <fs_idoc_data> assigning <fs_idoc_segment> where segnam = segment_
name.
    move <fs_idoc_segment>-sdata to source_e1edka1.
    if source_e1edka1-parvw = parvw_value.
      assign component 'VKORG' of structure cs_sidx_entry to <fs_kfld> .

   SELECT SINGLE vkorg INTO edsdc-vkorg FROM edsdc WHERE lifnr = source_e1edka1-
partn.
    IF sy-subrc = 0.
      <fs_kfld>  = edsdc-vkorg.
      condense <fs_kfld>.
      endif.
    endif.
  endloop.

ENDFUNCTION.
```

Listing 8.2 ABAP Function Module ZAIF_KEYFIELD_ORDERS_SORG

```
FUNCTION ZAIF_KEYFIELD_ORDERS_DESCR .
*"----------------------------------------------------------------------
*"*"Local Interface:
*"  IMPORTING
*"     REFERENCE(IS_RAW_STRUCTURE) TYPE  ANY OPTIONAL
*"     REFERENCE(IS_SAP_STRUCTURE) TYPE  ANY OPTIONAL
*"     REFERENCE(IS_KFLDS) TYPE  /AIF/T_INF_KFLDS
*"  CHANGING
*"     REFERENCE(CS_SIDX_ENTRY) TYPE  ANY
*"----------------------------------------------------------------------
TABLES: EDPAR.
TABLES: KNA1.

constants: segment_name    type edilsegtyp value 'E1EDKA1',
           parvw_value    type edi_qualfo     value 'AG'.

  data: source_e1edka1 type e1edka1.

  field-symbols: <fs_ifname>        type /aif/ifname,
                 <fs_idoc_data>     type edidd_tt,
                 <fs_idoc_segment> type edidd,
                 <fs_kfld>          type any.
```

```
  assign component 'IFNAME' of structure cs_sidx_entry to <fs_ifname>.
  if <fs_ifname> is not assigned.
    return.
  endif.

  assign component 'IDOC_DATA' of structure is_raw_structure to <fs_idoc_data>.
  if <fs_idoc_data> is not assigned.
    return.
  endif.

  loop at <fs_idoc_data> assigning <fs_idoc_segment> where segnam = segment_
name.
    move <fs_idoc_segment>-sdata to source_e1edka1.
    if source_e1edka1-parvw = parvw_value.
      assign component 'NAME1' of structure cs_sidx_entry to <fs_kfld> .

   SELECT SINGLE inpnr INTO edpar-inpnr FROM edpar WHERE expnr = source_e1edka1-
lifnr AND parvw = 'AG'.
    IF sy-subrc = 0.

         SELECT SINGLE name1 INTO kna1-name1 FROM kna1 WHERE kunnr = edpar-
inpnr.

      <fs_kfld>  = kna1-name1.
      condense <fs_kfld>.
      endif.
    endif.
  endloop.

ENDFUNCTION.
```

Listing 8.3 ABAP Function Module ZAIF_KEYFIELD_ORDERS_DESCR

```
FUNCTION ZAIF_KEYFIELD_ORDRSP_VKORG .
*"----------------------------------------------------------------------
*"*"Local Interface:
*"  IMPORTING
*"     REFERENCE(IS_RAW_STRUCTURE) TYPE  ANY OPTIONAL
*"     REFERENCE(IS_SAP_STRUCTURE) TYPE  ANY OPTIONAL
*"     REFERENCE(IS_KFLDS) TYPE  /AIF/T_INF_KFLDS
*"  CHANGING
*"     REFERENCE(CS_SIDX_ENTRY) TYPE  ANY
*"----------------------------------------------------------------------
```

```
constants: segment_name    type edilsegtyp value 'E1EDK14',
           qualifier_value    type edi_qualfo     value '008'.

  data: source_e1edk14 type e1edk14.

  field-symbols: <fs_ifname>        type /aif/ifname,
                 <fs_idoc_data>     type edidd_tt,
                 <fs_idoc_segment> type edidd,
                 <fs_kfld>          type any.

  assign component 'IFNAME' of structure cs_sidx_entry to <fs_ifname>.
  if <fs_ifname> is not assigned.
    return.
  endif.

  assign component 'IDOC_DATA' of structure is_raw_structure to <fs_idoc_data>.
  if <fs_idoc_data> is not assigned.
    return.
  endif.

  loop at <fs_idoc_data> assigning <fs_idoc_segment> where segnam = segment_
name.
    move <fs_idoc_segment>-sdata to source_e1edk14.
    if source_e1edk14-qualf = qualifier_value.
      assign component 'VKORG' of structure cs_sidx_entry to <fs_kfld> .
      <fs_kfld>  = source_e1edk14-ORGID.
      condense <fs_kfld>.
    endif.
  endloop.

ENDFUNCTION.
```

Listing 8.4 ABAP Function Module ZAIF_KEYFIELD_ORDRSP_VKORG

```
FUNCTION ZAIF_KEYFIELD_INVOIC_CC .
*"----------------------------------------------------------------------
*"*"Local Interface:
*"  IMPORTING
*"     REFERENCE(IS_RAW_STRUCTURE) TYPE  ANY OPTIONAL
*"     REFERENCE(IS_SAP_STRUCTURE) TYPE  ANY OPTIONAL
*"     REFERENCE(IS_KFLDS) TYPE  /AIF/T_INF_KFLDS
*"  CHANGING
*"     REFERENCE(CS_SIDX_ENTRY) TYPE  ANY
*"----------------------------------------------------------------------
```

```
constants: segment_name    type edilsegtyp value 'E1EDK14',
           qualifier_value    type edi_qualfo     value '003'.

  data: source_e1edk14 type e1edk14.

  field-symbols: <fs_ifname>        type /aif/ifname,
                 <fs_idoc_data>     type edidd_tt,
                 <fs_idoc_segment> type edidd,
                 <fs_kfld>          type any.

  assign component 'IFNAME' of structure cs_sidx_entry to <fs_ifname>.
  if <fs_ifname> is not assigned.
    return.
  endif.

  assign component 'IDOC_DATA' of structure is_raw_structure to <fs_idoc_data>.
  if <fs_idoc_data> is not assigned.
    return.
  endif.

  loop at <fs_idoc_data> assigning <fs_idoc_segment> where segnam = segment_
name.
    move <fs_idoc_segment>-sdata to source_e1edk14.
    if source_e1edk14-qualf = qualifier_value.
      assign component 'COMP_CODE' of structure cs_sidx_entry to <fs_kfld> .
      <fs_kfld>  = source_e1edk14-ORGID.
      condense <fs_kfld>.
    endif.
  endloop.

ENDFUNCTION.
```

Listing 8.5 ABAP Function Module ZAIF_KEYFIELD_INVOIC_CC

Define Changeable Fields

One additional feature that we enable for more efficient error handling is a possibility to change a specific field. By default, you can't change any data in the IDoc visible in the SAP Application Interface Framework monitoring tool. With the configuration as shown in Figure 8.37, we allow that the IDoc field that contains the material number can be changed. This is applicable for the sales order interface, and the intention is to support the common case in which we receive an EDI order for materials that are not known or not active within the given sales organization. This configuration piece enables order desk employees to modify the material number in an incoming IDoc message and continue with the order processing. We will demonstrate how that works in practice in the following section.

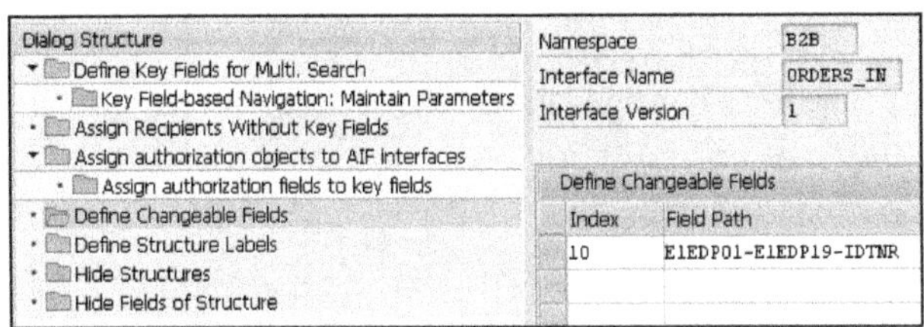

Figure 8.37 Define Changeable Fields

To define changeable fields, follow these steps:

1. Navigate to **Error Handling • Define Interface-Specific Features** and choose the **ORDERS_IN** interface from the B2B namespace.
2. Then choose **Define Changeable Field** on the left-hand side. Click **New Entries**.
3. Add "10" in the **Index** field.
4. In **Field Path**, maintain the exact field you want to enable for change via SAP Application Interface Framework monitoring. One option is to type the full path to the IDoc field, but our recommendation is to click at the end of the line and open the view of the raw structure, from which we can choose the exact field. The latter approach is less error-prone.

The structure would look like the one shown in Figure 8.38.

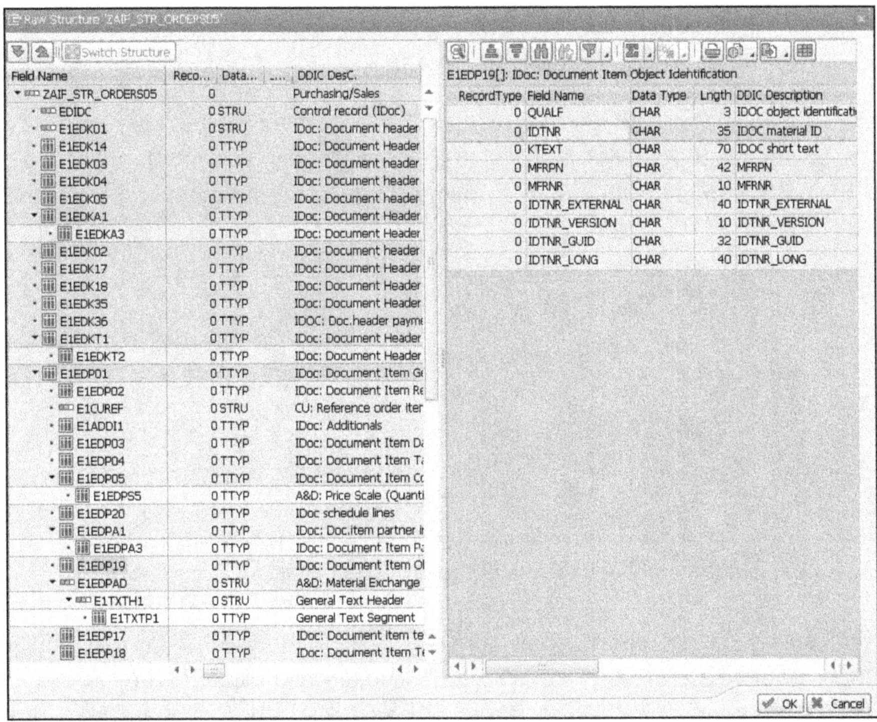

Figure 8.38 Raw Structure Used for Selection of Changeable Fields

8.2.4 System Configuration

In this section, we will perform some final configuration related to recipient groups, users within these groups, and interface determination.

Content-Based Recipient Determination

Here we assign users to the appropriate recipient groups. Whenever there is an alert trigger out of SAP Application Interface Framework monitoring, it is forwarded to one or multiple recipient groups. We can have an almost unlimited number of users assigned to each recipient group. Each user has its own set of parameters.

What we implemented earlier is dynamic recipient determination. The exact group is determined based on the message content and precisely based on the value of the key fields. The concept is that we have a global enterprise with multiple company codes, sales organizations, and plants. Whenever there is a message dedicated to one of these organizational structures, we notify only the related entity. We also defined the SUPPORT_IT group, which receives all alerts.

In our use case, we only have a single organization, but we define a future-proof solution that can support the concept. This also allows us to explain the technical details behind the concept.

We maintain the following recipient assignment tables:

- ZAIF_ALRT_BUKRS, used for invoice interface
- ZAIF_ALRT_VKORG, used for sales order and sales order response interface
- ZAIF_ALRT_WERKS, used for outbound delivery interface

Figure 8.39 shows the content of table ZAIF_ALRT_BUKRS, where we link company code 1710 to recipient group BUSINESS_BUKRS_1710. The second recipient group, SUPPORT_IT, has no specific company code assigned; this will work as if all company codes are assigned.

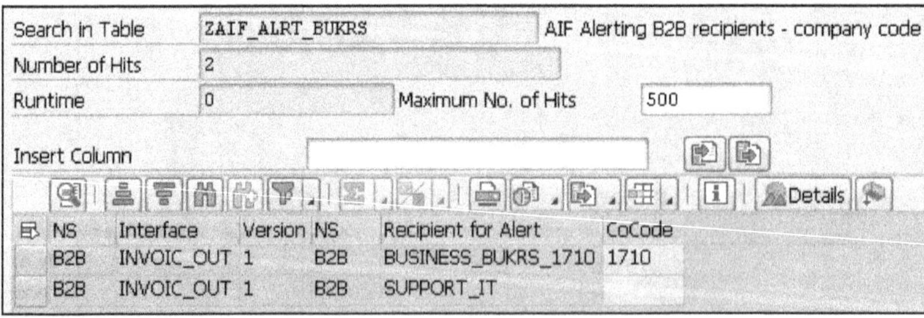

NS	Interface	Version	NS	Recipient for Alert	CoCode
B2B	INVOIC_OUT	1	B2B	BUSINESS_BUKRS_1710	1710
B2B	INVOIC_OUT	1	B2B	SUPPORT_IT	

Figure 8.39 Recipient Assignment Table for Company Code Value

Figure 8.40 shows the content of table ZAIF_ALRT_VKORG, where we link sales organization 1710 to recipient group BUSINESS_VKORG_1710. The second recipient group, SUPPORT_IT, has no specific sales organization assigned; this will work as if all sales organizations are assigned.

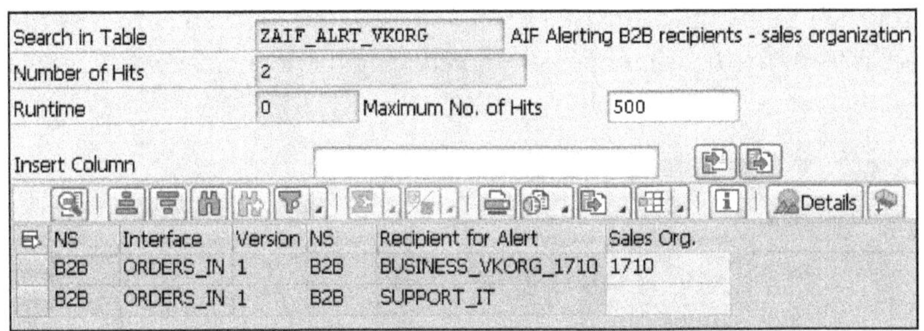

Figure 8.40 Recipient Assignment Table for Sales Organization Value

Figure 8.41 shows content of table ZAIF_ALRT_WERKS, where we link plant 1710 to recipient group BUSINESS_WERKS_1710. The second recipient group, SUPPORT_IT, has no specific plant assigned; this will work as if all plants are assigned.

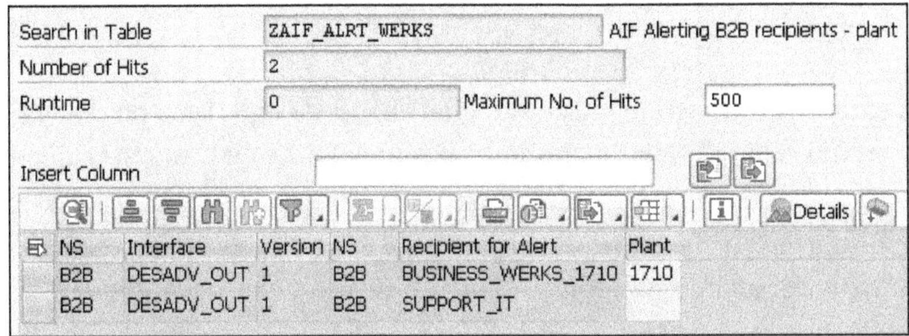

Figure 8.41 Recipient Assignment Table for Plant Value

Assigning Recipients

While we already configured the whole template for sending alerts based on the message content, including creation of the different recipient groups, we still do not have any end users added to the groups. You can add users to recipient groups by following menu path **System Configuration • Recipients • Assign Recipients**. After that, follow these steps:

1. On the initial selection screen, like that shown in Figure 8.42, enter "B2B" in **Namespace**.

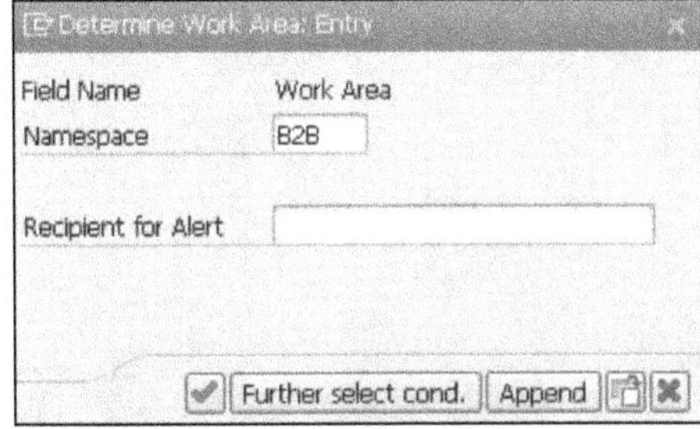

Figure 8.42 Initial Selection Screen for Recipient Assignment

2. Enter an appropriate recipient group in the **Recipient for Alert** field. You can type the name or select it from the list that opens upon clicking the icon at the end of this field. The selection screen is shown in Figure 8.43.

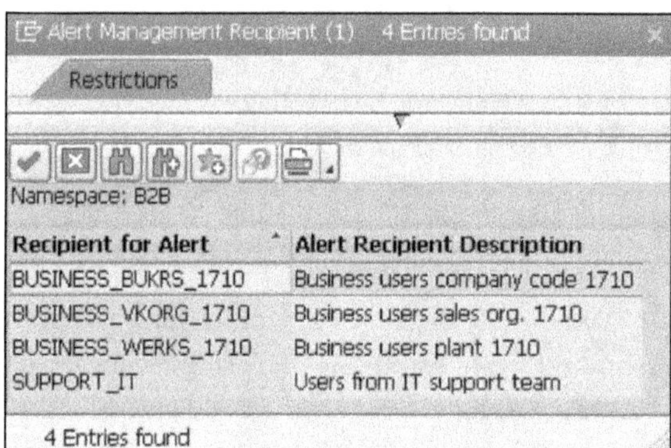

Figure 8.43 Selection List for Recipient Groups

3. As soon you enter the transaction, make sure you choose **Assign Users** on the left-hand side. Then you can click **New Entries** at the bottom of the screen to start adding users.

4. You will be adding users to the SUPPORT_IT group in the B2B namespace. Maintain the fields as follows, as shown in Figure 8.44:

   - **User Number: 10**
     Of course, this is just an example; you can maintain whatever number you need. The whole point of this is to keep track of all the future entries. It is fully up to you what kind of approach you use.
   - **User Name: PIASECZNYM**
     This is the user ID as created in your SAP S/4HANA system. This example contains a user ID from the demo system.
   - **Message Type: Application Error or Technical Error**
     This selection allows the user to receive alerts on any message for an application or technical error. Note that it is not the only option. There are many more, which are self-explanatory (see Figure 8.45).
   - **Include on Overview**
     Needs to be set. This allows the user to see the messages in the Interface Monitoring transaction.
   - **Tech. User**
     Needs to be set. This allows the user to see the messages in *technical error* and *in process* statuses.

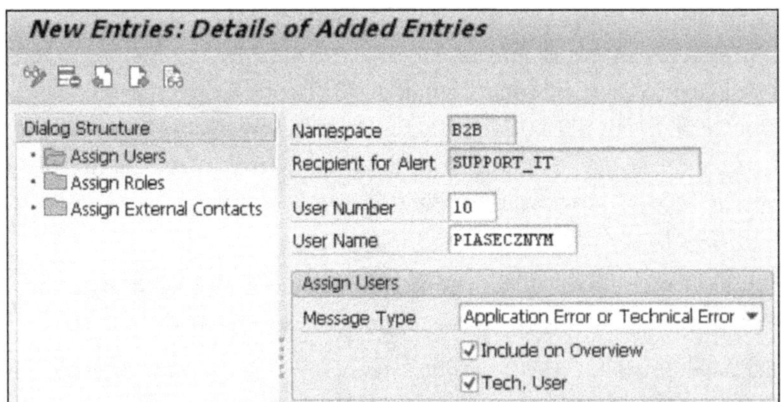

Figure 8.44 User Assignment to SUPPORT_IT Recipient Group

The **Message Type** selection list determines for which message status this specific user should be notified by the alert mechanism. Figure 8.45 shows all the available options.

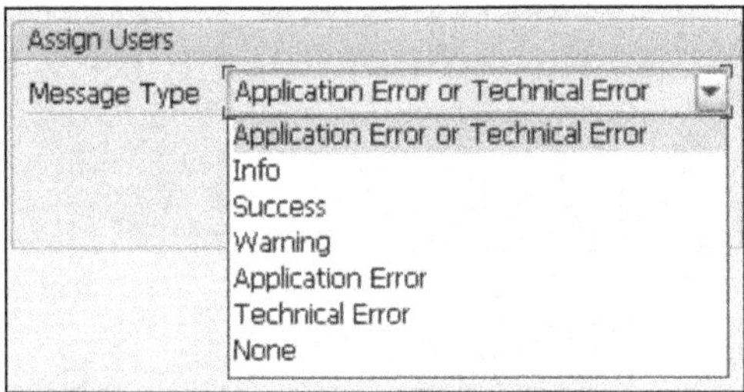

Figure 8.45 Selection of Message Type to Be Included in Alert

Interface Determination

In Chapter 7, we explained the decision to make use of the message variant (MESCOD) field from the IDoc control record to group and trace all B2B EDI-related messages. Here we will configure SAP Application Interface Framework interface determination, which is used to specify which SAP Application Interface Framework interface should be used to process a message. In our EDI solution, we apply a common rule for all four message types. Each IDoc related to our B2B EDI processing will have value B2B in the MESCOD field; we combine this with the basic type and message type and use these three elements to configure the interface determination.

Note that our rules are pretty simple. In our demo system, we don't have too many active IDoc-based interfaces. When you do the setup in a more complex and heavily used system, you will probably need more specific conditions to avoid conflicts among the different flows. If you look at the transaction itself, you'll see that it offers many more options than the ones used in this example.

Let's take the sales order interface as the first case. We already have preconfigured the SAP Application Interface Framework interface named ORDERS_IN. Now, with the configuration steps listed below, we will indicate that all IDocs of a basic type ORDERS05 and message type ORDERS should be identified as belongs to this SAP Application Interface Framework interface. As this is still quite a generic rule, we will additionally make use of the MESCOD field. Having the MESCOD field with value B2B will be our third condition for assigning IDoc to the ORDERS_IN interface.

To configure this, follow menu path **System Configuration • Interface Determination • Define Interface Determination for IDoc Interfaces** and perform these steps:

1. Before you click **New Entries** at the top of the screen, make sure you choose **Define Determination Key** in the left-hand list.
2. Now you can define the rules for interface determination. Note that these are on the IDoc level. The values required for our case are as follows, as shown in Figure 8.46:
   - **Basic Type: ORDERS05**
   - **Message Type: ORDERS**
   - **Field Category:** Choose **IDoc Control Record**
   - **Field Name: MESCOD**

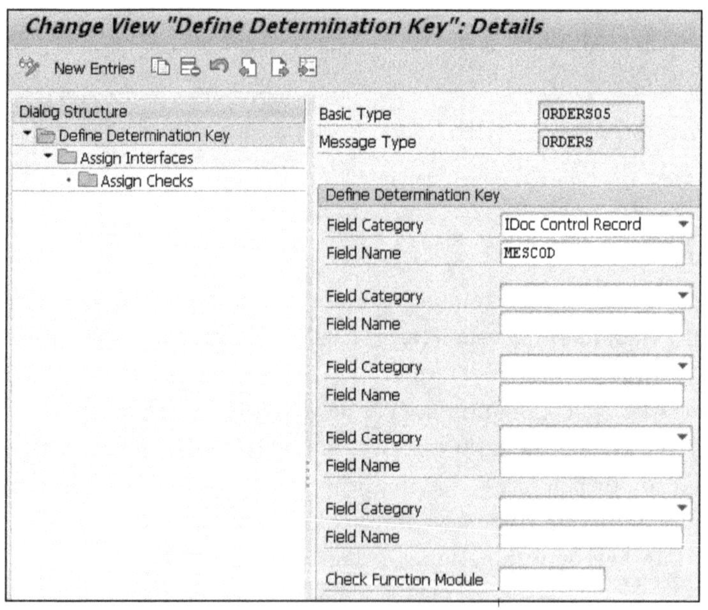

Figure 8.46 Defining Determination Key for Sales Order IDoc

3. We already have the rules we need to maintain the conditions for our case. When you are still on the **Define Determination Key** screen for ORDERS, choose **Assign Interfaces** from the left hand-side list and enter the values as below:

- **Value Number: 10**
- **Operator: Equal**
- **Value for Interface Determination: B2B**
- **Namespace: B2B**
- **Interface Name: ORDERS_IN**
- **Interface Version: 1**

The end results are shown in Figure 8.47.

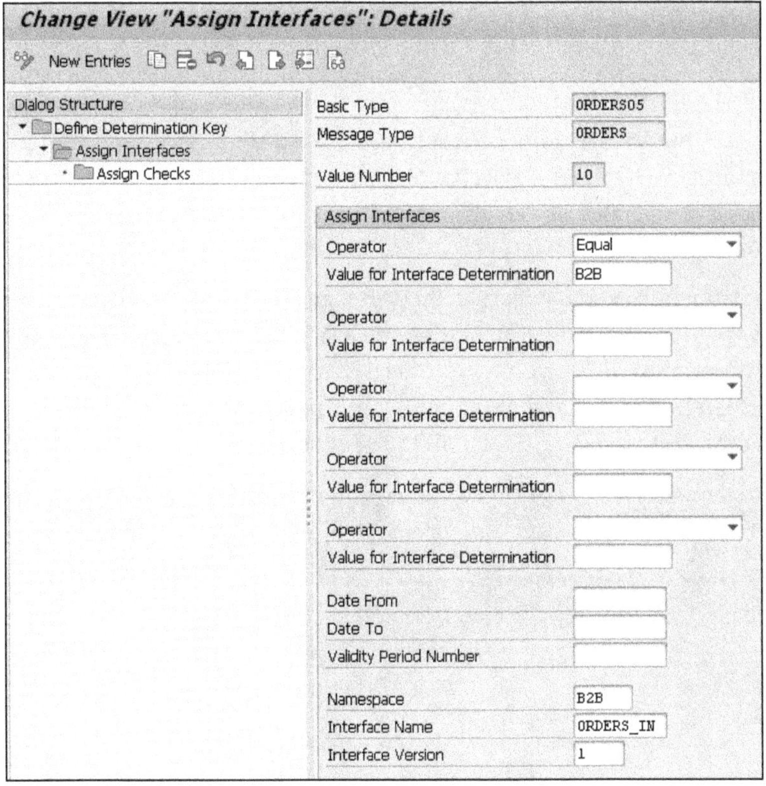

Figure 8.47 Assigning ORDERS_IN Interface for ORDERS.ORDERS05 IDoc and B2B Message Code

In the same way, we maintain the conditions for sales order response. In the first step we define interface determination key for IDoc basic type ORDERS05 and message type ORDRSP, which is the `MESCOD` field from the IDoc control record. In the second step, we add an interface determination entry on this key to indicate that when the `MESCOD` field is equal to `B2B`, the IDoc ORDRSP.ORDERS05 should be assigned to SAP Application Interface Framework interface ORDRSP_OUT in version 1.

Next is the outbound delivery interface. In the first step, we define the interface determination key for IDoc basic type DELVRY07 and message type DESADV. This is also the

MESCOD field from the IDoc control record. In the second step, we add an interface determination entry on this key, to indicate that when the MESCOD field is equal to B2B, the IDoc DESADV.DELVRY07 should be assigned to the SAP Application Interface Framework interface DESADV_OUT in version 1.

Finally, we have the invoice interface. Again, in the first step we define interface determination key for the IDoc basic type INVOIC02 and message type INVOIC which is again the MESCOD field from the IDoc control record. In the second step, we add an interface determination entry on this key, to indicate that when MESCOD field is equal to B2B, the IDoc INVOIC.INVOIC02 should be assigned to the SAP Application Interface Framework interface INVOIC_OUT in version 1.

SAP Fiori Message Monitoring

It wasn't an easy task, but we are done with setting up the SAP Application Interface Framework monitoring. We still have to wait a bit to see the results as we did not process any actual messages yet. But we can give you a little hint of what we have created and what awaits us in the future.

Start the SAP Fiori launchpad and navigate to the **Message Monitoring** space. There we can find the **Message Monitoring** tile (the first on the left in Figure 8.48). Click it to see the overview of interfaces to which we are assigned (see the recipient group configuration). In this transparent and flexible dashboard, you can see all four interfaces and their different statuses; messages will appear in the next chapters.

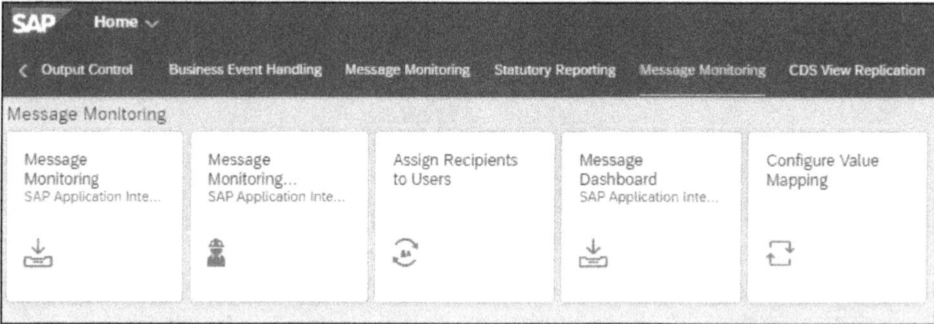

Figure 8.48 Message Monitoring Space

After starting Message Monitoring Overview, you can see a dashboard that gives a transparent and real-time view of the message processing status. See Figure 8.49 for a sample; in our example, we have no messages yet.

From this dashboard, it is possible to navigate to the details of specific message types or statuses. A sample view is shown in Figure 8.50, but again, without any data yet.

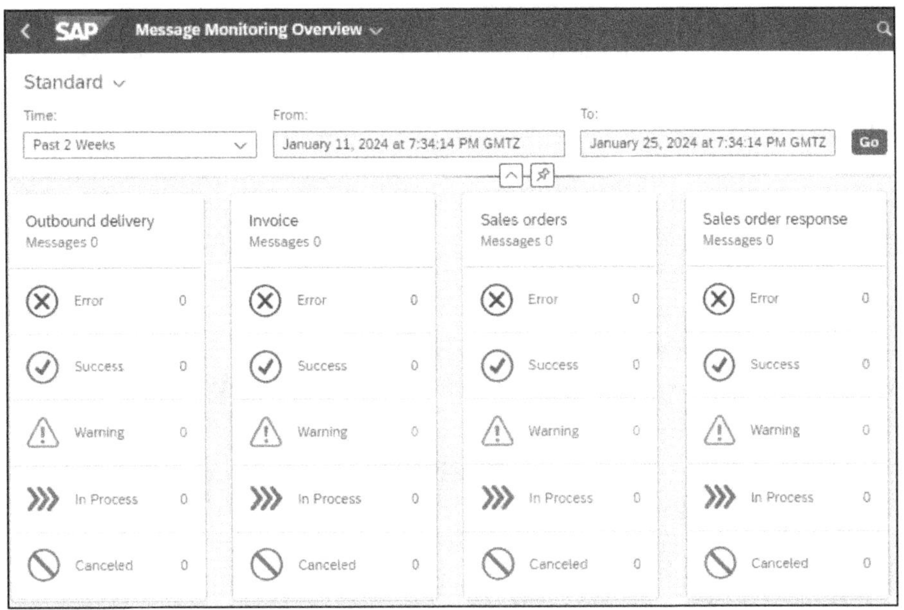

Figure 8.49 Message Monitoring Overview

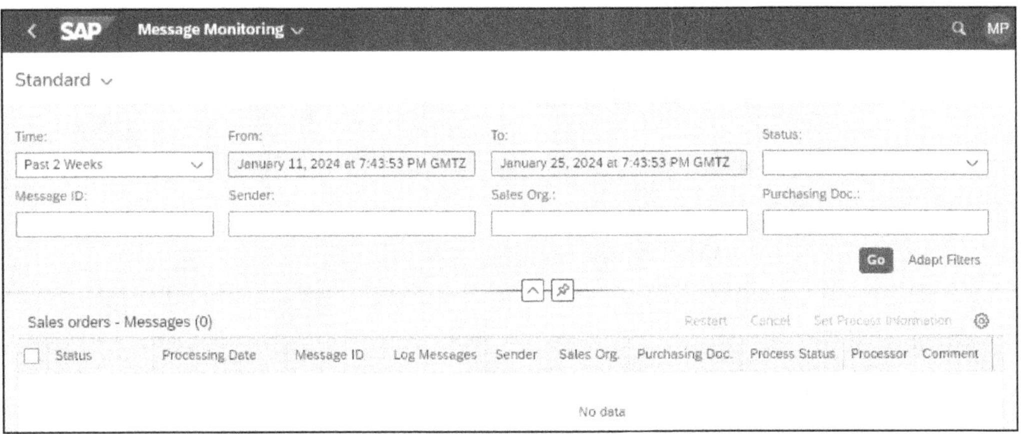

Figure 8.50 Message Monitoring Message List

8.3 Summary

In this chapter, we have done a lot of customizing and development related to enablement of IDoc processing and monitoring via SAP Application Interface Framework. We started by determining the key and most useful information within each of the interface, in the next steps we defined and used these fields and data via the SAP Application Interface Framework key fields. There are multiple use cases behind these key fields, and you will see some examples when we test our EDI solution. For each of the interfaces, we created a dedicated index table, each of which stores key information related to the processed message, along with all kinds of technical details and the values of our preselected key fields. During the whole process, we concentrated on our four interfaces, but you can easily make use of the given instructions for activating any additional flow on top of the given four. There are also two features that would have a significant, positive impact on the overall EDI solution in the long term. The first is the content-based recipient determination: although we have made the configuration for a single organization, the whole setup can easily support a complex SAP S/4HANA system with several or even hundreds of organizations. Literally, only entries are missing in the tables to enable new recipient groups. Now that you understand the concept in detail, you can also adjust it per your needs; you are not limited to company code, sales organization, and plant.

The SAP Fiori user experience is a game-changer, and you can find dedicated SAP Fiori applications for SAP Application Interface Framework message monitoring. These should increase the efficiency and transparency of monitoring and error handling activities. In the next chapter, we'll take a closer look at SAP Integration Suite and its Trading Partner Management component and describe how it facilitates classic EDI and B2B transactions.

Chapter 9
SAP Integration Suite

This chapter explains how classic EDI and B2B transactions can be realized on SAP Integration Suite and its latest Trading Partner Management component. We will elaborate on each configuration and development step required to position SAP Integration Service in the center of our SAP S/4HANA landscape.

In this chapter, we will move from SAP S/4HANA to SAP Business Technology Platform and its core integration service, SAP Integration Suite. In Chapter 6, we already performed activities required for the initial setup and we connected these two systems. Now it is time to work on the EDI scenario itself. This will be the longest chapter of this book, not necessarily because of the complexity of SAP Integration Suite but rather due to its comprehensiveness. We will make use of a few SAP Integration Suite components that cooperate with each other bring into being the EDI scenarios themselves but also deliver a set of tools to maintain it. A short introduction to SAP Integration Suite and its EDI capabilities was already made in Chapter 4, so we'll start here with a closer look at the concept of B2B EDI on this SAP BTP service. After that, we will continue with the necessary activities on each of the components.

9.1 B2B EDI on SAP Integration Suite

In Chapter 4, we provided a short theoretical introduction to the role of SAP Integration Suite and its components. Let's go step by step through all the elements that we will look at in this chapter to get a clear picture of the overall concept and function of each element:

- **Integration Advisor**
  We will define message implementation guidelines (MIGs) for each of our own interfaces and for each interface of our trading partner. A MIG is a definition of the message, its content, its elements, the requirements of the specific party, and other conditions. We start with the selection of the *type system*, which is in fact a B2B standard like ASC X12, UN/EDIFACT, SOAP, or IDoc. After that, we define the segments and data elements that should appear in the message and with which attributes and conditions (mandatory, optional, number of occurrences, qualifiers, and other dependencies). We define MIGs in a graphical tool that offers machine learning capa-

bilities and automatic generation of documentation and runtime artifacts. For our EDI scenario, we need to create eight MIGs; these will be the four IDocs as seen from SAP S/4HANA view and then four UN/EDIFACT messages looking from the trading partner perspective. Once MIGs are created, they can be reused across multiple different scenarios.

With the MIGs, the definitions of unique messages, we can go ahead with the second step, which is the creation of mapping guidelines (MAGs). These are simply links and mappings between two different MIGs. As an example, when we have a MIG definition for UN/EDIFACT ORDERS from our trading partner and then a MIG definition for IDoc ORDERS.ORDERS05, we need to define a MAG to specify how the UN/EDIFACT message transforms into an IDoc message. Integration Advisor supports us in MAG creation by once again offering a graphical tool with machine learning capabilities. Like with the MIGs definition, there are also options to describe and document everything within the tool itself. There is no need for parallel maintenance of any additional documentation outside the system. As you probably noticed, the MIGs are the input for the MAGs, so having good MIGs definition is key for reliable and transparent MAGs, while their definition does not require high effort.

- **Trading Partner Management**
  This is the actual core where not only the details of involved parties but also the B2B scenarios with all their transactions are defined. You will see how to configure your own company profile and trading partner profile. Then we will define an agreement template to set the framework for B2B within our company. In the last step, the framework will be copied to an agreement that incorporates the trading partner details and defines the final, functioning B2B scenario.
- **Cloud Integration**
  This will be the runtime engine for the B2B scenario. Apart from the iFlows that we already deployed during the initial setup, described in Chapter 6, Section 6.2, there is no need for additional configuration. As you will learn later, there is an automated mechanism to synchronize the configuration from Integration Advisor and Trading Partner Management via Cloud Integration.

9.2 Integration Advisor

This will be quite a long section. As you may recall from Chapter 4 and Figure 4.8, we will be creating four MIGs for IDoc messages and four more for UN/EDIFACT messages. In the next step, these eight MIGs will be used to define four MAGs. The technical steps are not very complex as the MIG and MAG objects can be created with almost no coding, but the overall effort is still quite high as we also need to decide what information to use and map in each EDI transaction. The following sections detail the entire process down to the smallest detail.

9.2.1 Defining Message Implementation Guidelines for IDoc Messages

First let's look at the EDI scenario from the SAP S/4HANA point of view. There are four IDoc messages involved. We will define the structure and content of these IDoc messages so that whenever there is any IDoc-based communication with our SAP S/4HANA it will be done with an IDoc structure defined as a MIG.

Sales Order IDoc

To define a MIG for the sales order IDoc, follow these steps:

1. Start SAP Integration Suite and navigate to the MIGs located in the **Design** section on the left-hand menu. If this is the first MIG you're creating, you will see an empty screen as shown in Figure 9.1. 9

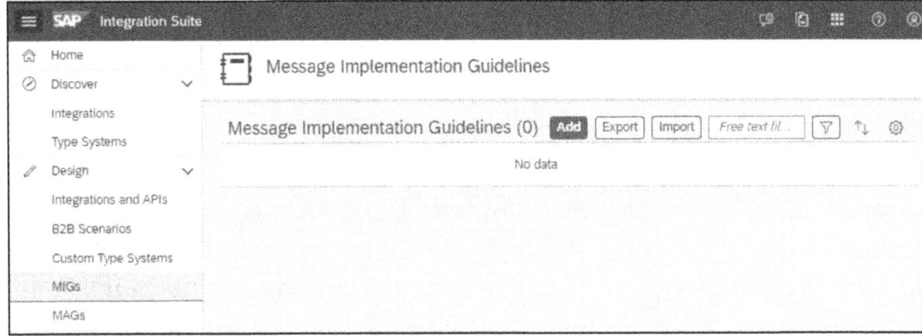

Figure 9.1 Initial Screen for Message Implementation Guidelines

2. Clicking the **Add** button at the top of the screen will trigger a wizard to guide you through all the necessary steps for defining the key attributes of the MIG. After this is done, you can continue with the structure definition within the graphical editor. First follow the steps given by the wizard:
   - The first step is type systems selection. This is nothing more than selection of the B2B standard for the message. You will see options like **UN/EDIFACT**, **cXML**, **Odette**, **GS1 XML**, **Automotive CL: VDA**, and many others. For this example, select **SAP S/4HANA On Premise IDoc**. You can make use of the search option located in the top-right corner, as shown in Figure 9.2.

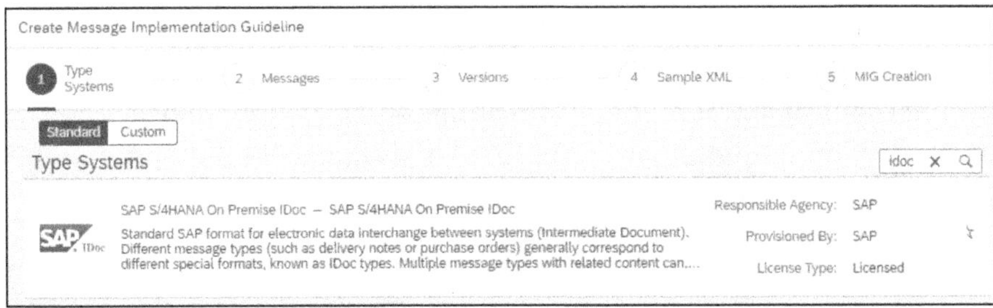

Figure 9.2 MIG Creation: Selection of SAP S/4HANA On-Premise IDoc for Type System

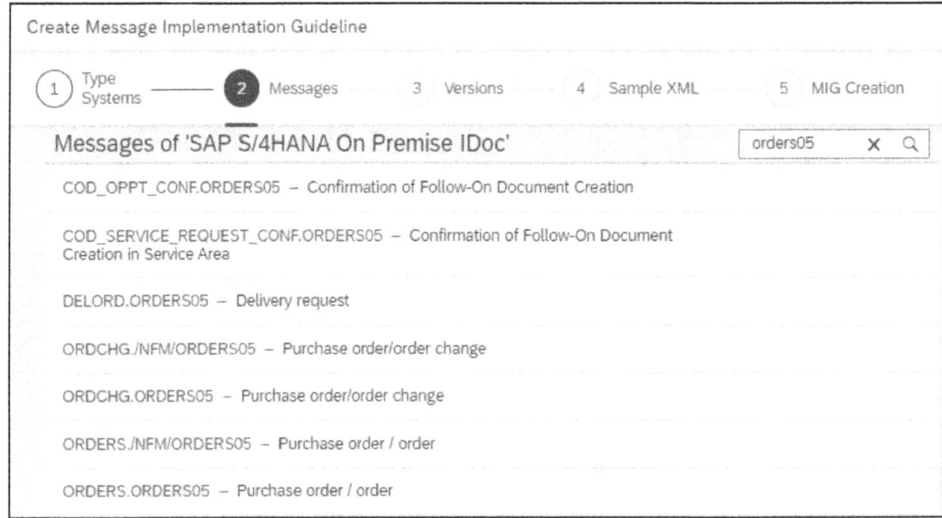

Figure 9.3 MIG Creation: Selection of Message from SAP S/4HANA On-Premise IDoc

- Next is the selection of a specific message type from the B2B standard selected in the previous step. In this case, there is a list of IDocs available within SAP S/4HANA on-premise, from which you should choose **ORDERS.ORDERS05—Purchase order/order**. As there are hundreds of IDocs available, the search function may be very useful here also to find the specific one you need (see Figure 9.3).
- Next is the selection of the version for the type system selected in the first step. In this case, as you chose **SAP S/4HANA On-Premise**, the selection options are limited. Simply choose the most recent version as that is the one that matches the demo SAP S/4HANA system. This is shown in Figure 9.4.

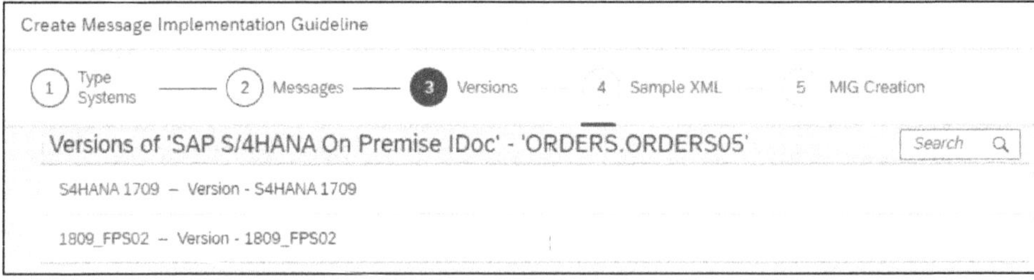

Figure 9.4 MIG Creation: Selection of SAP S/4HANA On-Premise IDoc Version

- If you have a sample XML message, you can upload it at this step and then use the sample data for MIG definition and testing. For this example, there is no sample XML for **ORDERS.ORDERS05**, so skip this step (see Figure 9.5).

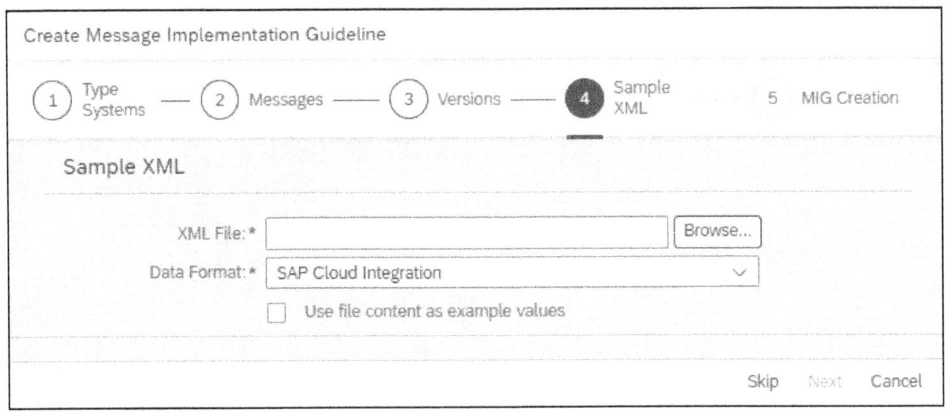

Figure 9.5 MIG Creation: Optional Step to Upload Sample XML

- The final wizard step is about maintaining general information related to the MIG and also the business context from your own and your partner's perspectives. Maintain the key information as follows (shown in Figure 9.6):
  - **Name: IDoc sales order**
  - **Direction: In**
  - **Version**, **Status** and information related to **Message Type** and **Type Systems** are filled automatically
  - Optionally, in **Documentation**, you can add a short description—for example, summarizing the purpose and usage of the MIG itself

For **Own Business Context**, maintain the following values (see Figure 9.7):

- **Business Process: Manage Purchase Order**
- **Business Process Role: Supplier**
- **Industry Classification: Sports and Entertainment**
- **Product Classification: Sports and Recreational Equipment and Supplies and Accessories**
- **Country/Region: United States**

There is one additional point to note about **Direction**. The inbound direction (**In**) is the flow in which a message goes from the trading partner to your own system—that is, to your company.

The outbound direction (**Out**) means that the message goes from the company to the trading partner.

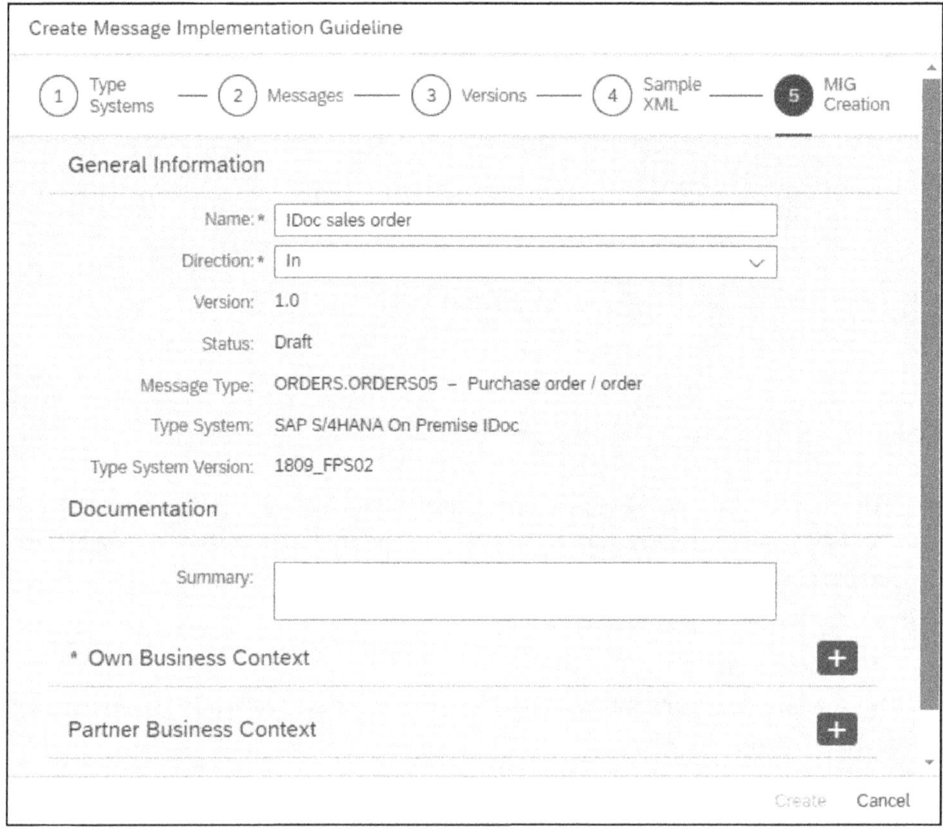

Figure 9.6 MIG Creation: Maintaining General Information

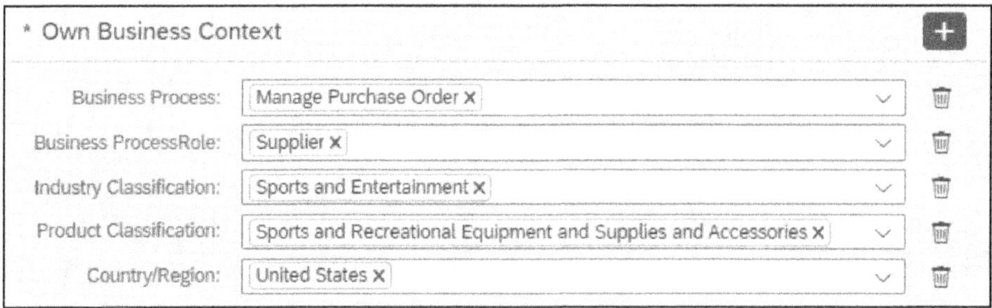

Figure 9.7 MIG Creation: Maintaining Own Business Context

3. After finalizing all the steps of the wizard, you will see the MIG editor, as shown in Figure 9.8.

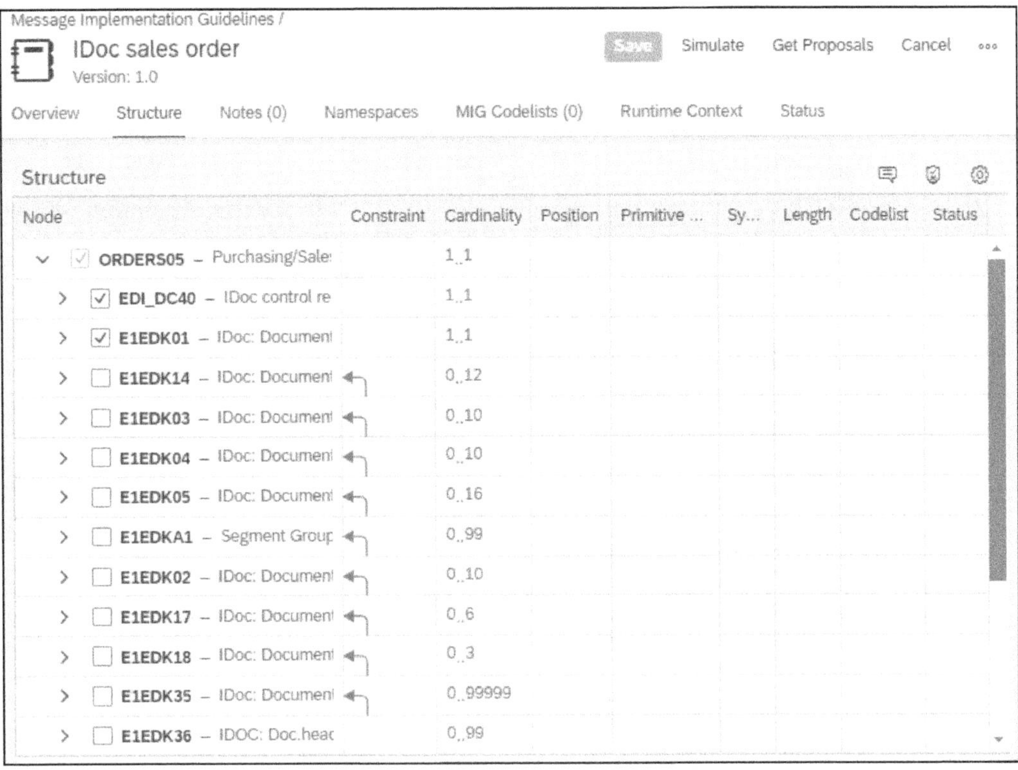

Figure 9.8 MIG Editor with Sales Order IDoc

We are done with the creation of the MIG itself, but the structure is not fully useable yet. As you will recall, we skipped the step related to uploading a sample XML message as we did not give any input on the expected structure; only the mandatory segments and fields are enabled by default. Our next step is to define what additional segments and fields are relevant for the sales order transaction. Before continuing with the structure itself, we will take a quick look at the key functions of the MIG editor.

Looking once again at Figure 9.8 and the tabs located in the top part of the screen, you can see the following:

- **Overview**
  On this tab you find the general information related to the MIG as maintained in the wizard. Most of this can be still edited at a later stage. There is also administrative data showing who created and changed this specific MIG and when.
- **Structure**
  This is the main tab where you actually define and modify the MIG.
- **Notes**
  This tab gives you an option to maintain your own notes related to the MIG. Each note has to be categorized and numbered.

- **Namespaces**
  If your message has specific requirements related to a namespace (like prefixes for its elements or attributes), you can make use of this tab. This is not applicable for our IDocs and UN/EDIFACT messages.
- **MIG Codelists**
  On this tab, you can add MIG-specific codelists. We will not use this option for our scenario.
- **Runtime Context**
  On this tab, you have an option to modify the supported XSD version for Cloud Integration. We keep the default values in this case. The parameter related to the envelope namespace is used for SAP Process Integration and not applicable for our scenario.
- **Status**
  You have an option to set the status of the MIG artifact from the predefined list. This can be a useful feature to support your change and/or quality assurance process.

Looking at the top right-hand corner of the screen (Figure 9.8), you can see the following functions:

- **Edit/Save**
  Switch from display to edit mode.
- **Simulate**
  Allows for a basic test of the MIG with the use of MIG example data (loaded in the wizard) or payload data. which is simply an external file.
- **Get Proposals**
  This function gives a proposal for the MIG definition based on the data from the central integration knowledge base (crowd source–based). We will define a sales order MIG manually without AI support, but we will make use of this option for the definition of MIGs related to outbound IDocs.
- **Cancel**
  Abort the edit process.
- **Activate**
  Share a MIG to the knowledge base.
- **Export**
  With this function, you can generate documentation and runtime artifacts for Cloud Integration or SAP Process Integration.
- While in display mode, there is a **Copy** function that creates a copy of a MIG and self-explanatory **Delete** function.

Now let's explain what we will be doing within the structure definition task. There will be (almost) no coding; the activities we will perform will be similar to a functional specification in which we customize and configure the structure. We will implement our own and our trading partner's requirements by selecting which node and which data elements should be used (optionally or mandatorily). There are also options for defining semantic qualification, properties of the nodes and data elements, operations for the codelist, and much more. In this case, we will not make use of all the features, just briefly demonstrate the key ones and the overall concept. All the aspects and details of the MIG editor could be probably a topic for a separate book.

After this brief look at the key functions of the MIG editor, we can now practice a bit. The prerequisite is that we know what the structure should look like. Providing sample XML is a significant simplification of the MIG definition process, but to have better understanding of the task we will maintain the ORDERS.ORDERS05 structure without loading the sample file. Luckily, we know exactly what the IDoc should look like and which data elements (fields) must be there. Based on this information, we can prepare a sample message, as in Figure 9.9, which shows a sample XML IDoc. As mentioned earlier, we do not try to build an extensive structure that can cover all aspects and variants of the purchasing process. Our sales order IDoc will be simple enough to create sales order within our dummy sales organization. We can already summarize what needs to be defined for the sales order IDoc structure:

- `EDI_DC40`
  This is the IDoc control record. There is no need to handle all technical details as Trading Partner Management will handle most of it (you will learn about that in Section 9.3). To make it a bit more transparent and simple, we assign a few constant values. In an actual implementation, any additional values or modifications related to the IDoc control record should be handled in a dynamic way (e.g., via global parameters).
- `E1EDK01`
  This is a mandatory segment. It must appear in the structure even if we do not have any fields within it.
- `E1EDKA1`
  We can see three occurrences of this segment, each with data related to different partner roles.
- `E1EDP01`
  The item segment in the sample transaction is also simplified to a bare minimum; only the item quantity and unit of measure appear on the top level.
- `E1EDP19`
  This is a child of `E1EDP01` with item identification data. We expect the material number used by the vendor and (optionally) a description.

```
<ORDERS05 xmlns="urn:sap-com:document:sap:idoc:soap:messages">
    <IDOC BEGIN="1">
        <EDI_DC40 SEGMENT="1">
        <E1EDK01 SEGMENT="1">
        </E1EDK01>
        <E1EDKA1 SEGMENT="1">
            <PARVW>AG</PARVW>
            <LIFNR>BEST_BIKES_HQ</LIFNR>
        </E1EDKA1>
        <E1EDKA1 SEGMENT="1">
            <PARVW>WE</PARVW>
            <LIFNR>BEST_BIKES_WHS</LIFNR>
        </E1EDKA1>
        <E1EDKA1 SEGMENT="1">
            <PARVW>LF</PARVW>
            <PARTN>COMPANY_1710_VN</PARTN>
        </E1EDKA1>
        <E1EDK02 SEGMENT="1">
            <QUALF>001</QUALF>
            <BELNR>TST PO number01</BELNR>
        </E1EDK02>
        <E1EDP01 SEGMENT="1">
            <MENGE>10</MENGE>
            <E1EDP19 SEGMENT="1">
                <QUALF>002</QUALF>
                <IDTNR>MZ-FG-R300</IDTNR>
                <KTEXT>R300 BIKE</KTEXT>
            </E1EDP19>
        </E1EDP01>
    </IDOC>
</ORDERS05>
```

Figure 9.9 Sample Sales Order IDoc in XML Format

After getting familiar with the expected end results, we can go back to the structure definition. The basic rule does not change: effort spent on MIG definition will pay off during MAG definition. We should aim to define the MIG as precisely as possible. One of the most important actions while defining the MIG is node qualification. Segment E1EDKA1 from the sales order IDoc is a perfect example to explain this concept. We expect the E1EDKA1 segment to occur three times, each time with a different partner role, which is indicated by a different PARVW value. The latter is the qualifier. In the structure, we define three unique E1EDKA1 segments, each with a predefined and fixed qualifier (PARVW) value. Follow these steps to qualify the E1EDKA1 segment:

1. Open the context menu for the E1EDKA1 segment by right-clicking it. Choose the **Qualify Node** option at the bottom (see Figure 9.10).
2. Choose **AG Sold-to-Party** as the value for the qualifier (see Figure 9.11).
3. The end results should appear as shown in Figure 9.12.
4. You also can maintain additional properties related to the E1EDKA1 segment, like **Cardinality** and **Name**, and can add a description in the **Documentation** area (see Figure 9.13).

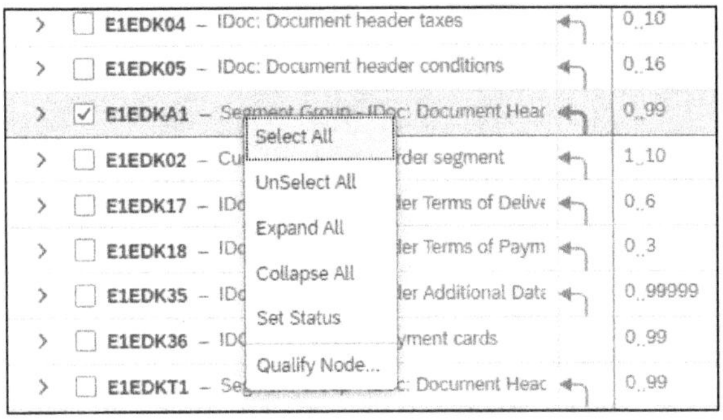

Figure 9.10 Qualify Node Option in Context Menu

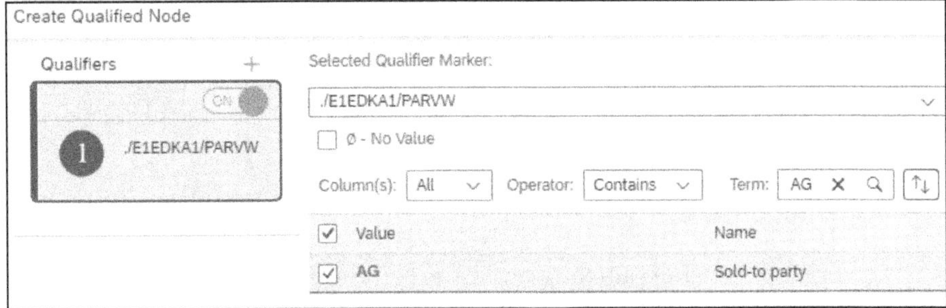

Figure 9.11 Selecting Qualifier Marker AG for ./E1EDKA/PARVW

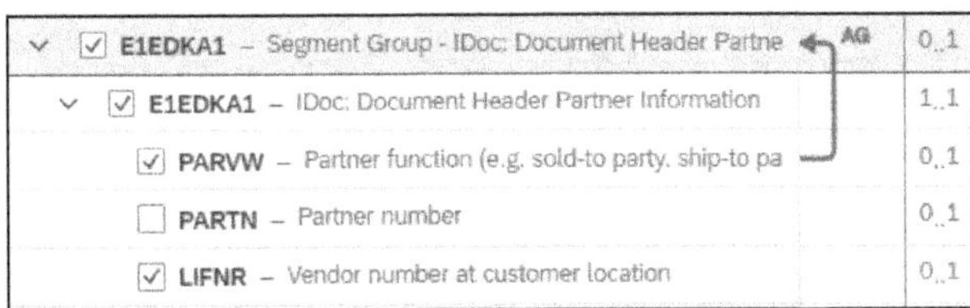

Figure 9.12 View of E1EDKA1 Segment with AG Qualifier on PARVW

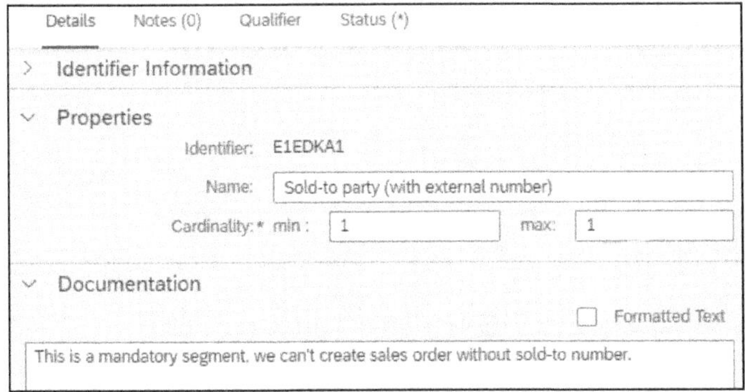

Figure 9.13 Properties Related to E1EDKA1 Segment

5. You've qualified one E1EDKA1 segment with the **AG Sold-to-Party** value for PARVW. To do the next two, make use of **Duplicate Qualified Node** option that now appears in the context menu (see Figure 9.14).

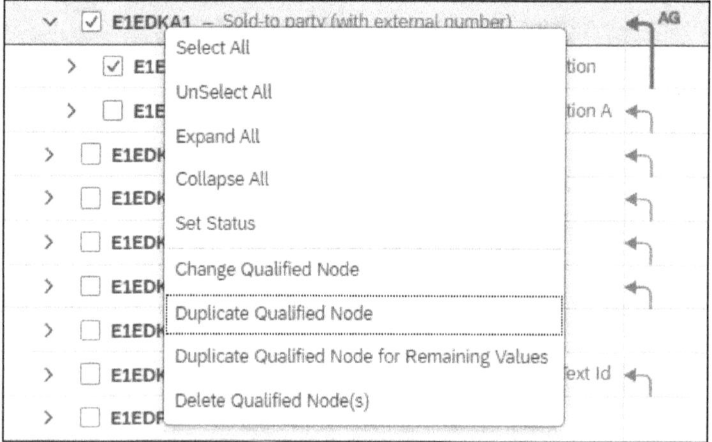

Figure 9.14 Context Menu to Duplicate Qualified Node

6. Maintain the duplicated qualified node in the same way as for the first qualified node. Choose **WE Ship-to Party** for the first one and **LF Vendor Party** for the second one. At the end, you should have the partner-related segment as shown in Figure 9.15.

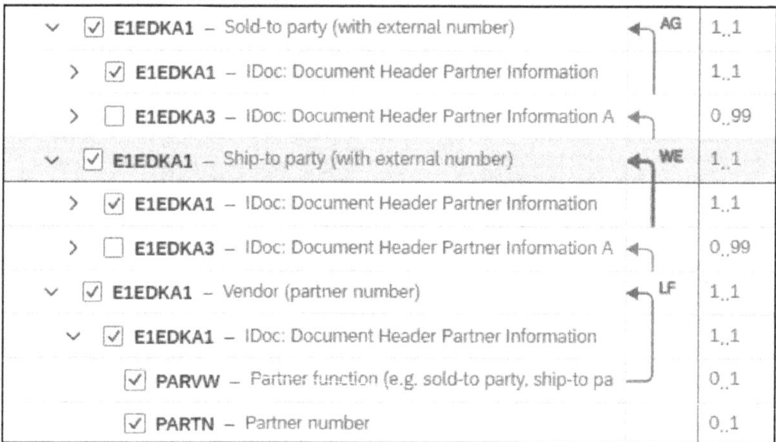

Figure 9.15 All Three Qualified E1EDKA1 Segments

In addition to qualifying the E1EDKA1 segment we also qualified the E1EDP02 001 and E1EDP19 segments by linking it to QUALF with value 002. And we also ticked the few fields expected in the IDoc. You can also see that the name was adjusted slightly for added transparency. The final structure of the MIG IDoc sales order looks as shown in Figure 9.16.

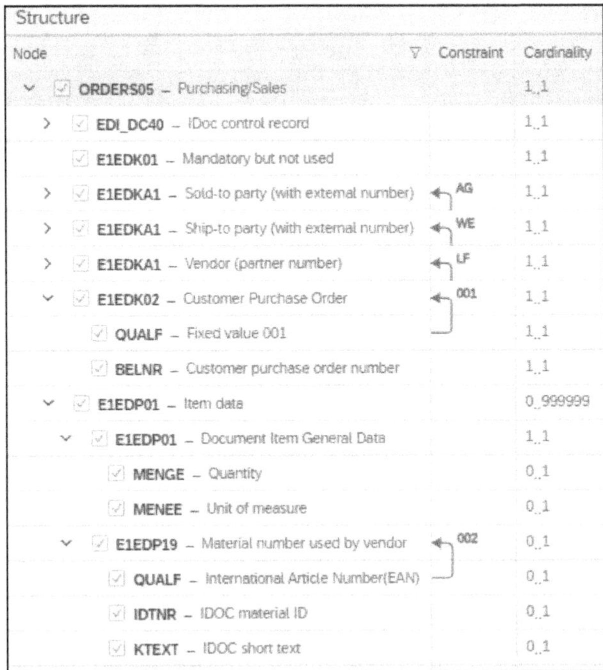

Figure 9.16 Final Structure of MIG IDoc Sales Order

Sales Order Response IDoc

Next we will discuss MIG creation for the outbound sales order response IDoc. To create a MIG related to the sales order response IDoc, we will make use of the sample XML IDoc. As you will see, it speeds up the structure definition by including in the initial structure all the nodes and data elements that were present in the sample XML. We start by performing the same steps as we did for the sales order IDoc: go to **Design • MIGs** in SAP Integration Suite and choose the **Add** button. The provided details for the general information will appear as shown in Figure 9.17.

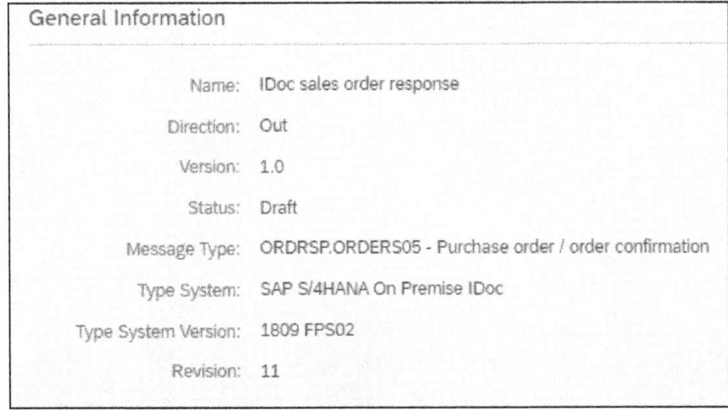

Figure 9.17 General Information Related to MIG IDoc Sales Order Response

This time, do not skip the **Sample XML** step. Load an XML file that contains the sample ORDRSP.ORDERS05 IDoc in XML format. Also select the **Use file content as example values** option (see Figure 9.18).

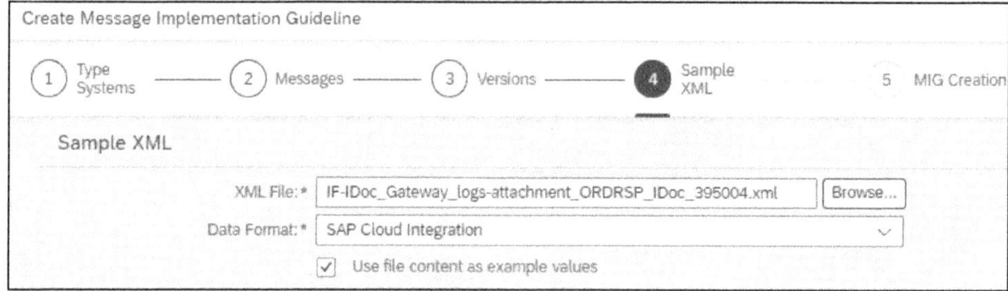

Figure 9.18 Creating Message Implementation Guideline Based on Sample XML IDoc

After going through all the steps of the MIG creator, you can see that the initially created structure has all the segments and fields enabled as per the sample XML (see Figure 9.19). This is a good starting point, and a good way to demonstrate the functionality, but we will apply a few adjustments to make the definition more transparent and to simplify the MAG creation in the future.

Message Implementation Guidelines /

IDoc sales order response

Version: 1.0

Overview | Structure | Notes (0) | Namespaces | MIG Codelists (0) | Runtime Context | Status

Structure

Node	Constraint	Cardinality
ORDERS05 – Purchasing/Sales		1..1
EDI_DC40 – IDoc control record for the interface to the external system		1..1
E1EDK01 – IDoc: Document header general data		1..1
E1EDK14 – IDoc: Document Header Organizational Data		0..12
E1EDK03 – IDoc: Document header date segment		0..10
E1EDK02 – IDoc: Document header reference data		0..10
E1EDK17 – IDoc: Document Header Terms of Delivery		0..6
E1EDK18 – IDoc: Document Header Terms of Payment		0..3
E1EDP01 – Segment Group - IDoc: Document Item General Data		0..999999
E1EDP01 – IDoc: Document Item General Data		1..1
E1EDP02 – IDoc: Document Item Reference Data		0..10
E1EDP03 – IDoc: Document Item Date Segment		0..10
E1EDP05 – Segment Group - IDoc: Document Item Conditions		0..16
E1EDP19 – IDoc: Document Item Object Identification		0..5
E1EDS01 – IDoc: Summary segment general		0..5

Figure 9.19 Initial MIG Structure Created Based on Sample XML IDoc

Table 9.1 lists the exact data elements and correlated IDoc elements from ORDRSP.ORDERS05 that we want to use when communicating a sales order response to a trading partner.

Data Element	IDoc Field
Currency	E1EDK01-CURCY
Supplier (SAP) sales order number	BELNR from E1EDK02 segment with QUALF='002'
Customer purchase order number	BELNR from E1EDK02 segment with QUALF='001'
Requested delivery data	DATUM from E1EDK03 segment with IDDAT='002'
Item level—quantity	E1EDP01-MENGE
line item amount (net)	E1EDP01-NETWR
Item price (net)	E1EDP01-VPREI
Material number	IDTNR from E1EDP19 segment with QUALF='002'
Material description	KTEXT from E1EDP19 segment with QUALF='002'
Sales order value (net)	SUMME from E1EDS01 segment where SUMID ='002'

Table 9.1 Data Elements from ORDRSP.ORDERS05 IDoc Planned for Sales Order Response Transaction

We approach this MIG structure in such a way that we keep all the segments and fields enabled by default, but on top of that we qualify the segments we are about to use. In addition, we apply more self-explantory descriptions of the data elements. Whenever possible, we encourage you to make use of the available options to describe a MIG and its elements directly in Integration Advisor. There are huge benefits to this, not to mention that if you do this, there is no need to maintain any additional documentation outside the system. Integration Advisor is able to generate transparent and ready-to-use documentation in a few formats like Excel, PDF, and Word. We will take a quick look at this feature later on. As this book has certain limitations and we've prioritized the key functionality, we will not emphasize the documentation and description aspects too much here.

There is one useful feature that helps edit extended structures. When you click the **Node** column just below the structure, you will see a popup window where you can apply a filter on the structure; see the sample in Figure 9.20. You can enter the name of a specific data element here, or you can make use of the **Show Selected Nodes** option; the latter is a good choice when you want to review or adjust the existing definition. We make use of that option for adjustments in ORDRSP.ORDERS05. Also keep in mind that the other columns offer similar filter functionality. You can explore the available options by clicking each of them.

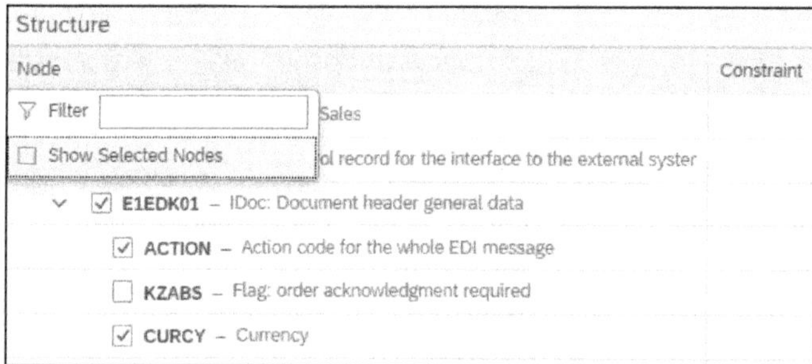

Figure 9.20 Filtering Structure Nodes by Selection

As you already know how to create qualified nodes, you may go ahead and create the following ones:

- Segment **E1EDK03** identifier **IDDAT**—value: **002** Requested delivery date
- Segment **E1EDK03** identifier **IDDAT**—value: * Remaining values. This covers all the values of **IDDAT** identifier, including those that are already explicitly defined (like **002** above).
- Segment **E1EDK02** identifier **QUALF**—value: **002** Supplier order
- Segment **E1EDK02** identifier **QUALF**—value: **001** Supplier order
- Segment **E1EDK02** identifier **QUALF**—value: * Remaining values
- Segment **E1EDS01** identifier **SUMID**—value: **002** Net document value
- Segment **E1EDS01** identifier **SUMID**—value: * Remaining values

The final structure should look as shown in Figure 9.21.

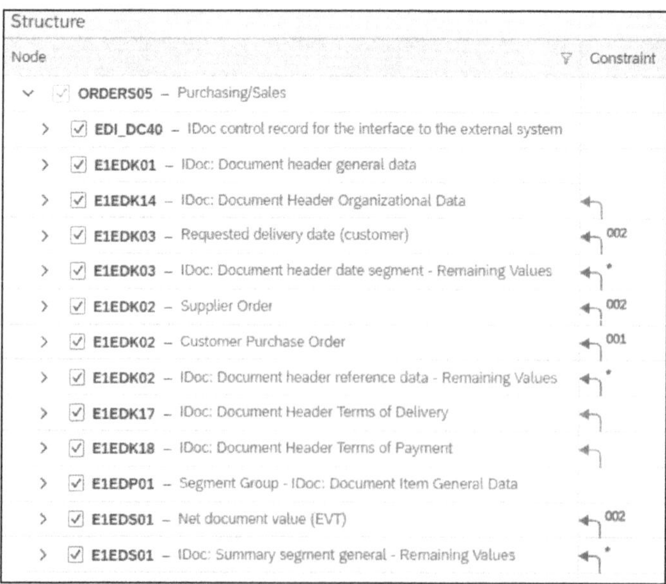

Figure 9.21 Final MIG Structure for Sales Order Response

The benefit of using a sample XML message when creating a MIG is that the payload was also stored as MIG example data. You now have the option to check the MIG against this preloaded data. Choose function **Simulate • Simulate with MIG Example Data** as shown in Figure 9.22.

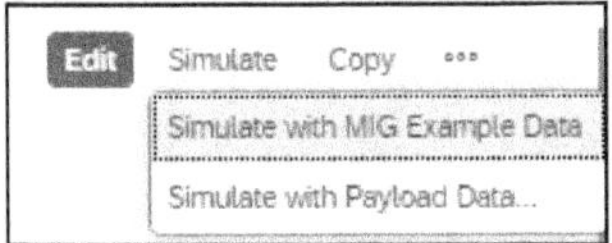

Figure 9.22 Simulate with MIG Example Data

After triggering the simulation function, you will see an additional column called **Simulation Data**, which is where you can see values mapped from the sample data to the specific data element. You can also see how many times each node will be created. Notice what the nodes qualified for *** Remaining Values** look like. They were created for each value of the qualifier, even those qualified directly. In the example shown in Figure 9.23, we have segment E1EDK03 002 occurring once based on the 002 qualifier. And then we also have two more occurrences of E1EDK03 with values 002 and 012 based on the * qualifier.

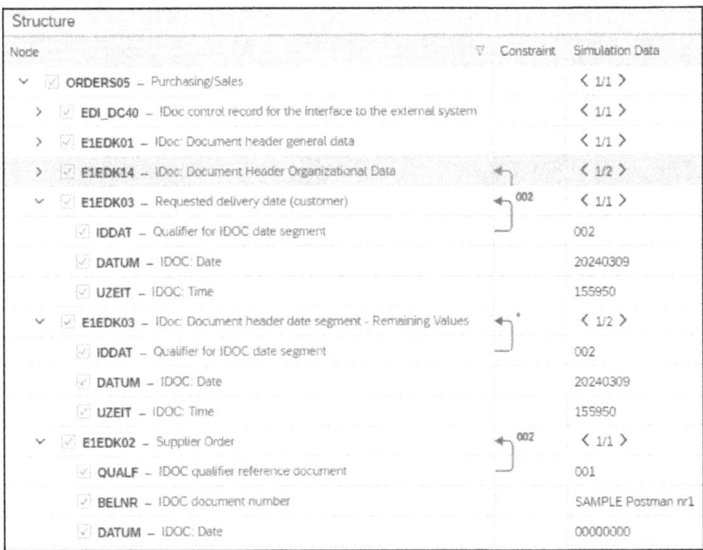

Figure 9.23 Results of MIG Simulation with MIG Example Data

Advanced Shipping Notification IDoc

After creating MIGs related to the sales order and sales order response, you should be able to follow up with the third MIG related to outbound IDocs, for advanced shipping notifications (ASNs). We will not be describing each unique step necessary for the MIG creation, but to help you with the task and allow you to fully recreate our demo scenario, you can find the exact nodes and data elements used for the ASN definition. The generic information maintained for the ASN should look as shown in Figure 9.24.

General Information

Name: * IDoc advance shipping notice
Direction: * Out
Version: 1.0
Status: Draft
Message Type: DESADV.DELVRY07 – Delivery: Shipping notification
Type System: SAP S/4HANA On Premise IDoc
Type System Version: 1809_FPS02

Figure 9.24 General Information Related to Advanced Shipping Notification IDoc MIG

Table 9.2 shows the exact data elements we need to enable in this MIG for the demo transaction. In the first column, you can find the name of the data element; the second column contains the exact data location in the IDoc.

Data Element	IDoc Field
Delivery document number	E1EDL20-VBELN
Total weight	E1EDL20-BTGEW
Net weight	E1EDL20-NTGEW
Unit of weight	E1EDL20-GEWEI
Sold-to party	Data from E1ADRM1 segment with PARVW='AG'
Ship-to	Data from E1ADRM1 segment with PARVW='WE'
Shipping point	Data from E1ADRM1 segment with PARVW='OSP'
Good issue date	NTANF from E1EDT13 segment with QUALF='006'
Outbound delivery date	NTANF from E1EDT13 segment with QUALF='007'
Item number	E1EDL24-POSNR
Material number	E1EDL24-MATNR
Material description	E1EDL24-ARKTX
Quantity delivered	E1EDL24-LFIMG

Table 9.2 Data Elements from DESADV.DELVRY07 IDoc Planned for Advanced Shipping Notification Transaction

Data Element	IDoc Field
Sales order number	BELNR from E1EDL43 segment with QUALF='C'
Sales order item number	POSNR from E1EDL43 segment with QUALF='C'
Sales order date	DATUM from E1EDL43 segment with QUALF='C'
Purchase order number	BELNR from E1EDL41 segment with QUALF='001'
Purchase order item number	POSNR from E1EDL41 segment with QUALF='001'

Table 9.2 Data Elements from DESADV.DELVRY07 IDoc Planned for Advanced Shipping Notification Transaction (Cont.)

The final structure for our ASN should look as shown in Figure 9.25. If so, you can save it and move to the last one, the invoice-related MIG.

Node	Constraint	Cardinality
DELVRY07 – Delivery Interface		1..1
EDI_DC40 – IDoc control record for the interface to the externa		1..1
E1EDL20 – Segment Group - Delivery header		1..9999
E1EDL20 – Delivery header		1..1
E1EDL22 – Delivery Header Descriptions		0..1
E1EDL21 – Segment Group - Delivery Header Additional Dat		0..1
E1EDL18 – IDOC: Controlling (Delivery)		0..99
E1ADRM1 – Sold-to party	AG	0..1
E1ADRM1 – Ship-to party	WE	0..1
E1ADRM1 – Shipping Point	OSP	0..1
E1ADRM1 – Segment Group - Central Address Segment Grc	*	0..1
E1EDT13 – Deadline (delivery) - Goods Issue	006	0..1
E1EDT13 – Deadline (delivery) - (Outbound) Delivery	007	0..1
E1EDT13 – Segment Group - IDoc: Deadline (delivery) - Ren	*	0..1
E1EDL33 – Segment Group - Export Data Delivery Header		0..1
E1EDL24 – Segment Group - Delivery Item		0..99999
E1EDL24 – Delivery Item		1..1
E1EDL25 – Delivery Item Descriptions		0..1
E1EDL26 – Segment Group - Delivery Item Additional Dat		0..1
E1EDL43 – Sales Order	C	0..1
E1EDL41 – Purchase order data of customer	001	0..1

Figure 9.25 Final MIG Structure for Advanced Shipping Notification

Invoice IDoc

Now it is time for the last MIG object, one related to invoice IDocs. For the creation of the invoice-related MIG, we will make use of the knowledge graph in the Integration Advisor. At the start, we will create the MIG as usual, maintaining all the general information as shown in Figure 9.26.

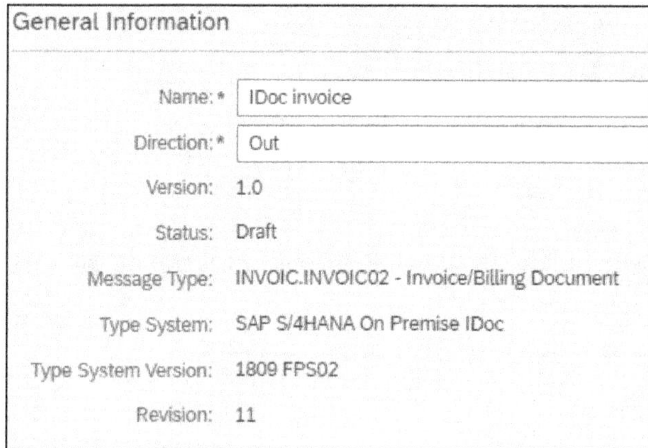

Figure 9.26 General Information Related to Invoice IDoc MIG

Optionally, you can again load sample XML, so your starting point would be a structure as shown in Figure 9.27.

Structure

Node	Constraint	Cardinality
INVOIC02 – Invoice/Billing document		1..1
EDI_DC40 – IDoc control record for the interface to the external system		1..1
E1EDK01 – IDoc: Document header general data		1..1
E1EDKA1 – IDoc: Document Header Partner Information	↰	0..99
E1EDK02 – IDoc: Document header reference data	↰	0..20
E1EDK03 – IDoc: Document header date segment	↰	0..20
E1EDK17 – IDoc: Document Header Terms of Delivery	↰	0..10
E1EDK18 – IDoc: Document Header Terms of Payment	↰	0..10
E1EDK28 – IDoc: Document Header Bank Data		0..20
E1EDK14 – IDoc: Document Header Organizational Data	↰	0..10
E1EDP01 – Segment Group - IDoc: Document Item General Data		0..999999
E1EDS01 – IDoc: Summary segment general	↰	0..30

Figure 9.27 Initial MIG Structure for Invoice

In the next step, click the **Get Proposals** function located in the top-right corner of the screen. Remember that you need to switch from display to edit mode to have this function available (see Figure 9.28).

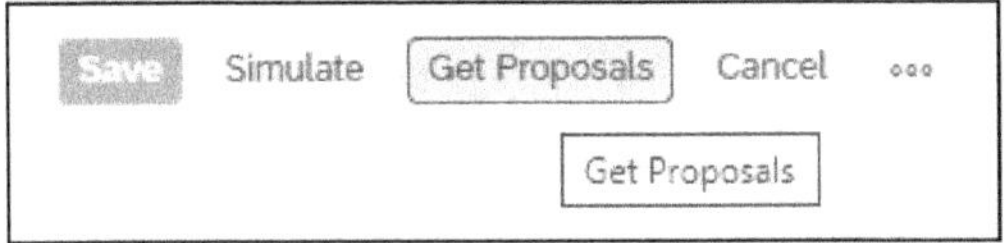

Figure 9.28 Actions Related to MIG Objects

As a result, you will see the structure view with one additional column, called **Confidence**. If you hover over the information icon, you will see additional information—in this example, **Proposal successfully calculated based on 30 best-fitting MIGs** (see Figure 9.29).

Structure

Node	Constraint	Confidence	Cardinality	Position
INVOIC02 – Invoice/Billing document			Proposal successfully calculated based on 30 best-fitting MIGs.	
EDI_DC40 – IDoc control record for the interface to the external system			1..1	
E1EDK01 – IDoc: Document header general data			1..1	
E1EDKA1 – IDoc: Document Header Partner Information			0..99	
E1EDK02 – IDoc: Document header reference data			0..20	
E1EDK03 – IDoc: Document header date segment			0..20	
E1EDK05 – IDoc: Document header conditions			0..99	
E1EDK04 – IDoc: Document header taxes			0..999	
E1EDK17 – IDoc: Document Header Terms of Delivery			0..10	
E1EDK18 – IDoc: Document Header Terms of Payment			0..10	

Figure 9.29 MIG Structure with Calculated Proposal: Result of Get Proposal Action

The confidence rating indicates the usage of the specific node or data element based on the knowledge graph content. If you hover over the blue bar, you will see the exact percentage, as shown in Figure 9.30.

Structure

Node	Constraint	Confidence	Cardinality	P
INVOIC02 – Invoice/Billing document			1..1	
EDI_DC40 – IDoc control record for the interface to the external system			1..1	
E1EDK01 – IDoc: Document header general data			1..1	
ACTION – Action code for the whole EDI message			0..1	
KZABS – Flag: order acknowledgment required			0..1	
CURCY – Currency			0..1	
HWAER – EDI local currency			Confidence Value: 77%	
WKURS – Exchange rate			Highest Confidence Value in the group: 77%	

Figure 9.30 Result of Proposal Service: Confidence Value for E1EDK01-CURCY Field

The proposal also gives you the confidence rating related to the usage of qualifiers. In Figure 9.31, you can see segment E1EDK01 and that the most commonly used qualifiers are RE, AG, and WE. In this view, you can choose the qualifiers that will result in automatically created qualified nodes.

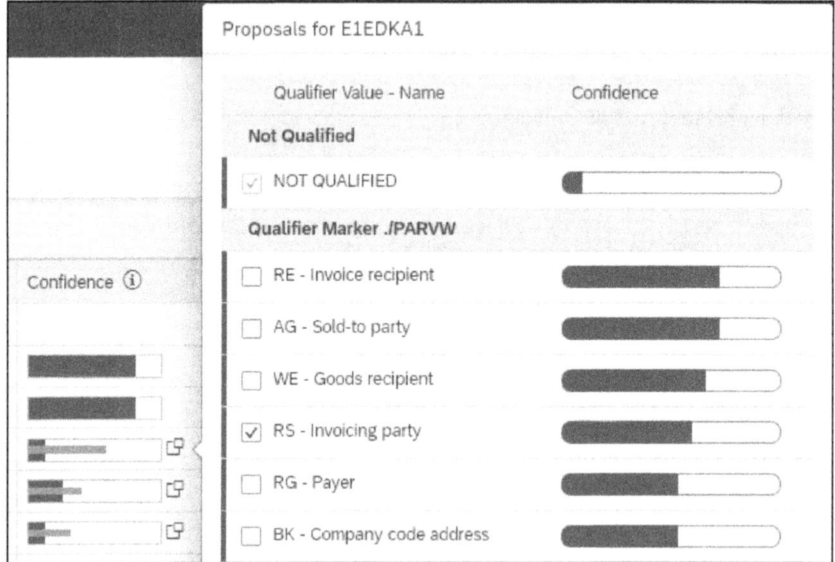

Figure 9.31 Detailed View of Proposal for E1EDKA1 Segment

Selecting **RS—Invoicing Party** on this screen results in the creation of a qualified node, as shown in Figure 9.32.

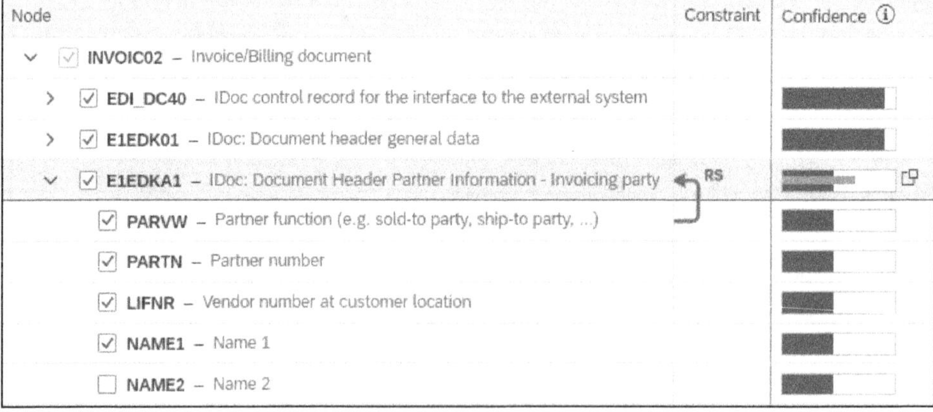

Figure 9.32 Accepted RS Invoicing Party Qualifier for E1EDKA1 Segment

In a similar way, you can go through the proposed elements and choose the ones applicable for your scenario. Figure 9.33 shows another example for the E1EDK02 segment, where you choose qualifiers 001, 002, and 009.

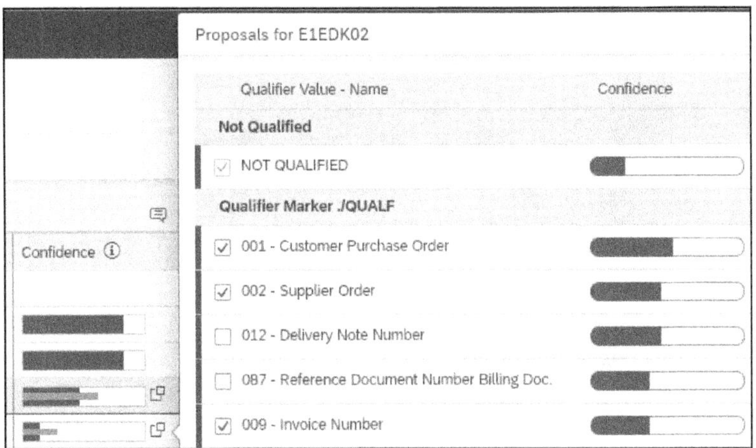

Figure 9.33 Accepting Qualifiers Proposed for E1EDK02

The results of these selections are qualified nodes, as shown in Figure 9.34.

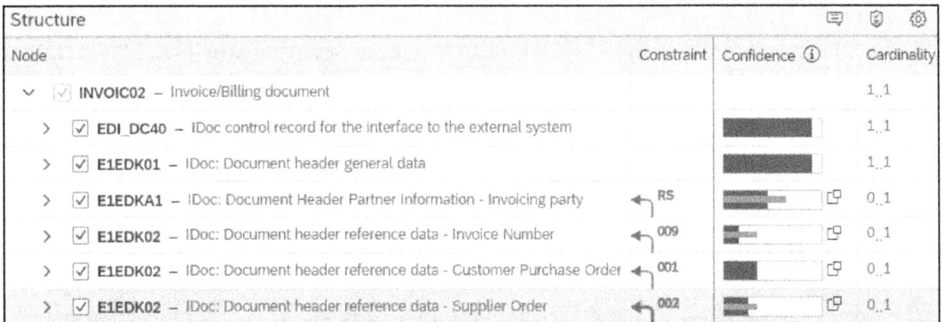

Figure 9.34 Accepted Proposal Related to E1EDKA1 and E1EDK02 Qualifiers

In addition to the proposals, you can still maintain your own requirements. For example, you can still trigger the **Duplicate Qualified Node for Remaining Values** option, as shown in Figure 9.35.

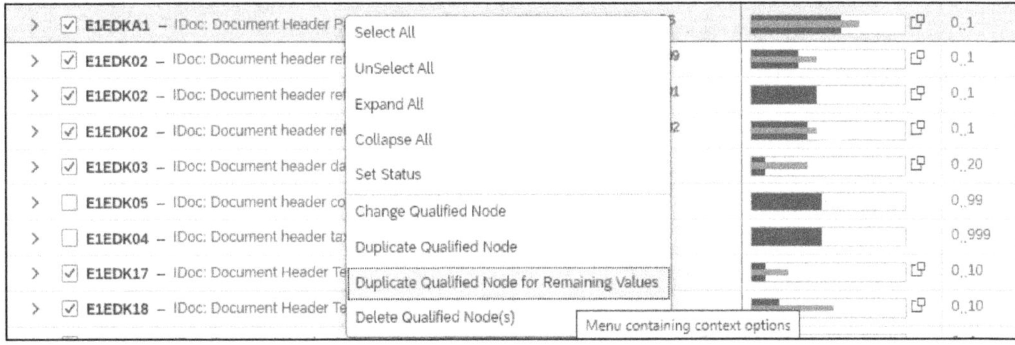

Figure 9.35 Context Menu: Duplicate Qualified Node for Remaining Values

The header of your invoice structure should look as shown in Figure 9.36.

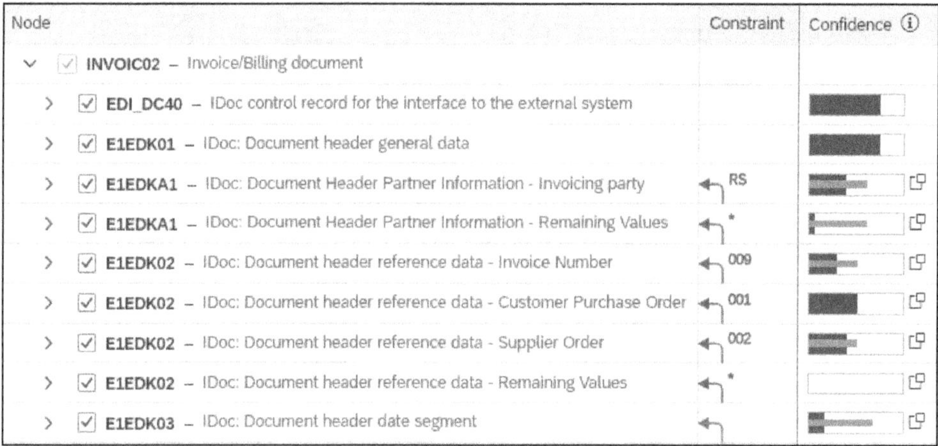

Figure 9.36 Final MIG Structure for Invoice: Header Level

In Figure 9.37, you can see the final structure defined for item-related segment E1EDP01. After all the practice you've had so far, you should be able to recreate it without detailed instruction.

Node	Constraint	Cardinality
E1EDP01 – Segment Group - IDoc: Document Item General Data		0..999
E1EDP01 – IDoc: Document Item General Data		1..1
E1EDP02 – IDoc: Document Item Reference Data - Customer Purchase Orde	001	0..1
E1EDP02 – IDoc: Document Item Reference Data - Number of Preceding Doc	016	0..1
E1EDP02 – IDoc: Document Item Reference Data - Supplier Order	002	0..1
E1EDP03 – IDoc: Document Item Date Segment		0..25
E1EDP19 – IDoc: Document Item Object Identification - Material number usec	002	0..1
E1EDP26 – IDoc: Document Item Amount Segment - Gross price	001	0..1
E1EDP26 – IDoc: Document Item Amount Segment - Gross value	002	0..1
E1EDP26 – IDoc: Document Item Amount Segment - Net value	003	0..1
E1EDP26 – IDoc: Document Item Amount Segment - Amount qualifying for ca	004	0..1
E1EDP26 – IDoc: Document Item Amount Segment - Subtotal 1 from calculat	010	0..1
E1EDP26 – IDoc: Document Item Amount Segment - Subtotal 2 from calculat	011	0..1

Figure 9.37 Final MIG Structure for Invoice: Item Level

Figure 9.38 shows the summary section of the invoice structure, with the qualified nodes on the E1EDS01 segment.

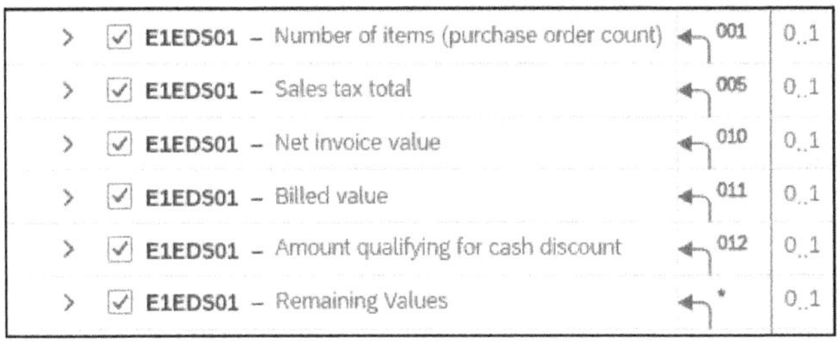

> ☑ **E1EDS01** – Number of items (purchase order count)	001	0..1
> ☑ **E1EDS01** – Sales tax total	005	0..1
> ☑ **E1EDS01** – Net invoice value	010	0..1
> ☑ **E1EDS01** – Billed value	011	0..1
> ☑ **E1EDS01** – Amount qualifying for cash discount	012	0..1
> ☑ **E1EDS01** – Remaining Values	*	0..1

Figure 9.38 Final MIG Structure for Invoice: Summary Level

Table 9.3 gives an overview of the key data elements that you should enable on the MIG structure related to the invoice IDoc. This is the data that you will be mapping to the target EDIFACT message.

Data Element	IDoc Field
Billing document number	E1EDK01-BELNR
Date	DATUM from E1EDK02 segment with QUALF='009'
Date	DATUM from E1EDK02 segment with QUALF='002'
Reference purchase order	BELNR from E1EDK02 segment with QUALF='001'
Reference sales order	BELNR from E1EDK02 segment with QUALF='002'
Supplier party	PARTN from E1EDKA1 segment with PARVW='RS'
	NAME1 from E1EDKA1 segment with PARVW='RS'
	STRAS from E1EDKA1 segment with PARVW='RS'
	ORT01 from E1EDKA1 segment with PARVW='RS'
	PSTLZ from E1EDKA1 segment with PARVW='RS'
	COUNC from E1EDKA1 segment with PARVW='RS'
Item number	E1EDP01-POSEX
Item quantity	E1EDP01-MENGE
Material ID	IDTNR from E1EDP01-E1EDP19 segment with QUALF='002'
Material description	KTEXT from E1EDP01-E1EDP19 segment with QUALF='002'
Total value	BETRG from E1EDP01-E1EDP26 segment with QUALF='010'
Item value	BETRG from E1EDP01-E1EDP26 segment with QUALF='001'
Total sales tax	SUMME from E1EDS01 segment with SUMID='005'
Total value	SUMME from E1EDS01 segment with SUMID='010'

Table 9.3 Data Elements from INVOIC.INVOIC02 IDoc Planned for Invoice Document

We have completed the creation of the first set of objects in Integration Advisor. This is a good time to have a look at the extended lifecycle management feature for the artifacts managed within Integration Advisor. One of the additional attributes of a MIG is its status, and each can have one of three statuses:

- **Draft**
  This is the initial status of a newly created MIG. The version is set to 1.0.
- **Active**
  When you activate a MIG, it gets this status. At the same time, the MIG is populated to the public knowledge base. SAP respects your privacy and copyrights, so there are some restrictions in regard to the data being shared. You will see an appropriate message when activating a MIG, as shown in Figure 9.39.
- **Deprecated**
  When you create the next version of the currently active MIG, the old version will get the deprecated status.

For our scenario, we keep all our MIGs in draft status. At this stage, we can't confirm yet that these are correct and final definitions; we keep everything as shown in Figure 9.40. Next, in Section 9.2.2, we will continue with the MIG definitions for EDIFACT messages.

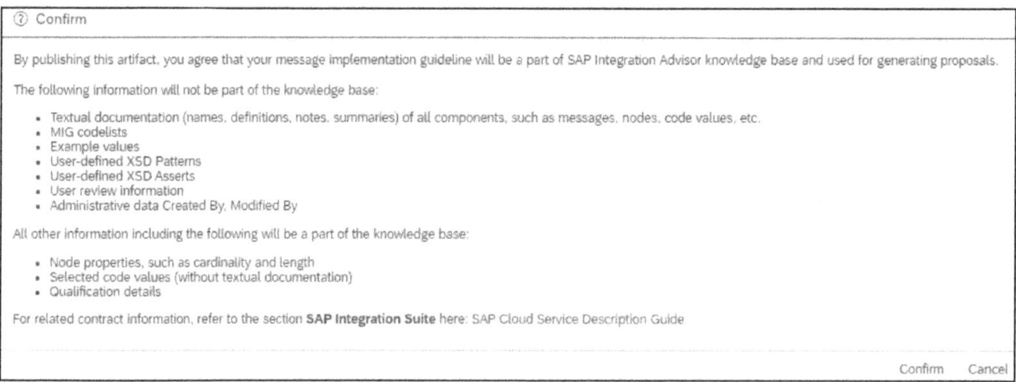

Figure 9.39 Additional Confirmation Requested when Activating MIGs

Let's take the sales order IDoc MIG as an example. The starting point is that all our IDoc-related MIGs are in **Draft Version: 1.0**, as shown in Figure 9.40.

You can activate a MIG by opening it and then choosing the **Activate** option located in the top-right-hand corner, as shown in Figure 9.41. This will change the **Status** of the MIG to **Active** (see Figure 9.42).

Message Implementation Guidelines (4)

IDoc sales order	Message Type: ORDERS.ORDERS05 Type System: SAP S/4HANA On Premise IDoc Type System Version: 1809 FPS02 Direction: In	Draft Version: 1.0 History	
IDoc advance shipping notice	Message Type: DESADV.DELVRY07 Type System: SAP S/4HANA On Premise IDoc Type System Version: 1809 FPS02 Direction: Out	Draft Version: 1.0 History	
IDoc invoice	Message Type: INVOIC.INVOIC02 Type System: SAP S/4HANA On Premise IDoc Type System Version: 1809 FPS02 Direction: Out	Draft Version: 1.0 History	
IDoc sales order response	Message Type: ORDRSP.ORDERS05 Type System: SAP S/4HANA On Premise IDoc Type System Version: 1809 FPS02 Direction: Out	Draft Version: 1.0 History	

Figure 9.40 Message Implementation Guidelines Related to IDoc Messages in EDI Scenario

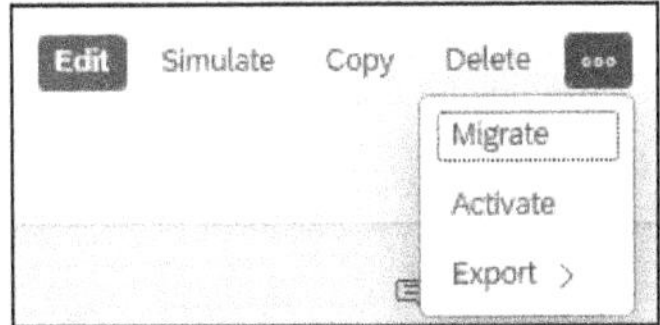

Figure 9.41 Action Menu for MIG Object

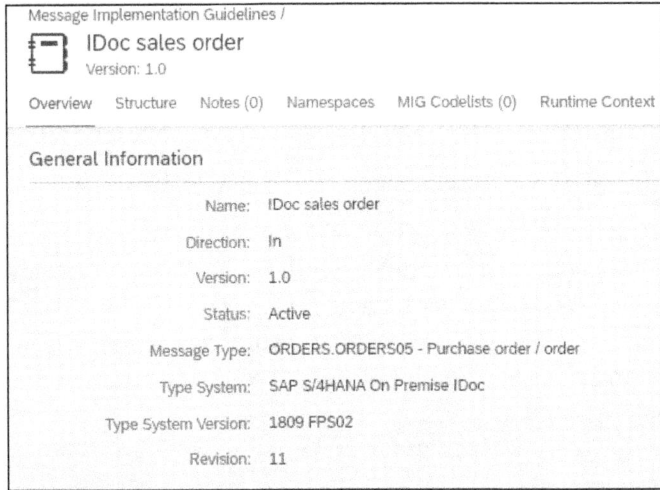

Figure 9.42 MIG Overview Screen for Sales Order IDoc

Let's go through a small experiment to make a change in an already active MIG. When we look at a MIG's history, our starting point is as shown in Figure 9.43. When you edit an active structure, there will be a new version created in a draft state, as shown in Figure 9.44. After implementing the additional changes and activating the new version, draft version 2.0 will be converted to active version 2.0. The previous active version 1.0 will be moved to deprecated version 1.0, as shown in Figure 9.45.

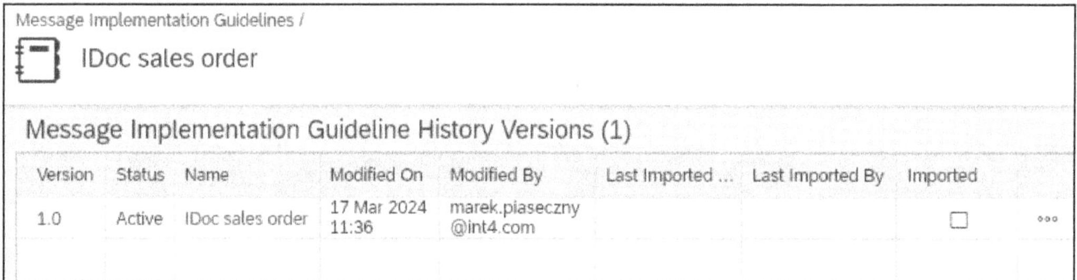

Message Implementation Guidelines /

IDoc sales order

Message Implementation Guideline History Versions (1)

Version	Status	Name	Modified On	Modified By	Last Imported ...	Last Imported By	Imported	
1.0	Active	IDoc sales order	17 Mar 2024 11:36	marek.piaseczny @int4.com			☐	ooo

Figure 9.43 Sales Order IDoc MIG: Active Version 1.0

Message Implementation Guideline History Versions (2)

Version	Status	Name	Modified On	Modified By	Last Imported On	Last Imported By	Imported	
2.0	Draft	IDoc sales order	17 Mar 2024 11:47	marek.piaseczny @int4.com			☐	ooo
1.0	Active	IDoc sales order	17 Mar 2024 11:36	marek.piaseczny @int4.com			☐	ooo

Figure 9.44 Sales Order IDoc MIG: Draft Version 2.0

Message Implementation Guideline History Versions (2)

Version	Status	Name	Modified On	Modified By	Last Imported On	Last Imported By	Imported	
2.0	Active	IDoc sales order	17 Mar 2024 11:51	marek.piaseczny @int4.com			☐	ooo
1.0	Deprecated	IDoc sales order	17 Mar 2024 11:51	marek.piaseczny @int4.com			☐	ooo

Figure 9.45 Sales Order IDoc MIG: Active Version 2.0

9.2.2 Defining Message Implementation Guidelines for EDIFACT Messages

From a technical point of view, use of the MIG editor and definition of MIGs for different type systems (message standards) does not differ significantly. What you already learned when creating IDoc-based MIGs will be also applicable for UN/EDIFACT-based MIGs. We will concentrate on the differences here and share some additional tips, but we'll skip some of the basic steps and activities that should be easily handled at this point even without detailed instructions.

EDIFACT Sales Order

Before we continue with MIG creation, let's look at the sample EDIFACT ORDERS message. This is an exact message that we expect from a trading partner for their purchase order. It was defined based on the mutual agreement between our own company and the trading partner. We will make use of this sample EDIFACT message while defining the MIG: simply load it in the MIG wizard in similar way as we did for outbound IDocs. A minor problem arises from the fact that so far, we only have flat EDIFACT, while the wizard expects the XML EDIFACT format. There is a quick solution for that. Examine the sample EDIFACT ORDERS D96A message shown in Figure 9.46.


```
UNB+UNOC:3+EDI_BEST_BIKES:ZZZ+EDI_COMPANY_1710:ZZZ+240301:1245+10001'
UNH+1+ORDERS:D:96A:UN'
BGM+220+BB_PO_12345+9'
DTM+137:20240301:102'
NAD+VN+COMPANY_1710_VN::ZZZ'
NAD+BY+BEST_BIKES_HQ::ZZZ'
NAD+DP+BEST_BIKES_WHS::ZZZ'
LIN+1++MZ-FG-R300:SA'
IMD+F++:::R300 BIKE'
QTY+21:5.00:PCE'
UNS+S'
CNT+2:1'
CNT+1:5.00'
UNT+13+1'
UNZ+1+10001'
```

Figure 9.46 Sample UN/EDIFACT ORDERS D96A Message

The trick is to build a simple iFlow that needs neither any actual senders nor receivers. You can see the full iFlow in Figure 9.47. Self-explanatory descriptions of the different functions should already give you an idea of what it does. We trigger this iFlow using the **Timer Start—Run Once** option.

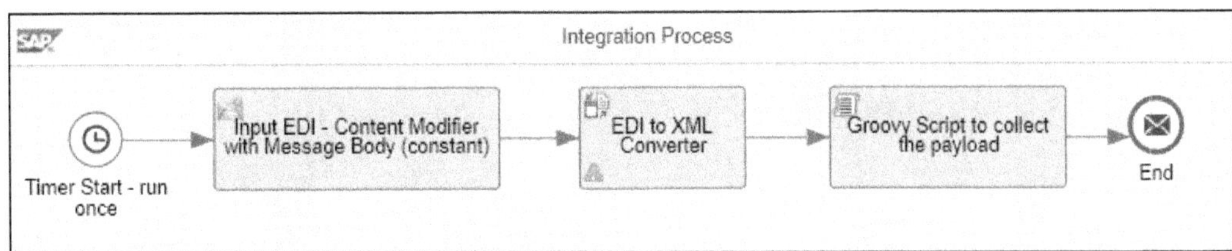

Figure 9.47 Sample iFlow to Convert EDI Message to XML Format

In a second step, we make use of the content modifier that stores the EDIFACT message and pushes it to the EDI-to-XML converter defined in a third step. The exact data for the EDI-to-XML conversion is stored as a constant message body in the content modifier; see Figure 9.48.

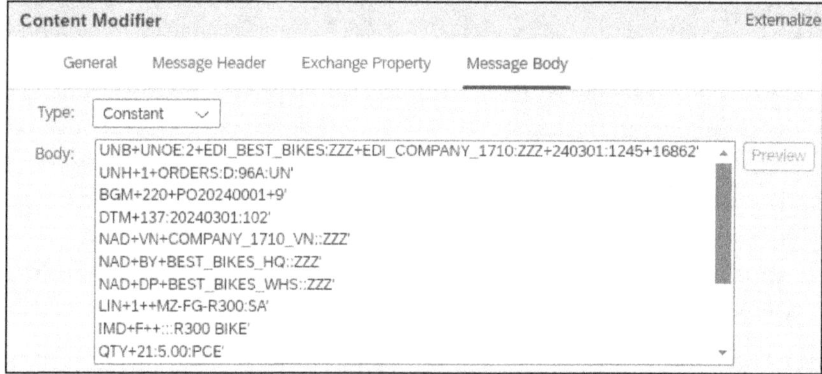

Figure 9.48 Content Modifier Step in Sample iFlow to Convert EDI Message to XML Format

The last important piece is that in the EDI converter, we need to add schemas related to the converted EDI format. In this case, we need a schema related to UN/EDIFACT ORDERS D96A, and the related XSD file can be exported from the MIG editor (see Figure 9.49 for a reference).

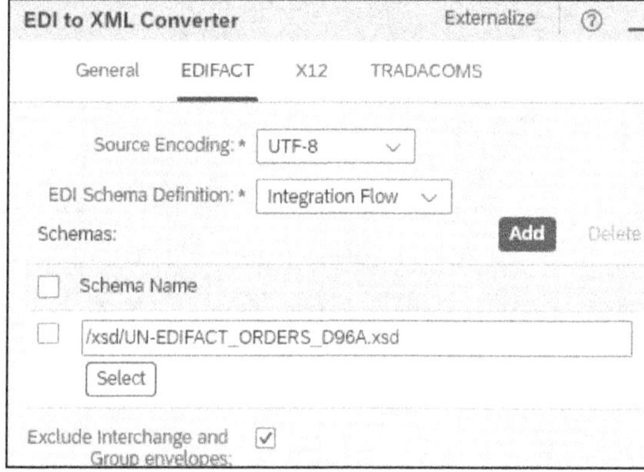

Figure 9.49 EDI to XML Converter Step in Sample iFlow to Convert EDI Message to XML Format

The task of the Groovy script located at the end of the iFlow is to collect the XML file generated by the EDI-to-XML converter. In our example, starting the iFlow will result in XML output, as in Listing 9.1.

```
<?xml version="1.0" encoding="ISO-8859-1"?><M_ORDERS>
    <S_UNH>
        <D_0062>1</D_0062>
        <C_S009>
            <D_0065>ORDERS</D_0065>
            <D_0052>D</D_0052>
            <D_0054>96A</D_0054>
```

```
        <D_0051>UN</D_0051>
    </C_S009>
</S_UNH>
<S_BGM>
    <C_C002>
        <D_1001>220</D_1001>
    </C_C002>
    <D_1004>BB_PO_12345</D_1004>
    <D_1225>9</D_1225>
</S_BGM>
<S_DTM>
    <C_C507>
        <D_2005>137</D_2005>
        <D_2380>20240301</D_2380>
        <D_2379>102</D_2379>
    </C_C507>
</S_DTM>
<G_SG2>
    <S_NAD>
        <D_3035>VN</D_3035>
        <C_C082>
            <D_3039>COMPANY_1710_VN</D_3039>
            <D_3055>ZZZ</D_3055>
        </C_C082>
    </S_NAD>
</G_SG2>
<G_SG2>
    <S_NAD>
        <D_3035>BY</D_3035>
        <C_C082>
            <D_3039>BEST_BIKES_HQ</D_3039>
            <D_3055>ZZZ</D_3055>
        </C_C082>
    </S_NAD>
</G_SG2>
<G_SG2>
    <S_NAD>
        <D_3035>DP</D_3035>
        <C_C082>
            <D_3039>BEST_BIKES_WHS</D_3039>
            <D_3055>ZZZ</D_3055>
        </C_C082>
    </S_NAD>
</G_SG2>
```

```
    <G_SG25>
        <S_LIN>
            <D_1082>1</D_1082>
            <C_C212>
                <D_7140>MZ-FG-R300</D_7140>
                <D_7143>SA</D_7143>
            </C_C212>
        </S_LIN>
        <S_QTY>
            <C_C186>
                <D_6063>21</D_6063>
                <D_6060>5.00</D_6060>
                <D_6411>PCE</D_6411>
            </C_C186>
        </S_QTY>
    </G_SG25>
    <S_UNS>
        <D_0081>S</D_0081>
    </S_UNS>
    <S_CNT>
        <C_C270>
            <D_6069>2</D_6069>
            <D_6066>1</D_6066>
        </C_C270>
    </S_CNT>
    <S_CNT>
        <C_C270>
            <D_6069>1</D_6069>
            <D_6066>5.00</D_6066>
        </C_C270>
    </S_CNT>
    <S_UNT>
        <D_0074>13</D_0074>
        <D_0062>1</D_0062>
    </S_UNT>
</M_ORDERS>
```

Listing 9.1 Sample EDIFACT ORDERS D96A Message in XML Format

To create a MIG for UN/EDIFACT messages, we need to follow the same steps as for IDoc messages. In the first step, we create a new MIG, choosing appropriate options as follows:

1. Go to **Design • MIGs** via the left-hand menu and click **Add**.
2. Choose **UN/EDIFACT** in the **Type Systems** section (see Figure 9.50).

Figure 9.50 Creation of MIG for EDIFACT ORDERS D96A: Type Systems Selection

3. Choose **ORDERS—Purchase Order Message** in the **Messages** view (see Figure 9.51).

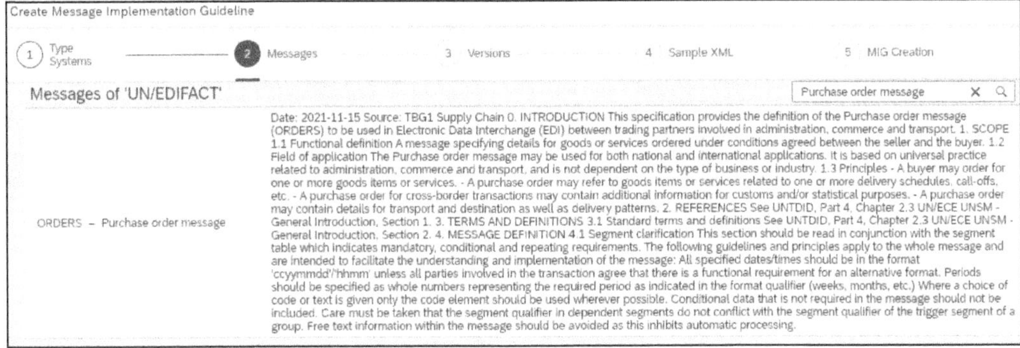

Figure 9.51 Creation of MIG for EDIFACT ORDERS D96A: Message Selection

4. Choose the **D.96A S3** version for **UN/EDIFACT ORDERS** (see Figure 9.52).

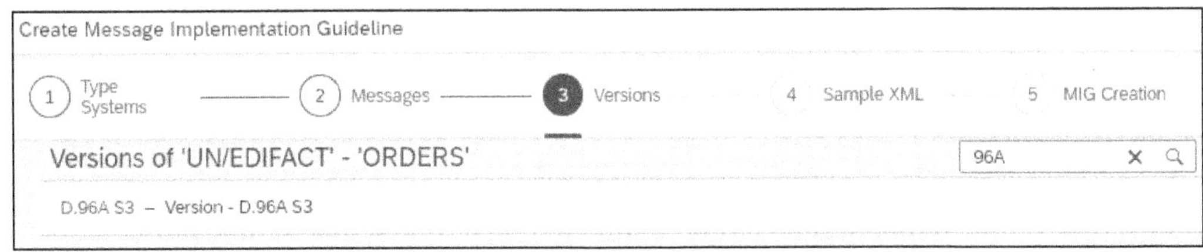

Figure 9.52 Creation of MIG for EDIFACT ORDERS D96A: Message Version Selection

5. Upload the sample XML file by browsing to the appropriate location in your file system. Keep **Data Format** set to **SAP Cloud Integration**. Select the **Use file content as sample values** option, which will help with the message simulation and testing (see Figure 9.53).
6. Maintain the general information. For **Name**, enter "EDIFACT sales order". For **Direction**, select **In**. Maintain your own business context values in the same way as for the other messages (see Figure 9.54).

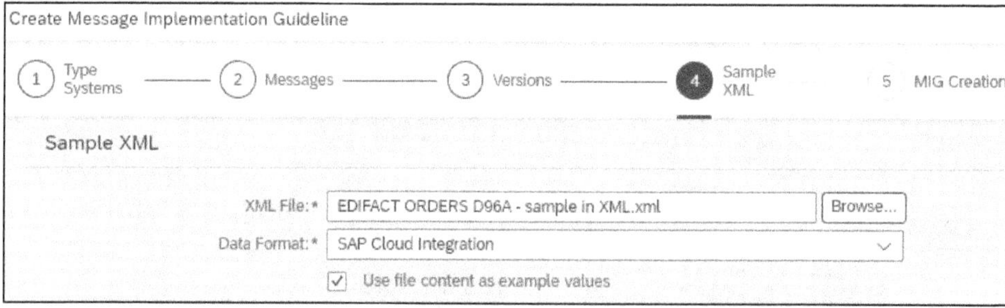

Figure 9.53 Creation of MIG for EDIFACT ORDERS D96A: Loading Sample XML

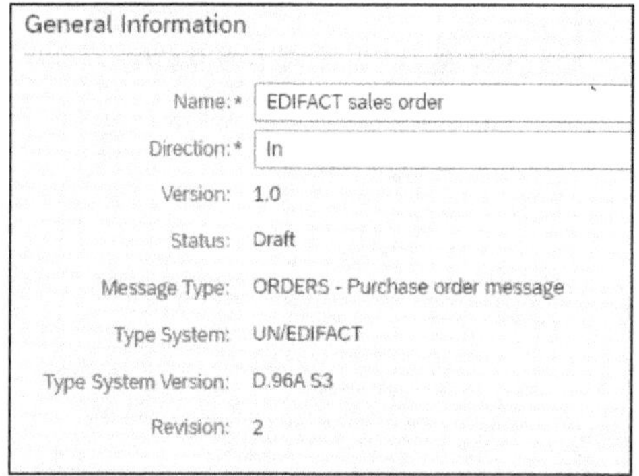

Figure 9.54 EDIFACT Sales Order MIG: General Information

If you open the MIG structure created from these steps, you can see that using the sample EDIFACT XML message gives two benefits. The most important is that you can find segments existing in the sample message that are already defined in the MIG structure. Looking at the structure shown in Figure 9.55, you can see segments like `BGM`, `DTM`, `SG2/NAD`, `SG25/LIN`, and so on. That definitely decreases the required effort for MIG maintenance. The second benefit is that all the data from the sample XML was also taken into the appropriate segments/fields as example values.

Structure

Node	Constraint	Cardinality	Position
ORDERS – Purchase order message		1..1	
UNH – MESSAGE HEADER		1..1	0010
BGM – Beginning of message		1..1	0020
DTM – Date/time/period	↰	1..35	0030
SG2 – Segment Group 2: NAD-LOC-FII-SG3-SG4-SG5	↰	0..99	0110
SG25 – Segment Group 25: LIN-PIA-IMD-MEA-QTY-PCD-ALI-DTM-MOA		0..200000	0930
UNS – SECTION CONTROL		1..1	2090
CNT – Control total	↰	0..10	2110
UNT – MESSAGE TRAILER		1..1	2160

Figure 9.55 MIG Structure Defined Based on Sample XML Message

Once again, you should qualify the necessary segments. Table 9.4 gives you an overview of the nodes and data elements enabled for the sales order transaction.

Data Element	EDIFACT Segment Field
Supplier's purchase order number	BGM segment
Supplier's purchase order date	DTM segment with qualifier 137
Vendor party	SG2 segment with qualifier VN
Delivery party	SG2 segment with qualifier DP
Buyer	SG2 segment with qualifier BY
Item	LIN segment
Line item number	LIN data element 1082
Item identifier	LIN C212 node with 7140 and 7143 data elements
Material description	IMD segment
Item quantity	QTY segment with qualifier 21

Table 9.4 Data Elements from UN/EDIFACT ORDERS D96A Message

Figure 9.56 shows the final structure for the UN/EDIFACT sales order.

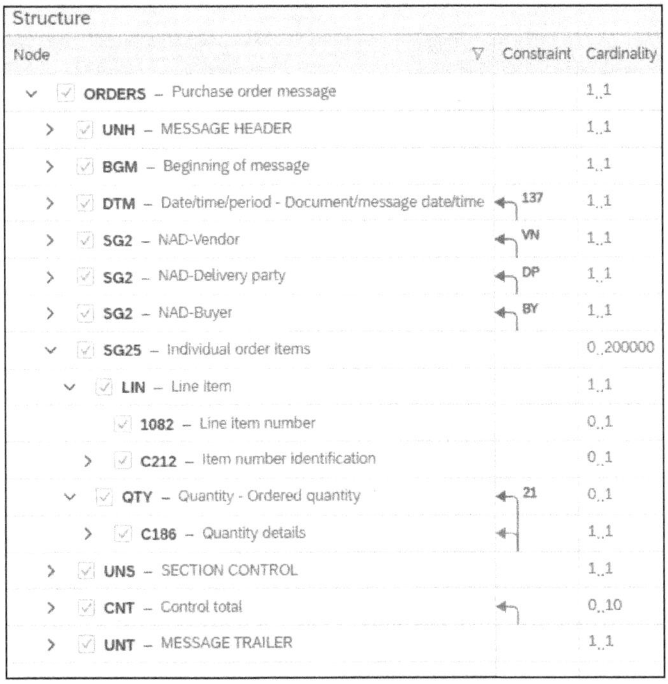

Structure

Node	Constraint	Cardinality
ORDERS – Purchase order message		1..1
UNH – MESSAGE HEADER		1..1
BGM – Beginning of message		1..1
DTM – Date/time/period - Document/message date/time	137	1..1
SG2 – NAD-Vendor	VN	1..1
SG2 – NAD-Delivery party	DP	1..1
SG2 – NAD-Buyer	BY	1..1
SG25 – Individual order items		0..200000
LIN – Line item		1..1
1082 – Line item number		0..1
C212 – Item number identification		0..1
QTY – Quantity - Ordered quantity	21	0..1
C186 – Quantity details		1..1
UNS – SECTION CONTROL		1..1
CNT – Control total		0..10
UNT – MESSAGE TRAILER		1..1

Figure 9.56 MIG EDIFACT Sales Order: Final Structure

EDIFACT Sales Order Response

Create the MIG structure for the UN/EDIFACT ORDRSP message with the **General Information** settings shown in Figure 9.57.

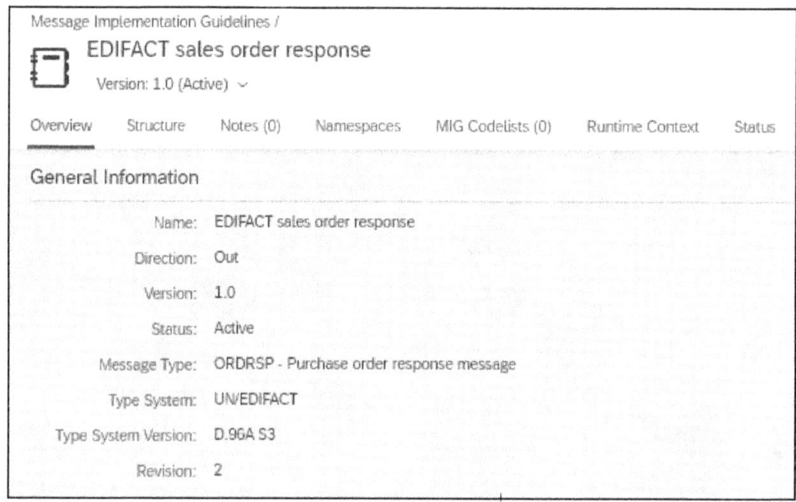

Figure 9.57 EDIFACT Sales Order Response MIG: General Information

Table 9.5 summarizes all the data elements that you should customize in the MIG structure related to the EDIFACT sales order response.

Data Element	EDIFACT Segment Field
Supplier (SAP) sales order number	BGM segment
Supplier (SAP) sales order date	DTM segment with qualifier 137
Customer purchase order number	RFF segment with qualifier ON
	RFF segment qualified for * remaining values
Currency	CUX segment
Material number	LIN segment
Material description	IMD segment
Item level—quantity	QTY segment with qualifier 21
Requested delivery data	DTM segment with qualifier 2
Line item amount (net)	MOA segment with qualifier 203
Item price (net)	PRI segment
Sales order value (net)	MOA segment with qualifier 139

Table 9.5 Data Elements from UN/EDIFACT ORDRSP D96A Message

Figure 9.58 shows the final structure of the sales order response MIG.

Figure 9.58 EDIFACT Sales Order Response MIG: Final Structure

EDIFACT Advanced Shipping Notification

Create the MIG structure for the EDIFACT DESADV message with the **General Information** settings as shown in Figure 9.59.

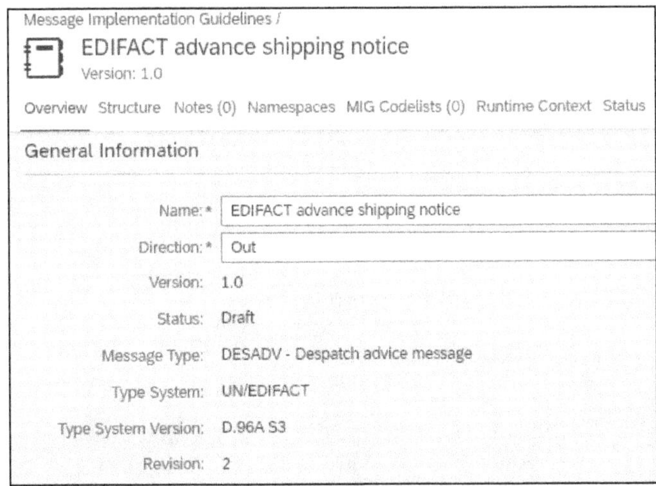

Figure 9.59 EDIFACT Advanced Shipping Notification MIG: General Information

Figure 9.60 shows the final structure.

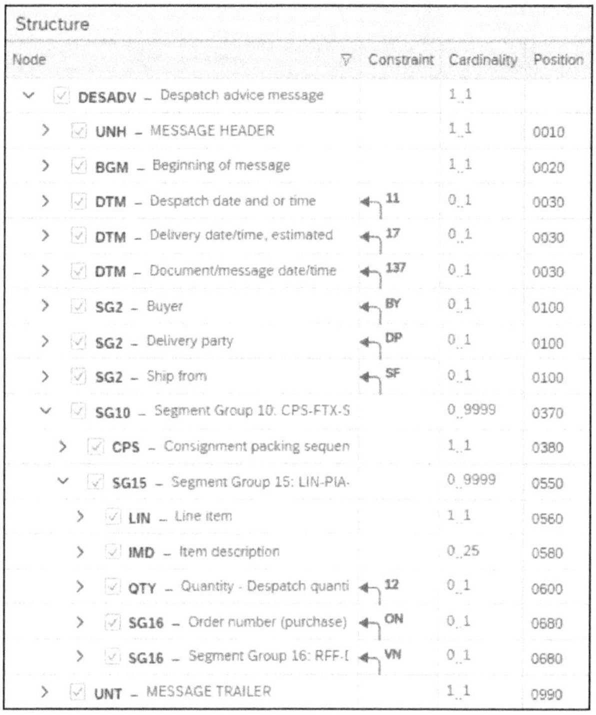

Structure

Node	Constraint	Cardinality	Position
DESADV – Despatch advice message		1..1	
UNH – MESSAGE HEADER		1..1	0010
BGM – Beginning of message		1..1	0020
DTM – Despatch date and or time	11	0..1	0030
DTM – Delivery date/time, estimated	17	0..1	0030
DTM – Document/message date/time	137	0..1	0030
SG2 – Buyer	BY	0..1	0100
SG2 – Delivery party	DP	0..1	0100
SG2 – Ship from	SF	0..1	0100
SG10 – Segment Group 10: CPS-FTX-S		0..9999	0370
CPS – Consignment packing sequen		1..1	0380
SG15 – Segment Group 15: LIN-PIA-		0..9999	0550
LIN – Line item		1..1	0560
IMD – Item description		0..25	0580
QTY – Quantity - Despatch quanti	12	0..1	0600
SG16 – Order number (purchase)	ON	0..1	0680
SG16 – Segment Group 16: RFF-I	VN	0..1	0680
UNT – MESSAGE TRAILER		1..1	0990

Figure 9.60 EDIFACT Advanced Shipping Notification MIG: Final Structure

Table 9.6 summarizes all the data elements you should customize in the MIG structure related to the EDIFACT advanced shipping notification.

Data Element	IDoc Field
Delivery document number	BGM segment
Goods issue date	DTM segment with qualifier 11 (NTANF from E1EDT13 segment with QUALF='006')
Outbound delivery date	DTM segment with qualifier 17 (NTANF from E1EDT13 segment with QUALF='007')
Outbound delivery date	DTM segment with qualifier 137 (NTANF from E1EDT13 segment with QUALF='007')
Sold-to party	SG2 segment with qualifier BY; enable all the date elements within this node
Ship-to party	SG2 segment with qualifier DP; enable all the date elements within this node

Table 9.6 Data Elements from UN/EDIFACT DESADV D96A

Data Element	IDoc Field
Shipping point	SG2 segment with qualifier SF; enable and all the date elements within this node
Consignment packing sequence	CPS+1
Item position and material number	LIN segment
Material description	IMD segment
Quantity delivered	QTY segment with qualifier 12
Sales order number	RFF segment with qualifier VN
Purchase order reference	RFF segment with qualifier ON

Table 9.6 Data Elements from UN/EDIFACT DESADV D96A (Cont.)

EDIFACT Invoice

Create the MIG structure for the EDIFACT invoice message with the **General Information** settings as shown in Figure 9.61.

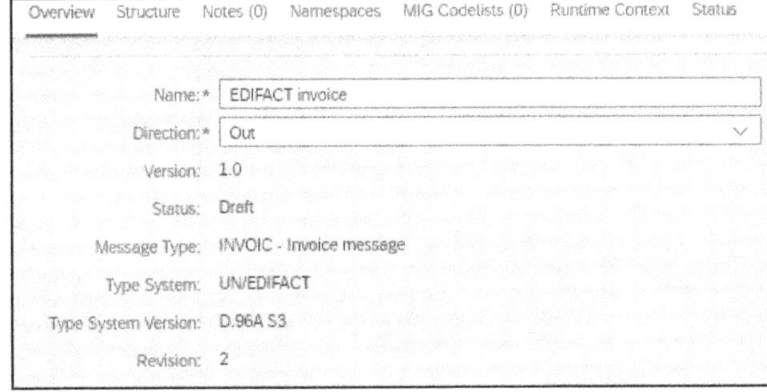

Figure 9.61 EDIFACT Invoice MIG: General Information

Table 9.7 contains all key elements to enable in the structure. If applicable, the related qualifiers also are mentioned.

Data Element	IDoc Field
Billing document number	BGM segment
Document date	DTM segment with qualifier 137
Delivery date	DTM segment with qualifier 35
Purchase order number	RFF segment with qualifier ON
Sales order number	RFF segment with qualifier VN

Table 9.7 Data Elements from UN/EDIFACT INVOIC D96A

Data Element	IDoc Field
Vendor party	NAD segment with qualifier SU
Item line number and ID	LIN segment
Material description	IMD segment
Item quantity	QTY segment with qualifier 47
Line item amount	MOA segment with qualifier 203
Item price	PRI segment with qualifier AAA
Total tax amount	MOA segment with qualifier 124
Total line items amount	MOA segment with qualifier 79

Table 9.7 Data Elements from UN/EDIFACT INVOIC D96A (Cont.)

Figure 9.62 shows the final MIG structure for the EDIFACT invoice.

Node	Constraint	Cardinality
INVOIC – Invoice message		1..1
UNH – MESSAGE HEADER		1..1
BGM – Beginning of message		1..1
DTM – Document issue date time	137	1..1
DTM – Delivery date/time, actual	35	1..1
SG1 – Order document identifier, buyer assigned	ON	0..1
RFF – Reference		1..1
SG1 – Delivery note number	DQ	0..1
SG1 – Order number (vendor)	VN	0..1
SG2 – Supplier	SU	0..1
SG25 – Item line		0..9999999
UNS – SECTION CONTROL		1..1
SG48 – Tax amount	124	1..1
MOA – Monetary amount		1..1
SG48 – Segment Group 48: MOA-SG49 - Total line items amount	79	1..1
UNT – MESSAGE TRAILER		1..1

Figure 9.62 MIG EDIFACT Invoice: Final Structure

At the end, we activate all the MIG structures related to EDIFACT messages. There are four of them, as shown in Figure 9.63. Now we can move to the next step and make use of these newly created MIGs as source and target structures in MAG definitions.

Message Implementation Guidelines (4)

EDIFACT advance shipping notice	Message Type: DESADV Type System: UN/EDIFACT Type System Version: D.96A S3 Direction: Out	Active Version: 1.0 History
EDIFACT invoice	Message Type: INVOIC Type System: UN/EDIFACT Type System Version: D.96A S3 Direction: Out	Active Version: 1.0 History
EDIFACT sales order	Message Type: ORDERS Type System: UN/EDIFACT Type System Version: D.96A S3 Direction: In	Active Version: 1.0 History
EDIFACT sales order response	Message Type: ORDRSP Type System: UN/EDIFACT Type System Version: D.96A S3 Direction: Out	Active Version: 1.0 History

Figure 9.63 Overview of UN/EDIFACT-Related MIGs

9.2.3 Defining Mapping Guidelines

The next step after creating MIGs is to define MAGs. There will be four of them in total. During these activities, you will already see some benefits of the effort invested in detailing the MIG structures, especially the qualified nodes. We will start with MAGs related to the inbound flow, where we map the source EDIFACT sales order MIG to the sales order IDoc MIG. One additional note is that we will not make use of the proposal service: we'll keep our messages very simple as we know exactly what the target messages should look like and what they should contain.

Let me introduce you to how we will structure this section. We will start with mappings related to sales order transactions. In this first example, you will read how to create a MAG and get familiar with the key features of the MAG editor. We will go through a few examples for the most typical use cases you may encounter during the mapping definition. This should give a solid background to start practicing with the tool and create basic mappings. To complement this topic, in the following section you will find a mapping overview for all four mappings to use for the B2B scenario. Based on this information, you can recreate the exact MAGs we use in this book.

Mapping the EDIFACT Sales Order to the IDoc Sales Order

Start by navigating to **Design • MAGs** via the left-hand menu in the SAP Integration Suite cockpit. We are welcomed by an overview screen, which in this case is empty as there are no MAGs created yet. Make use of the **Add** option located at the top-right corner to create a new MAG (see Figure 9.64).

Figure 9.64 MAG Creation: Maintenance Options for MAG Objects

You will see a new window listing all the available MIGs; you should see the eight created earlier. In the top-left corner, there is a hint to the required action. The first step is selection of the source MIG. Choose the appropriate structure from the list and click the **Next** button located in the bottom-right corner. This button will become active as soon you choose the source structure; in this case it is **EDIFACT Sales Order**, the third item in the list, as shown in Figure 9.65.

Select Source MIG

Source: EDIFACT sales order   Target: ...

Search

Name	Type System	Version	Message Type Namespace	Message	Direction	Version	Status
EDIFACT advance shipping notice	UN/EDIFACT	D.96A S3		DESADV	Out	1.0	Active
EDIFACT invoice	UN/EDIFACT	D.96A S3		INVOIC	Out	1.0	Active
EDIFACT sales order	UN/EDIFACT	D.96A S3		ORDERS	In	1.0	Active
EDIFACT sales order response	UN/EDIFACT	D.96A S3		ORDRSP	Out	1.0	Active
IDoc advance shipping notice	SAP S/4HANA On Premise IDoc	1809 FPS02		DESADV.DELVRY07	Out	1.0	Active
IDoc invoice	SAP S/4HANA On Premise IDoc	1809 FPS02		INVOIC.INVOIC02	Out	1.0	Active
IDoc sales order	SAP S/4HANA On Premise IDoc	1809 FPS02		ORDERS.ORDERS05	In	2.0	Active
IDoc sales order response	SAP S/4HANA On Premise IDoc	1809 FPS02		ORDRSP.ORDERS05	Out	1.0	Active

Next Create Cancel

Figure 9.65 MAG Creation: Selection of Source MIG

After confirming the selection of the source MIG by clicking **Next**, you go to the next step, selection of the target MIG. In this case, choose **IDoc Sales Order** as shown in Figure 9.66 and confirm by clicking **Create**.

You just created a MAG. In the new window that opens automatically, you can see the two MIG structures as shown in Figure 9.67. On the left-hand side, there is a source EDIFACT sales order and on the right-hand side there is an IDoc sales order. While browsing through the hierarchy of the MIG, you can see that all the details defined earlier in the MIG editor are taken into account. At the bottom part of the screen, you can see an area called **Mapping List**. It contains a summary of all the mapping rules already applied for the given MAG. At the beginning, you can see there are several entries related to constant values defined in the target MIG; all the other mapping rules related to the transformation from EDIFACT to IDoc will be added in the following parts of this section. As mentioned earlier, we will not make use of the AI-based proposal service. It is definitely a good option that may speed up your developments, but because we know well what to map, we choose to practice with the manual maintenance of the MAG.

Select Target MIG

Source: EDIFACT sales order **Target:** IDoc sales order

Search

Name	Type System	Version	Message Type Namespace	Message	Direction	Version	Status
EDIFACT advance shipping notice	UN/EDIFACT	D.96A S3		DESADV	Out	1.0	Active
EDIFACT invoice	UN/EDIFACT	D.96A S3		INVOIC	Out	1.0	Active
EDIFACT sales order	UN/EDIFACT	D.96A S3		ORDERS	In	1.0	Active
EDIFACT sales order response	UN/EDIFACT	D.96A S3		ORDRSP	Out	1.0	Active
IDoc advance shipping notice	SAP S/4HANA On Premise IDoc	1809 FPS02		DESADV.DELVRY07	Out	1.0	Active
IDoc invoice	SAP S/4HANA On Premise IDoc	1809 FPS02		INVOIC.INVOIC02	Out	1.0	Active
IDoc sales order	SAP S/4HANA On Premise IDoc	1809 FPS02		ORDERS.ORDERS05	In	2.0	Active
IDoc sales order response	SAP S/4HANA On Premise IDoc	1809 FPS02		ORDRSP.ORDERS05	Out	1.0	Active

Next Create Cancel

Figure 9.66 MAG Creation: Selection of Target MIG

Figure 9.67 Initial Screen of MAG Mapping Editor

Before we define our mapping rules, let's briefly go through the key functions of the mapping editor. While working with a single MAG, there are five tabs available at the top of the screen:

- **Overview**
  On this tab, you can see all the key information related to the MAG. Most of the attributes can be maintained: the name, source, and target business context, and the

documentation summary and definition. You can also change the source and target MIGs or choose different versions. One of powerful feature here is pretransformation of the given MIG structure. Its purpose is to support scenarios where direct mapping from the source to a target structure may be very complex or even not possible. We will not make use of it in our scenario, but as an example by using it we could perform EDIFACT ORDERS D96A to EDIFACT ORDERS D96A premapping (with some kind of sorting, filtering, and so on) before performing the transformation as per the MAG—that is, EDIFACT ORDERS D96A to IDoc ORDERS.ORDERS05 mapping.

- **Mapping**
  This we will study in detail separately. First we will go through all the available options, and shortly thereafter you will find the detailed description of how to define the mapping itself.
- **Global Parameter**
  This allows you to define global parameters that can be reused within a MAG. For educational purposes, we will make use of this within the MAG related to ASNs.
- **Shared Code**
  In this tab, you can create reusable code snippets.
- **Notes**
  In this tab, you can add any notes related to the MAG or to changes performed, or just add descriptions that you expect in the documentation.

Going back to the **Mapping** tab, in the right-hand corner there are options related to actions on the MAG object. These are similar to the ones known from the MIG editor:

- **Edit/Save**
  Switch to edit mode/save the changes.
- **Simulate**
  An option to test the mapping using sample data loaded earlier in the MIG or external payload.
- **Copy**
  Create a copy of the MAG.
- **Delete**
  Delete the MAG.
- **Activate**
  This is related to versioning features. It's also worth remembering that during activation, objects are shared with the knowledge graph.
- **Inject**
  Push artifacts to Cloud Integration.
- **Export**
  Here you can export the runtime artifacts or documentation to a local file system. We will look at the features related to documentation in a later part of this section, and the exporting function for the runtime artifacts will be reviewed in Section 9.2.4.

You can navigate through the source and target structures, expand/collapse nodes, and review any data element. When working with the mapping editor, the option to search for a specific value can be useful. This is available for a view with source structure, target structure, and mapping list. In Figure 9.68, you can see an example where we search for "E1EDP01" in the target IDoc structure. The value was found in two places and highlighted in orange/yellow.

What you probably also notice when looking at Figure 9.68 are the multiple signs (#) located on the left-hand side and linked to several fields of the target structure. This is how the constant values are represented; they are populated in the MAG from the MIG definition. You can still adjust or even delete these in the MAG editor.

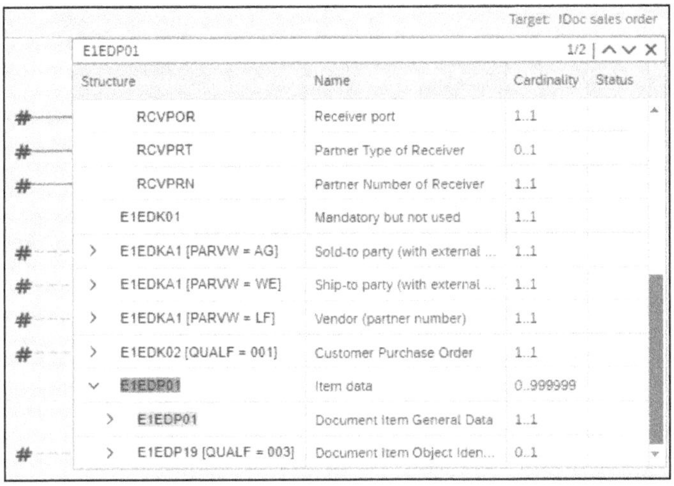

Figure 9.68 MAG Editor: Searching for E1EDP01 within Target Structure

If you choose one of the fields with a constant value, the details are displayed in the mapping list located at the bottom of the screen. Figure 9.69 shows the constant value AG after choosing the PARVW field in the E1EDKA1 segment qualified with AG.

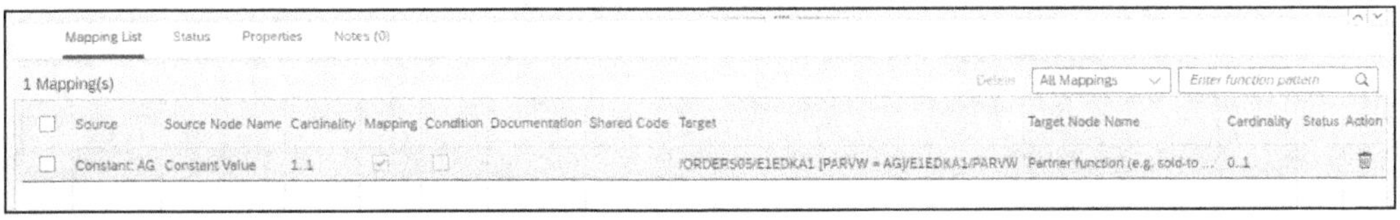

Figure 9.69 Mapping List in MAG Editor with Sample Constant Value

Let's start with a simple case of how to map a source element to a target element. We will take the buyer party identification data element as an example. The simplest approach is that you draw a mapping rule by dragging the source element and dropping it onto the target element, and that's the exact situation to see the benefit of qualified nodes. If you look at Figure 9.70, we dragged the EDIFACT party identification data element 3039 from the source to the target IDoc field LIFNR. As an outcome of that action, a line linking these two elements appeared. Thanks to the qualified nodes in the source and target MIGs, we do not have to add any additional conditions or functions. The intentions are very clear to the mapping editor as we chose a very specific source element from the appropriate node qualified as buyer party SG2 [3035=BY] and placed it within the target node qualified for the same data, sold-to party E1EDKA1 [PARVW=AG].

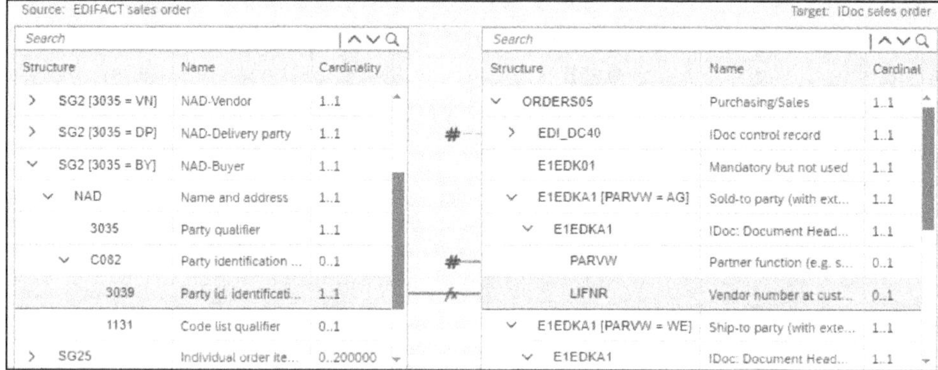

Figure 9.70 MAG Editor: Mapping between UN/EDIFACT 3039 and SAP IDoc LIFNR Data Elements

When you look closely at the line, you can see that there is a function icon placed on the line next to the target structure. As an automated action in the background, there was indeed a default function applied; you can find it in the bottom **Mapping List** section in the **Function** tab, as shown in Figure 9.71.

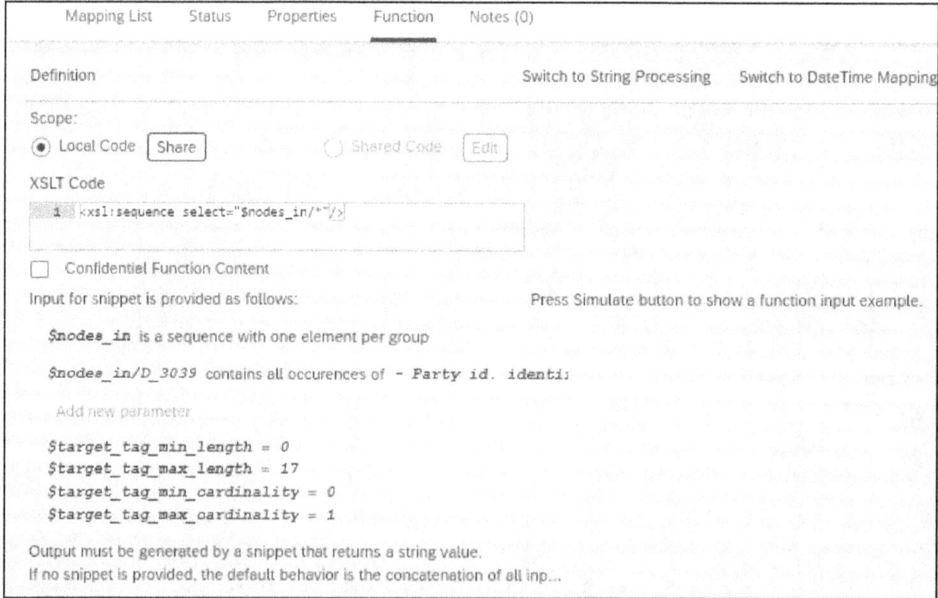

Figure 9.71 MAG Editor: XSLT Function Created by Dragging and Dropping

For the straightforward 1:1 mapping in this case where you just dragged and dropped the appropriate element, you don't need to worry about it; there is no need to modify anything. What you see in the **Function** tab is the XSLT code editor, which gives you almost unlimited options to define mapping rules based on XSLT V2.0 or XPath 2.0; all functions of these two styling languages are supported.

> **XSLT and XPath**
>
> As soon as you go into more complex scenarios and requirements, it becomes inevitable that you will need to make use of some XSLT or XPath expressions. The syntax and usage of these two languages is a topic much beyond this book, but it is something definitely worth learning, understanding, and practicing. The XSLT 2.0 specification is published by the World Wide Web Consortium (W3C) on their official webpage at *http://s-prs.co/v573605*.

The drag-and-drop action also generates an entry in the **Mapping List** tab. In Figure 9.72, you can see the source element and target elements represented as XPath expressions.

Mapping List | Status | Properties | Function | Notes (0)

1 Mapping(s) — Delete — All Mappings — Enter funct

	Source	Source Node N...	Cardinality	Mapping	Condition	Documentation	Shared Code	Target
☐	/ORDERS/SG2 [3035 = BY]/NAD/C082/3039	Party id. identif...	1..1	☑	☐			/ORDERS05/E1EDKA1 [PARVW = AG]/E1EDKA1/LIFNR

Figure 9.72 MAG Editor: Mapping List Section with Details for Buyer Identification Mapping

If you are still looking at the **Mapping List** tab, there is one selection option in the middle of the screen that may grab your attention. If you tick the **Condition** box, you will create an additional tab with yet another XSLT code editor in which you can implement additional expressions and conditions related to the preselected mapping rule. This is shown in Figure 9.73.

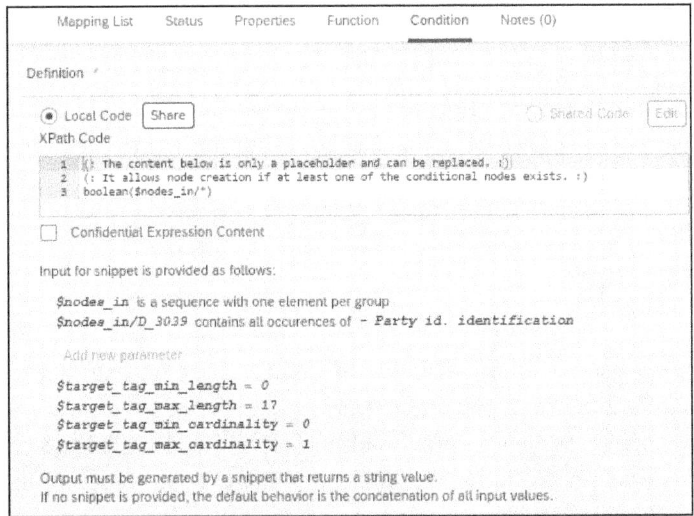

Figure 9.73 MAG Editor: Mapping List and Condition Tab

When you are back to the top section with the main MIG structures, you can go to the target structure and hover over any element located there. You will see three dots on the right-hand side next to the name of the segment or field. This context menu gets you different options depending on which element it is related to:

- For nodes, you will see the **Duplicate** option, which allows you to duplicate the given segment in the target structure and then apply specific mapping rules to it (different from the original occurrence of this segment).
- For data elements, you will see the three options shown in Figure 9.74. The **Duplicate** option is also there, but there are two more: **Set Constant** and **Set Function.** These are self-explanatory: it is not always the source structure or the drag-and-drop approach that defines the target value. In some cases, this is a fixed value; in others, the starting point could be an XLST function without a source node/data element.
- You probably noticed that selecting a node or data element in the source structure does not give you any options.

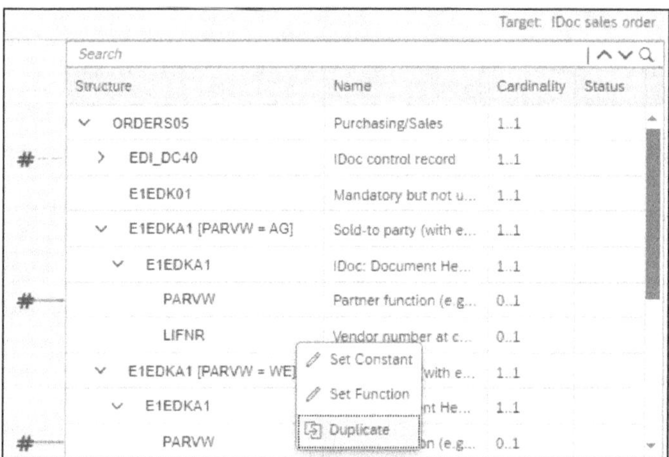

Figure 9.74 MAG Editor: Context Menu for Field in Target Structure

We will consciously omit a detailed description of the functions related to value transformations as we will not make use of these in our scenario—but it is worth mentioning the options, the names of which give an idea of their purposes:

- **Date Time Conversion**
- **String Processing Type**
- **Value Mapping**

When the drag-and-drop approach and other graphical functions are not enough to create the required transformation rules, you can make use of XSLT code. This option is available under the **Set Function** name and it gives almost unlimited possibilities to transform the source and target nodes. In the ASN MAG, we will make use of the simple functions to demonstrate the functionality.

We briefly went through the key functions, which should be enough to have a basic understanding of the capabilities and to handle our scenario, where we only make use of the basic features. To discover and practice with all the available options definitely would require more time, and a pretty simple scenario with four messages may not be enough.

Now that we've taken a very brief look at the key functions related to MAG maintenance, we can go ahead with a mapping definition for the sales order structures. All the details we already listed and explained in Section 9.2.1 and Section 9.2.2 now become very handy. Later on in this section, in Figure 9.77 and Figure 9.79, we will summarize what the mapping will look like. We skip the IDoc control record (EDI_DC40) for now; you may define some mapping rules for its data element, but that's not necessary as the configuration of B2B transaction itself will take care of populating the proper values. You will learn about this in Section 9.3.

After MAG creation, you may still need to do some updates in the mapping itself and the related MIG objects. In the MAG editor, there is a possibility to work with the different MIG versions. This option is available on the **Overview** screen of the MAG object in the **Source and Target MIGs** section. For both the source and target MIG, you can find an **Edit** icon located next to the **Version (Status)** label. As shown in Figure 9.75, you have the option to switch among all available MIG versions. Confirm your choice by clicking **Change Version.**

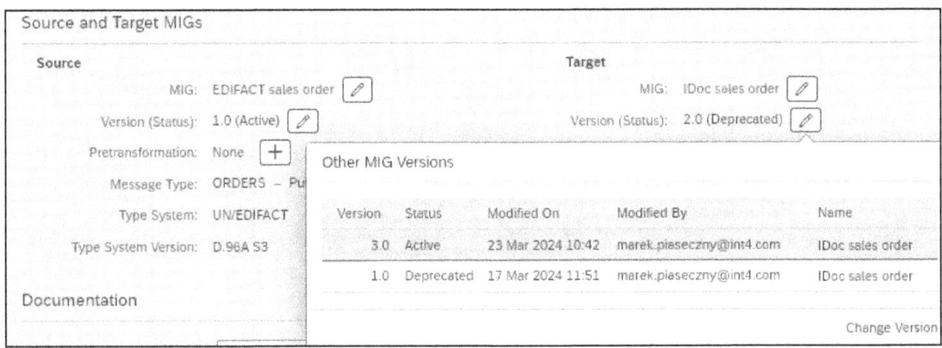

Figure 9.75 Selection of Target MIG for Sales Order MAG

By using the information from the sections related to MAG definitions and then following the drag-and-drop approach, we can complete the whole mapping required for sales order processing. There is no need to implement any additional functions, and there are no self-written XSLT/XPath expressions. Thanks to all the built-in features in Integration Advisor related to documentation and notes, we are now in the comfortable position of not needing to spend additional time in any external text editors or spreadsheet applications trying to collect all the activities and mapping definitions. Let's now look at the export functions that can generate documentation based on the configuration and development. Go to the options located in the top-right-hand corner. After clicking the three dots, choose **Export • Documentation (Excel)**, as shown in Figure 9.76.

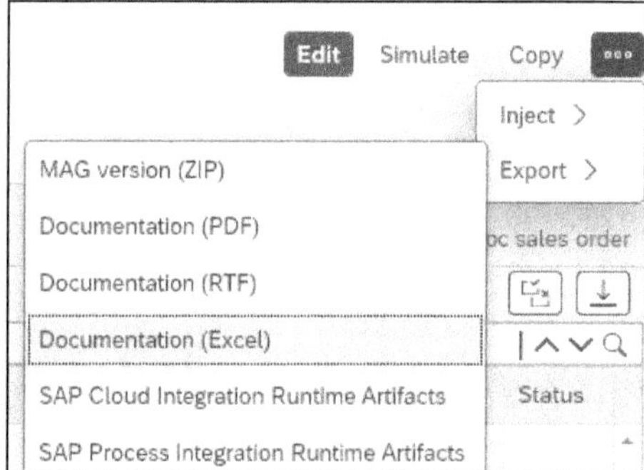

Figure 9.76 MAG Editor: Functionality to Export Documentation (Excel)

The result of this function will be a file in Excel format that contains all the information related to the mapping—not only the source and target elements but their full descriptions, names, and notes maintained by us. It is enough to mention that in our example the spreadsheet has 39 columns. After hiding most of the columns and limiting the view to the key information, we can see all the mapping rules included in the mapping (see Figure 9.77; you can make use of this figure while recreating the scenario). This is really only a small snapshot of what's in the spreadsheet.

SAP Mapping Documentation
Mapping Guideline
Mapping EDIFACT sales order to IDoc sales order
Document Version: 1.0 –

Target						**Source**
No #	**Structure Level 1**	**Lev. 2**	**Lev. 3**	**Lev. 4**	**Name**	**Path**
00.	ORDERS05				Purchasing/Sales	
01.		EDI_DC40			IDoc control record	
02.		E1EDK01			Mandatory but not used	
03.		E1EDKA1 [PARVW = “AG”]			Sold-to party (with external number)	
03.01.			E1EDKA1		IDoc Document Header Partner Information	
03.01.01.				PARVW	Partner function (e.g. sold-to party, ship-to party, ...)	
03.01.02.				LIFNR	Vendor number at customer location	/ORDERS/SG2[./NAD/3035=BY]/NAD/C082/3039
04.		E1EDKA1 [PARVW = “WE”]			Ship-to party (with external number)	
04.01.			E1EDKA1		IDoc Document Header Partner Information	
04.01.01.				PARVW	Partner function (e.g. sold-to party, ship-to party, ...)	
04.01.02.				LIFNR	Vendor number at customer location	/ORDERS/SG2[./NAD/3035=DP]/NAD/C082/3039
05.		E1EDKA1 [PARVW = “LF”]			Vendor (partner number)	
05.01.			E1EDKA1		IDoc Document Header Partner Information	
05.01.01.				PARVW	Partner function (e.g. sold-to party, ship-to party, ...)	
05.01.02.				PARTN	Partner number	/ORDERS/SG2[./NAD/3035=VN]/NAD/C082/3039
06.		E1EDK02 [QUALF = “001”]			Customer Purchase Order	
06.01.			QUALF		Fixed value 001	
06.02.			BELNR		Customer purchase order number	/ORDERS/BGM/1004
07.		E1EDP01			Item data	/ORDERS/SG25
07.01.			E1EDP01		Document Item General Data	
07.01.01.				MENGE	Quantity	/ORDERS/SG25/QTY[./C186/6063=21]/C186/6060
07.01.02.				MENEE	Unit of measure	/ORDERS/SG25/QTY[./C186/6063=21]/C186/6411
07.02.			E1EDP19 [QUALF = “003”]		Document Item Object Identification - International Article Number(EAN)	
07.02.01.				QUALF	International Article Number(EAN)	
07.02.02.				IDTNR	IDOC material ID	/ORDERS/SG25/LIN/C212/7140
07.02.03.				KTEXT	IDOC short text	/ORDERS/SG25/LIN/C212/7143

Figure 9.77 MAG Documentation: Excel File Generated from MAG Editor for Mapping EDIFACT Sales Order to IDoc Sales Order

In similar way, by choosing **Export • Documentation (PDF)**, we can generate a file in a more readable format, but also much more extensive. For our simple mapping, the generated PDF document has 30 pages; all the details are collected and presented in a very transparent way. Figure 9.78 shows the table of contents of the automatically generated PDF document, which should give you some idea of where the mentioned 30 pages come from. The document is truly complete and describes all aspects of the mapping.

TABLE OF CONTENTS

1	OVERVIEW	3
1.1	Mapping Overview	4
1.2	Mapping Elements	5
2	DETAILS	7
2.0	Namespaces	
2.1	ORDERS05 — Purchasing/Sales	7
2.2	EDI_DC40 — IDoc control record	8
2.3	E1EDK01 — Mandatory but not used	12
2.4	E1EDKA1 [PARVW = "AG"] — Sold-to party (with external number)	13
2.5	E1EDKA1 — IDoc: Document Header Partner Information	14
2.6	E1EDKA1 [PARVW = "WE"] — Ship-to party (with external number)	16
2.7	E1EDKA1 — IDoc: Document Header Partner Information	17
2.8	E1EDKA1 [PARVW = "LF"] — Vendor (partner number)	19
2.9	E1EDKA1 — IDoc: Document Header Partner Information	20
2.10	E1EDK02 [QUALF = "001"] — Customer Purchase Order	22
2.11	E1EDP01 — Item data	24
2.12	E1EDP01 — Document Item General Data	25
2.13	E1EDP19 [QUALF = "003"] — Document Item Object Identification - International Article Number(EAN)	27
3	COPYRIGHT STATEMENTS	29
3.1	Copyright Statement for MAG	29
3.2	Copyright Statement for Type System ASC X12	29
3.3	Copyright Statement for Type System UN/EDIFACT	29
3.4	Copyright Statement for Type System ISO Codelists	29
3.5	Copyright Statement for Type System UN/CEFACT	29
INDEX		30

Figure 9.78 Table of Contents of Automatically Generated Sales Order Documentation

Figure 9.79 shows the mapping overview as generated by the **Export • Documentation (PDF)** function. As you can see, this single page could be already enough to get an idea of the functionality of this interface. The following pages of this PDF document describe all additional details, and the document covers all possible aspects of the MAG objects.

1.1 MAPPING OVERVIEW

Source					Target	
Path	**Name**	**Card.**	**Prim.Type**	**Length**	**Node**	**Card.**
					ORDERS05 — Purchasing/Sales	1 .. 1
					├ **EDI_DC40** — IDoc control record	1 .. 1
					├ **TABNAM** — Name of Table Structure	1 .. 1
Constant STRING value: I					├ **DIRECT** — Direction	1 .. 1
Constant STRING value: ORDERS05					├ **IDOCTYP** — Name of basic type	1 .. 1
Constant STRING value: ORDERS					├ **MESTYP** — Message type	1 .. 1
Constant STRING value: B2B					├ **MESCOD** — Message code	0 .. 1
Constant STRING value: SAPCI					├ **SNDPOR** — Sender port (SAP System, external subsystem)	1 .. 1
Constant STRING value: KU					├ **SNDPRT** — Partner type of sender	1 .. 1
					├ **SNDPRN** — Partner Number of Sender	1 .. 1
Constant STRING value: SAPS42					├ **RCVPOR** — Receiver port	1 .. 1
Constant STRING value: LS					├ **RCVPRT** — Partner Type of Receiver	0 .. 1
Constant STRING value: S42CLNT400					└ **RCVPRN** — Partner Number of Receiver	1 .. 1
					├ **E1EDK01** — Mandatory but not used	1 .. 1
					├ **E1EDKA1** — Sold-to party (with external number)	1 .. 1
					└ **E1EDKA1** — IDoc: Document Header Partner Information	1 .. 1
Constant STRING value: AG					├ **PARVW** — Partner function (e.g. sold-to party, ship-to party, ...)	0 .. 1
/ORDERS/SG2[./NAD/3035=BY]/NAD/C082/3039	Party id. identification	1 .. 1	String	1 .. 35	└ **LIFNR** — Vendor number at customer location	0 .. 1
					├ **E1EDKA1** — Ship-to party (with external number)	1 .. 1
					└ **E1EDKA1** — IDoc: Document Header Partner Information	1 .. 1
Constant STRING value: WE					├ **PARVW** — Partner function (e.g. sold-to party, ship-to party, ...)	0 .. 1
/ORDERS/SG2[./NAD/3035=DP]/NAD/C082/3039	Party id. identification	1 .. 1	String	1 .. 35	└ **LIFNR** — Vendor number at customer location	0 .. 1
					├ **E1EDKA1** — Vendor (partner number)	1 .. 1
					└ **E1EDKA1** — IDoc: Document Header Partner Information	1 .. 1
Constant STRING value: LF					├ **PARVW** — Partner function (e.g. sold-to party, ship-to party, ...)	0 .. 1
/ORDERS/SG2[./NAD/3035=VN]/NAD/C082/3039	Party id. identification	1 .. 1	String	1 .. 35	└ **PARTN** — Partner number	0 .. 1
					├ **E1EDK02** — Customer Purchase Order	1 .. 1
Constant STRING value: 001					├ **QUALF** — Fixed value 001	1 .. 1
/ORDERS/BGM/1004	Document/message number	0 .. 1	String	1 .. 35	└ **BELNR** — Customer purchase order number	1 .. 1
/ORDERS/SG25	Individual order items	0 .. 200000			└ **E1EDP01** — Item data	0 .. 99999[illegible]
					├ **E1EDP01** — Document Item General Data	1 .. 1
/ORDERS/SG25/QTY[./C186/6063=21]/C186/6060	Quantity	1 .. 1	Decimal	1 .. 15	├ **MENGE** — Quantity	0 .. 1
/ORDERS/SG25/QTY[./C186/6063=21]/C186/6411	Measure unit qualifier	0 .. 1	String	1 .. 3	└ **MENEE** — Unit of measure	0 .. 1
					└ **E1EDP19** — Document Item Object Identification - International Article Number(EAN)	0 .. 1
Constant STRING value: 003					├ **QUALF** — International Article Number(EAN)	0 .. 1
/ORDERS/SG25/LIN/C212/7140	Item number	0 .. 1	String	1 .. 35	├ **IDTNR** — IDOC material ID	0 .. 1
/ORDERS/SG25/IMD/C273/7009	Item description identification	0 .. 1	String	1 .. 17	└ **KTEXT** — IDOC short text	0 .. 1

Figure 9.79 MAG Documentation: Mapping Overview for Mapping EDIFACT Sales Order to IDoc Sales Order

We are now done with the mapping definition, but it is not recommended to leave without at least a quick unit test of the MAG. To verify the MAG, we will use of the simulation option and the UN/EDIFACT XML file that we created and loaded while working on the EDIFACT sales order MIG. The sample file is already stored within the MIG, so you can trigger **Edit • Simulate with MIG Example Data** from the MAG editor, as shown in Figure 9.80.

Figure 9.80 MAG Function to Simulate with Payload Data

After the successful load of the sample XML file, we will see two additional elements within the view related to source data (see Figure 9.81):

- **Simulation Data: From MIG Example Data** offers an additional icon to rerun the simulation. This is located between the name of the source structure and the structure view itself.

- The **Simulation Data** column appears in the structure view.

Source: EDIFACT sales order

Simulation Data: From MIG example data

Search

Structure	Name	Cardinality	Simulation Data
∨ ORDERS	Purchase order message	1..1	< 1/1 >
> UNH	MESSAGE HEADER	1..1	< 1/1 >

Figure 9.81 Simulation Data Based on MIG Example Data

Alternatively, we can also load any other external XML file using **Edit • Simulate with Payload Data**. In such a case, the functionality will work the same; the only small difference is that you will see the file name in the source structure view (see Figure 9.82).

Source: EDIFACT sales order

Simulation Data: sample EDIFACT ORDERS XML In - two items.xml

Search

Structure	Name	Cardinality	Simulation Data
∨ ORDERS	Purchase order message	1..1	< 1/1 >
> UNH	MESSAGE HEADER	1..1	< 1/1 >

Figure 9.82 Simulation Data Based on External Payload

It may happen that your sample data that you load for simulation contains more data than is defined in the MIG structure. In such a case, you will see a warning message like that shown in Figure 9.83.

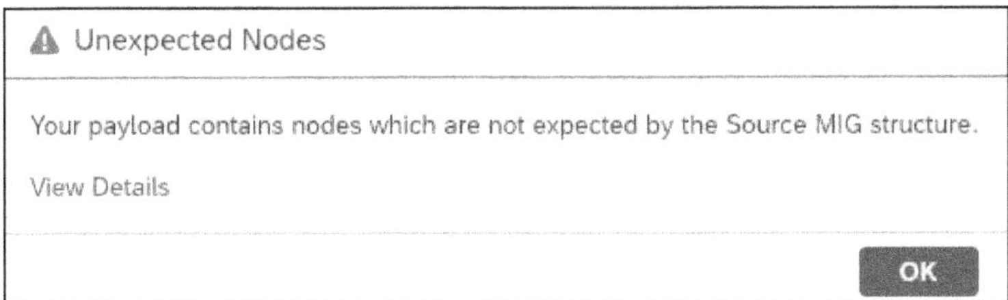

Figure 9.83 MAG Simulation: Warning Message when Source Contains Unexpected Nodes

You can also look at the details to see which nodes/data elements are triggering this warning. It does not necessarily mean that there is a problem with the MIG or MAG, but you should double-check if these are redundant elements that can be indeed ignored or perhaps missed elements that should be enabled in the MIG and following MAG. A sample warning message related to unexpected nodes is shown in Figure 9.84.

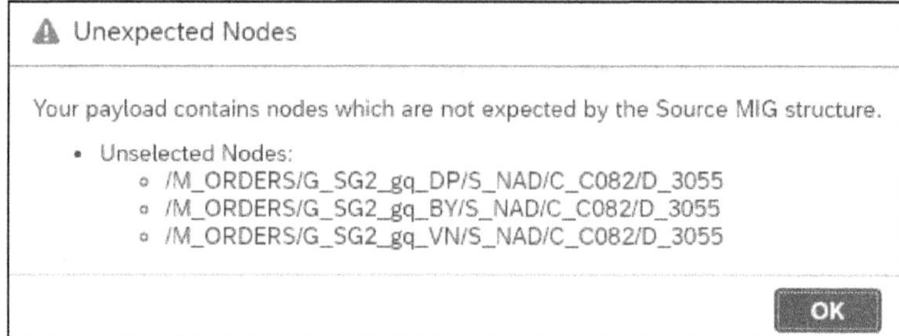

Figure 9.84 MAG Simulation: Detail View of Warning Message when Source Contains Unexpected Nodes

In Figure 9.85, you can see the results of the simulation run. There is an additional **Simulation Data** column in both views, for the source and target structures. On the left side, you can see the input value related to each node/data element, and on the right side, you can see the output value related to each node/data element within the target structure. In our example, you can see that the quantity and unit of measure values from the QTY segment with qualifier 21 were mapped to IDoc fields MENGE and MENEE—which is exactly how we were expecting the mapping to work.

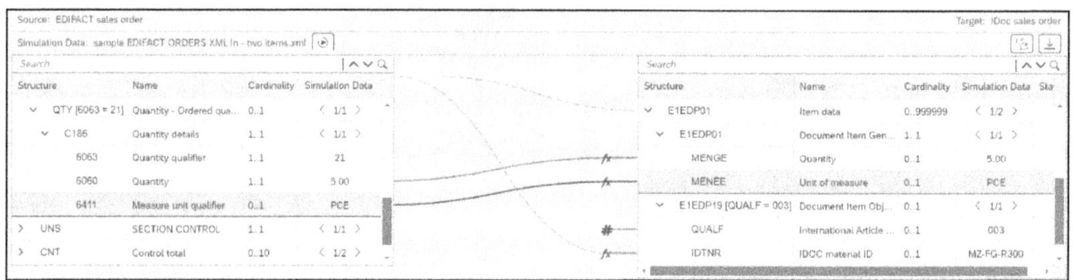

Figure 9.85 MAG Editor: Simulation Results

Mapping Sales Order Response IDoc to EDIFACT Sales Order Response

At this point, you should already have the basic understanding and skills to work with the MAG editor. We will not go through all the activities necessary for creation of the sales order response MAG. Instead, we will share the applied mapping rules with the use of documentation generated from the MAG object. It is worth mentioning that the documentation generated from this MAG object is 67 pages in length. Figure 9.86 shows the overview page that summarizes the key information related to the MAG object.

Next, we'll show the key elements from the documenation and its mapping overview pages. (The original pages are wider and contain additional columns, but to keep it readable, we omitted the less important parts.) Figure 9.87 shows the first part of the sales order response mapping overview.

1 OVERVIEW

General Information

Name	Mapping IDoc sales order response to EDIFACT sales order response
Version	3.0
Status	Active

Source and Target MIGs

	Source	Target
Name	IDoc sales order response	EDIFACT sales order response
Version	2.0	1.0
Status	Active	Active
Message Type	ORDRSP.ORDERS05	ORDRSP
Type System	SAP_IDoc	UNEDIFACT
Version	1809_FPS02	D.96A S3
External Category	Message Type	Message Type

Documentation

Summary	UID-85687cd1-a035-478d-82a8-7d335753964a
Definition	

Business Context Data

	Source Business Context	Target Business Context
Country/Region	United States	United States
Product Classification	Sports and Recreational Equipment and Supplies and Accessories	Sports and Recreational Equipment and Supplies and Accessories
Industry Classification	Sports and Entertainment	Sports and Entertainment
Business ProcessRole	Supplier	Supplier
Business Process	Manage Purchase Order	Manage Purchase Order

Figure 9.86 Overview Page from PDF Document related to MAG: Mapping IDoc Sales Order Response to EDIFACT Sales Order Response

Source						
Path	**Name**	**Card.**	**Prim.Type**	**Length**		**Node**
						ORDRSP — Purchase order response message
						├ **UNH** — MESSAGE HEADER
						| ├ **0062** — Message reference number
						| └ **S009** — MESSAGE IDENTIFIER
Constant STRING value: ORDRSP						| ├ **0065** — Message type identifier
Constant STRING value: D						| ├ **0052** — Message type version number
Constant STRING value: 96A						| ├ **0054** — Message type release number
Constant STRING value: UN						| └ **0051** — Controlling agency
/ORDERS01/E1EDK02[./QUALF=002]	Supplier Order	0 .. 1				├ **BGM** — Beginning of message
/ORDERS01/E1EDK02[./QUALF=002]/BELNR	IDOC document number	0 .. 1	String	0 .. 35		| ├ **C002** — Document/message name
Constant STRING value: 231						| | └ **1001** — Document/message name, coded
/ORDERS01/E1EDK02[./QUALF=002]/BELNR	IDOC document number	0 .. 1	String	0 .. 35		| └ **1004** — Document/message number
						├ **DTM** — Date/time/period - Document/message date/time
						| └ **C507** — Date/time/period
Constant STRING value: 137						| ├ **2005** — Date/time/period qualifier
/ORDERS01/E1EDK02[./QUALF=002]/DATUM	IDOC: Date	0 .. 1	String	0 .. 8		| ├ **2380** — Date/time/period
Constant STRING value: 102						| └ **2379** — Date/time/period format qualifier
						├ **SG1** — Segment Group 1: RFF-DTM - Order number (purchase
						| ├ **RFF** — Reference
						| | └ **C506** — Reference
Constant STRING value: ON						| | ├ **1153** — Reference qualifier
/ORDERS01/E1EDK02[./QUALF=001]/BELNR	IDOC document number	0 .. 1	String	0 .. 35		| | └ **1154** — Reference number

Figure 9.87 Mapping Overview for Mapping SAP IDoc ORDRSP to UN/EDIFACT ORDRSP, Part 1

9

Figure 9.88 shows the second part of the sales order response mapping overview.

Path	Name	Card.	Prim.Type	Length	Node
/ORDERS01/E1EDK01/CURCY	Currency	0 .. 1	String	0 .. 3	| └ **CUX** — Currencies
					| └ **C504** — Currency details
Constant STRING value: 2					| ├ **6347** — Currency details qualifier
/ORDERS01/E1EDK01/CURCY	Currency	0 .. 1	String	0 .. 3	| └ **6345** — Currency, coded
/ORDERS01/E1EDP01	Segment Group - IDoc: Document Item General Data	0 .. 999999			├ **SG26** — Segment Group 26: LIN-PIA-IMD-MEA-QTY-DTM-MOA-GIN-GIR-QVR-DOC-PAI-FTX-SG53-SG54
					| ├ **LIN** — Line item
/ORDERS01/E1EDP01/E1EDP01/POSEX	Item number	0 .. 1	String	0 .. 6	| | ├ **1082** — Line item number
/ORDERS01/E1EDP01/E1EDP01/ACTION	Action code for the item	0 .. 1	String	0 .. 3	| | ├ **1229** — Action request/notification, coded
/ORDERS01/E1EDP01/E1EDP19[./QUALF=002]	IDoc: Document Item Object Identification - Material number used by vendor	0 .. 1			| | └ **C212** — Item number identification
/ORDERS01/E1EDP01/E1EDP19[./QUALF=002]/IDTNR	IDOC material ID	0 .. 1	String	0 .. 35	| | ├ **7140** — Item number
Constant STRING value: NA					| | └ **7143** — Item number type, coded
/ORDERS01/E1EDP01/E1EDP19[./QUALF=002]	IDoc: Document Item Object Identification - Material number used by vendor	0 .. 1			| ├ **IMD** — Item description
Constant STRING value: F					| | ├ **7077** — Item description type, coded
					| | ├ **7081** — Item characteristic, coded
					| | └ **C273** — Item description
/ORDERS01/E1EDP01/E1EDP19[./QUALF=002]/KTEXT	IDOC short text	0 .. 1	String	0 .. 70	| | └ **7009** — Item description identification
					| ├ **QTY** — Quantity - Ordered quantity
					| | └ **C186** — Quantity details
Constant STRING value: 21					| | ├ **6063** — Quantity qualifier
/ORDERS01/E1EDP01/E1EDP01/MENGE	Quantity	0 .. 1	String	0 .. 15	| | ├ **6060** — Quantity
					| | └ **6411** — Measure unit qualifier

Figure 9.88 Mapping Overview for Mapping SAP IDoc ORDRSP to UN/EDIFACT ORDRSP, Part 2

Figure 9.89 shows the third part of the sales order response mapping overview.

Source					
Path	**Name**	**Card.**	**Prim.Type**	**Length**	**Node**
/ORDERS01/E1EDP01/E1EDP01	IDoc: Document Item General Data	1 .. 1			| ├ **DTM** — Date/time/period - Delivery date/time, requested
					| | └ **C507** — Date/time/period
Constant STRING value: 2					| | ├ **2005** — Date/time/period qualifier
/ORDERS01/E1EDK03[./IDDAT=002]/DATUM	IDOC: Date	0 .. 1	String	0 .. 8	| | ├ **2380** — Date/time/period
Constant STRING value: 102					| | └ **2379** — Date/time/period format qualifier
/ORDERS01/E1EDP01/E1EDP01	IDoc: Document Item General Data	1 .. 1			| ├ **MOA** — Monetary amount - Line item amount
					| | └ **C516** — Monetary amount
Constant STRING value: 203					| | ├ **5025** — Monetary amount type qualifier
/ORDERS01/E1EDP01/E1EDP01/NETWR	Item value (net)	0 .. 1	String	0 .. 18	| | └ **5004** — Monetary amount
					| └ **SG30** — Segment Group 30: PRI-CUX-APR-RNG-DTM
/ORDERS01/E1EDP01/E1EDP01	IDoc: Document Item General Data	1 .. 1			| └ **PRI** — Price details
					| └ **C509** — Price information
Constant STRING value: AAA					| ├ **5125** — Price qualifier
/ORDERS01/E1EDP01/E1EDP01/VPREI	Price (net)	0 .. 1	String	0 .. 15	| └ **5118** — Price
					├ **UNS** — SECTION CONTROL
					| └ **0081** — Section identification
/ORDERS01/E1EDS01[./SUMID=002]	Net document value (EVT)	0 .. 1			├ **MOA** — Monetary amount - Total payment amount
					| └ **C516** — Monetary amount
Constant STRING value: 139					| ├ **5025** — Monetary amount type qualifier
/ORDERS01/E1EDS01[./SUMID=002]/SUMME	Total value of sum segment	0 .. 1	String	0 .. 18	| └ **5004** — Monetary amount
					└ **UNT** — MESSAGE TRAILER
					├ **0074** — Number of segments in a message
					└ **0062** — Message reference number

Figure 9.89 Mapping Overview for Mapping SAP IDoc ORDRSP to UN/EDIFACT ORDRSP, Part 3

Mapping IDoc Advanced Shipping Notification to EDIFACT Advanced Shipping Notification

We will continue with the same approach as in previous section; that is, we will skip the step-by-step instruction and only share the mapping overview for IDoc DESADV.DELVRY07 to UN/EDIFACT DESADV. This is again in the PDF documentation generated from the MAG editor. Figure 9.90 shows the first part of the mapping overview.

Path	Name	Node
		DESADV — Despatch advice message
		UNH — MESSAGE HEADER
		0062 — Message reference number
		S009 — MESSAGE IDENTIFIER
Constant STRING value: DESADV		0065 — Message type identifier
Constant STRING value: D		0052 — Message type version number
Constant STRING value: 96A		0054 — Message type release number
Constant STRING value: UN		0051 — Controlling agency
		BGM — Beginning of message
/DELVRY01/E1EDL20/E1EDL20/VBELN	Sales and Distribution Document Number	C002 — Document/message name
Constant STRING value: 351		1001 — Document/message name, coded
		1131 — Code list qualifier
/DELVRY01/E1EDL20/E1EDL20/VBELN	Sales and Distribution Document Number	1004 — Document/message number
Constant STRING value: 9		1225 — Message function, coded
		DTM — Despatch date and or time
		C507 — Date/time/period
Constant STRING value: 11		2005 — Date/time/period qualifier
/DELVRY01/E1EDL20/E1EDT13[./E1EDT13/QUALF=006]/E1EDT13/NTANF	Constraint for activity start (Basic)	2380 — Date/time/period
Constant STRING value: 102		2379 — Date/time/period format qualifier
		DTM — Delivery date/time, estimated
		C507 — Date/time/period
Constant STRING value: 17		2005 — Date/time/period qualifier
/DELVRY01/E1EDL20/E1EDT13[./E1EDT13/QUALF=007]/E1EDT13/NTANF	Constraint for activity start (Basic)	2380 — Date/time/period
Constant STRING value: 102		2379 — Date/time/period format qualifier
		DTM — Document/message date/time
		C507 — Date/time/period
Constant STRING value: 137		2005 — Date/time/period qualifier
/DELVRY01/EDI_DC40/CREDAT	Created On	2380 — Date/time/period
Constant STRING value: 102		2379 — Date/time/period format qualifier

Figure 9.90 Mapping Overview for Mapping IDoc ASN to UN/EDIFACT DELVRY, Part 1

Figure 9.91 shows the second part of the mapping overview.

		SG2 — Buyer
		NAD — Name and address
Constant STRING value: BY		3035 — Party qualifier
		C082 — Party identification details
/DELVRY01/E1EDL20/E1ADRM1[./E1ADRM1/PARTNER_Q=AG]/E1ADRE1/EXTEND_D	Additional data, e.g. ILN or D&B no.	3039 — Party id. identification
		1131 — Code list qualifier
Constant STRING value: ZZZ		3055 — Code list responsible agency, coded
		C058 — Name and address
		3124 — Name and address line
		C080 — Party name
/DELVRY01/E1EDL20/E1ADRM1[./E1ADRM1/PARTNER_Q=AG]/E1ADRM1/NAME1	Name or address line	3036 — Party name
		C059 — Street
		3042 — Street and number/p.o. box
/DELVRY01/E1EDL20/E1ADRM1[./E1ADRM1/PARTNER_Q=AG]/E1ADRM1/CITY1	Town or city	3164 — City name
		3229 — Country sub-entity identification
/DELVRY01/E1EDL20/E1ADRM1[./E1ADRM1/PARTNER_Q=AG]/E1ADRM1/POSTL_COD1	Postal code	3251 — Postcode identification
/DELVRY01/E1EDL20/E1ADRM1[./E1ADRM1/PARTNER_Q=AG]/E1ADRM1/COUNTRY1	Country indicator (ISO alphanumeric)	3207 — Country, coded
		SG2 — Delivery party

Figure 9.91 Mapping Overview for Mapping IDoc ASN to UN/EDIFACT DELVRY, Part 2

Figure 9.92 shows the third part of the mapping overview.

Path	Name	Node
		NAD — Name and address
Constant STRING value: DP		**3035** — Party qualifier
		C082 — Party identification details
/DELVRY01/E1EDL20/E1ADRM1[./E1ADRM1/PARTNER_Q= WE]/E1ADRE1/EXTEND_D	Additional data, e.g. ILN or D&B no.	**3039** — Party id. identification
		1131 — Code list qualifier
Constant STRING value: ZZZ		**3055** — Code list responsible agency, coded
		C058 — Name and address
		3124 — Name and address line
		C080 — Party name
/DELVRY01/E1EDL20/E1ADRM1[./E1ADRM1/PARTNER_Q= WE]/E1ADRM1/NAME1	Name or address line	**3036** — Party name
		C059 — Street
/DELVRY01/E1EDL20/E1ADRM1[./E1ADRM1/PARTNER_Q= WE]/E1ADRM1/STREET1	Street and house number	**3042** — Street and number/p.o. box
/DELVRY01/E1EDL20/E1ADRM1[./E1ADRM1/PARTNER_Q= WE]/E1ADRM1/CITY1	Town or city	**3164** — City name
		3229 — Country sub-entity identification
/DELVRY01/E1EDL20/E1ADRM1[./E1ADRM1/PARTNER_Q= WE]/E1ADRM1/POSTL_COD1	Postal code	**3251** — Postcode identification
/DELVRY01/E1EDL20/E1ADRM1[./E1ADRM1/PARTNER_Q= WE]/E1ADRM1/COUNTRY1	Country indicator (ISO alphanumeric)	**3207** — Country, coded
		SG2 — Ship from
		NAD — Name and address
Constant STRING value: SF		**3035** — Party qualifier
		C082 — Party identification details
Function		**3039** — Party id. identification
		1131 — Code list qualifier
Constant STRING value: ZZZ		**3055** — Code list responsible agency, coded
		C058 — Name and address
		3124 — Name and address line
		C080 — Party name
		3036 — Party name
		C059 — Street
/DELVRY01/E1EDL20/E1ADRM1[./E1ADRM1/PARTNER_Q= OSP]/E1ADRM1/STREET1	Street and house number	**3042** — Street and number/p.o. box
/DELVRY01/E1EDL20/E1ADRM1[./E1ADRM1/PARTNER_Q= OSP]/E1ADRM1/CITY1	Town or city	**3164** — City name
		3229 — Country sub-entity identification
/DELVRY01/E1EDL20/E1ADRM1[./E1ADRM1/PARTNER_Q= OSP]/E1ADRM1/POSTL_COD1	Postal code	**3251** — Postcode identification
/DELVRY01/E1EDL20/E1ADRM1[./E1ADRM1/PARTNER_Q= OSP]/E1ADRM1/COUNTRY1	Country indicator (ISO alphanumeric)	**3207** — Country, coded

Figure 9.92 Mapping Overview for Mapping IDoc ASN to UN/EDIFACT DELVRY, Part 3

Figure 9.93 shows the fourth part of the mapping overview.

		SG10 — Segment Group 10: CPS-FTX-SG11-SG15
		CPS — Consignment packing sequence
Constant STRING value: 1		**7164** — Hierarchical id. number
		7166 — Hierarchical parent id.
/DELVRY01/E1EDL20/E1EDL24	Segment Group - Delivery Item	**SG15** — Segment Group 15: LIN-PIA-IMD-MEA-QTY-ALI-GIN-GIR-DLM-DTM-FTX-MOA-SG16-SG17-SG20-SG23
		LIN — Line item
/DELVRY01/E1EDL20/E1EDL24/E1EDL24/POSNR	Item number of the SD document	**1082** — Line item number
		C212 — Item number identification
/DELVRY01/E1EDL20/E1EDL24/E1EDL24/MATNR	Material Number (18 Characters)	**7140** — Item number
Constant STRING value: EN		**7143** — Item number type, coded
		IMD — Item description
Constant STRING value: F		**7077** — Item description type, coded
		C273 — Item description
Constant STRING value:		**7009** — Item description identification
/DELVRY01/E1EDL20/E1EDL24/E1EDL24/ARKTX	Short Text for Sales Order Item	**7008** — Item description

Figure 9.93 Mapping Overview for Mapping SAP IDoc ASN to UN/EDIFACT DELVRY, Part 4

Figure 9.94 shows the fifth part of the mapping overview.

Path	Name	Node
		└ 7008 — Item description
		├ QTY — Quantity - Despatch quantity
		└ C186 — Quantity details
Constant STRING value: 12		├ 6063 — Quantity qualifier
/DELVRY01/E1EDL20/E1EDL24/E1EDL24/LGMNG	Actual quantity delivered in stockkeeping units	└ 6060 — Quantity
		├ SG16 — Order number (purchase)
		└ RFF — Reference
		└ C506 — Reference
Constant STRING value: ON		├ 1153 — Reference qualifier
/DELVRY01/E1EDL20/E1EDL24/E1EDL41[./E1EDL41/QUALF=001]/E1EDL41/BSTNR	Customer Reference	└ 1154 — Reference number
		└ SG16 — Segment Group 16: RFF-DTM - Order number (vendor)
		├ RFF — Reference
		└ C506 — Reference
Constant STRING value: VN		├ 1153 — Reference qualifier
/DELVRY01/E1EDL20/E1EDL24/E1EDL43[./QUALF=C]/BELNR	IDOC document number	└ 1154 — Reference number
		└ DTM — Date/time/period
		└ C507 — Date/time/period
		├ 2005 — Date/time/period qualifier
		├ 2380 — Date/time/period
		└ 2379 — Date/time/period format qualifier
		└ UNT — MESSAGE TRAILER
		├ 0074 — Number of segments in a message
		└ 0062 — Message reference number

Figure 9.94 Mapping Overview for Mapping SAP IDoc ASN to UN/EDIFACT DELVRY, Part 5

At this stage, we would like to pause for a moment and look at two examples for educational purposes in the MAG related to ASNs:

- **Sample XSLT function to extend the mapping condition between source and target data elements**
  The use case is that there is an issue with the IDoc field containing the postal code related to the shipping point. The length of this source field is 10 characters, and the value coming from SAP S/4HANA master data contains a dash character (-). The target postcode identification field in UN/EDIFACT is nine characters long. This type of situation sometimes happens during EDI implementation, and it should be always analyzed on a case-by-case basis starting from the functional side. To demonstrate the XLST function, we applied a simple code that removes the dash from the source field. The result is that the 10-character-long value with a dash after XSLT processing becomes a nine-character-long value and fits into the target field in the UN/EDIFACT message (see Figure 9.95).
- **Sample function using global parameters on MAG level**
  In this use case, we need to map the ship-from party identifier, but the value is not available in the source IDoc. Navigate to the **Global Parameters** tab and create a parameter called **OWN_IDENTIFIER_CC_1710** with the default value **COMPANY_1710_VN** (see Figure 9.96).

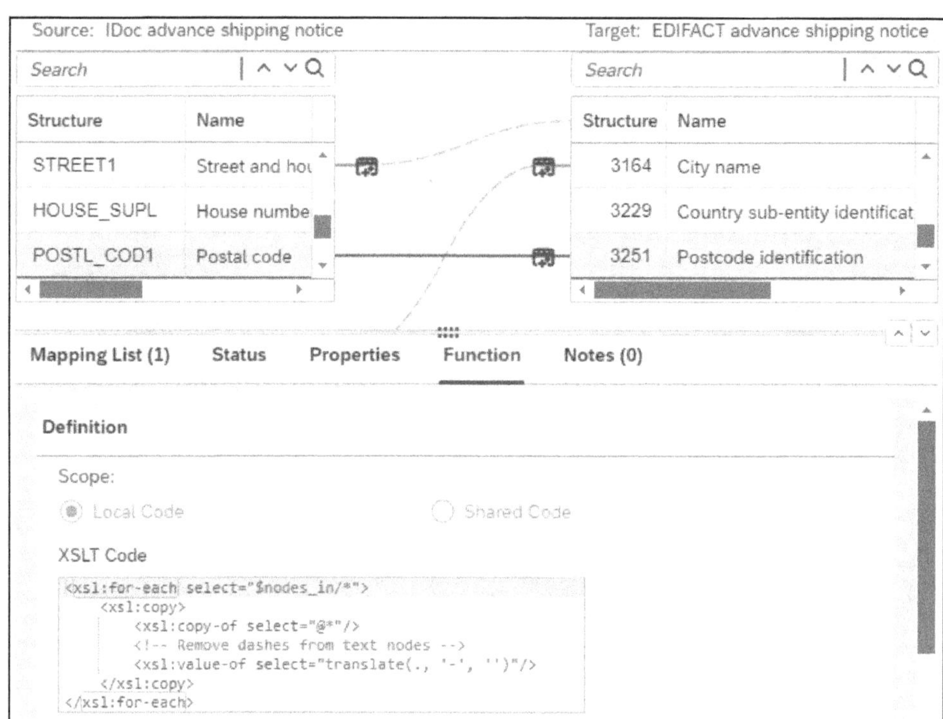

Figure 9.95 XSLT Function Applied on Postcode Identification in Mapping IDoc Advanced Shipping Notification to EDIFACT Advanced Shipping Notification

> **Note**
>
> This example is only invented for educational purposes. It's not ideal to store and populate properties like party identifiers within the middleware—especially when all the other party identifiers are already coming from the business system.

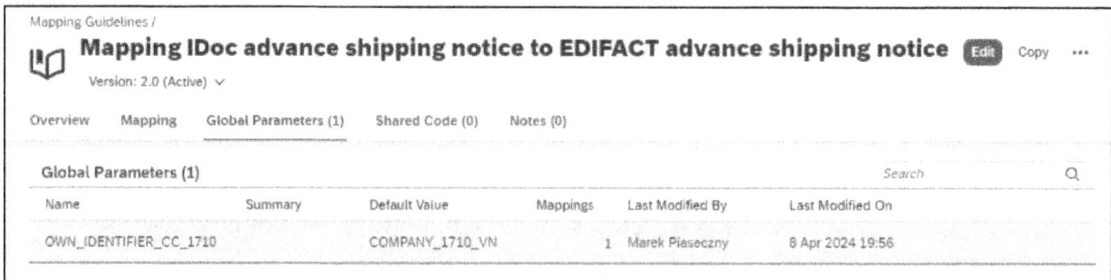

Figure 9.96 Global Parameter OWN_IDENTIFIER_CC_1710

In the next step, navigate to the appropriate data element (3039) and set **Function** (see Figure 9.97).

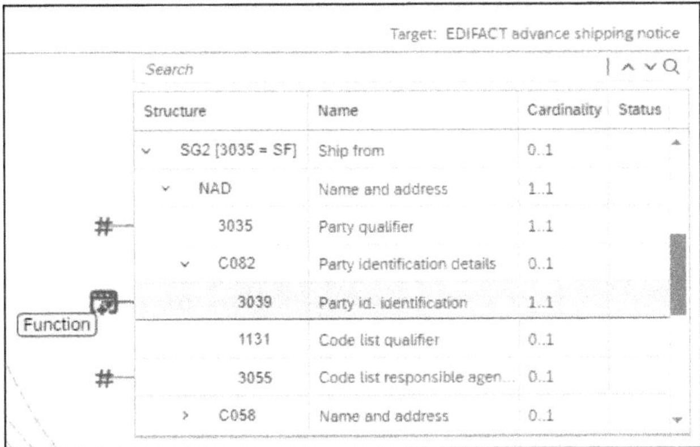

Figure 9.97 Function Set on Party Identification Data Element

Navigate to the **Function** tab. There is no XSLT code yet; before maintaining it, you need to define the parameter. Click **Add New Function Parameter**, which is located below the XSLT code editor (see Figure 9.98).

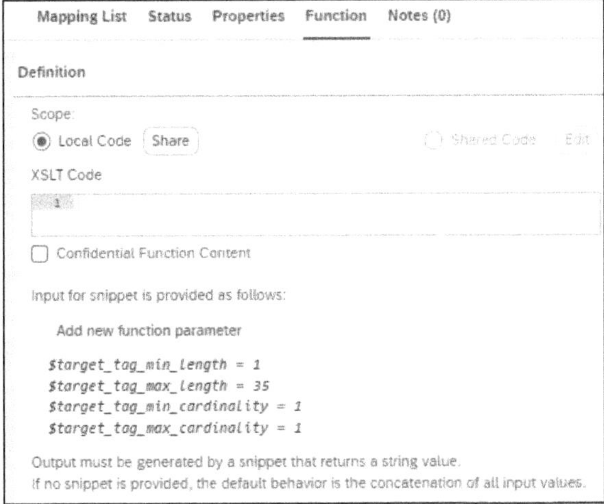

Figure 9.98 Function Tab with XSLT Editor

Create a new function parameter by giving it a name; in this example, it is again **OWN_IDENTIFIER_CC_1710** (see Figure 9.99).

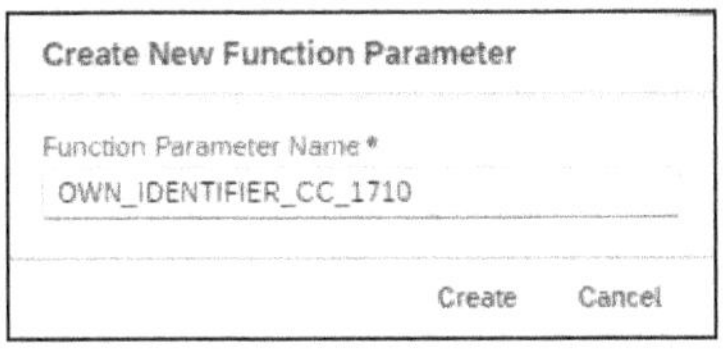

Create New Function Parameter

Function Parameter Name *

OWN_IDENTIFIER_CC_1710

Create Cancel

Figure 9.99 Creating New Function Parameter

After creating the new parameter, choose a global parameter to be assigned to the newly created parameters (see Figure 9.100).

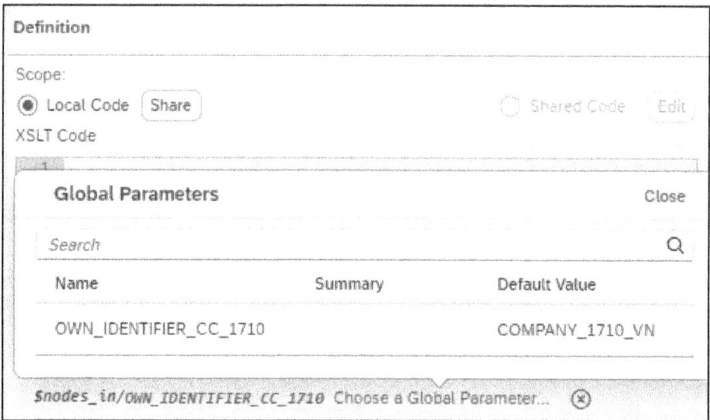

Figure 9.100 Assigning Global Parameter to Function Parameter

At this stage, the newly created function parameter OWN_IDENTIFIER_CC_1710 can be used in the XSLT code. It contains value of the OWN_IDENTIFIER_CC_1710 global parameter. Although in our example these two are named the same, that does not have to be the case. In Figure 9.101, you can see the XSLT code to read function parameter OWN_IDENTIFIER_CC_1710.

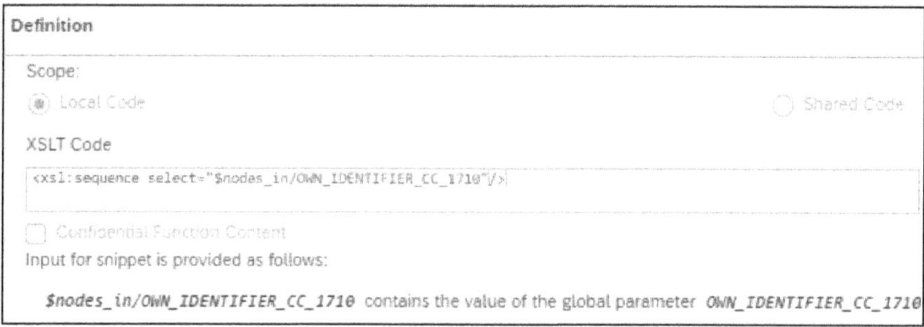

Figure 9.101 XSLT Code to Read Function Parameter OWN_IDENTIFIER_CC_1710

Mapping IDoc Invoice to EDIFACT Invoice

For the last MAG object related to mapping the IDoc invoice to the EDIFACT invoice, we again limit the approach to share the end results from the PDF documentation. There are two figures where you can see all the details of the applied mapping rules. Figure 9.102 shows the first part of the mapping overview.

Path	Name	Node
		INVOIC — Invoice message
		├ **UNH** — MESSAGE HEADER
		│ ├ **0062** — Message reference number
		│ └ **S009** — MESSAGE IDENTIFIER
Constant STRING value: INVOIC		│ ├ **0065** — Message type identifier
Constant STRING value: D		│ ├ **0052** — Message type version number
Constant STRING value: 96A		│ ├ **0054** — Message type release number
Constant STRING value: UN		│ └ **0051** — Controlling agency
		├ **BGM** — Beginning of message
/INVOIC01/E1EDK01/BELNR	IDOC document number	│ ├ **C002** — Document/message name
Constant STRING value: 380		│ │ └ **1001** — Document/message name, coded
/INVOIC01/E1EDK01/BELNR	IDOC document number	│ ├ **1004** — Document/message number
Constant STRING value: 9		│ └ **1225** — Message function, coded
		├ **DTM** — Date/time/period - Document issue date time
		│ └ **C507** — Date/time/period
Constant STRING value: 137		│ ├ **2005** — Date/time/period qualifier
/INVOIC01/E1EDK02[./QUALF=009]/DATUM	IDOC: Date	│ ├ **2380** — Date/time/period
Constant STRING value: 102		│ └ **2379** — Date/time/period format qualifier
		├ **DTM** — Date/time/period - Delivery date/time, actual
		│ └ **C507** — Date/time/period
Constant STRING value: 35		│ ├ **2005** — Date/time/period qualifier
/INVOIC01/E1EDK02[./QUALF=002]/DATUM	IDOC: Date	│ ├ **2380** — Date/time/period
Constant STRING value: 102		│ └ **2379** — Date/time/period format qualifier

Figure 9.102 Mapping Overview for Mapping IDoc Invoice to EDIFACT Invoice, Part 1

Figure 9.103 shows the second part of the mapping overview.

		├ **SG1** — Segment Group 1: RFF-DTM - Delivery note number
		│ └ **RFF** — Reference
		│ └ **C506** — Reference
Constant STRING value: DQ		│ ├ **1153** — Reference qualifier
		│ └ **1154** — Reference number
		├ **SG1** — Segment Group 1: RFF-DTM - Order number (vendor)
		│ └ **RFF** — Reference
		│ └ **C506** — Reference
Constant STRING value: VN		│ ├ **1153** — Reference qualifier
/INVOIC01/E1EDK02[./QUALF=002]/BELNR	IDOC document number	│ └ **1154** — Reference number
		├ **SG2** — Segment Group 2: NAD-LOC-FII-SG3-SG4-SG5 - Supplier
		│ └ **NAD** — Name and address
Constant STRING value: SU		│ ├ **3035** — Party qualifier
		│ ├ **C082** — Party identification details
/INVOIC01/E1EDKA1[./PARVW=RS]/PARTN	Partner number	│ │ ├ **3039** — Party id. identification
Constant STRING value: ZZZ		│ │ └ **1131** — Code list qualifier
		│ ├ **C058** — Name and address
/INVOIC01/E1EDKA1[./PARVW=RS]/NAME1	Name 1	│ │ ├ **3124** — Name and address line
		│ │ └ **3124** — Name and address line
		│ ├ **C059** — Street
/INVOIC01/E1EDKA1[./PARVW=RS]/STRAS	Street and house number 1	│ │ └ **3042** — Street and number/p.o. box
/INVOIC01/E1EDKA1[./PARVW=RS]/ORT01	City	│ ├ **3164** — City name
		│ ├ **3229** — Country sub-entity identification
/INVOIC01/E1EDKA1[./PARVW=RS]/PSTLZ	Postal code	│ ├ **3251** — Postcode identification

Figure 9.103 Mapping Overview for Mapping IDoc Invoice to EDIFACT Invoice, Part 2

Figure 9.104 shows third part of the mapping overview.

Path	Name	Node
/INVOIC01/E1EDKA1[./PARVW=RS]/COUNC	County code	└ 3207 — Country, coded
/INVOIC01/E1EDP01	Segment Group - IDoc: Document Item General Data	├ SG25 — Segment Group 25: LIN-PIA-IMD-MEA-QTY-PCD-ALI-DTM-GIN-GIR-QVR-EQD-FTX-SG26-SG46-SG47
		├ LIN — Line item
/INVOIC01/E1EDP01/E1EDP01/POSEX	Item number	├ 1082 — Line item number
		├ 1229 — Action request/notification, coded
		└ C212 — Item number identification
/INVOIC01/E1EDP01/E1EDP19[./QUALF=002]/IDTNR	IDOC material ID	├ 7140 — Item number
Constant STRING value: SA		└ 7143 — Item number type, coded
		├ IMD — Item description
Constant STRING value: F		├ 7077 — Item description type, coded
		├ 7081 — Item characteristic, coded
		└ C273 — Item description
/INVOIC01/E1EDP01/E1EDP19[./QUALF=002]/KTEXT	IDOC short text	└ 7009 — Item description identification
		├ QTY — Quantity - Invoiced quantity
		└ C186 — Quantity details
Constant STRING value: 47		├ 6063 — Quantity qualifier
/INVOIC01/E1EDP01/E1EDP01/MENGE	Quantity	├ 6060 — Quantity
Constant STRING value: EA		└ 6411 — Measure unit qualifier
		├ SG26 — Segment Group 26: MOA-CUX - Line item amount (inclusive of allowances and charges)
		└ MOA — Monetary amount
		└ C516 — Monetary amount
Constant STRING value: 203		├ 5025 — Monetary amount type qualifier
/INVOIC01/E1EDP01/E1EDP26[./QUALF=010]/BETRG	Total value of sum segment	├ 5004 — Monetary amount
/INVOIC01/E1EDK01/CURCY	Currency	├ 6345 — Currency, coded
		└ 6343 — Currency qualifier

Figure 9.104 Mapping Overview for Mapping IDoc Invoice to EDIFACT Invoice, Part 3

Figure 9.105 shows fourth part of the mapping overview.

		└ SG28 — Segment Group 28: PRI-APR-RNG-DTM - Reference price
		└ PRI — Price details
		└ C509 — Price information
Constant STRING value: AAA		├ 5125 — Price qualifier
/INVOIC01/E1EDP01/E1EDP26[./QUALF=001]/BETRG	Total value of sum segment	├ 5118 — Price
		└ 5375 — Price type, coded
		├ UNS — SECTION CONTROL
		└ 0081 — Section identification
		├ SG48 — Segment Group 48: MOA-SG49 - Tax amount
		└ MOA — Monetary amount
		└ C516 — Monetary amount
Constant STRING value: 124		├ 5025 — Monetary amount type qualifier
/INVOIC01/E1EDS01[./SUMID=005]/SUMME	Total value of sum segment	└ 5004 — Monetary amount
		├ SG48 — Segment Group 48: MOA-SG49 - Total line items amount
		└ MOA — Monetary amount
		└ C516 — Monetary amount
Constant STRING value: 79		├ 5025 — Monetary amount type qualifier
/INVOIC01/E1EDS01[./SUMID=010]/SUMME	Total value of sum segment	└ 5004 — Monetary amount
		└ UNT — MESSAGE TRAILER
		├ 0074 — Number of segments in a message
		└ 0062 — Message reference number

Figure 9.105 Mapping Overview for Mapping IDoc Invoice to EDIFACT Invoice, Part 4

To summarize this section, Figure 9.106 shows all the MAGs created for our EDI scenario. We are almost ready to move to Trading Partner Management and start configuration of the EDI transactions. Before that, we will have one short section where you can learn about exporting MAG-related artifacts.

		Source	Target		
Mapping EDIFACT sales order to IDoc sales order	MIG Name: Message Type: Type System: Type System Version:	EDIFACT sales order ORDERS – Purchase order message UN/EDIFACT D.96A S3	IDoc sales order ORDERS.ORDERS05 – Purchase order / order SAP S/4HANA On Premise IDoc 1809 FPS02	Active Version: 1.0 History	UID-a5590ad6-a1eb-4d67-9874-a63974b75b2a
		Source	Target		
Mapping IDoc advance shipping notice to EDIFACT advance shipping notice	MIG Name: Message Type: Type System: Type System Version:	IDoc advance shipping notice DESADV.DELVRY07 – Delivery: Shipping notifi... SAP S/4HANA On Premise IDoc 1809 FPS02	EDIFACT advance shipping notice DESADV – Despatch advice message UN/EDIFACT D.96A S3	Active Version: 1.0 History	UID-299cec40-723d-4de4-b00b-389dc11d274b
		Source	Target		
Mapping IDoc invoice to EDIFACT invoice	MIG Name: Message Type: Type System: Type System Version:	IDoc invoice INVOIC.INVOIC02 – Invoice/Billing Document SAP S/4HANA On Premise IDoc 1809 FPS02	EDIFACT invoice INVOIC – Invoice message UN/EDIFACT D.96A S3	Active Version: 1.0 History	UID-08983842-134f-4b8a-b457-3e40237859da
		Source	Target		
Mapping IDoc sales order response to EDIFACT sales order response	MIG Name: Message Type: Type System: Type System Version:	IDoc sales order response ORDRSP.ORDERS05 – Purchase order / order ... SAP S/4HANA On Premise IDoc 1809 FPS02	EDIFACT sales order response ORDRSP – Purchase order response message UN/EDIFACT D.96A S3	Active Version: 2.0 History	UID-85687cd1-a035-478d-82a8-7d335753964a

Figure 9.106 Final List of Four MAGs Created for B2B Scenario

9.2.4 Generating Runtime Artifacts

There are two additional options available for each MAG that we did not mention yet. You can see both options in Figure 9.76, shown earlier in this chapter:

- **Inject • SAP Cloud Integration Flow Resource**
- **Export • SAP Cloud Integration Runtime Artifacts**

The export option generates ZIP packages that contain the definitions of our MIG and MAG artifacts in XSD and XSLT format, and this allows for custom implementation in Cloud Integration or SAP Process Integration. In our case, we do not need to make use of this functionality. We also do not need to use the inject option to push artifacts to Cloud Integration. Trading Partner Management, which we will examine in the next section, eliminates the need for manual injection or export of the artifacts created in Integration Advisor. The different SAP Integration Suite components cooperate with each other, and Trading Partner Management is able to pull/inject the necessary artifacts based on the configured B2B scenarios. You will see exactly how this is done in the next section.

9.3 Trading Partner Management

Trading Partner Management is the second core component of SAP Integration Suite, and we will configure it in this section. There are four key elements to configure:

- Company profile
- Trading partner profile
- Trading partner agreement template
- Trading partner agreement

9

The sequence in which we configure these elements is important: it needs to be exactly in the listed order. First we start with defining our own company and its characteristics, and then we can do the same for the trading partner. When we have both actors of the B2B integration defined in profile form, we can define the trading partner agreement, but instead of doing this directly we make use of the trading partner agreement template. At the end in the trading partner agreement, we add a specific trading partner and adjust the other parameters related to B2B transactions.

9.3.1 Creating the Company Profile

There is a single company profile, and creating it is a one-time action performed at the very beginning of Trading Partner Management configuration. To start, use the left-hand menu to navigate to **Design • B2B Scenarios • Company Profile**. When you start this for the first time you are welcomed by a screen informing you that there is no company profile set yet, exactly as shown in Figure 9.107.

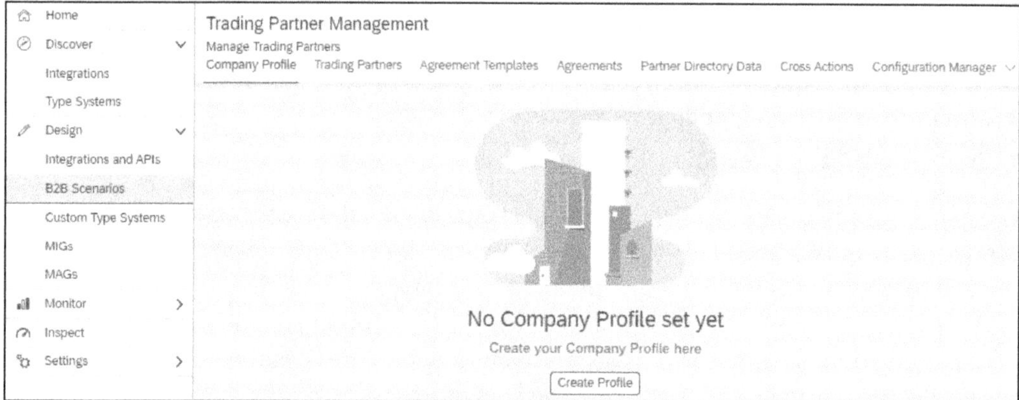

Figure 9.107 Initial Screen of Trading Partner Management

To start the company profile creation, click the **Create Profile** button located at the bottom of the screen. The first screen you see is **Company Profile Overview** with only two mandatory fields:

- **Company Name**
  Maintain value **Company Code 1710**
- **Company Short Name**
  Maintain value **1710**

We'll skip the **URL** and **Logo**, but fill in a few address-related details in line with the company code details in SAP S/4HANA, as shown in Figure 9.108. At the top of the screen, you can see a few additional tabs, each related to different properties. We will follow these exactly in this order.

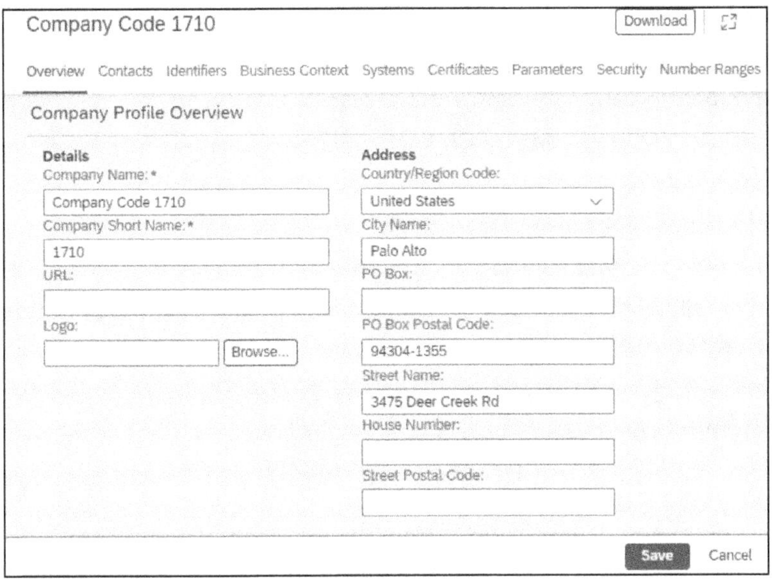

Figure 9.108 Company Profile Overview Maintained with Company Data

The next tab gives you the opportunity to fill in contact information related to the company. It is not a mandatory step; you do not have to add any contact information. If you decide to do so, which we recommend, note that you can create multiple entries. Usually these would include the details of a person or people who support the B2B EDI solution within your organization. There are three mandatory fields: **First Name**, **Last Name**, and **Alias**. Figure 9.109 shows sample data maintained in our example.

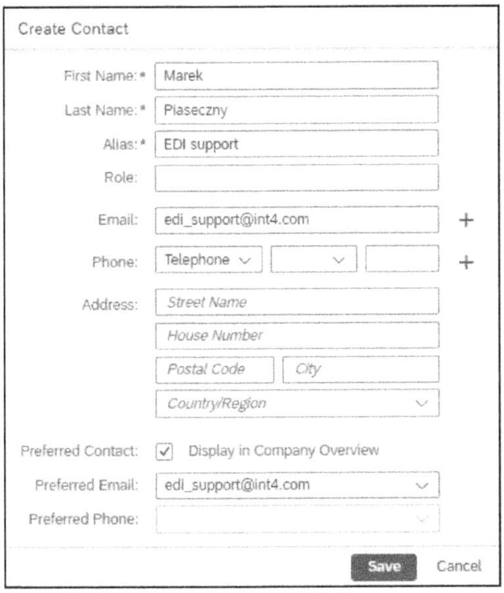

Figure 9.109 Creating Contact Details in Company Profile

The next tab is critical for the overall solution. It is related to the identifiers we elaborated on in Chapter 5. We need to create two entries:

- One for the identifier used in UN/EDIFACT messages:
  - **Identification**—maintain value **EDI_COMPANY_1710**
  - **Alias**—maintain value **B2B-Identifier**
  - **Type System**—choose **UN/EDIFACT** from the list
  - **Scheme Name**—choose **Mutually defined** from the list
  - **Scheme Code**—will be filled automatically based on the above selection. In this case with value **ZZZ**
  - **Agency Name** and **Agency Code** you can ignore
- One for the identifier used in IDocs:
  - **Identification**—maintain value **S42CLNT400**
  - **Alias**—maintain value **S42-Identifier**
  - **Type System**—choose **SAP S/4HANA On-Premise IDoc** from the list
  - **Scheme Name**—choose **N/A** from the list
  - **Scheme Code**—will be filled automatically based on the above selection. In this case with value **N/A**
  - **Agency Name** and **Agency Code** you can ignore

After finishing, the two entries should look like as shown in Figure 9.110.

Identifiers (2)

Identification	Alias	Type System	Scheme Name	Scheme Code	Agency Name	Agency Code
EDI_COMPANY_1710	B2B-Identifier	UNEDIFACT	Mutually defined	ZZZ		
S42CLNT400	S42-Identifier	SAP_IDoc	N/A	N/A		

Figure 9.110 Company Profile: Identifiers

We can say that the next properties are less important; they are not critical for the overall functionality, but they do give the business context in which our company operates. We continue with the story that we are a US-based supplier operating within the sports and entertainment industry. Figure 9.111 shows the sample values.

On the next tab, **Systems**, we maintain systems related to the company—that is, SAP S/4HANA—using the following values:

- **Name**—value **S4HANA_S42_DEV**.
- **Alias**—**S42_DEV**.
- **Type**—choose **S/4HANA** from the list.
- **Purpose**—choose **Prod**. Additional explanation is needed here. There are two other options, **Dev** and **Test**, which may seem to be better choices. But the reason for not

selecting them is that they influence the value of the test indicator in the IDoc control record. When the purpose is set as **Dev** or **Test**, it also sets the test indicator within the IDoc, and that limits the IDoc processing within SAP S/4HANA.

- **Link**—can be ignored.
- **Description**—freeform description, here, **Development environment of the SAP S/4HANA**.
- **Status**—this value will be filled automatically.

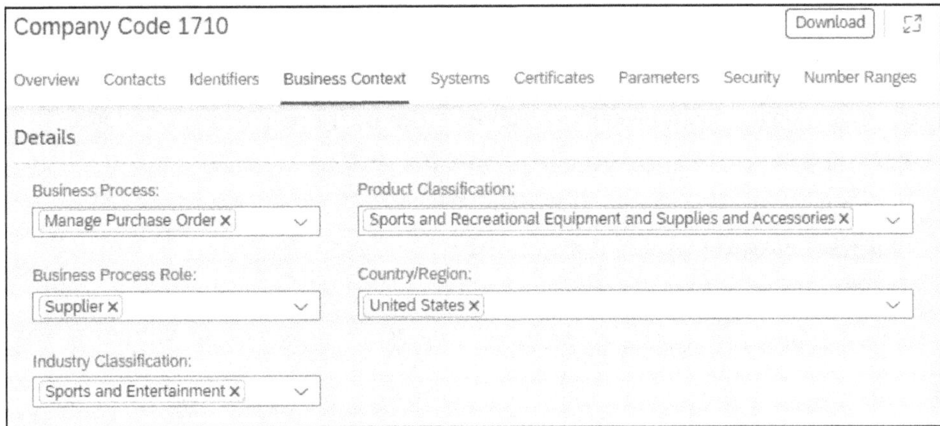

Figure 9.111 Business Context of Company

Finally we should see the system on the list like in Figure 9.112.

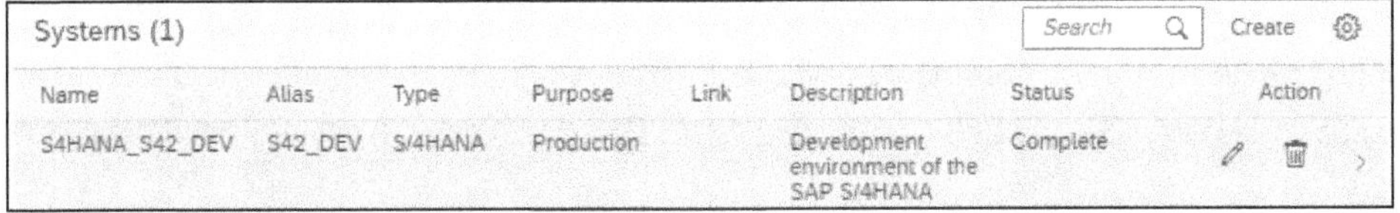

Figure 9.112 Company Profile: Systems

After creating the system, we must maintain two more sections related to type systems and communication. To start the process, click on the system, which will trigger an additional screen as shown in Figure 9.113. We'll start with the **Type Systems** tab. Click the **Create** button located on the right-hand side. In the popup window, there are three parameters to be filled in:

- **Name**—choose **SAP S/4HANA On Premise IDoc** from the given selection list
- **Version**—choose the version applicable for your system; in this case, the most recent version, **1809_FPS02**
- **Input version manually**—gives an option to maintain the version value with any value

Figure 9.113 shows the type systems created in our case.

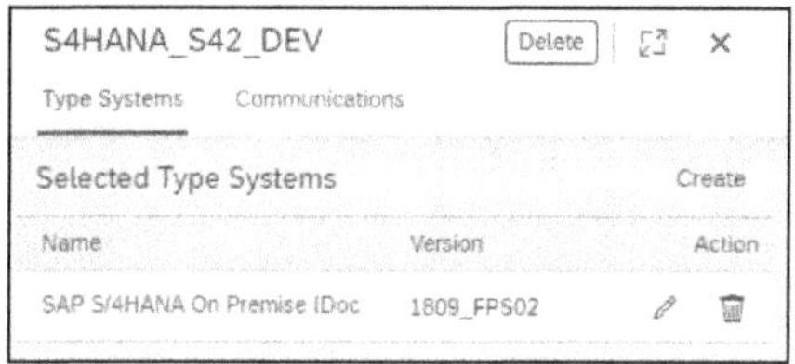

Figure 9.113 Company Profile: System: Type Systems

Now move to the **Communications** tab. Click the **Create** button located on the right-hand side. We will create two entries: one for sending data to SAP S/4HANA (receiver IDoc adapter) and one for receiving IDocs from SAP S/4HANA (sender IDoc adapter). To create communication for the receiver IDoc adapter, maintain the parameters as follows:

- **Name** and **Alias**—**IDoc_Receiver**
- **Description**—**IDocs towards S42 system**
- **Direction**—**Receiver**
- **Adapter**—**IDOC**

For the **Connection Details** section:

- **Address**—**http://int4saps42.int4sap.com:44301/sap/bc/srt/idoc?sap-client=400**
- **Proxy Type**—**On-Premise**
- **Location ID**—**S42**
- **Authentication**—**Basic**
- **Credential Name**—**B2B_S4HANA_DEV**

Figure 9.114 shows the complete configuration.

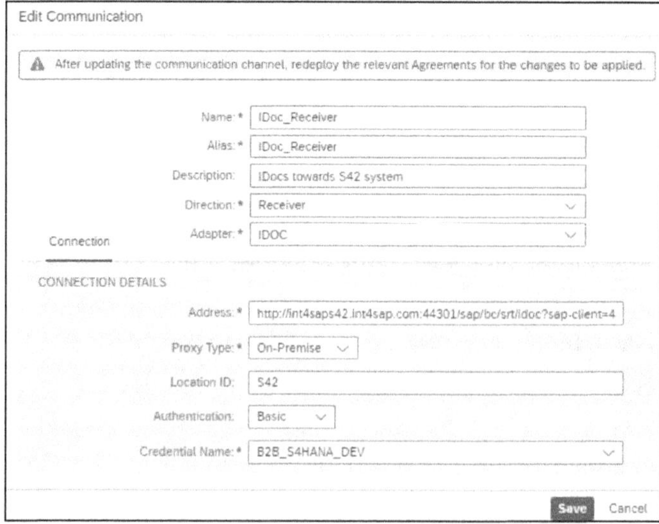

Figure 9.114 Configuration of Receiver Communication IDoc

To create communication for the sender IDoc adapter, maintain the parameters as follows:

- **Name—IDoc_Sender**
- **Alias—IDoc_Sender**
- **Description—IDocs from S42 system**
- **Direction—Sender**
- **Adapter—IDOC**

The complete configuration is shown in Figure 9.115.

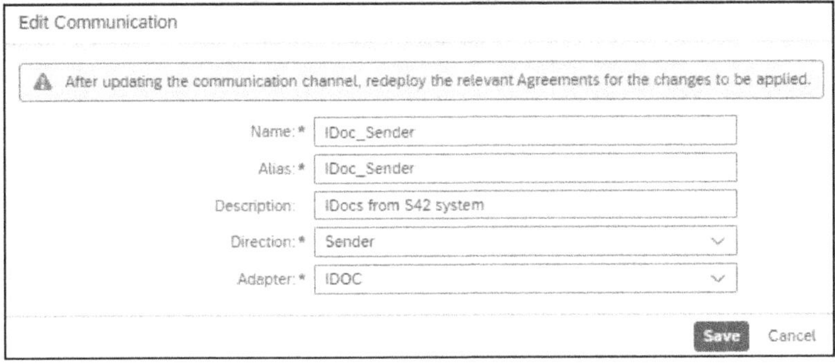

Figure 9.115 Company Profile: Configuration of Sender Communication IDoc

After maintaining the **Communication** section, go back to the top level of the company profile. The next two sections, **Certificates** and **Parameters**, can be skipped for our basic scenario.

In the **Security** section, maintain configuration related to the AS2 inbound decryption. We need one entry as follows:

- **User Account** and **Alias—AS2_COMPANY_1710**, the AS2 identifier.
- **Private Key Alias—company_code_1710**, the security material on Cloud Integration under which we loaded our AS2 private key

The last step is about assigning a number range. Click the **Add** button located on the right-hand side of the screen and add an entry with the **ICN_DEFAULT** value in the **Number Range** and **Alias** fields (see Figure 9.116).

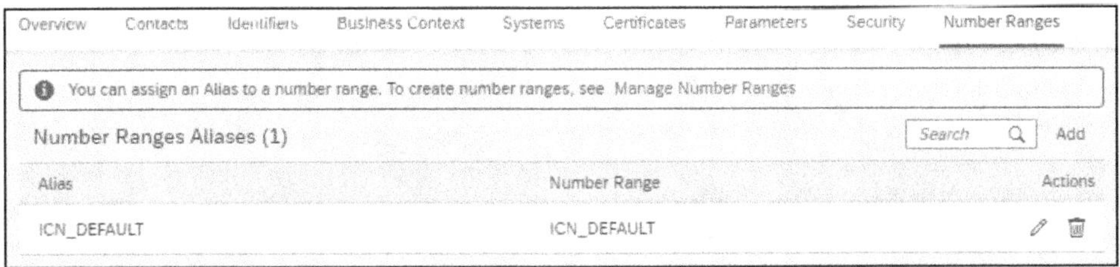

Figure 9.116 Company Profile: Number Ranges

9.3.2 Creating a Trading Partner Profile

From a technical point of view, creating a profile for a trading partner does not differ from creating the company profile. The main difference is that you can only have one company profile, and it is part of each B2B scenario (each agreement, as you will learn soon), but there can be hundreds profiles related to trading partners and each usually has its own scenario. We will quickly go through the creation of the trading partner profile, concentrating mainly on the key parameters:

- **Overview**
  On this screen, maintain the key information related to the trading partner. The most important elements are the mandatory **Company Name** and **Company Short Name**. See the data related to our scenario in Figure 9.117.

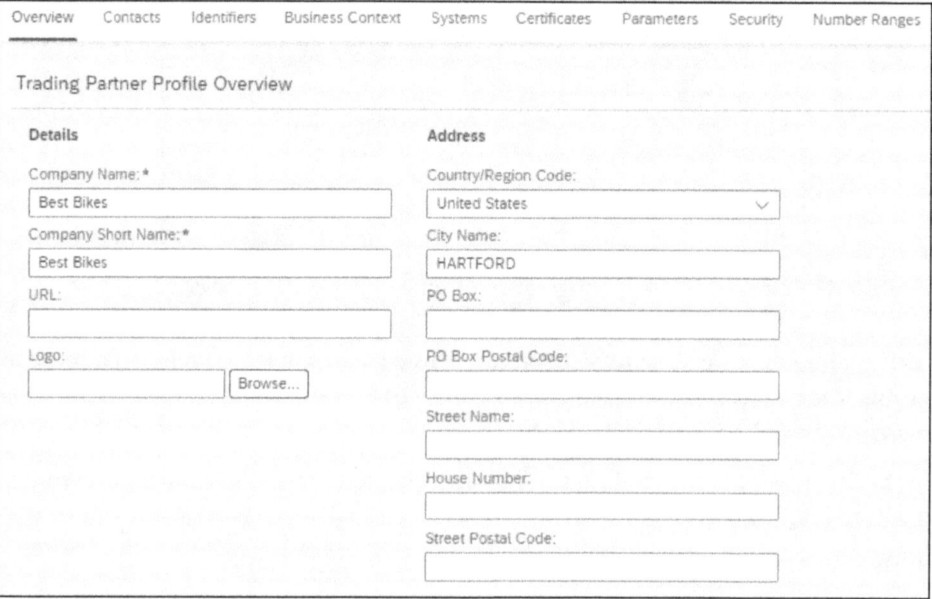

Figure 9.117 Trading Partner Profile: Overview

- **Contacts**
  This is a self-explanatory screen. Simply add the contact details. This is not relevant to the processing itself but definitely a must-have for support and operational aspects.
- **Identifiers**
  Similar to the company profile, we need to maintain two identifiers:
  - One for the identifier used in UN/EDIFACT messages:
    - **Identification**—maintain value **EDI_BEST_BIKES**
    - **Alias**—maintain value **B2B-Identifier**
    - **Type System**—choose **UN/EDIFACT** from the list

    - **Scheme Name**—choose **Mutually defined** from the list
    - **Scheme Code**—will be filled automatically based on the above selection, in this case with value **ZZZ**
    - **Agency Name** and **Agency Code** you can ignore
  - One for the identifier used in IDocs:
    - **Identification**—maintain value **USCU_S05**, the business partner defined within SAP S/4HANA for this trading partner and the value used for EDI partner profile definition within the SAP backend system
    - **Alias**—maintain value **USCU_S05**
    - **Type System**—choose **SAP S/4HANA On Premise IDoc** from the list
    - **Scheme Name**—choose **N/A** from the list
    - **Scheme Code**—will be filled automatically based on the previous selection, in this case with value **N/A**
    - **Agency Name** and **Agency Code** you can ignore

  Figure 9.118 shows the two identifiers created for our Best Bikes customer.

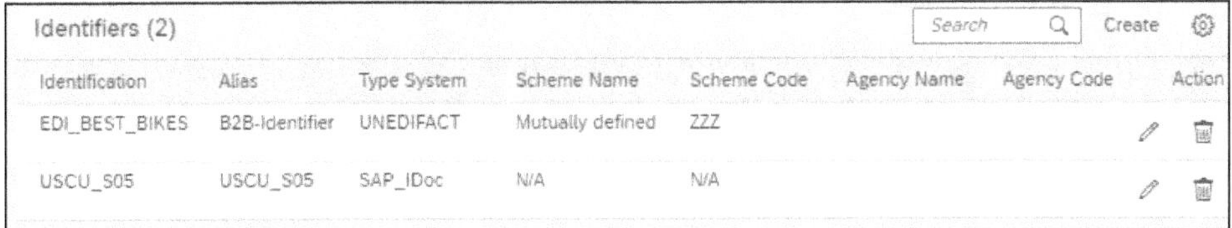

Identifiers (2)

Identification	Alias	Type System	Scheme Name	Scheme Code	Agency Name	Agency Code	Action
EDI_BEST_BIKES	B2B-Identifier	UNEDIFACT	Mutually defined	ZZZ			
USCU_S05	USCU_S05	SAP_IDoc	N/A	N/A			

Figure 9.118 Trading Partner Profile: Identifiers

- **Business Context**

  This is a self-explanatory screen. We maintain the business context as already known from the MIG and MAG definitions.

- **Systems**
  - **Name**—value **TP_BestBikes_ERP**
  - **Alias**—**TP_ERP**
  - **Type**—choose **B2B System** from the list
  - **Purpose**—choose **Prod.** An additional explanation is needed here. There are two other options, **Dev** and **Test**, which may seem to be a better choice—but the reason for not selecting them is that they influence the test indicator within the EDIFACT message. That brings some limitations that we want to bypass.
  - **Link**—can be ignored
  - **Description**—leave empty or add any useful description
  - **Status**—this will be filled automatically after you save the entry

  The complete entry is shown in Figure 9.119.

Systems (1) Search Create

Name	Alias	Type	Purpose	Link	Description	Status	Action
TP_BestBikes_ERP	TP_ERP	B2B System	Production			Complete	

Figure 9.119 Trading Partner Profile: Systems

- **Type Systems**
  When creating this entry, there are two properties that we fill by choosing the values from a selection list:
  - **Name**—**UN/EDIFACT**
  - **Version**—**D.96A S3**

  The **Selected Type Systems** screen should look as shown in Figure 9.120.

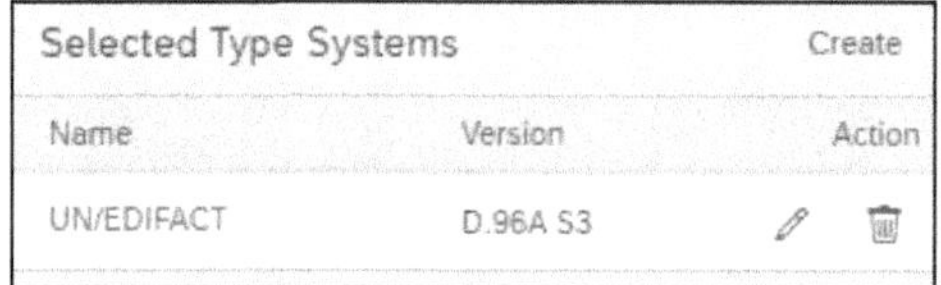

Selected Type Systems Create

Name	Version	Action
UN/EDIFACT	D.96A S3	

Figure 9.120 Trading Partner Profile: Systems: Type Systems

- **Communications**
  Here we define how we connect to the trading partner's system. In our case, it is by the AS2 protocol, and we need both an AS2 receiver adapter and an AS2 sender adapter. We will start with the AS2 receiver adapter, which handles the outbound connection from Cloud Integration to the third-party system. Fill in all the fields from the top to the bottom, noting that the section related to AS2 settings appears only after selection of an appropriate direction and adapter:
  - **Name**—**AS2_Receiver**
  - **Alias**—**AS2_Receiver**
  - **Description**—can stay empty, but feel free to add any meaningful and useful text
  - **Direction**—**Receiver**
  - **Adapter**—**AS2**

  After maintaining the parameters from the top part, we will have four additional sections dedicated to the AS2 receiver adapter:
  - **Connection**
  - **Recipient URL**—the AS2 entry point for the trading partner, who here has a dummy AS2 server running on a trial instance of SAP BTP/SAP Integration Suite
  - **Proxy Type**—**Internet**
  - **Authentication Type**—**Basic Authentication**
  - **Credential Name**—**B2B_TP_BESTBIKES_DEV**

Figure 9.121 shows all these parameters entered as needed.

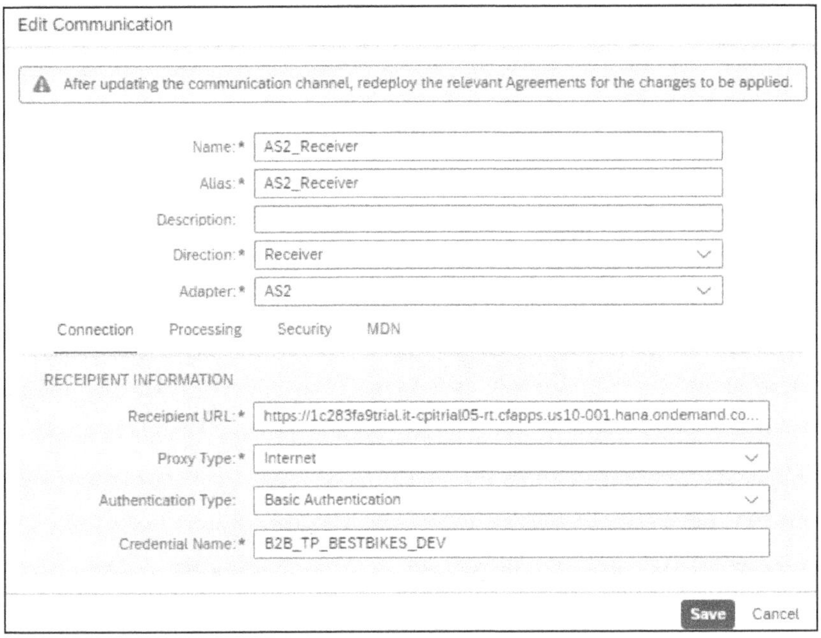

Figure 9.121 Trading Partner Profile: AS2 Receiver Adapter

We stay on the same maintenance view for the AS2 receiver adapter and only switch to the next section, **Processing**, which is located in the middle of the screen. The parameters to enter there are as follows:

- **Append Timestamp**—set this parameter
- **Own AS2 ID—AS2_COMPANY_1710**
- **Partner AS2 ID—AS2_BEST_BIKES**
- **Message Subject—EDI message**
- **Own email address**—maintain appropriate email address
- **Content-Type—application/EDI-X12**
- **Content Transfer Encoding—binary**

There is no need to maintain the other parameters. The complete screen should look as shown in Figure 9.122.

Staying on the AS2 receiver adapter, move to the next tab, **Security**, and maintain values as follows:

- **Compress Message**—do not set this parameter
- **Sign Message**—set this parameter
- **Algorithm—SHA512**

- **Private Key Alias—company_code_1710**
- **Encrypt Message**—set this parameter
- **Algorithm—AES256**
- **Public Key Alias—bestbikes**

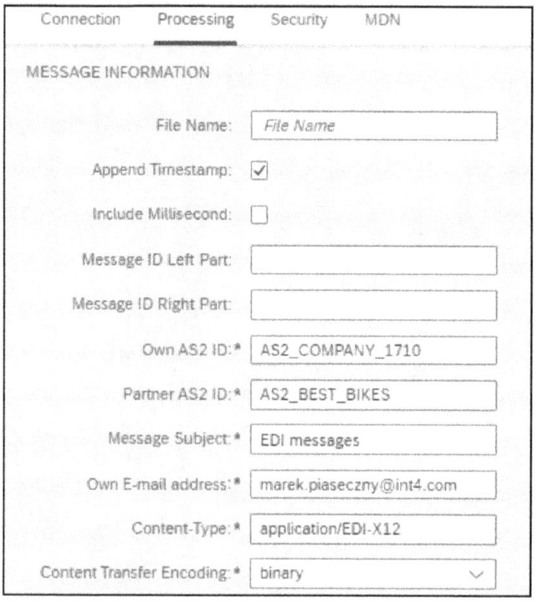

Connection Processing Security MDN
MESSAGE INFORMATION
File Name: File Name
Append Timestamp: ☑
Include Millisecond: ☐
Message ID Left Part:
Message ID Right Part:
Own AS2 ID:* AS2_COMPANY_1710
Partner AS2 ID:* AS2_BEST_BIKES
Message Subject:* EDI messages
Own E-mail address:* marek.piaseczny@int4.com
Content-Type:* application/EDI-X12
Content Transfer Encoding:* binary

Figure 9.122 Trading Partner Profile: AS2 Receiver Adapter: Processing

The section should now be filled in as shown in Figure 9.123.

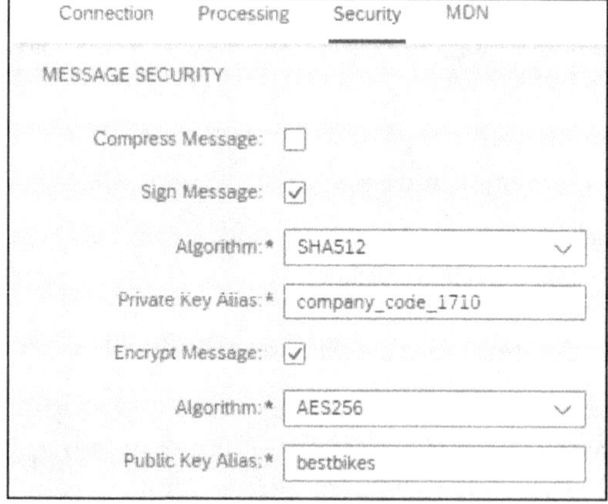

Figure 9.123 Trading Partner Profile: AS2 Receiver Adapter: Security

The last section we need to maintain still for the AS2 receiver adapter is called **MDN**, and the parameters are as follows:

- **Type—Synchronous**
- **Request Signing**—set this parameter
- **Verify Signature**—set this parameter
- **Public Key Alias—bestbikes**
- **Verify MIC**—set this parameter
- **Message Failure on Negative MDN**—set this parameter

The complete section for should look as shown in Figure 9.124.

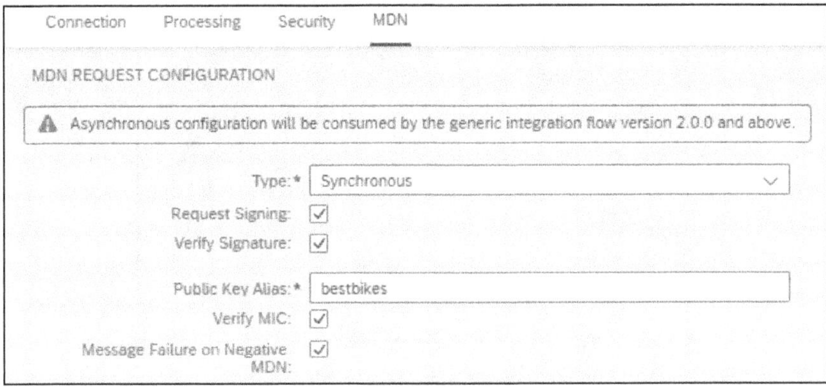

Figure 9.124 Trading Partner Profile: AS2 Receiver Adapter: MDN

After maintaining all the details related to the AS2 receiver adapter, we can create the configuration related to the Cloud Integration inbound connection, which will be an AS2 sender adapter maintained within the **Communications** section. Again, fill in all the fields from the top to the bottom, and the section related to AS2 settings appears only after selection of the appropriate direction and adapter:

- **Name—AS2_Sender**
- **Alias—AS2_Sender**
- **Description**—can stay empty, but feel free to add any meaningful and useful text
- **Direction—Sender**
- **Adapter—AS2**
- **Security Configuration Mode—Channel**
- **User Account**—this references the credentials used by the trading partner to post messages on our AS2 entry point. In this example, it is the **clientid** value taken from the service key that was generated on the SAP Process Integration runtime service. This is maintained on the subaccount on which SAP Integration Suite is subscribed.

After entering the parameters in the top part, we will have two sections in the middle of the screen dedicated to the AS2 sender adapter. The AS2 sender adapter parameters to be maintained within the first section called **MDN** are as follows:

- **Private Key Alias for Signature**—**company_code_1710**
- **Signature Encoding**—**binary**
- **Proxy Type**—**Internet**
- **Authentication**—**Basic Authentication**
- **Credential Name**—**B2B_TP_BESTBIKES_DEV**

The complete configuration of the AS2 sender adapter and its **MDN** section should look as shown in Figure 9.125.

Figure 9.125 Trading Partner Profile: AS2 Sender Adapter

The final section related to security of the AS2 sender adapter should be filled in as follows:

- **Decrypt Message**—set this parameter
- **Private Key Alias**—**company_code_1710**
- **Verify Signature**—**Not Required**

The complete **Security** tab for the AS2 sender adapter should look as shown in Figure 9.126.

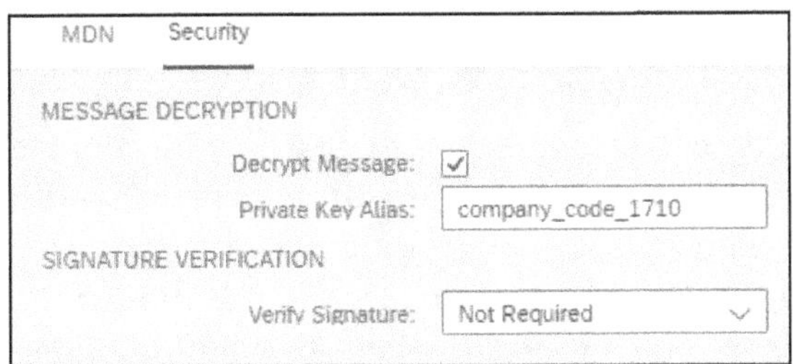

Figure 9.126 Trading Partner Profile: AS2 Sender: Security

After maintaining the **Type Systems** and **Communications** sections, we can go back to the top level of the partner profile maintenance view. There are two sections we can skip in our scenario: **Certificates** and **Parameters**. In the next section, **Security**, we need to maintain signature verification configurations. We keep this part simple by not enabling signatures and MIC verification (although it is recommended to use both options in a production setup):

- **AS2 Partner ID—AS2_BEST_BIKES**
- **Alias—AS_BEST_BIKES**
- **Signature—Not Required**
- **Verify MIC (MDN Only)**—not set

The complete entry in **Security** section should look like in Figure 9.127.

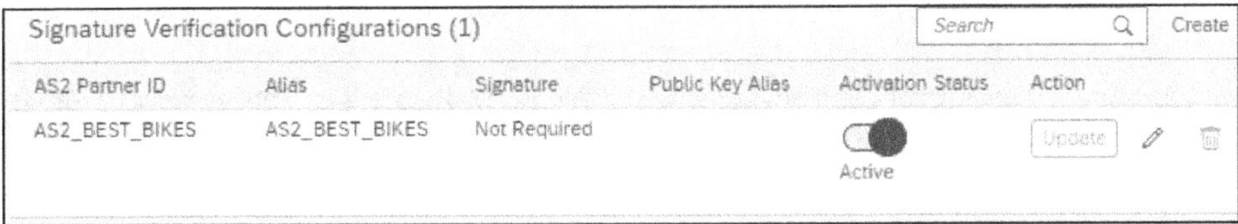

Figure 9.127 Trading Partner Profile: Security

In **Number Ranges**, add one entry with values as follows:

- **Number Ranges—ICN_DEFAULT**
- **Alias—ICN_DEFAULT**

The complete entry in the **Number Ranges** section should look as shown in Figure 9.128.

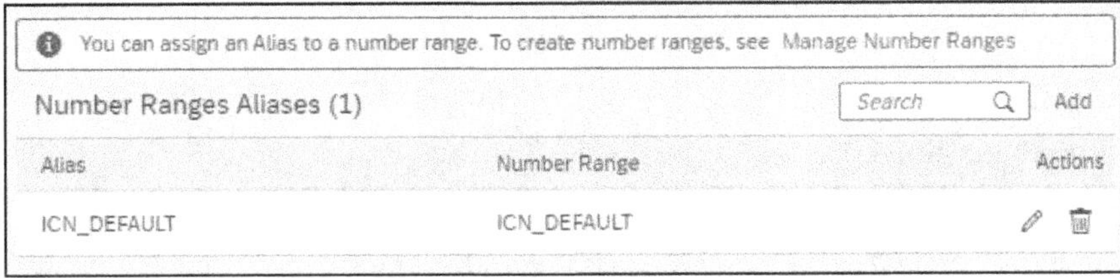

Figure 9.128 Trading Partner Profile: Number Ranges

9.3.3 Creating Agreement Templates

In the previous sections, we created MIGs and MAGs and defined the company profile and trading partner profile. Now it is time to combine them all in a fully functioning B2B scenario. This is done via an agreement, which not only pairs our company with the respective trading partner but more importantly defines the set of EDI transactions between the preselected parties and the role of each party. During this process, we choose the previously predefined type systems, type system versions, identifiers, communication adapters, and so on. The agreement defines all the small details of the B2B scenario between two indicated parties. Because one party is always our own company, the concept is extended by the agreement template, which must be created before the agreement itself. The former helps not only with harmonizing the setup but also speeding up the configuration itself. An agreement template does not define any aspects related to trading partners, but it does define all the aspects related to processing within our own company. It defines not only the type systems, versions, identifiers, and so on but it also sets of transactions. It acts as a template that can be reused when we have an active specific trading partner via the agreement. We only need to maintain the part related to trading partner; optionally we can also adjust the other parameters. The benefit is that we don't have to define agreements for each unique trading partner from scratch.

As you probably noticed earlier, even for our own company we can define multiple systems and then multiple following type systems, versions, and communication adapters. That means that there is no limitation to run B2B scenarios on a single ERP system or on a single system type/version within our own landscape. The same holds for trading partner systems. All these options and variables are handled by agreement templates and agreements. It is also possible to create multiple agreement templates. In our scenario, we only have one system within our own company, and the same holds for our trading partner. A single agreement template will be enough.

In the following steps, we will create the agreement template, which will define that our own company runs on SAP S/4HANA system and makes use of IDocs for processing four different UN/EDIFACT transactions. Most of the values come from the previously defined profiles, so the maintenance is mainly about selecting the appropriate values from a selection list.

Follow menu path **Design • B2B Scenarios • Agreement Templates**, then maintain the **Overview** screen as follows:

- **Detail**
  - **Name**—add value **Company 1710 UN/EDIFACT template**
  - **Description**—add value **Template for SAP S/4HANA IDocs and UN/EDIFACT**
  - **Version**—add value **1.0**
- **My Company Details**
  - **Name: Company Code 1710** (this field is not editable and filled automatically)
  - **System**—choose **S4HANA_S42_DEV**

- **Purpose**—choose **PROD**
- **Type Systems**—choose **SAP S/4HANA On Premise IDoc**
- **Type Systems Version**—choose **1809_FPS02**
- **Identifier in Company Type Systems**—choose **S42CLNT400**
- **Identifier in Trading Partner Type System**—choose **EDI_COMPANY_1710** (note: to be able to maintain this value we need to define that on trading partner side we will have a UN/EDIFACT type system, mentioned ahead)

- **Trading Partner Details**
  - **Type Systems**—choose **UN/EDIFACT**

The complete maintenance view for the agreement template overview should look as shown in Figure 9.129.

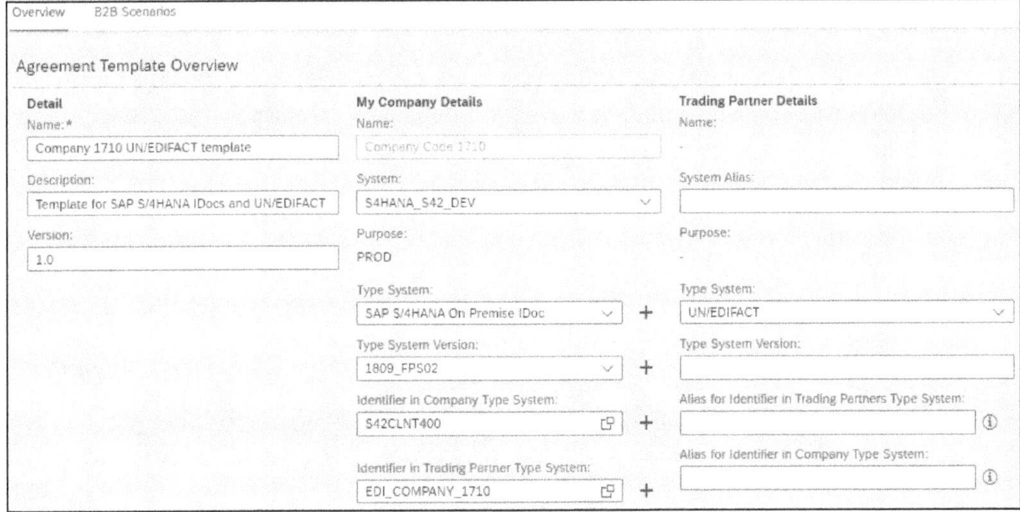

Figure 9.129 Maintenance View for Agreement Template Overview

After maintaining the **Overview** section go to the **B2B Scenarios** tab and click the **Create Business Transaction** button located at the bottom of the screen. The first of the four transactions is related to purchase order documents sent by a customer, and we will create it as follows:

- **Name**—**Sales order**
- **Business Transaction Pattern**—**One Way Notification**
- **My Company Role**—**Reactor**

At the end we should have values maintained as shown in Figure 9.130.

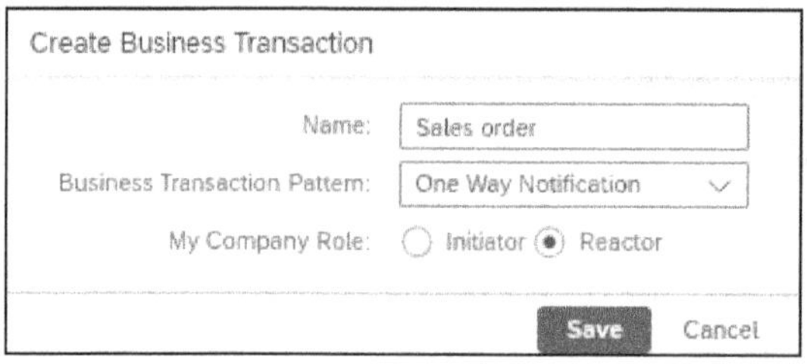

Create Business Transaction

Name: Sales order

Business Transaction Pattern: One Way Notification

My Company Role: Initiator Reactor

Save Cancel

Figure 9.130 Agreement Template: Create Business Transaction

After confirming creation of the transaction by clicking **Save**, we will see a new screen with boxes representing the key (technical) steps and their sequence within the transaction. We will work with the sales order transaction here, but the technical steps are exactly the same for each transaction. The direction of the message can be the opposite, but even that does not change the steps; only the arrows go in the opposite direction.

You can see all the steps of the sales order transaction in Figure 9.131. On the top level of the transaction, we have an option to maintain **Activity Parameters** and **Custom Search Attributes**; these are self-explanatory, and at this stage we will not make use of this functionality.

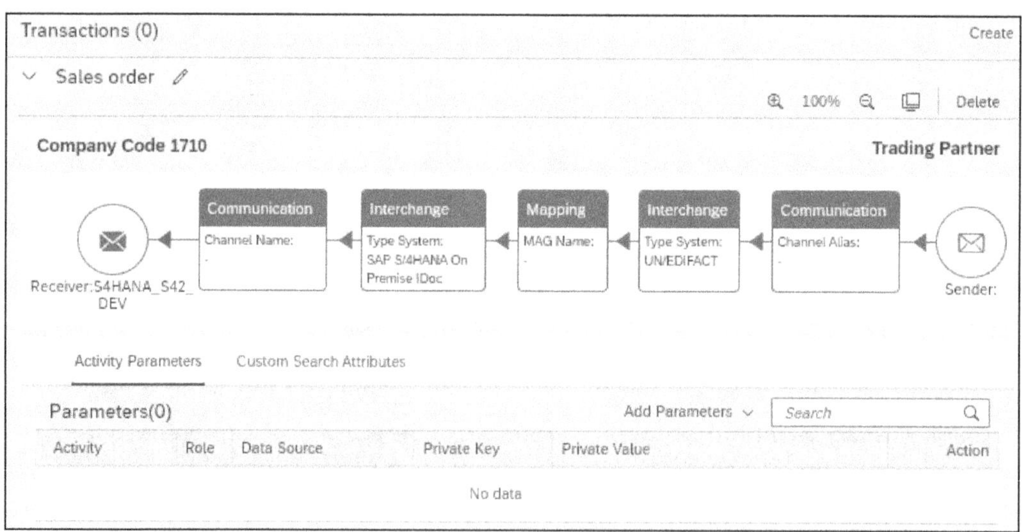

Figure 9.131 Agreement Template: Sales Order Transaction

The task here is to go through all the steps of the given transaction and review and maintain the appropriate parameters. We will start from the sender side, so click the circle at the farthest right, **Sender**. At the bottom of the screen, you will see parameters related to the selected step. And as we said before, the agreement template does not involve any trading partner. That's why although you can see some parameters after selecting the sender, there is no option to maintain any of them. The same happens for the **Communication** step related to the **Sender** (second from the right); it is also read-only. In Figure 9.132, you can see the details of the sender system. All information comes from agreement template overview and there is no option to change anything on this screen.

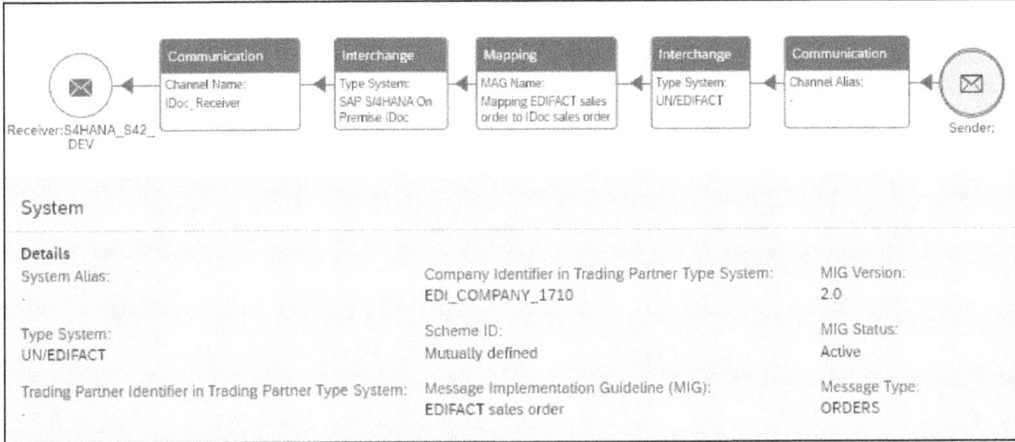

Figure 9.132 Agreement Template: Sales Order Transaction: Sender

The first element we need to maintain is the interchange on the sender side, which defines what kind of message type we receive and link to the MIG definition. This is also the source for the MAG defined in the next step. After choosing the **Interchange** step, you can see three sections at the bottom of the screen. The values to maintain there are as follows:

- **Details**
  - **Type Systems: UN/EDIFACT**— from the template agreement overview (not editable in this view)
  - **Type System Version: D.96A S3**—one of the few values we need to type, not choose from the list
  - **Message Implementation Guideline (MIG): EDIFACT sales order**—chose appropriate MIG and its version from the list
  - **MIG Status** and **Message Type**—values come from the MIG itself; not editable fields
  - **Create Functional Acknowledgement: Required**
- **Custom Integration Flow:** In this and in most of the other steps we have an option to hook a custom iFlow where we can address any additional, custom requirements. The link is done via a process direct adapter. We will not make use of this, but it is one of the features that makes Trading Partner Management solutions almost limitless.
- **Validation Options:** This step allows validation of the actual message against its definition in MIG. It is not only about the structure itself but also includes other aspects like cardinality, codelist, and so on. Keep it disabled for now.

Figure 9.133 shows the complete setup for the sender interchange.

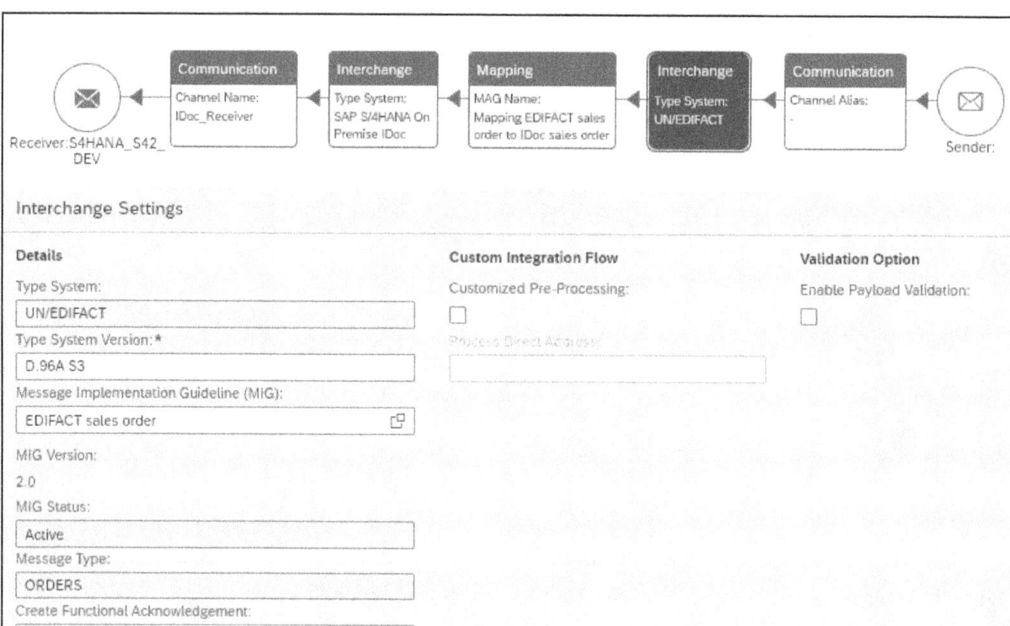

Figure 9.133 Agreement Template: Sales Order Transaction: Interchange on Sender Side

The **Message Implementation Guideline (MIG)** field offers a selection screen where you can search and filter on all available MIGs and their attributes. A sample **Select MIG** screen is shown in Figure 9.134.

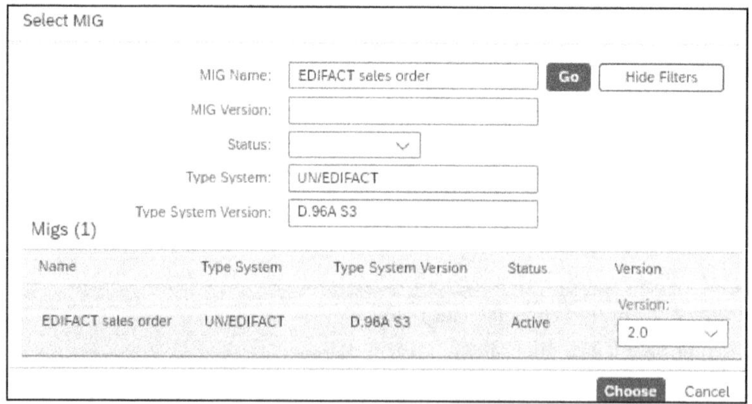

Figure 9.134 Agreement Template: Sales Order Transaction: Interchange: MIG Selection

The next step is related to a mapping used to transform the sender message into the receiver message. For choosing the right MAG, we have a dedicated selection screen available with multiple search and filter options. It is worth mentioning that we have the option to choose any MAG version—not only the most recent and active version but even the older, deprecated ones.

Figure 9.135 shows the **Mapping** step and its properties, then Figure 9.136 shows the mapping selection screen. Pay attention to the selection list located on the right-hand side of the screen where you can choose the MAG's version.

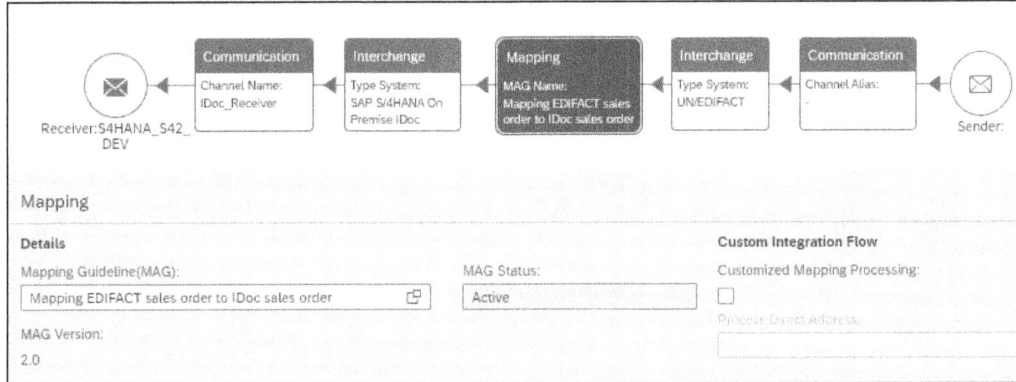

Figure 9.135 Agreement Template: Sales Order Transaction: Mapping

Filters

MAG Name: Mapping EDIFACT sales order to IDoc sales order | Go | Hide Filters

MAG Version: 2.0

MAG Status: Active

Name (Source MIG): EDIFACT sales order

Name (Target MIG): IDoc sales order

MAGs (1)

Name	Source MIG	Target MIG	Version
Mapping EDIFACT sales order to IDoc sales order	EDIFACT sales order	IDoc sales order	Version: 2.0

Choose Cancel

Figure 9.136 Agreement Template: Sales Order Transaction: Mapping Selection Screen

After configuring the **Mapping** step and moving to the left, we have yet another **Interchange** step. This one is related to the message that we finally send to the receiver system. The message defined here needs to match the target message from the MAG defined in the previous step. There are three key items we maintain here:

- **Type System Version—SAP S/4HANA 2022 02 (05/2024) FPS**
- **Message implementation Guideline (MIG)**—using the selection screen we choose **IDoc sales order** in the most recent, active version
- **Number Range**—choose **ICN_DEFAULT**

The other fields are greyed out and not editable. We don't make use of the custom iFlow or the validation option. Figure 9.137 shows the interchange settings with all the parameters.

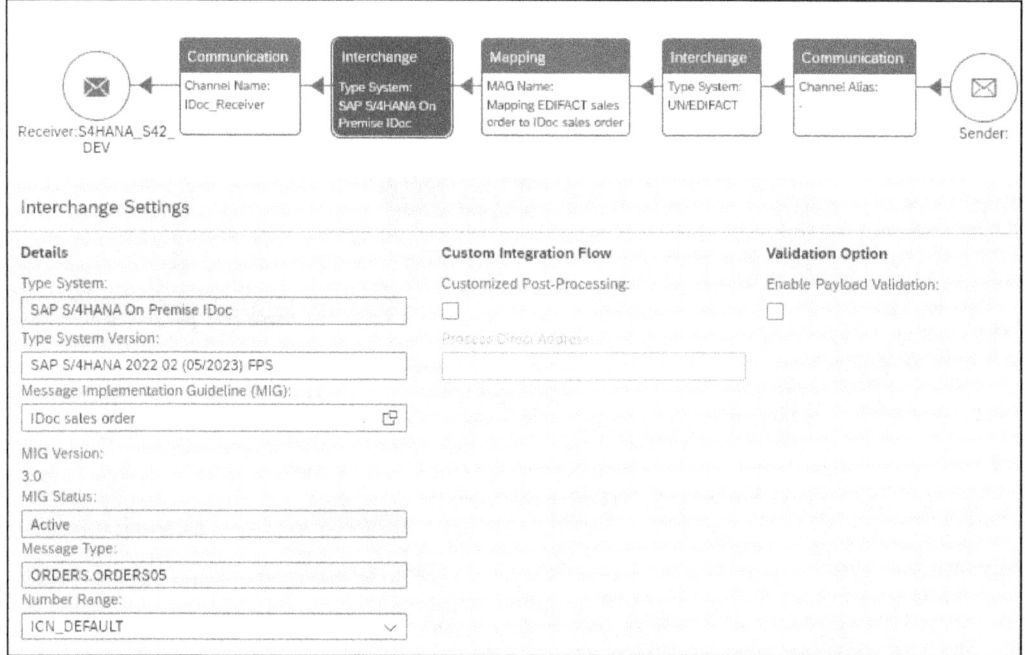

Figure 9.137 Agreement Template: Sales Order Transaction: Interchange on Receiver Side

The last step is related to the communication with the receiver system. From the selection list, choose **IDoc_Receiver**, the IDoc receiver adapter we defined earlier in the company profile for our SAP S/4HANA system. That's the only parameter available in this step. The complete sales order transaction and all its steps are shown in Figure 9.138.

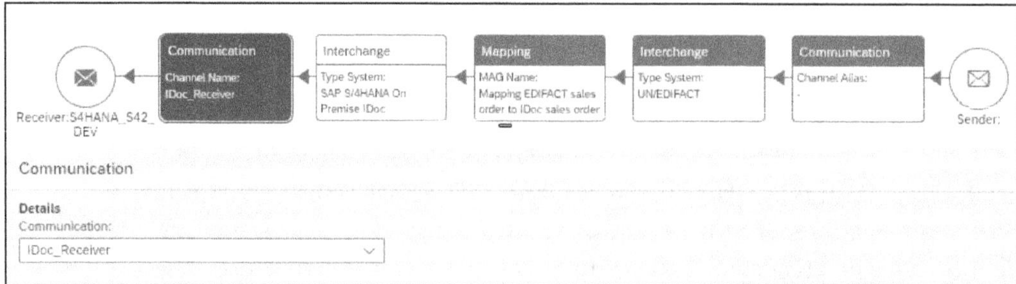

Figure 9.138 Agreement Template: Sales Order Transaction: Communication on Receiver Side

When you click **Receiver:S4HANA_S42_DEV**, you can only display data previously maintained in the company profile. There is nothing to maintain in this step; all information is read-only and coming from the template agreement overview (see Figure 9.139).

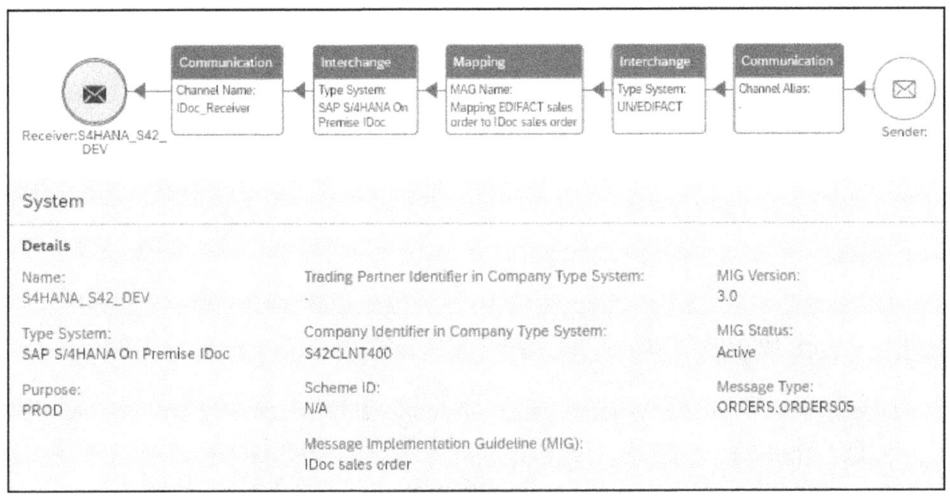

Figure 9.139 Agreement Template: Sales Order Transaction: Receiver

After saving all modifications, we have completed the first transaction that is part of the B2B scenario. You need to maintain three more, but the work does not differ much from the provided example. Your main doubts may be related to the fact that for the other three transactions, your own company is the initiator of the process. In addition, the messages go out from SAP S/4HANA toward the trading partner. Let's go through the sales order response to practice on one case.

While you are in **Design • B2B Scenarios • Agreement Template • B2B Scenarios**, click **Edit** to switch to edit mode and then click **Create**. Maintain values as follows to create the second transaction (see Figure 9.140):

- **Name—Sales order response**
- **Business Transaction Pattern—One Way Notification**
- **My Company Role—Initiator**

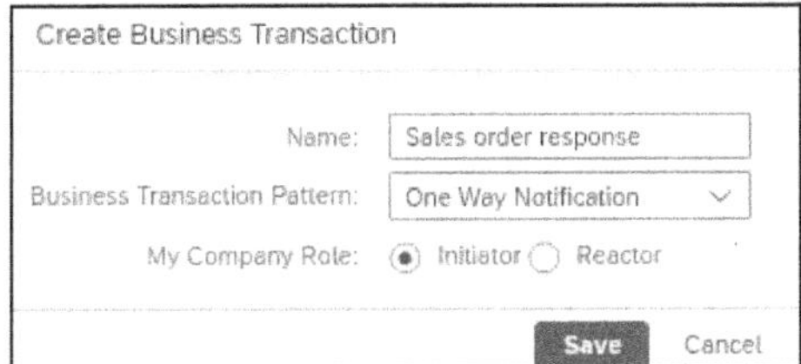

Figure 9.140 Agreement Template: Sales Order Response: Create Transaction

As you can see in Figure 9.141, the steps do not differ from the previous sales order transaction; only the direction of the arrows for the flow has changed. My recommendation is to maintain all the steps from the left side, starting from the sender communication channel. There is again only one parameter to be maintained via a selection list; choose **IDoc_Sender** as shown in Figure 9.141.

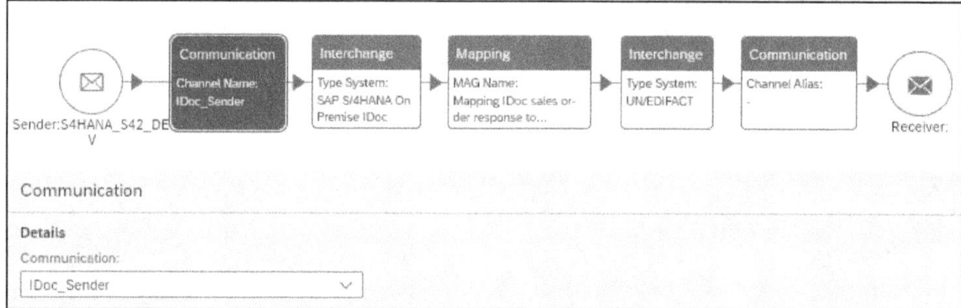

Figure 9.141 Agreement Template: Sales Order Response: Sender Communication

The next step is related to the interchange on sender side. At this step you need to configure the IDoc message coming from SAP S/4HANA. Set **Type System Version** as **1809_FPS02**. Using the selection screen, for **Message Implementation Guideline (MIG)**, choose **IDoc sales order response** in the most recent, active version. We don't make use of **Custom Integration Flow** or **Validation Option**. The complete configuration is shown in Figure 9.142.

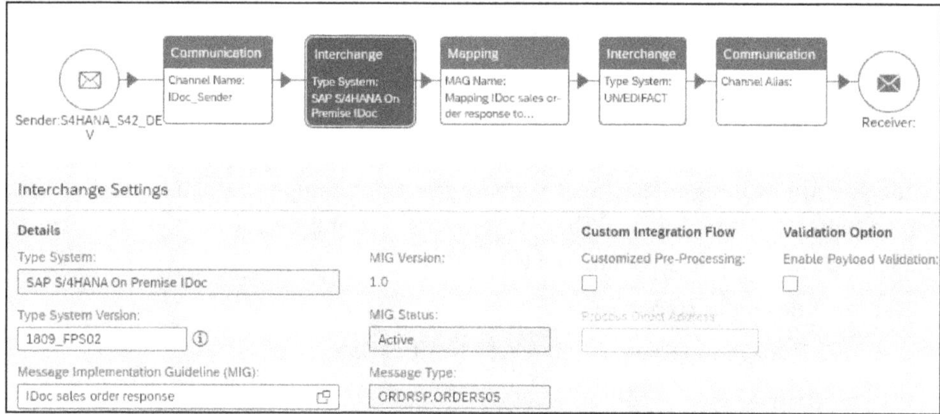

Figure 9.142 Agreement Template: Sales Order Response: Sender Interchange

In the **Mapping** step, we only need to choose the appropriate MAG: **Mapping IDoc sales order response to EDIFACT sales order response**, as shown in Figure 9.143.

The last step we need to maintain is the interchange on receiver side. As we aim to send UN/EDIFACT ORDRSP, the configuration should be as follows:

- **Type Systems**: **UN/EDIFACT**—value comes from the template agreement overview; not editable
- **Type System Version**: **D.96A S3**—one of the few values we need to type, not choose from the list
- **Message Implementation Guideline (MIG)**: **EDIFACT sales order response**—choose the appropriate MIG and its version from the list
- **MIG Status** and **Message Type**—values come from the MIG itself; not editable fields

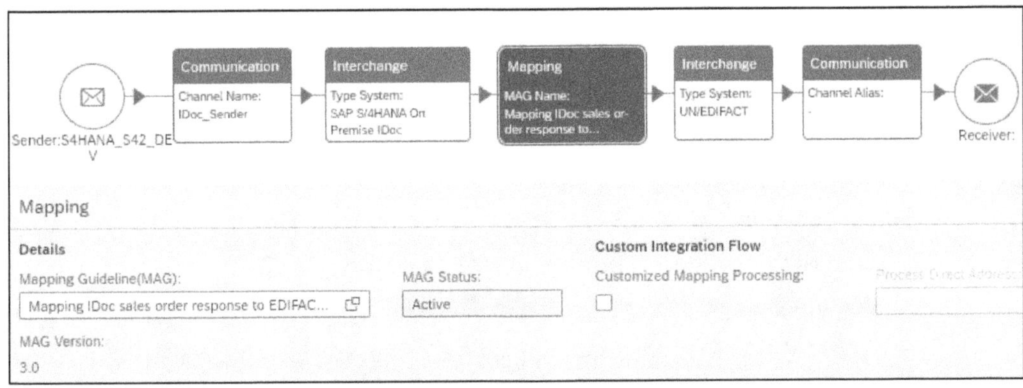

Figure 9.143 Agreement Template: Sales Order Response: Mapping

We will not use **Custom Integration Flow** or **Validation Options**. You may notice two additional options: **Receiver Functional Acknowledgment** and **Custom Separators**. The former enables additional functional acknowledgments for the message transmitted by us, but it requires additional configuration in the communication section. In **Custom Separators**, you can influence the separators used in the UN/EDIFACT message. We'll skip both options for now. Figure 9.144 shows the complete setup for the sender interchange.

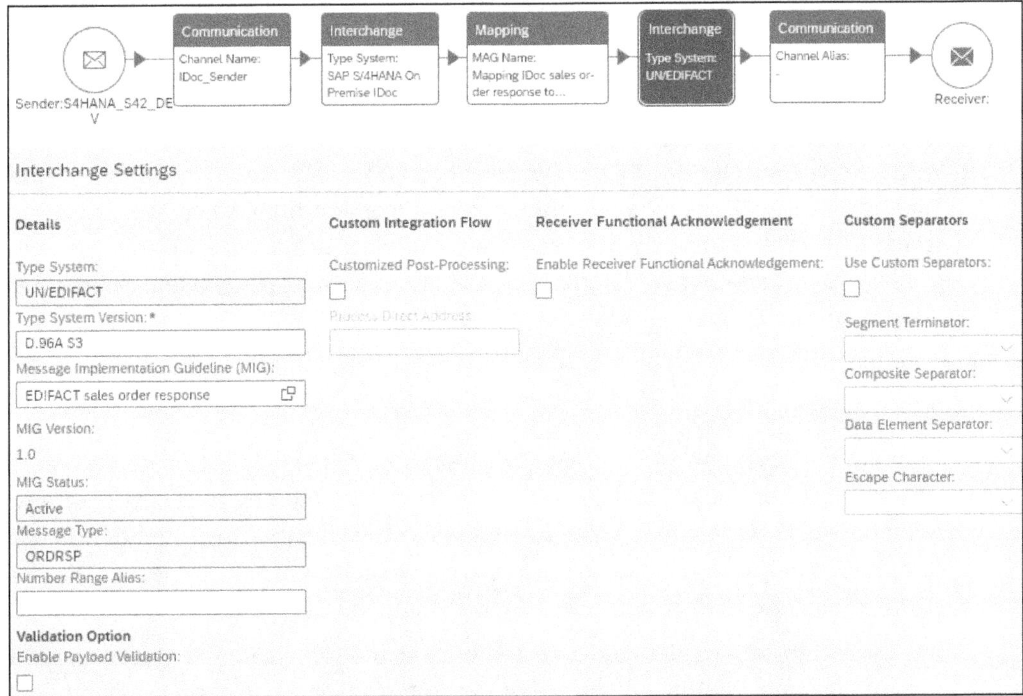

Figure 9.144 Agreement Template: Sales Order Response: Receiver Interchange

We keep the receiver communication and receiver system steps as is, without maintaining anything, and we are now done with the setup of the second transaction within our B2B scenario.

Based on the given examples it should be a straightforward task to maintain the other two transactions, for the ASN and invoice. To summarize:

- **Sender Communication**
  - For both these transactions should have value **IDoc_Sender**
- **Sender Interchange**
  - Choose MIG as **IDoc advance shipping notice** for ASN transaction
  - Choose MIG as **IDoc invoice** for ASN transaction
  - Remember to maintain **Type System Version** as **1809_FPS02**
- **Mappings**
  - Choose MAG as **Mapping IDoc advance shipping notice to EDIFACT advance shipping notice** for ASN transaction
  - Choose MAG as **Mapping IDoc invoice to EDIFACT invoice** for ASN transaction
- **Receiver Interchange**
  - Choose MIG as **EDIFACT advance shipping notice** for ASN transaction
  - Choose MIG as **EDIFACT invoice** for ASN transaction
  - Remember about maintaining **Type System Version** as **D96A S3** for both
- **Receiver Communication**
  - Does not need to be maintained

We have completed template agreement creation and can move to the final task within Trading Partner Management: creation of the agreement itself, where we enable the scenario for a specific trading partner. What's important to know is that while the template is a base for the agreement there is still option to adjust many of the parameters. We will look at this option in the next section.

9.3.4 Agreements

To create an agreement, go to **Design • B2B Scenarios • Agreements** and click the **Create** button located on the right-hand side of the screen. On the first screen, you need to select the existing agreement template that will be used for agreement creation. In our case, there is only one template to choose, as shown in Figure 9.145: **Company 1710 UN/EDIFACT template**. Confirm the selection by clicking **Next**.

On the next screen, we define the creation mode and can create the agreement as a copy of the template. This will give us the option to adjust some of the parameters as there will be no relation to the agreement template. That's the option to choose in this case. In addition, make sure that all the available transactions are selected and, as a last step, make use of the selection list at the bottom to choose Best Bikes as the trading partner. When that's in place, click on **Open Draft**. Figure 9.146 shows what the selections should look like.

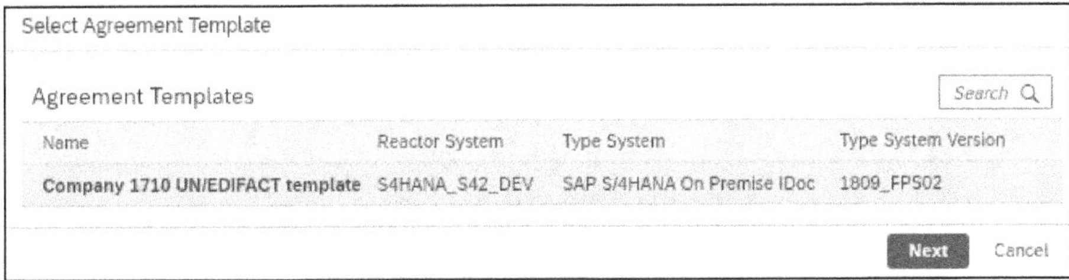

Figure 9.145 Agreement Creation: Selection of Agreement Template

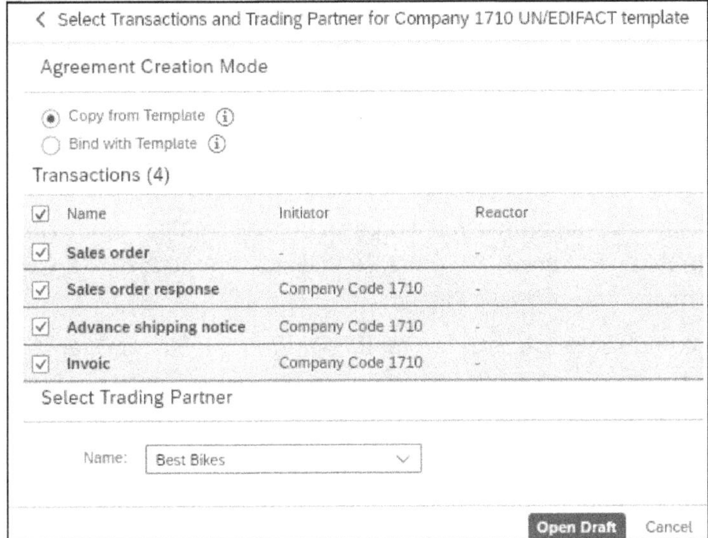

Figure 9.146 Agreement Creation: Selection of Transactions and Trading Partners

Now we can maintain the agreement itself. Similar to the agreement template, we also have two tabs: **Overview** and **B2B Scenarios**. We start with the **Overview** screen where we add the trading partner details that were not available in the template itself. The values we need to add are as follows:

- **System**—**TP_ERP**
- **Type System**—**UN/EDIFACT**
- **Type System Version**—**D.96A S3**
- **Identifier in Trading Partners Type System**—choose **B2B-Identifier** from the list
- **Identifier in Company Type System**—choose **USCU_S05** from the list

Figure 9.147 shows the complete setup.

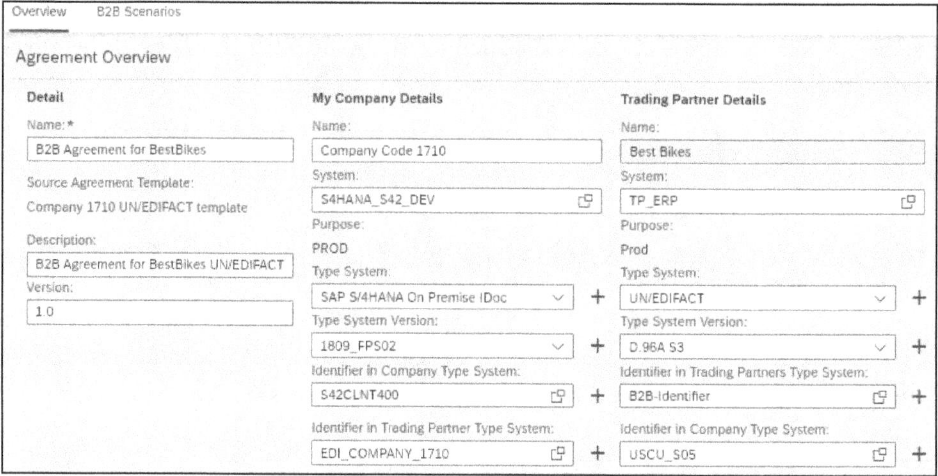

Figure 9.147 Agreement Overview

After saving the **Overview** screen, go to the **B2B Scenarios** tab. You will see there all four transactions as they were maintained in the agreement template. When you expand the first transaction, you will see the already known view, and you may notice that the sender and its system is already identified and assigned. But the configuration is not complete; there are still settings we need to maintain. It is also recommended to review each step and each parameter before final activation.

You need to maintain communication with the trading partner for each transaction, one of the elements that are not filled in automatically. For the sales order transaction, click the (sender) **Communication** step and maintain the two attributes as follows, making use of the selection list:

- **Communication—AS2_Sender**
- **Communication for Sender Functional Acknowledgment—AS2_Receiver**

This is the only action required for the sales order transaction. The final setup for communication is shown in Figure 9.148.

The next transaction in our agreement is the sales order response. Like in the sales order flow, we need to maintain communication with the trading partner. **AS2_Receiver** is the channel to choose from the selection list (see Figure 9.149).

There is one more parameter we need to maintain for the sales order response and the other two outbound transactions. This will also demonstrate that although we configured a specific value in the company profile and the agreement template, we still have an option to adjust it in the final agreement. The parameter we have to adjust is **Type System Version**. The reason behind this is that our SAP S/4HANA version generates and sends out IDocs with a release value as 758. That does not match the **Type System Version 1809_FPS02** we configured in the company profile nor does it match the agreement template. We will adjust this value directly in the agreement. This is not the recommended approach, but it is a good example to demonstrate the functionality of having the agreement created as a copy of the agreement template (see Figure 9.150).

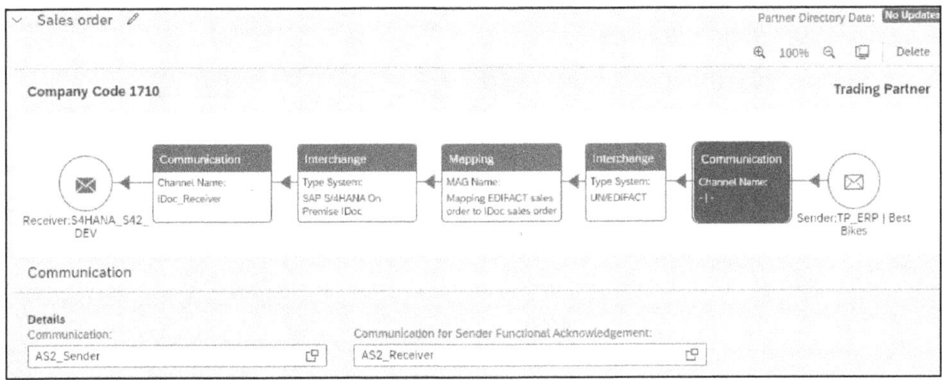

Figure 9.148 Agreement: Sales Order Transaction

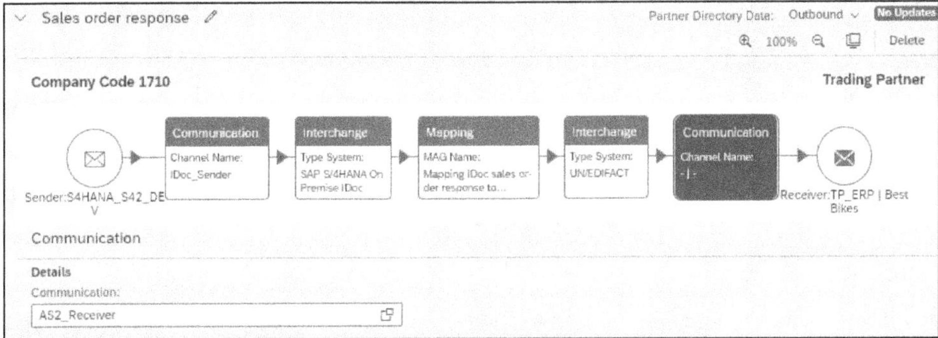

Figure 9.149 Agreement: Sales Order Response Transaction: Communication

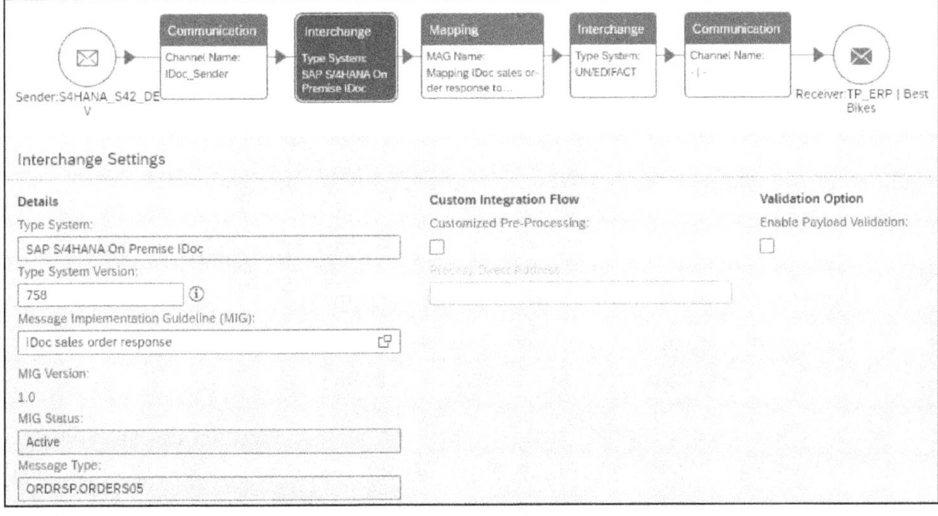

Figure 9.150 Agreement: Sales Order Response Transaction: Interchange

The third and the last property we maintain is **Number Range Alias** on the (receiver) **Interchange** step. **ICN_Default** is the value to maintain in the **Number Range Alias** field, as shown in Figure 9.151.

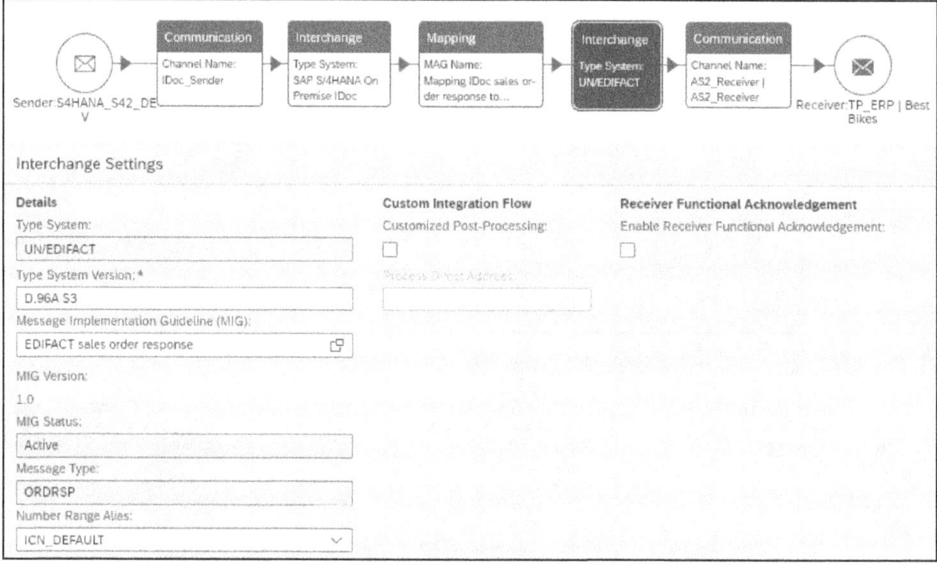

Figure 9.151 Agreement: Sales Order Response Transaction: Interchange

After these steps, we can proceed to the ASN transaction, where exactly the same adjustments are required:

- Click the **Interchange** step related to the sender and maintain value **758** in field **Type System Versions**.
- Click the **Interchange** step related to the receiver and choose **ICN_DEFAULT** in field **Number Range Alias** from the selection list at the bottom.
- Click the **Communication** step related to the receiver and choose **AS2_Receiver** channel from the selection list at the bottom, as shown in Figure 9.152.

When you move to the invoice, again, exactly the same adjustments are required:

- Click the **Interchange** step related to the sender and maintain value **758** in field **Type System Versions**.
- Click the **Interchange** step related to the receiver and choose **ICN_DEFAULT** in field **Number Range Alias** from the selection list at the bottom.
- Click the **Communication** step related to the receiver and choose **AS2_Receiver** from the selection list at the bottom, as shown in Figure 9.153.

The last step, when the overview and all transactions of the agreement are adjusted accordingly, is related to the activation of the object. When you look at the header of the agreement shown in Figure 9.154, you can see that **Activation Status** has value **Draft**.

In the upper-right corner, there is also an **Activate** button available. Click the button to activate the newly created agreement object.

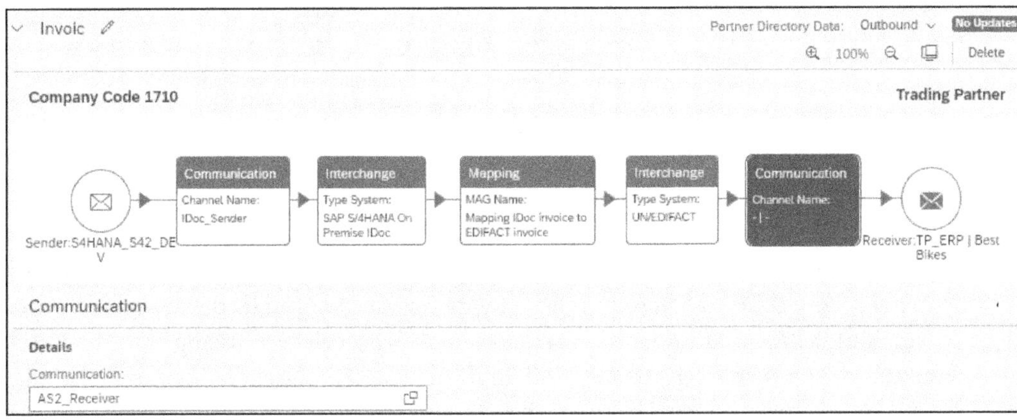

Figure 9.152 Agreement: Invoice: Communication with Trading Partner

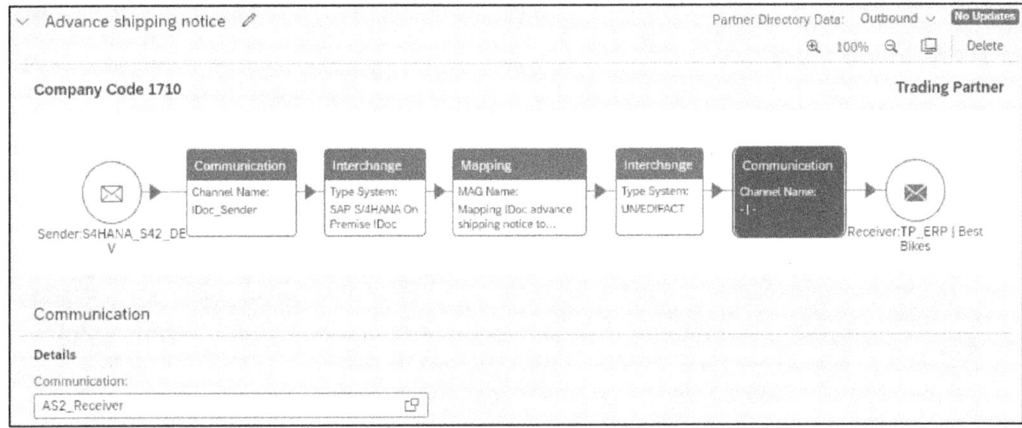

Figure 9.153 Agreement: ASN: Communication with Trading Partner

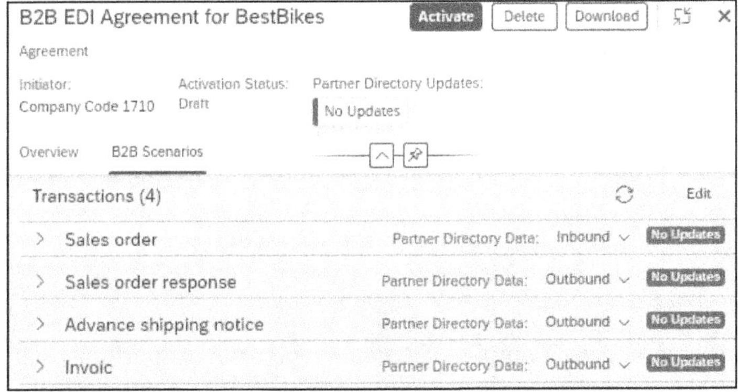

Figure 9.154 Header of B2B EDI Agreement for Best Bikes before Activation

The activation of an agreement requires an additional confirmation because the configuration is loaded into the partner directory. This is a standard process and known functionality; agree to it by clicking the **OK** button as shown in Figure 9.155.

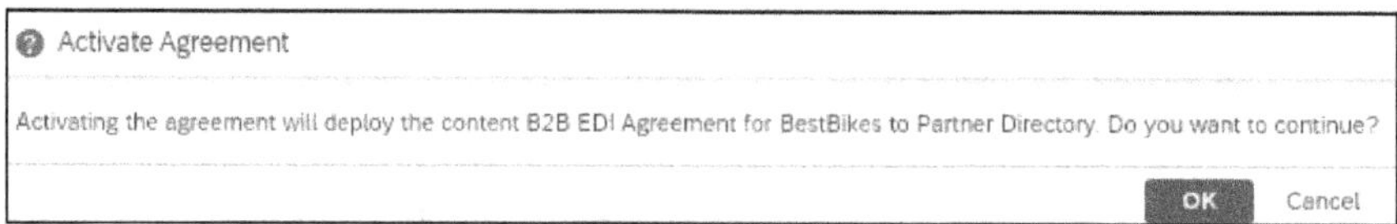

Figure 9.155 Activation of Agreement: Confirmation Screen

After successful activation of the agreement, you can see that the **Activation Status** changed from **Draft** to **Active** and the **Activate** button changed its description to **Deactivate**; after deactivation of the agreement, you can delete it. Using the **Download** option, you can export the whole configuration related to the agreement to a JSON file. Figure 9.156 shows the final status of the agreement.

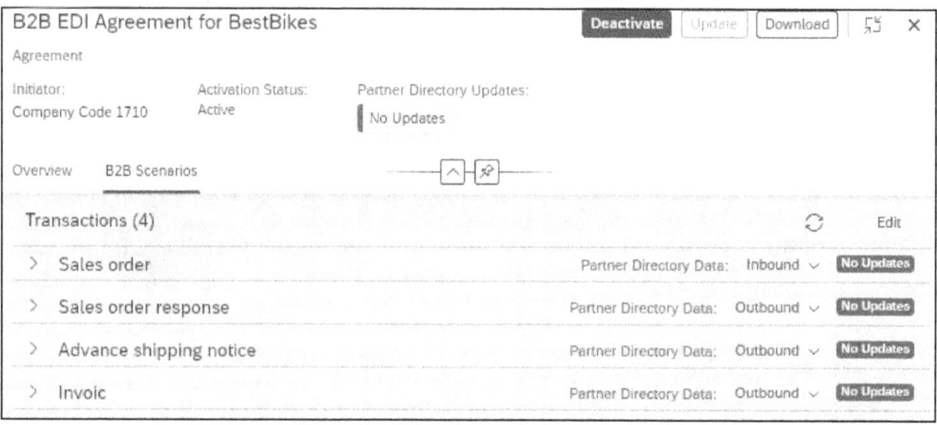

Figure 9.156 Header of B2B EDI Agreement for Best Bikes after Activation

The active agreement can be still changed, in both the overview and the transaction-related data. Be aware that (currently) there is no automated version management for the agreement object, so the value of the **Version** field on the **Overview** screen needs to be controlled and maintained manually. After changing an active agreement, you must trigger an update that will synchronize the changes into the partner directory. The update can be triggered by the **Update** button located in the top-right corner; the need for the update itself and the progress of the update can be traced with the **Status** fields located next to each transaction and in the header.

9.3.5 Partner Directory Data

With the activation of the agreement completed, there is one more important background activity to handle. All the parameters configured for the transactions are pushed into the partner directory. You can see the exact entries loaded into the partner directory also via a link from the agreement itself. When you display the list of transactions, the link to **Partner Directory Data** is located on the right-hand side of the transaction name. Click the link, which has a name that matches the direction of the given transaction (**Inbound** or **Outbound**). In the small popup menu that appears, choose **View Data**, as shown in Figure 9.157.

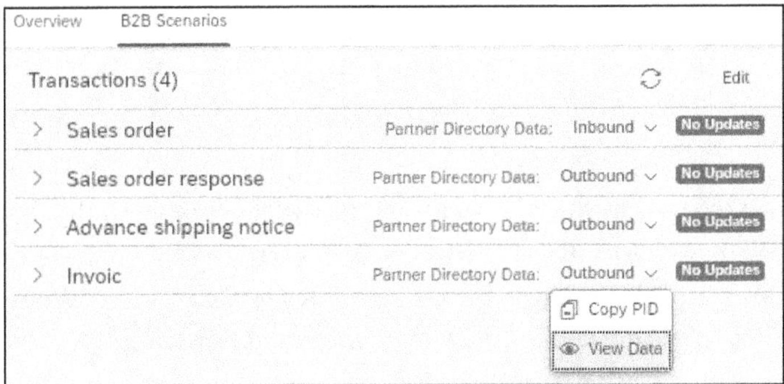

Figure 9.157 Agreement: Partner Directory Data: View Data

Choosing **View Data** will redirect you to a new view, as shown in Figure 9.158. The window is split into two parts. On the left-hand side you can see the available **Partner Directory IDs**. The one related to the invoice transaction is already selected as this is what we chose in the agreement view. On the right-hand side you can see a list of all **String Parameters** and **Binary Parameters** related to the selected partner directory ID.

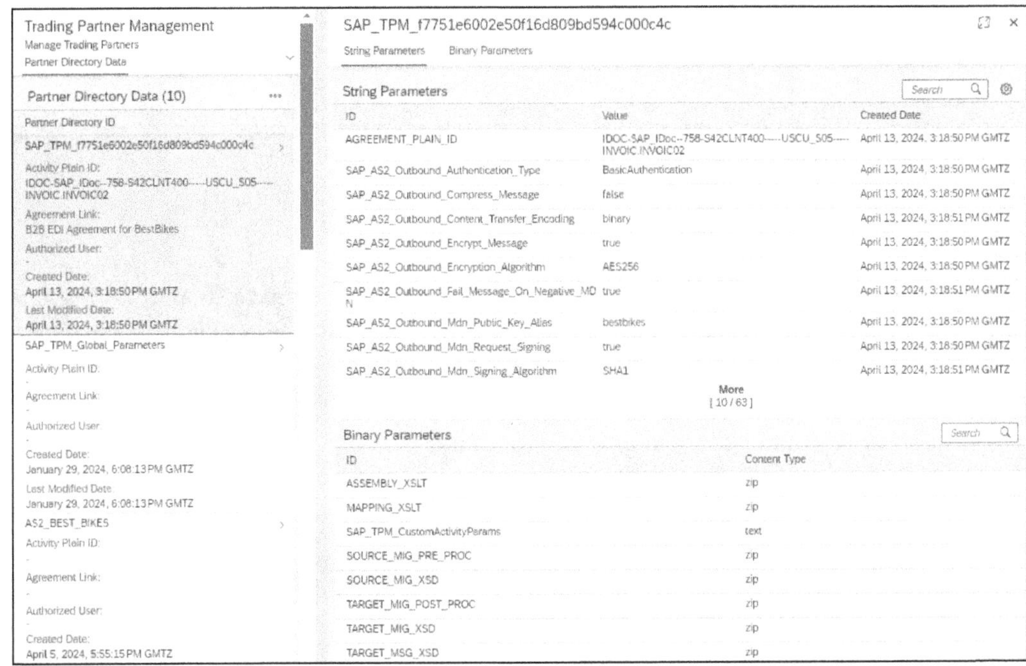

Figure 9.158 Partner Directory Data: View of Invoice Transaction

When you close the window with the parameters, you will see the overview of the partner directory data. This is the same view you can see directly via menu path **Design • B2B Scenarios • Partner Directory Data**. When you look at the overview in Figure 9.159, you can find the key elements related to the invoice transaction from the agreement:

- **Partner Directory ID—SAP_TPM_f7751e6002e50f16d809bd594c000c4c**
- **Agreement Link—B2B EDI Agreement for BestBikes**
- **Activity Plain ID—IDOC-SAP_IDoc--758-S42CLNT400-----USCU_S05-----INVOIC.INVOIC02**

Looking at the content of the partner directory, you should realize that it is critical to the functioning of the entire B2B solution on SAP Integration Suite. It links all the different components and defines how each transaction or even each step should be executed. The key rule is to always make sure that the content is well synchronized. When processing errors, especially for newly configured environments or flows, the first step is usually to check the content of the partner directory.

Company Profile | Trading Partners | Agreement Templates | Agreements | Partner Directory Data | Cross Actions | Configuration Manager

Partner Directory Data (10)

Search

Partner Directory ID	Activity Plain ID	Agreement Link	Authorized User	Created Date	Last Modified Date
SAP_TPM_f7751e6002e50f16d809bd594c000c4c	IDOC-SAP_IDoc--758-S42CLNT400-----USCU_S05-----INVOIC.INVOIC02	B2B EDI Agreement for BestBikes	-	April 13, 2024, 3:18:50 PM GMTZ	April 13, 2024, 3:18:50 PM GMTZ
SAP_TPM_Global_Parameters	-	-	-	January 29, 2024, 6:08:13 PM GMTZ	January 29, 2024, 6:08:13 PM GMTZ
AS2_BEST_BIKES	-	-	-	April 5, 2024, 5:55:15 PM GMTZ	April 5, 2024, 5:55:15 PM GMTZ
SAP_TPM_2f438e01d3314d5fa467805ac20cbc44	-	-	sb-f9ace2ab-ee06-4a11-8e1d-53c9998f279b!b152182|it-rt-int4-cloud-integration-cf-u0uji6hs!b117912	April 1, 2024, 3:26:34 PM GMTZ	April 1, 2024, 3:26:34 PM GMTZ
SAP_TPM_c4f9de2bde9b2be450671bb0b2b6214a	-	-	as2_company_1710	April 9, 2024, 8:49:37 PM GMTZ	April 9, 2024, 8:49:37 PM GMTZ
SAP_TPM_5be28d26bc802ae39def7f4625f1894d	IDOC-SAP_IDoc--758-S42CLNT400-----USCU_S05-----DESADV.DELVRY07	B2B EDI Agreement for BestBikes	-	April 13, 2024, 3:18:51 PM GMTZ	April 13, 2024, 3:18:51 PM GMTZ
SAP_TPM_676fc5f74e7692639a59e659ff8d2268	IDOC-SAP_IDoc--758-S42CLNT400-----USCU_S05-----ORDRSP.ORDERS05	B2B EDI Agreement for BestBikes	-	April 13, 2024, 3:18:52 PM GMTZ	April 13, 2024, 3:18:52 PM GMTZ
SAP_TPM_1f05e7fdaeb950c87902444882cafc22	AS2-UN-EDIFACT--D.96A S3-EDI_BEST_BIKES-ZZZ----EDI_COMPANY_1710-ZZZ----ORDERS	B2B EDI Agreement for BestBikes	-	April 13, 2024, 3:18:53 PM GMTZ	April 13, 2024, 3:18:53 PM GMTZ
SAP_TPM_28a0b554d3bec8996de3eb1948dfa323	-	-	-	-	-
SAP_TPM_e2271c3faaeda358a88ddbcd8b298783	-	-	-	-	-

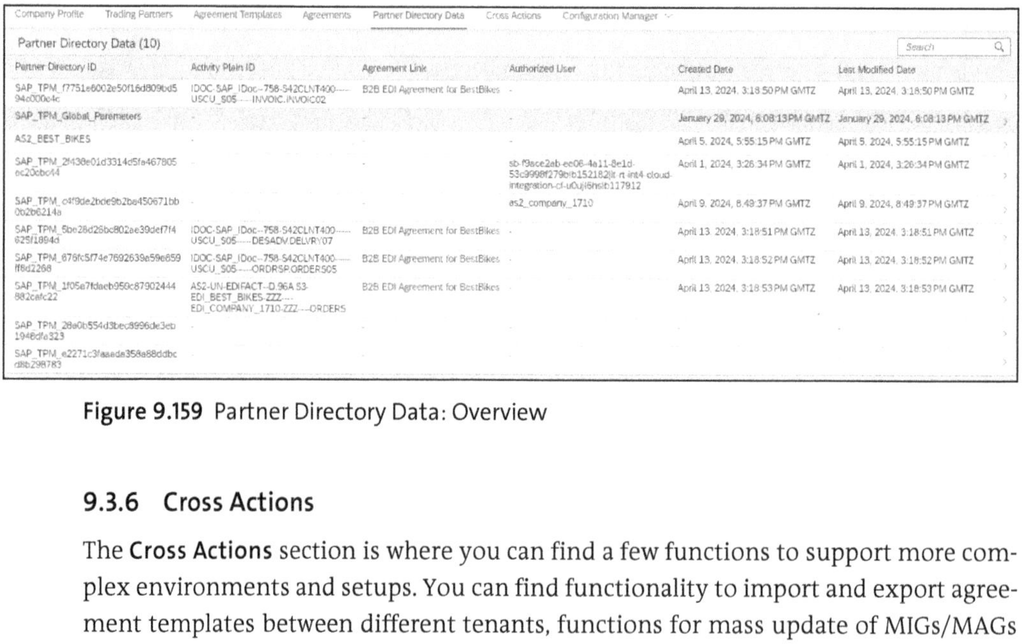

Figure 9.159 Partner Directory Data: Overview

9.3.6 Cross Actions

The **Cross Actions** section is where you can find a few functions to support more complex environments and setups. You can find functionality to import and export agreement templates between different tenants, functions for mass update of MIGs/MAGs for a group of agreements, or mass deactivation for a group of agreements. We will not be using any of these for our demo B2B scenario.

9.3.7 Configuration Manager

In the **Configuration Manager** section, you can find two useful features related to operational and error handling aspects of the B2B solution.

Custom Search Attributes

In the next chapter, we will take a close look at how to monitor B2B messages. A key part of this will be the monitor for B2B scenarios in Cloud Integration. In the **Configuration Manager** tab, you have the option to define custom fields that can later be added to the standard views and filters.

Navigate to **Design • B2B Scenarios • Configuration Manager** and using **Create** button located on the right-hand side of the screen add an entry with values as follows:

- **Name—CSA Sales order number**
- **Description—Custom Search Attribute on sales order number**

The results should match Figure 9.160.

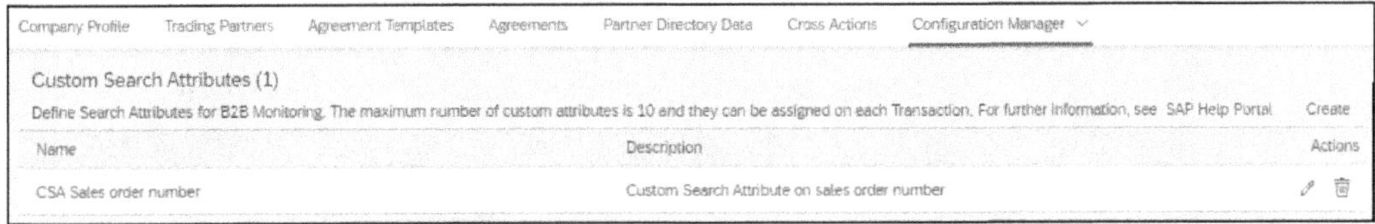

Figure 9.160 Configuration Manager: Adding Entry for Sales Order Number

When you have defined a custom attribute, you can assign it in the appropriate transaction and define the source of the data. Go to the agreement and switch to edit mode. Choose the sales order transaction and in the bottom part of the screen click the **Custom Search Attribute** tab. On the right-hand side of the screen, you can find an **Add** button. Click it to add one entry as follows:

- **Name—CSA Sales order number**
- **Source Type—Xpath**
- **Source Value—/M_ORDERS/S_BGM/D_1004**
- **Data Source—Sender Interchange**

All entries but the source value you will choose from the selection screen. The Xpath points to the location in the UN/EDIFACT ORDERS message from which you pull the sales order number. You can see the final setup in Figure 9.161.

As you probably recall, each change in the agreement needs to be updated so that the changes are synchronized to the partner directory. When you save the changes to the agreement, you can see that the need for an update is clearly indicated by the status of the agreement. Look at Figure 9.162, where you can see the **Update Pending** remarks located on the far right-hand side next to each transaction. Use the **Update** button in the top-right corner to push the changes to the partner directory. The process takes a few seconds.

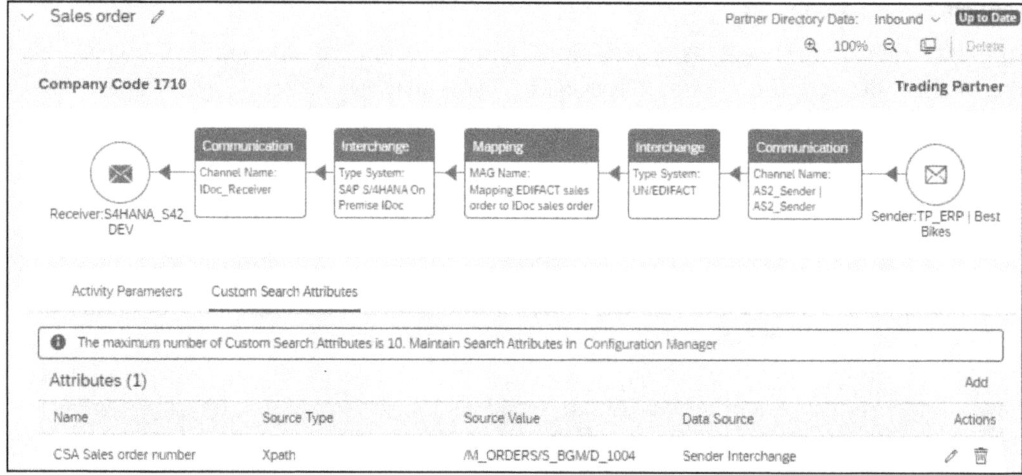

Figure 9.161 Agreement: Adding Customer Search Attribute to Sales Order Transaction

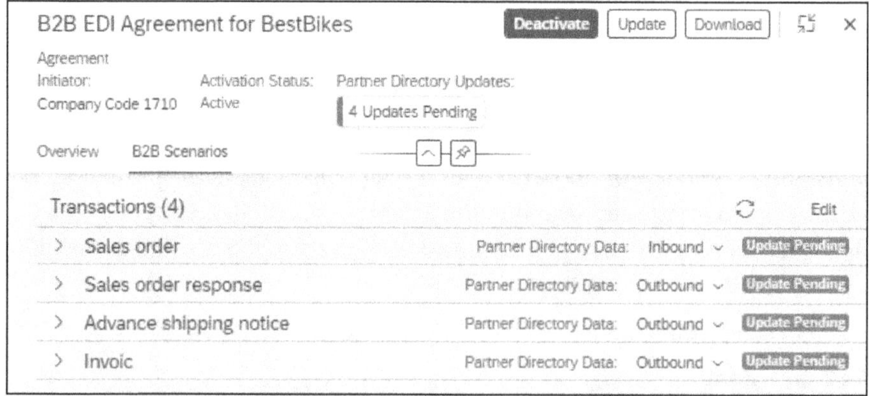

Figure 9.162 Agreement: Update Pending

To fully understand the process, you can open the partner directory ID related to this specific agreement and its sales order transaction. Among dozens of string parameters, you can find one that keeps all the details of the customer search attribute. Figure 9.163 shows that exact string parameter, called `SAP_TPM_Receiver_Custom_Search_Attribute_Values`, and its value.

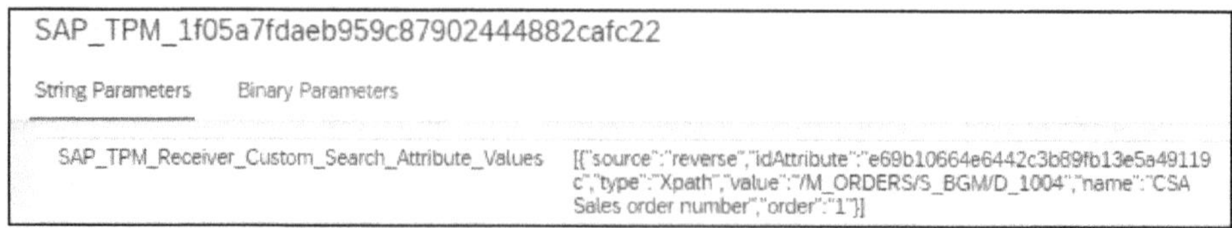

Figure 9.163 Partner Directory Entry for Custom Search Attribute

In a similar way, you can maintain up to 10 custom search attributes, including custom search attributes to handle reference document numbers for the other transactions, such as sales order response, advance shipping notice, and invoice.

Maximum Retry Count

This option is rather self-explanatory. There can be some temporary issues during interchange processing (the second iFlow from the template), and the same can happen with connectivity to the receiver. There is already inbuilt functionality for reprocessing, and within this section we can configure the number of retries. Each case has its own parameters, as follows:

- **Interchange Processing Flow**
- **Receiver Communication Flow**

The default value is 3 and it can be increased up to 25. The screen to set these values is shown in Figure 9.164.

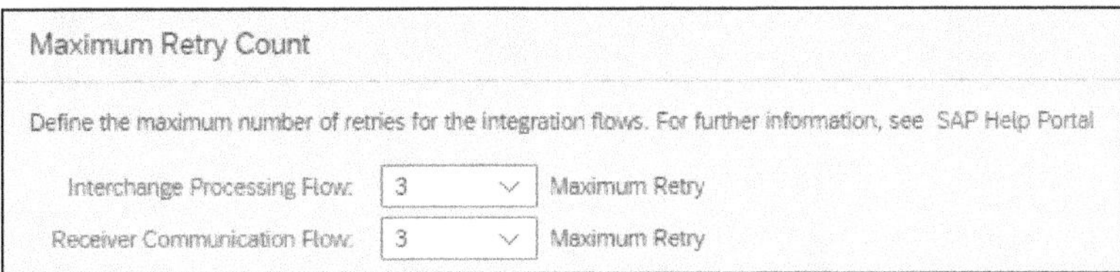

Figure 9.164 Configuration Manager: Maximum Retry Count

9.4 Summary

This is the longest chapter in the whole book, but it cannot be helped as Cloud Integration is the core of the B2B solution for the SAP S/4HANA landscape. But the length is not only due to the importance of the component; it's even more due to the number and complexity of the performed tasks. Keep in mind that there are actually three separate components that together create a uniform and harmonious processing environment for the B2B EDI area. We started the chapter with the Integration Advisor component, in which we defined MIGs for SAP IDocs and UN/EDIFACT messages. In the next step, still in Integration Advisor, we defined MAGs to transform the different types of messages. We also had a chance to look at the documentation generated automatically for our objects. With these artifacts created, we could move to the next component: Trading Partner Management. Defining your own company profile and trading partner profile was only an introduction to the definition of the B2B scenario. The latter was done in the form of an agreement object—created not directly, but as a copy of a reusable agreement template. In the agreement itself, we included all four transactions on which we've worked since the initial chapters. We went through all the processing steps for each transaction, maintaining the sender and receiver communication, assigning appropriate MIGs and MAGs, and finally activating the agreement to synchronize the whole configuration to the partner directory. One important point worth mentioning is that we managed to configure AS2 connectivity with the trading partner for both directions, to receive and send UN/EDIFACT messages in a secure manner.

In the next chapter, we will verify the configuration of the B2B scenario in the most practical way. We will go through the whole business process triggered by the sales order message received from the trading partner. You will have a chance to see how the processing can be monitored on each component. We will also look at B2B monitoring in Cloud Integration and then also at the SAP Fiori message monitoring applications in the SAP S/4HANA system.

Chapter 10
Checking the EDI Solution

The tough work related to configuration and development is finished, and now we will enjoy the initial execution of the EDI scenario. We will verify and look at each step related to inbound and outbound EDI transactions.

You cannot consider the work complete until the results are validated. This principle applies to our case as well. Despite having hundreds of pages detailing the setup of our B2B solution, we still lack proof that it is functioning correctly and as intended. The aim of this chapter is to validate the configuration of each component and the overall B2B setup by executing the business process with all its EDI transactions.

To start, we will assume the role of our trading partner and initiate an EDI sales order directed toward the newly configured environment. After verifying the EDI sales order processing on Cloud Integration and within SAP S/4HANA, we will proceed with the subsequent transactions from the purchasing process. You will have the opportunity to familiarize yourself with and observe in action the B2B monitoring in Cloud Integration and the SAP Fiori–based Message Monitoring - SAP Application Interface Framework application on SAP S/4HANA. Both tools provide more than just technical status updates; they also offer a user-friendly and transparent view of the overall status.

10.1 Verifying Inbound Processing

Although the EDI implementation project contains a phase where we make an agreement with our trading partner about the technical and functional aspects of each EDI transaction, there is always a question if these are fully applied. This question can be quickly answered by processing the initial EDI sales order from the trading partner. Any deviations and shortcomings from the agreement will be immediately visible in one or many processing steps. In this section, we will look at what the received EDI sales order message looks like in our system and how it is processed through the different steps.

10.1.1 Sales Order EDI Transaction

Before we simulate the trading partner's action, we need to know the AS2 endpoint of the B2B solution running on Cloud Integration. As you probably recall from Chapter 5, during the initial setup we were checking endpoints for the IDoc communication, and we configured a URL on the sender side, which in this case was the SAP S/4HANA system. To find the AS2 endpoint, navigate to the list of deployed integration content by following path **Monitor • Integrations and APIs • Manage Integration Content**. The sender AS2 channel is exposed in the iFlow **Step 1—Sender AS2 Communication Flow V2**; make use of the search option to find this specific iFlow. You can find the endpoint on the right-hand side in the tab of the same name (see Figure 10.1). In our example the URL is as follows, and this is what we share with our trading partner together with other parameters: *https://int4-cloud-integration-cf-uOujl6hs.it-cpiO18-rt.cfapps.eu10-003.hana.ondemand.com/as2/as2/tpm/b2b/as2.*

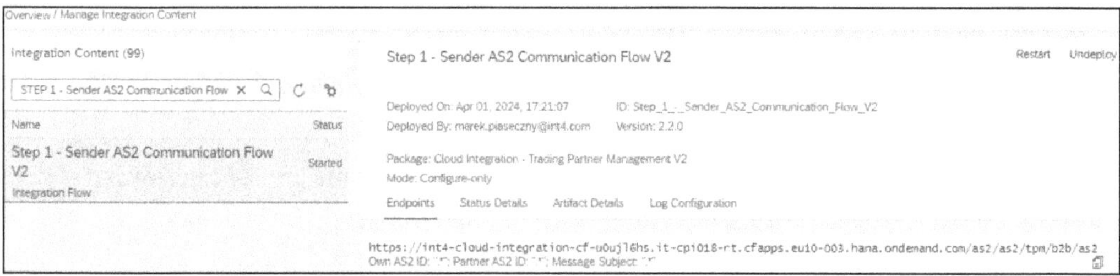

Figure 10.1 View of iFlow Exposing Sender AS2 Channel: AS2 Endpoint

When you went through the list of iFlows in the Trading Partner Management V2 package, you probably noticed that there is a second iFlow with a sender AS2 adapter. That's correct, but this one is dedicated to receiving MDN receipts. We deployed that iFlow earlier in Chapter 5, but in our scenario we don't make use of it. It is only applicable for asynchronous MDN receipt, where the message is exchanged in a separate HTTP(S) session. In our case, we configured synchronous MDN where the receipt is exchanged within the same session like the original message. Figure 10.2 shows the sender AS2 MDN endpoint.

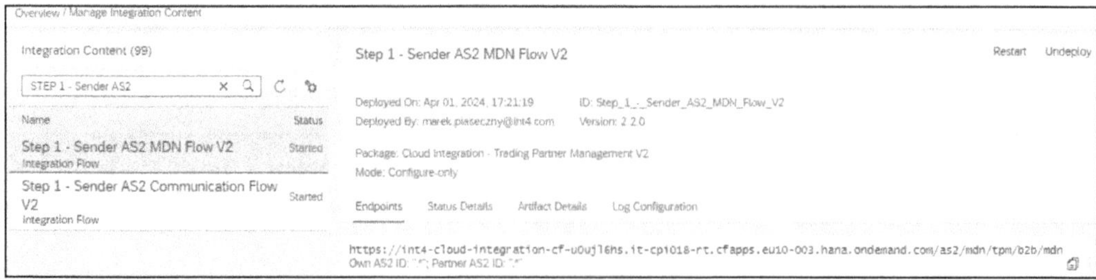

Figure 10.2 View of iFlow Exposing Sender AS2 Channel for MDN Purpose: AS2 Endpoint

Let's assume that our trading partner already has all the necessary details and that we agreed to start the EDI process. The whole process started on the side of the trading partner, the first EDI sales order was just sent, and we got an email with a copy of the message so that we can make the necessary checks and analyses within our IT system.

The initial EDI sales order is in line with our earlier agreement. The exact content is shown in Figure 10.3.

```
UNB+UNOC:3+EDI_BEST_BIKES:ZZZ+EDI_COMPANY_1710:ZZZ+240512:1020+24001'
UNH+1+ORDERS:D:96A:UN'
BGM+220+PO2400001+9'
DTM+137:20240505:102'
NAD+VN+COMPANY_1710_VN::ZZZ'
NAD+BY+BEST_BIKES_HQ::ZZZ'
NAD+DP+BEST_BIKES_WHS::ZZZ'
LIN+1++MZ-FG-R300:SA'
IMD+F++:::R300 BIKE'
QTY+21:5.00:PCE'
UNS+S'
CNT+2:1'
CNT+1:5.00'
UNT+13+1'
UNZ+1+24001'
```

Figure 10.3 Initial EDI Sales Order as Sent by Trading Partner

We will start by checking the message receipt and processing on Cloud Integration as this is the first component in the end-to-end flow. If you worked with Cloud Integration earlier, you know that any message processed by any artifact (iFlow) can be found in a dedicated section called **Monitor Message Processing**; you can navigate there via **Monitor • Integrations and APIs • Monitor Message Processing**. That's still an option, but the B2B processing is quite specific and as it goes through multiple processing steps (artifacts), finding appropriate relationships and sequencing the steps are quite challenging. Another aspect not seen out of the box in the standard monitoring is the B2B EDI-specific attributes like message type, sender, receiver, and so on. Of course it is possible to run the iFlow in trace mode and find all the details, but that would not be the most efficient option. Luckily there is a dedicated solution for B2B monitoring that is actually an integral part of Cloud Integration and that makes use of information maintained in Trading Partner Management and Integration Advisor. Its main advantage is that it offers a simplified view of B2B message processing across multiple iFlows and at the same time gives a full set of additional information related and critical to B2B processing.

Before we start looking for the message in Cloud Integration or SAP S/4HANA, let's look at what happens on the trading partner side after sending the message:

- On the AS2 communication layer, the trading partner receives the message disposition notification (MDN). This is the electronic receipt for a successfully transmitted AS2 message. Figure 10.4 shows the full content of MDN receipt sent by the newly configured EDI solution to the trading partner. Apart from the recipient ID and time stamp, you can find a unique message ID, which can be useful for message tracing.

```
Date: Sun, 12 May 2024 08:23:18 +0000 (UTC)

------=_Part_3_1386873734.1715502198977
Content-type: text/plain
Content-Transfer-Encoding: 7bit

Message received and processed at local time Sun May 12 08:23:18 UTC 2024, internal message ID is ID-490f760f-a020-4564-672a-b2d5-1715490981752-51-1
------=_Part_3_1386873734.1715502198977
Content-Type: message/disposition-notification
Content-Transfer-Encoding: 7bit

Reporting-UA: SAP AS2 Adapter
Original-Recipient: rfc822; AS2_COMPANY_1710
Final-Recipient: rfc822; AS2_COMPANY_1710
Original-Message-ID: <cc75c4d3-7045-4766-b2e9-ff69d4a2d755@AS2_BEST_BIKES>
Disposition: automatic-action/MDN-sent-automatically; processed

------=_Part_3_1386873734.1715502198977--
```

Figure 10.4 MDN Receipt Received by Trading Partner after Sending Initial EDI Sales Order

- After MDN receipt, the trading partner receives a technical acknowledgment from the newly configured EDI subsystem. This second confirmation is a CONTRL message from UN/EDIFACT standard and it confirms receipt and acceptance of a specific interchange. As you can see in Figure 10.5, in the third line there is a UCI interchange response segment that contains the following:
  - The reference interchange, 24001
  - The interchange sender, EDI_BEST_BIKES
  - The interchange receiver, EDI_COMPANY_1710
  - Action/acknowledgment code 7, which we can read as acknowledging the interchange, but some additional rejections may follow from the lower-level/next processing steps

```
1 UNB+UNOC:3+EDI_COMPANY_1710:ZZZ+EDI_BEST_BIKES:ZZZ+240512:0823+24001'
2 UNH+24001+CONTRL:D:3:UN'
3 UCI+24001+EDI_BEST_BIKES:ZZZ+EDI_COMPANY_1710:ZZZ+7'
4 UNT+3+24001'UNZ+1+24001'
```

Figure 10.5 UN/EDIFACT CONTRL Message: Functional Acknowledgment

10.1.2 Tracing the Message on Cloud Integration: B2B Scenarios Monitoring

From the perspective of the trading partner, the message has been successfully sent and its processing has been confirmed. Now it is time to navigate to **Monitor • B2B Scenarios** to check what the processing looks like in Cloud Integration. Figure 10.6 shows the initial screen that gives an overview of successfully identified **Interchanges** and **Unassigned Interchanges**. Click the first **Interchanges** box, where the known message type from the fully configured trading partner should end.

You should see a screen split into two parts: The top part offers search and filter options on multiple values, and visibility of specific attributes and their sequence can be configured. On the bottom part of the screen there is a list showing interchanges, which matches the selection criteria. Figure 10.7 shows the default selection criteria, with all ORDERS messages from the past 10 minutes.

We can already confirm that the message sent by the trading partner has been successfully received and processed by our B2B middleware. It has status **Completed**, but let's check in detail what that means and what exactly happened. But first, let's take a short look at the options available on this screen.

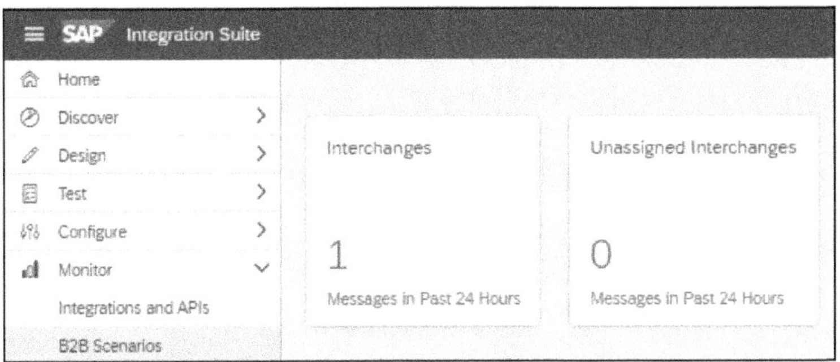

Figure 10.6 Monitoring for B2B Scenarios: Starting Screen

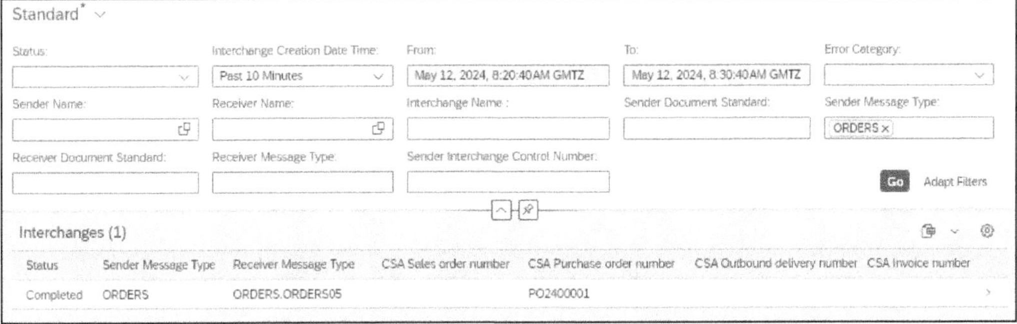

Figure 10.7 Monitoring for B2B Scenarios: View of Interchanges

Most of the options visible at the top of the screen and dedicated to filtering should be self-explanatory. In the next section of this chapter, when we have more message processes, we will apply some additional filters to demonstrate the functionality and usage of key filters. For now, let's go through the fields with selection options:

- **Interchange Creation Date Time**
  There is an option to choose all messages irrelevant of the creation date, and there are also a few predefined time frames and an option to define a custom time frame (see Figure 10.8).

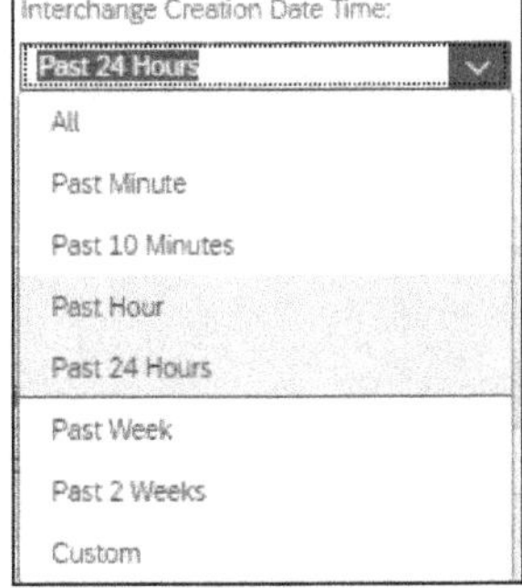

Figure 10.8 B2B Monitoring: Selection List for Interchange Creation Date Time

- **Status**
  You can filter the interchange selection based on its status. You have an option to choose one or multiple statuses, which are self-explanatory (see Figure 10.9).

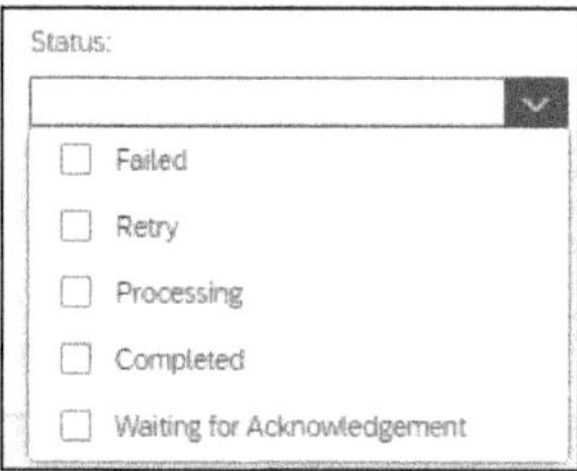

Figure 10.9 B2B Monitoring: Selection List for Interchange Status

- **Error Category**
  When the interchange is in error, the different errors are categorized and also can be used for filtering the interchange selection (see Figure 10.10).

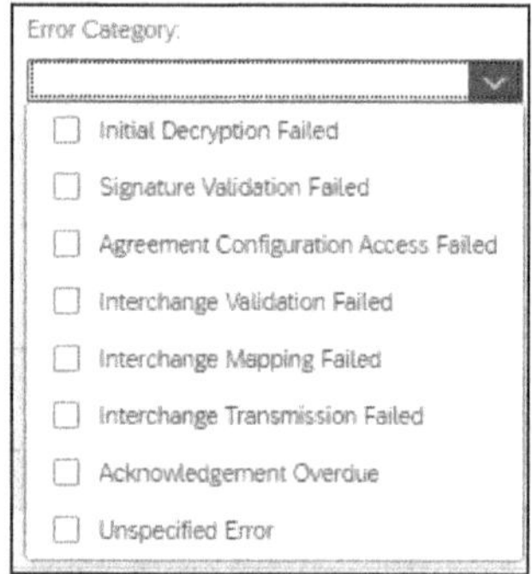

Figure 10.10 B2B Monitoring: Selection List for Interchange Error Category

- **Sender Name/Receiver Name**
  There are two selection options that allow filtering on the sender or receiver name. The exact values come from Trading Partner Management, and operating on the exact name is definitely a way to increase transparency (see Figure 10.11).

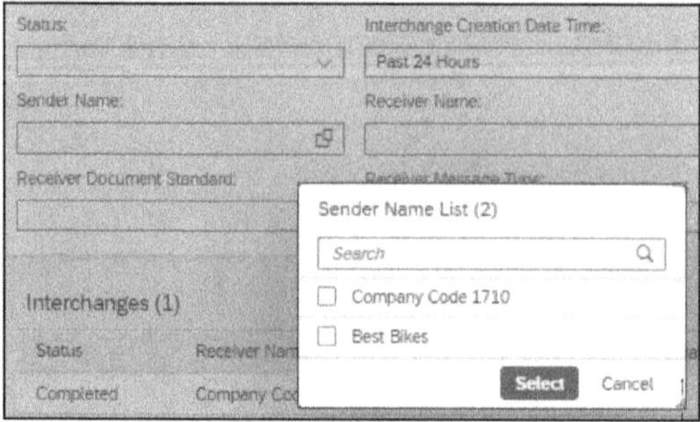

Figure 10.11 B2B Monitoring: Selection List for Sender Name

When you look at the top part, the overview list with interchanges as shown in Figure 10.7, there are two useful features available via icons located on the right-hand side of the screen:

- **Export to Spreadsheet**
  This option allows you to generate an Excel file with all the information available on the interchange list. While doing this, you can decide if filter criteria should be included or excluded.
- **Settings**
  You can define which data (columns) you want to display in the interchange list. As you can see in Figure 10.12, there are more than 10 fields available, including fields defined as custom search attributes. In our case, we have **CSA Sales order number**, which is a custom attribute defined in previous chapter.

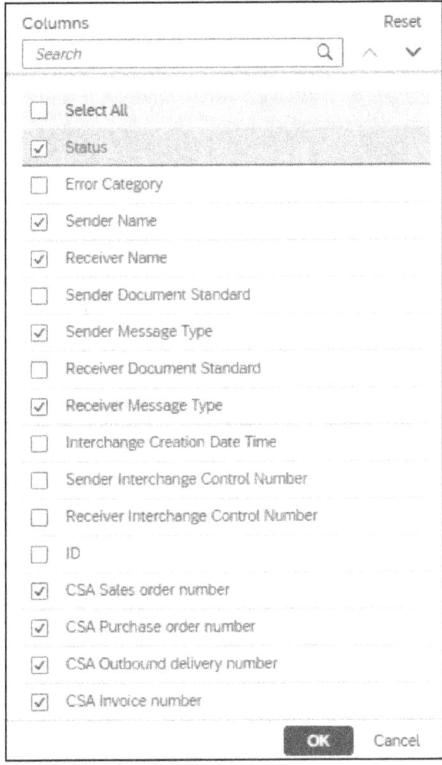

Figure 10.12 B2B Monitoring: Selecting Columns Visible on Interchange Results List

Now let's look at the details of the interchange related to the first EDI sales order. Each item shown in the bottom list as in Figure 10.7 can be clicked, and it will redirect to the detailed view. Results for our sales order message are shown in Figure 10.13.

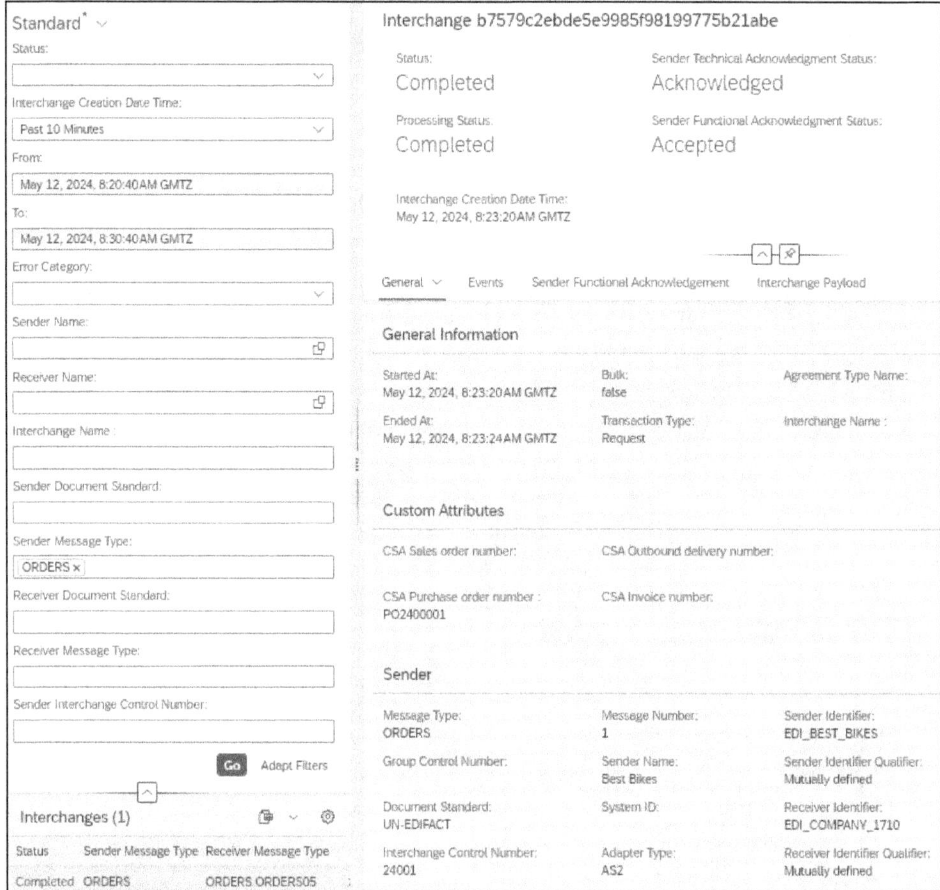

Figure 10.13 B2B Monitoring: Details of Interchange Related to First Incoming ORDERS

The screen is now split into two parts vertically. One the left-hand side you can still find the previous overview sections, but it is a kind of compressed view. On the right-hand side you can find a new view containing multiple sections with all details related to the selected interchange. Starting from the top, you can see the following sections:

- **Status overview**
  Next to the unique identifier of the interchange and its creation date and time, you can also find the overall status, which comes from the processing status. The two additional statuses seen on the right are **Sender Technical Acknowledgment** and **Sender Functional Acknowledgment**. Figure 10.14 shows a closer look at the **Status** section.
- **General tab**
  If you scroll down, you can find additional, generic information related to the interchange collected in four additional sections:
  - **General Information**—shows when the processing started and ended, if the interchange contains a single or multiple documents (bulk), and the transaction type (in our case, a request)

- **Custom Attributes**—shows all custom search attributes defined in the agreement (in our case, four custom search attributes)
- **Sender**—shows sender-related details determined from the UNB and UNH UN/EDIFACT segments
- **Receiver** —shows receiver-related details, like characteristics of the IDoc and SAP S/4HANA system

Figure 10.15 shows **General Information** and the three following sections related to the first received interchange with ORDERS message.

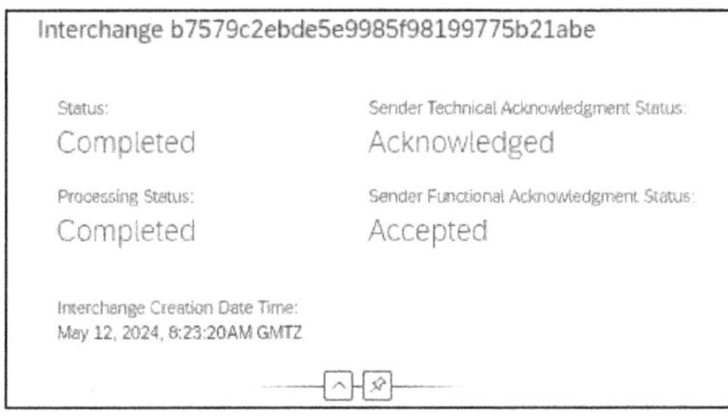

Figure 10.14 B2B Monitoring: Overview Section from Interchange Details

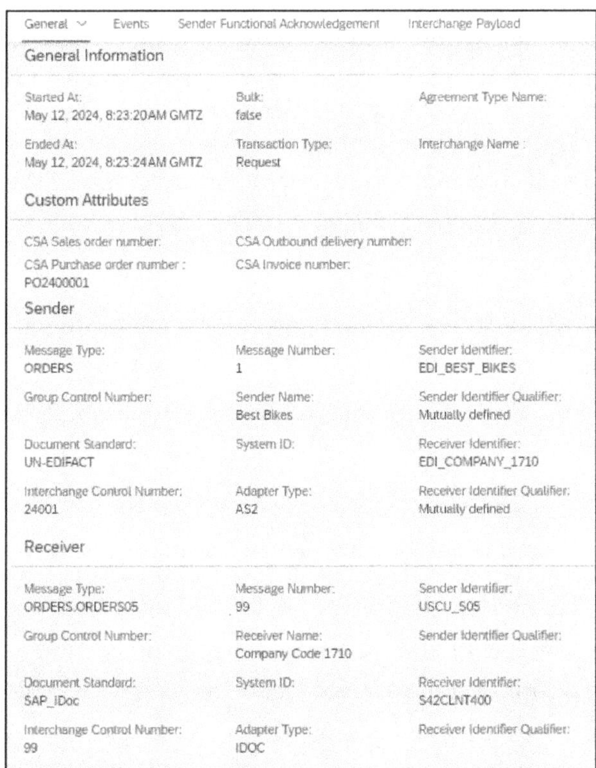

Figure 10.15 B2B Monitoring: General Section from Interchange Details

The next section in the **Interchange** details is called **Events** and it shows the processing steps related to our interchange. In the list you can find type of the event, exact time of the event, and unique identification number, which also acts as a link that redirects you to the standard message monitoring. As you need to remember, in the background each interchange is a multiple-step process that uses multiple iFlows. The B2B monitoring offers a simplified and more transparent view of the processing, but you always have the option to look at the standard logs. Figure 10.16 shows all the **Events** related to our initial ORDERS.

General ∨ Events Sender Functional Acknowledgement Interchange Payload

Events

Events (8)

Event Type	Date	Monitoring Reference
Sender Interchange Received	May 12, 2024, 8:23:20AM GMTZ	AGZAfHcyODDfNBZw_RtlI0k_
Technical Acknowledgement Sent (for Received Sender Interchange)	May 12, 2024, 8:23:20AM GMTZ	AGZAfHcyODDfNBZw_RtlI0k_
Sender Interchange Mapped	May 12, 2024, 8:23:21AM GMTZ	AGZAfHcyODDfNBZw_RtlI0k_
Functional Acknowledgement Assembled	May 12, 2024, 8:23:21AM GMTZ	AGZAfHcyODDfNBZw_RtlI0k_
Receiver Interchange Assembled	May 12, 2024, 8:23:21AM GMTZ	AGZAfHI4l-F6Doal-BNhz-1MC
Functional Acknowledgement Sent	May 12, 2024, 8:23:23AM GMTZ	AGZAfHkY0Ne8GgokBQn3dlrl
Technical Acknowledgement Received (for Sent Functional Acknowledgement)	May 12, 2024, 8:23:24AM GMTZ	AGZAfHkY0Ne8GgokBQn3dlrl
Receiver Interchange Sent	May 12, 2024, 8:23:24AM GMTZ	AGZAfHI4l-F6Doal-BNhz-1MC

Figure 10.16 B2B Monitoring: Events Section from Interchange Details

After the **Events** section, you can find **Sender Functional Acknowledgement**. This section only appears for interchanges received from trading partners as it is related to the functional acknowledgement we sent to confirm the receipt and correctness of the inbound message. What you see in Figure 10.17 is related to the UN/EDIFACT CONTRL message seen in Figure 10.5.

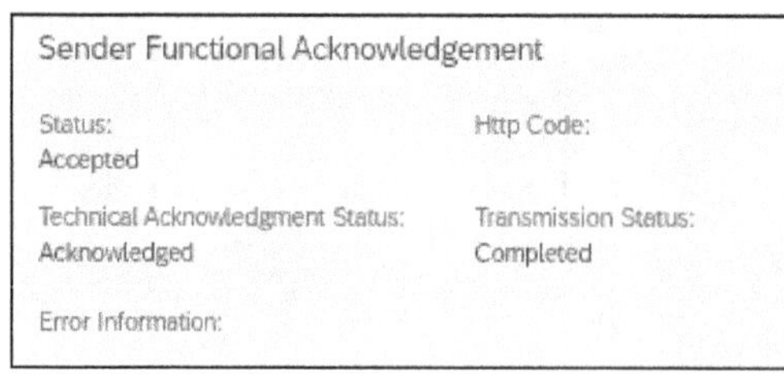

Figure 10.17 B2B Monitoring: Sender Functional Acknowledgement Section from Interchange Details

Last but not least is the **Interchange Payload** section, where we can find two versions of the payload. There is an option to download the payload, or we can simply copy the information from the text box:

- **Sender Interchange Received**—contains the original UN/EDIFACT ORDERS D96A message as sent by the trading partner (see Figure 10.18)

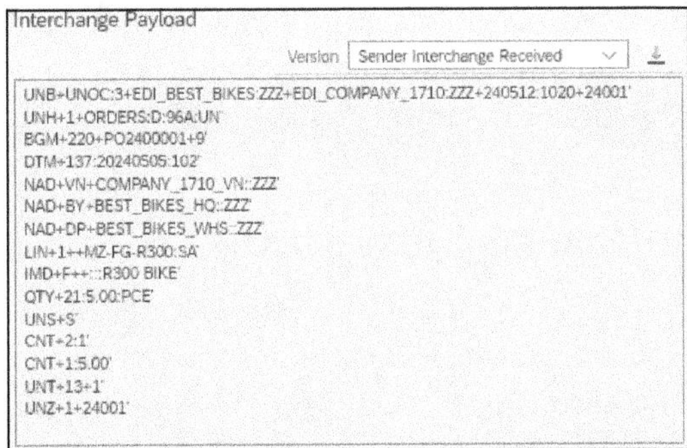

Figure 10.18 B2B Monitoring: Interchange Payload Section from Interchange Details: Sender Interchange Received

- **Receiver Interchange Assembled**—the IDoc message created after the UN/EDIFACT message processing via the appropriate mapping and the exact payload that will be pushed to the receiver system (see Figure 10.19)

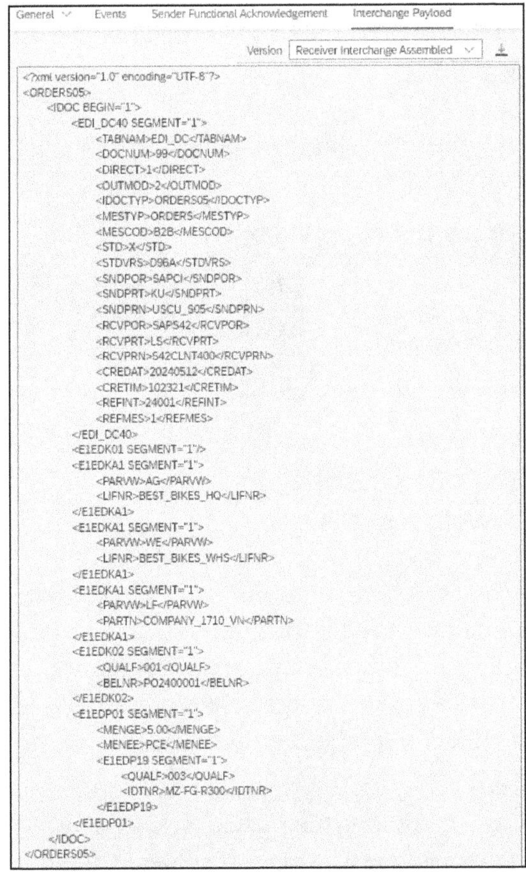

Figure 10.19 B2B Monitoring: Interchange Payload Section from Interchange Details: Receiver Interchange Assembled

We went through the B2B monitoring in Cloud Integration assuming that the initial UN/EDIFACT ORDERS was indeed received and processed correctly. Now we can go to the SAP S/4HANA system to check how the sales order processing went there.

10.1.3 Tracing the Message on SAP S/4HANA: Message Monitoring

To trace our first EDI sales order in SAP S/4HANA, we'll make use of the Message Monitoring applications. As you recall from Chapter 8, we enabled monitoring of all our interfaces in SAP Application Interface Framework. In the first step, after starting the SAP Fiori launchpad, you need to find the **Message Monitoring** space (see Figure 10.20).

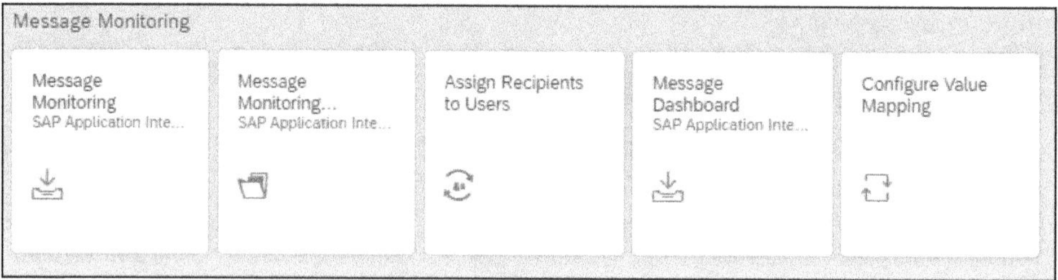

Figure 10.20 SAP Fiori: Message Monitoring space

The alternative is to search directly for a tile called **Message Monitoring Overview—SAP Application Interface Framework**, as shown in Figure 10.21.

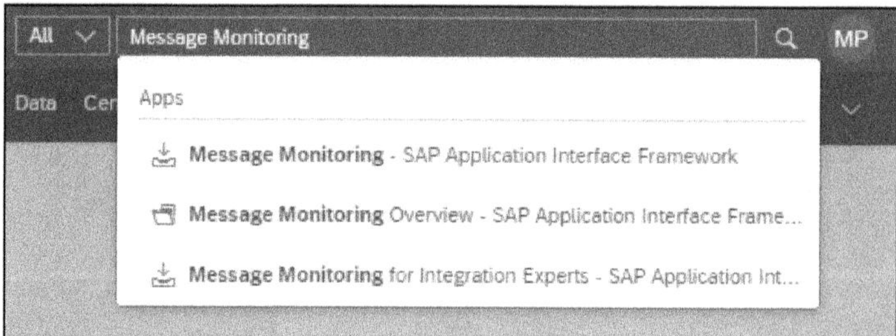

Figure 10.21 SAP Fiori: Searching for Message Monitoring Tiles

After starting the Message Monitoring Overview app, you will see a dashboard with four boxes, each representing a unique interface. Each box has a description coming from the customizing we performed earlier and shows a summary of the number of messages for each status category. In our case, we have one successfully processed sales order and one successfully processed sales order response (see Figure 10.22).

You can navigate to the details of the sales order by clicking the header text or the area around the green **Success** icon. The message monitoring transaction looks as shown in Figure 10.23, showing only single sales order message. The top part allows for filtering the results using multiple selection options; the bottom part displays the appropriate messages and their details. In both the top selection screen and the bottom list, you can find the fields we customized as key fields in SAP Application Interface Framework. You can look at these as custom search attributes, although the key fields can be also used for other purposes. The most important conclusion for now is that the first EDI sales order from Best Bikes not only arrived in the SAP S/4HANA system but was also successfully processed there.

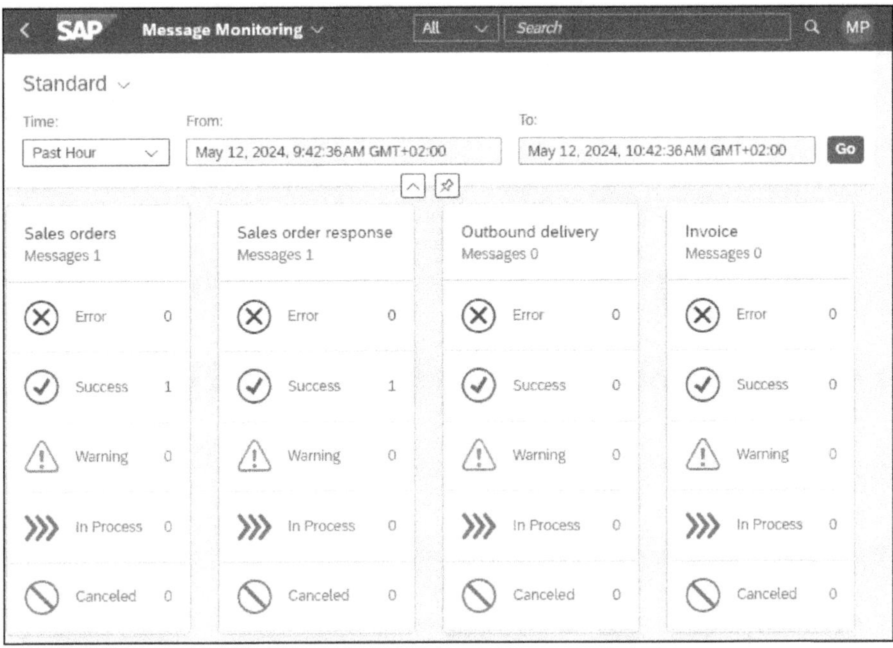

Figure 10.22 SAP Fiori: Message Monitoring Overview

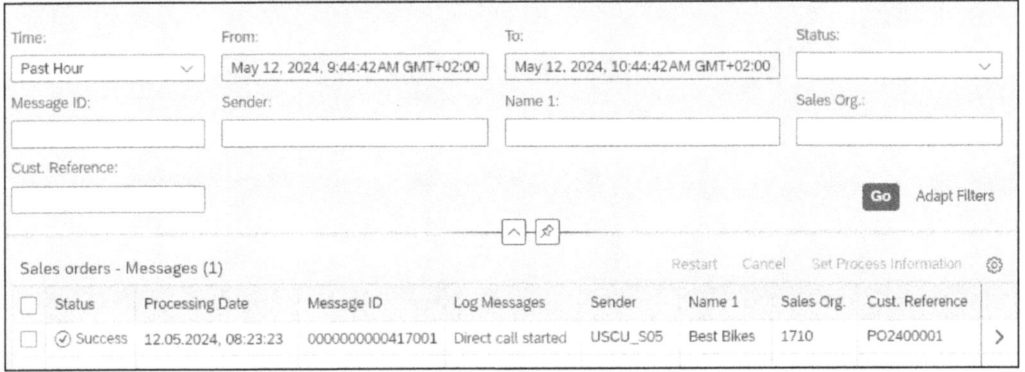

Figure 10.23 Message Monitoring: Listing Available Sales Order Messages

What you see in Figure 10.23 is only the generic information of the message. The additionally customized key fields add value and allow immediate identification of the message. If that's not enough, you can look at the exact content of a message by simply clicking it. Figure 10.24 shows the initial screen after navigating to the message details.

On the left-hand side you can see **Log Messages** related to the message processing. On the right-hand side you can navigate through the content of the message. The navigation starts from the **Key Information** section, which is once again a view of all keys fields. The second section is for **Control record (IDoc)**. The following sections correspond to the next IDoc segments.

The most important message is this: **Standard Order 76309 has been saved.**

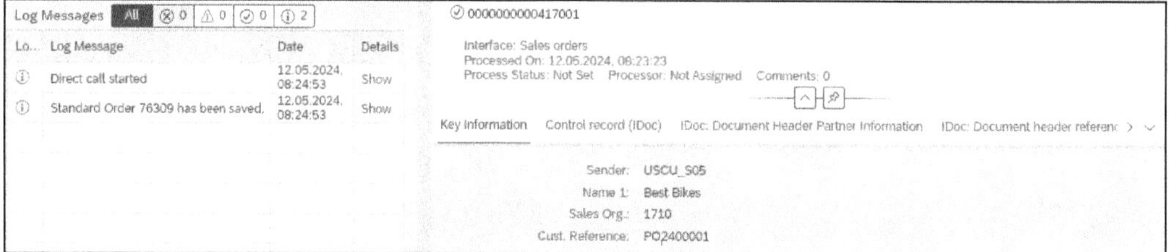

Figure 10.24 Message Monitoring: View of Sales Order Message Details

Figure 10.25 shows the content of the IDoc control record as seen in Message Monitoring.

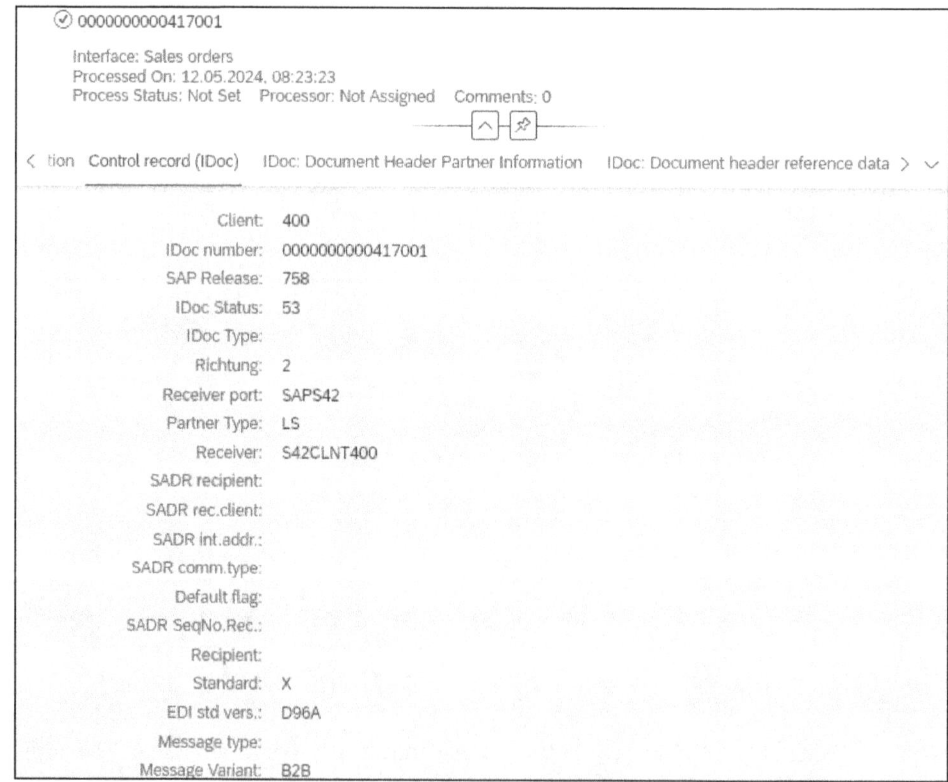

Figure 10.25 Message Monitoring: View of IDoc Control Records

Figure 10.26 shows the header partner information in our initial sales order messages.

0000000000417001

Interface: Sales orders
Processed On: 12.05.2024, 08:23:23
Process Status: Not Set Processor: Not Assigned Comments: 0

<)oc) IDoc: Document Header Partner Information IDoc: Document header reference data IDoc: Document Item Objec >

Pa...	Partner	Supplier	Name 1
AG		BEST_BIKES_HQ	
WE		BEST_BIKES_WHS	
LF	COMPANY_1710_VN		

Figure 10.26 Message Monitoring View of Header Partner Details

10.1.4 Checking the Sales Order Document via Manage Sales Orders App: Version 2

With the help of Message Monitoring, we have confirmed that the first EDI sales order has been successfully processed and that standard order 76309 has been created. In the following steps we will look at the sales order document. The starting point is the Manage Sales Orders—Version 2 app, and once again, you can use the search options as shown in Figure 10.27.

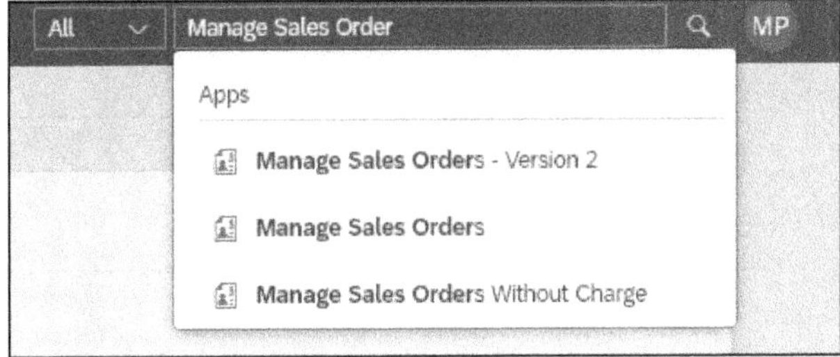

Figure 10.27 Searching for Manage Sales Orders Tiles

When you start the application for sales order management, there are two options to find the appropriate document:

- You can enter "76309" which is the SAP S/4HANA sales order number found earlier in the Message Monitoring application. Enter the value in the **Sales Order** search field located in the top part of the screen next to the other selection fields. This approach is shown in Figure 10.28.
- You can make use of the **Customer Reference** search field; this takes the purchase order number originating from the customer purchase order. Enter "PO2400001" in the mentioned field.

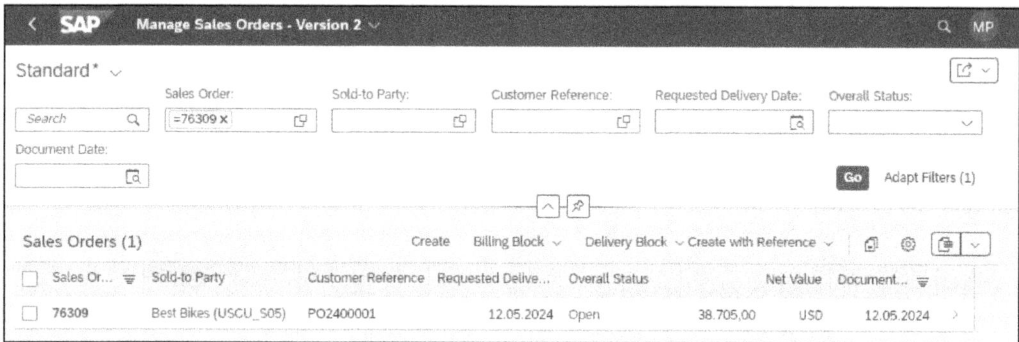

Figure 10.28 Manage Sales Orders—Version 2: Finding Initial EDI Sales Order

Click the sales order document to open its details. The initial view shows you **General Information** related to the document (see Figure 10.29).

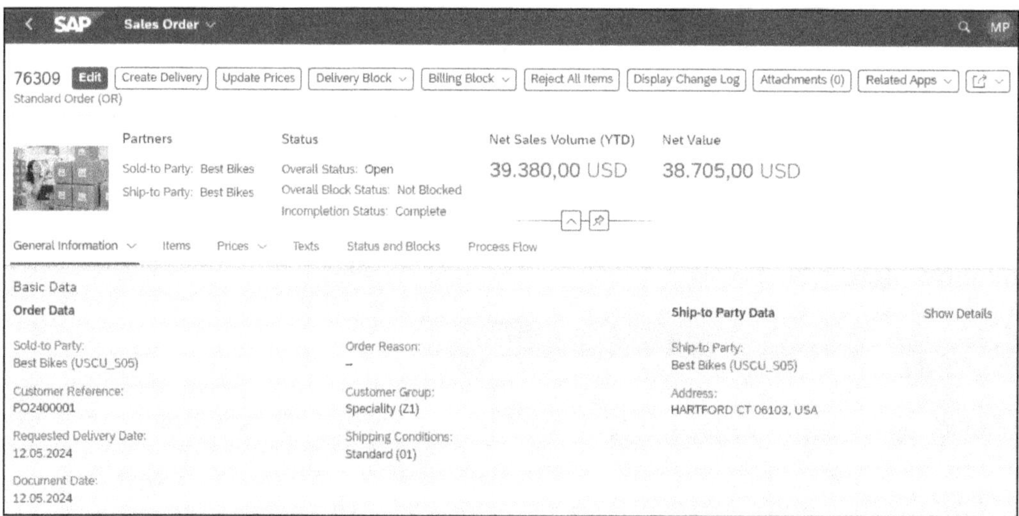

Figure 10.29 General Information about Initial EDI Sales Order

Now you should verify the header and item details of the sales order created based on the EDI process to make sure that it works per expectations. Cross-check if the data sent in the original message matches what you see in the ERP system. We will go through the key elements of the sample sales order as shown in Figure 10.30:

- The sales order has been created as a standard order (OR) in sales organization 1710, distribution channel 10, and product division 00. This matches the input we received in the EDI message (see Figure 10.31) and the configuration maintained in table EDSDC (see Figure 10.32).

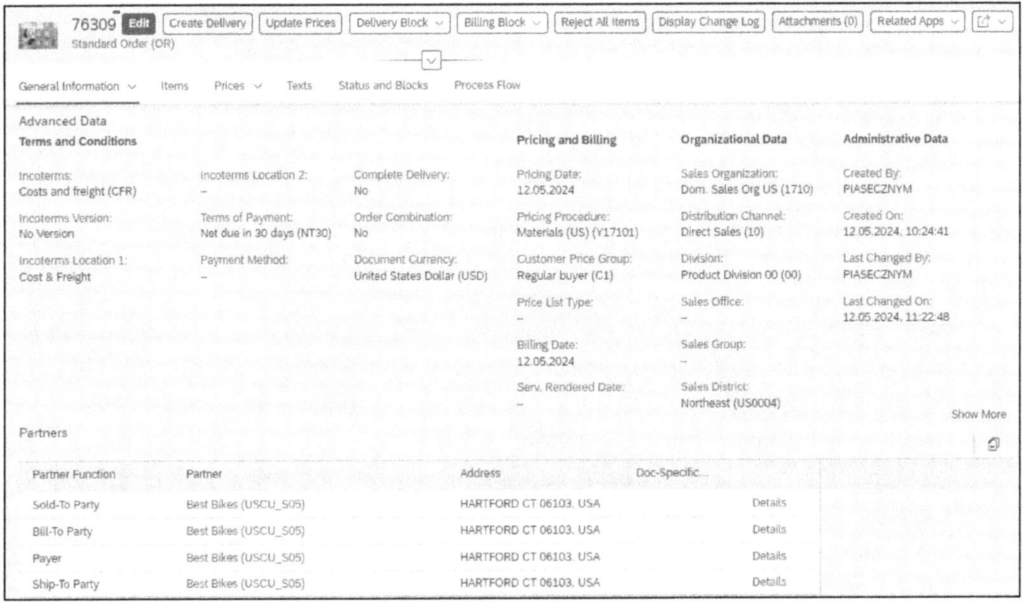

Figure 10.30 Organizational Data and Partners of Initial EDI Sales Order

```
NAD+VN+COMPANY_1710_VN::ZZZ'
NAD+BY+BEST_BIKES_HQ::ZZZ'
NAD+DP+BEST_BIKES_WHS::ZZZ'
```

Figure 10.31 Partner-Related Segments from Initial EDI Sales Order

Customer	Vendor number	SOrg.	DChl	Dv	SaTy	Ret. Order	DSP Allow.	SH Use
USCU_S05	COMPANY_1710_VN	1710	10	00	OR			

Figure 10.32 Table EDSDC Entries Maintained for Sales Order Processing

- The sold-to and ship-to parties have been determined as customer number USCU_S05. This matches the input received in the EDI message (Figure 10.31) and the configuration maintained in table EDPAR (Figure 10.33).

Customer	Funct	External Partner	Int. No.	Addr. No.
USCU_S05	SP	BEST_BIKES_HQ	USCU_S05	
USCU_S05	PI	COMPANY_1710_VN		
USCU_S05	SH	BEST_BIKES_WHS	USCU_S05	

Figure 10.33 Table EDPAR Entries Maintained for Sales Order Processing

- The purchase order number and document date can be confirmed as correctly mapped.
- The last key element would be the product and quantity. Figure 10.34 shows that both the product code **MZ-FG-R300** and quantity **5 PC** match the original EDI sales order message.

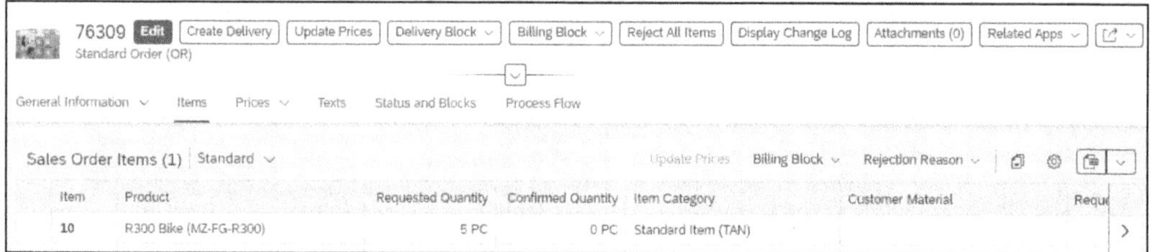

Figure 10.34 Item-Related Information about Initial EDI Sales Order

Although this was just a quick review, it is enough to confirm that the key elements are correct. In an actual EDI implementation, especially when the processing runs on a new solution or components, it would be recommended to make a thorough review. That would also include involving functional consultants and businesspeople to verify all details, not only of the sales orders document but also of all the following and related documents. In our model EDI scenario there is no actual risk of interrupting business operations, so we can limit our validation steps to the bare minimum and move to the next step, which is checking the sales order confirmation triggered from the sales order document.

10.2 Verifying Outbound Processing

We already have the initial sales order created in our SAP S/4HANA system based on the EDI transaction. Now it is time to continue with the business process, which will result in outbound EDI messages triggered out of our system. The difference between inbound and outbound processing is that in the latter full responsibility is on us. We need to not only trigger the proper output but also do it at the right time. Next to that is the aspect of including the proper data and transforming it into the agreed-upon format. Last but not least is the transmission of the EDI message to our trading partner. All these steps we will verify in this section. The expected results are that we will successfully generate and send sales order response, advanced shipping notification (ASN), and invoice messages.

10.2.1 Sales Order Response EDI Transaction

In Chapter 7 we configured the sales order confirmation output using Business Rules Framework plus (BRFplus). Before we check the result, we will look at the process flow to check the status of the sales order. As you can see in Figure 10.35, **Standard Order 76309** is **Not Shipped**; the delivery is planned for a future date. This is still correct and acceptable, but we can also see that there is a warning on unconfirmed quantities. The ordered quantity of 5 PC cannot be confirmed at this moment. First we will look at the sales order response message and will confirm the quantity before creating the following outbound delivery document.

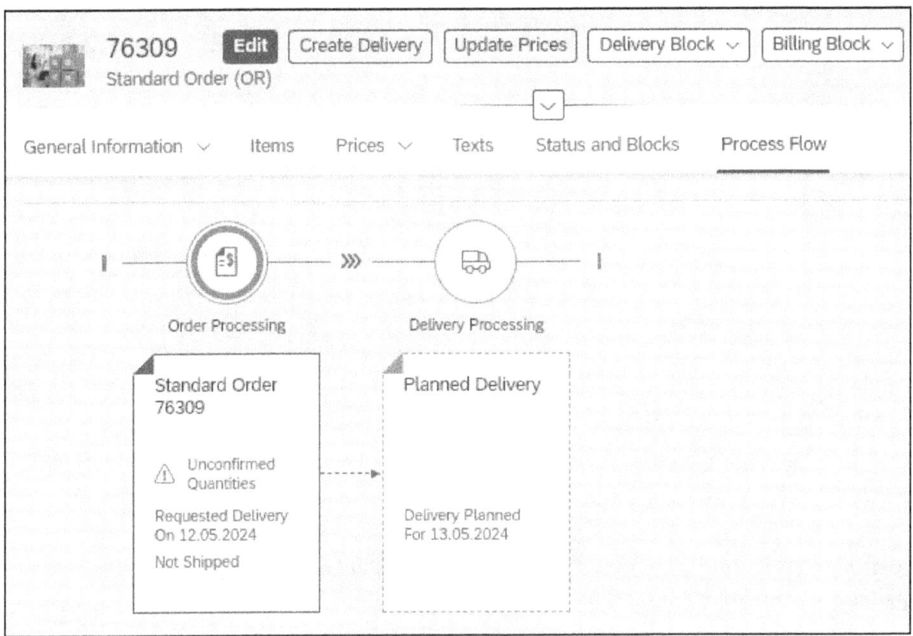

Figure 10.35 Process Flow after Sales Order Creation

To check the output generated from the sales order document, start the Manage Output Items app and search for application object ID 000076309. Figure 10.36 shows the output on the IDoc channel triggered from sales order 76309. The receiver of this output is **Best Bikes (USCU_S05)**, and the dispatch time is set to **Immediately**.

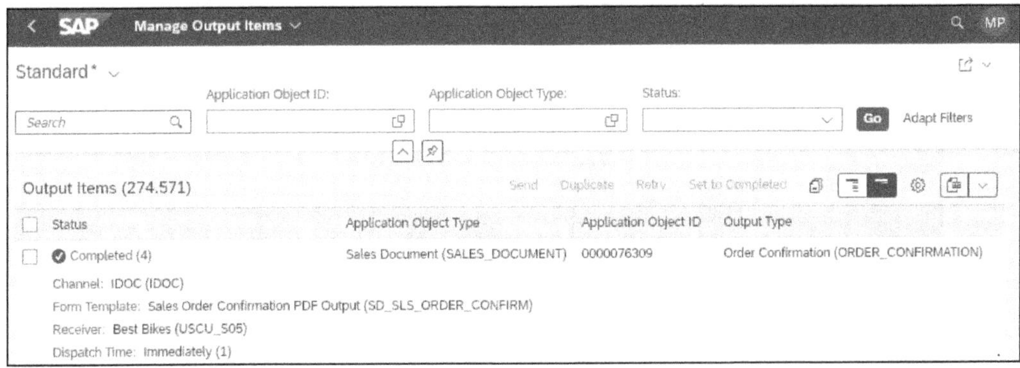

Figure 10.36 Manage Output Items: Sales Order Confirmation Output on IDoc Channel

After checking on the successful output processing, navigate to the Message Monitoring Overview app and choose **Sales Order Response** on the initial dashboard to display the list of messages. In Figure 10.37, you can see one sales order response message with customer reference PO2400001. It has been successfully processed.

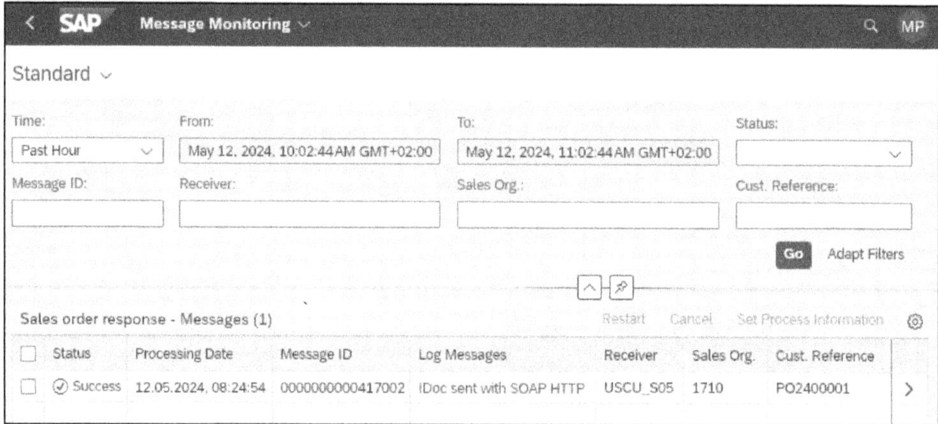

Figure 10.37 Message Monitoring: Sales Order Response Message with Customer Reference PO2400001

We can also navigate to the message details and look at **Log Message** to see a more specific diagnosis of what happened. See Figure 10.38, which confirms that our sales order response IDoc should be already in Cloud Integration.

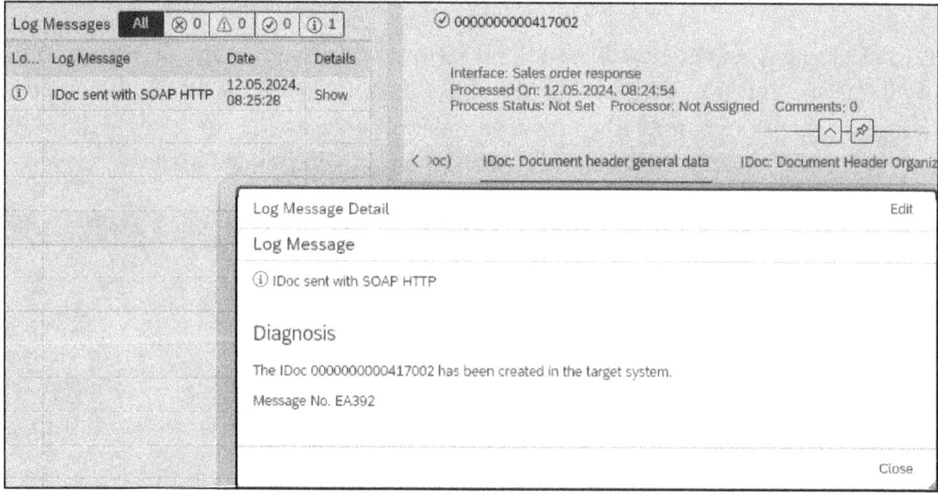

Figure 10.38 Message Monitoring: Sales Order Response Message and Its Log Message

Now we will leave SAP Fiori in the SAP S/4HANA system and go to Cloud Integration and its B2B monitoring. We will display all B2B messages processed in Cloud Integration in the past hour. Figure 10.39 shows two messages: **Sender Message Type ORDERS**, based on which we created a sales order in SAP S/4HANA, and **Receiver Message Type ORDRSP**, which is the confirmation triggered out of SAP S/4HANA. Thanks to the customer search attributes we customized earlier, we can quickly identify and confirm the message related to purchase order PO2400001.

Status:
Interchange Creation Date Time:
Past Hour
From:
May 12, 2024, 8:00:36AM GMTZ
To:
May 12, 2024, 9:00:36AM GMTZ
Error Category:
Sender Name:
Receiver Name:
Interchange Name :
Sender Document Standard:
Sender Message Type:
Receiver Document Standard:
Receiver Message Type:
Sender Interchange Control Number:
Go
Adapt Filters
Interchanges (2)
Status
Sender Message Type
Receiver Message Type
CSA Sales order number
CSA Purchase order number
CSA Outbound delivery number
CSA Invoice number
Completed
ORDRSP.ORDERS05
ORDRSP
0000076309
PO2400001
Completed
ORDERS
ORDERS.ORDERS05
PO2400001

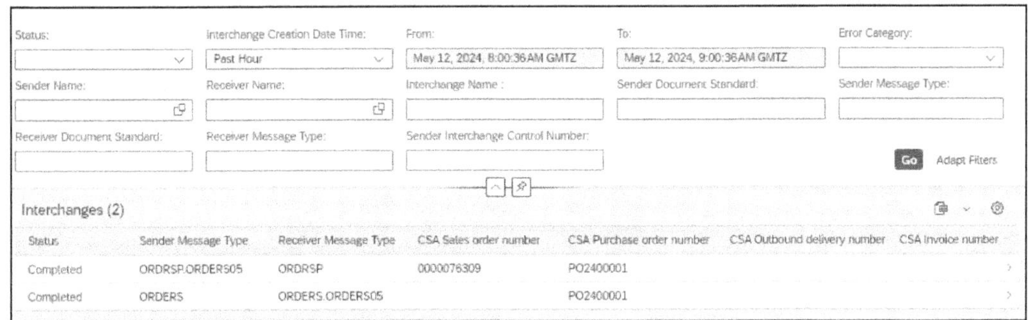

Figure 10.39 B2B Monitoring: ORDERS and ORDRSP Messages

Like for the inbound ORDERS message, we can go through all the details related to this interchange. Figure 10.40 shows the **General Information**, **Sender**, and **Receiver** details related to this interchange. At the top you can see the different statuses confirming that the processing has been not only successfully completed but also confirmed by a technical acknowledgement from the message receiver (trading partner).

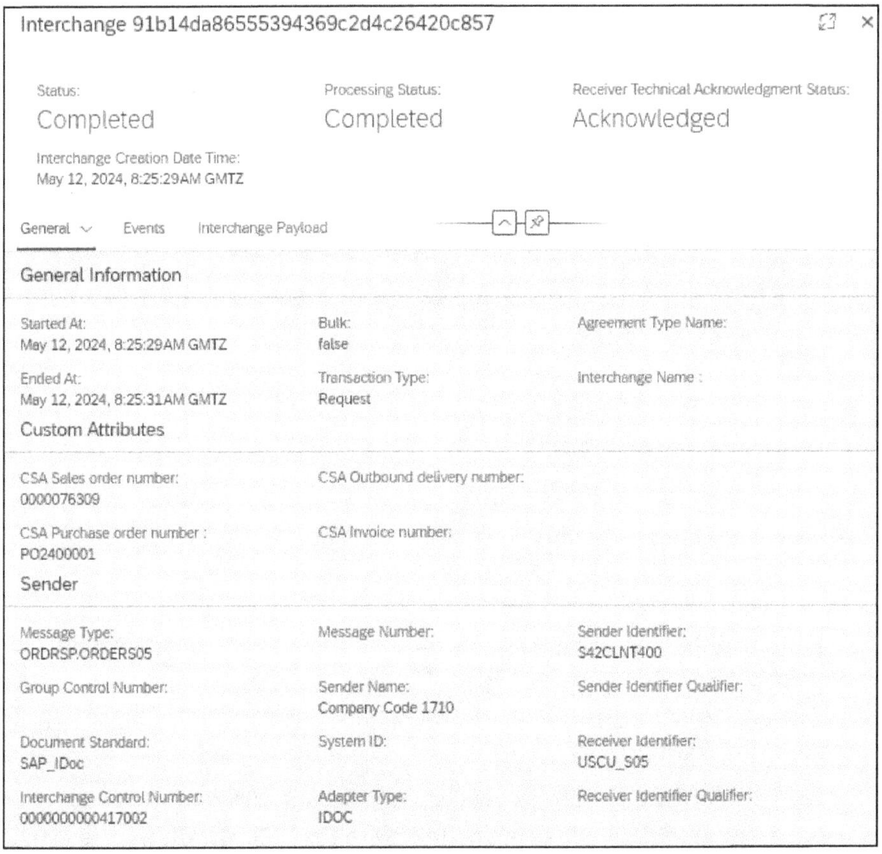

Interchange 91b14da86555394369c2d4c26420c857
Status:
Completed
Processing Status:
Completed
Receiver Technical Acknowledgment Status:
Acknowledged
Interchange Creation Date Time:
May 12, 2024, 8:25:29AM GMTZ
General
Events
Interchange Payload
General Information
Started At:
May 12, 2024, 8:25:29AM GMTZ
Bulk:
false
Agreement Type Name:
Ended At:
May 12, 2024, 8:25:31AM GMTZ
Transaction Type:
Request
Interchange Name :
Custom Attributes
CSA Sales order number:
0000076309
CSA Outbound delivery number:
CSA Purchase order number :
PO2400001
CSA Invoice number:
Sender
Message Type:
ORDRSP.ORDERS05
Message Number:
Sender Identifier:
S42CLNT400
Group Control Number:
Sender Name:
Company Code 1710
Sender Identifier Qualifier:
Document Standard:
SAP_IDoc
System ID:
Receiver Identifier:
USCU_S05
Interchange Control Number:
0000000000417002
Adapter Type:
IDOC
Receiver Identifier Qualifier:

Figure 10.40 B2B Monitoring: General Information Related to ORDRSP Message

After reviewing the general information, we will move to the **Interchange Payload** section to check the **Sender Interchange Received** version, which in this case is the ORDRSP.ORDERS05 IDoc in XML format. As shown in Figure 10.41, the IDoc control record matches the configuration done earlier and our expectations. In segment E1EDK01, you can see field BELNR with value 0000076309, which matches the number of the sales order created based on the first EDI message.

Figure 10.41 B2B Monitoring: Sales Order Response Interchange: Sender Interchange Received

After changing the **Interchange Payload** version to **Receiver Interchange Assembled**, you can see the UN/EDIFACT message ORDRSP generated for our trading partner (see Figure 10.42). We can quickly go through the key segments to confirm if the message is in line with our expectations:

- The UNB segment contains the proper sender identifier EDI_COMPANY_1710 and proper receiver identifier EDI_BEST_BIKES, both defined with the ZZZ code, which means that these are mutually agreed-upon values.

- The UNH segment confirms that we have the UN/EDIFACT ORDRSP D93A message.
- This sales response is related to document 0000076309, which is the correct supplier's sales order number.
- The RFF+ON segment contains the customer's reference number PO2400001, which is also correct.
- In the following part of the message, we can see that there is a single item with quantity 5 PC confirmed. There is also the line and document total amount shared. This is all as per our expectations.

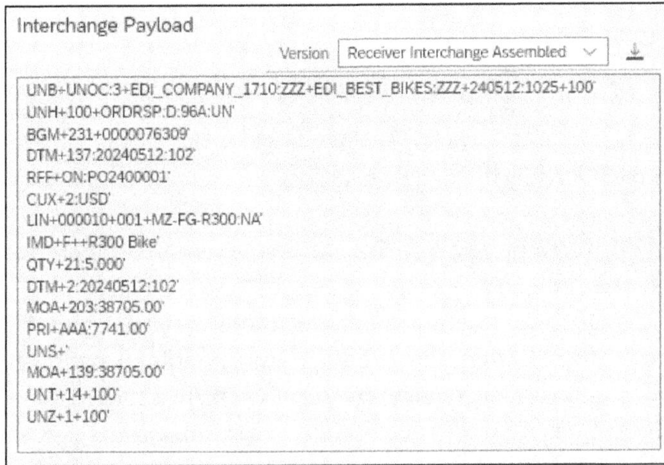
Interchange Payload

Version: Receiver Interchange Assembled

UNB+UNOC:3+EDI_COMPANY_1710:ZZZ+EDI_BEST_BIKES:ZZZ+240512:1025+100'
UNH+100+ORDRSP:D:96A:UN'
BGM+231+0000076309'
DTM+137:20240512:102'
RFF+ON:PO2400001'
CUX+2:USD'
LIN+000010+001+MZ-FG-R300:NA'
IMD+F++R300 Bike'
QTY+21:5.000'
DTM+2:20240512:102'
MOA+203:38705.00'
PRI+AAA:7741.00'
UNS+'
MOA+139:38705.00'
UNT+14+100'
UNZ+1+100'

Figure 10.42 B2B Monitoring: Sales Order Response Interchange: Receiver Interchange Assembled

At this stage, we can conclude that the EDI sales order response functionality is working properly, on both sides: the trigger in SAP S/4HANA and the EDI processing in Cloud Integration. Let's move on to the next step, where we will simulate shipment of the goods by creating an outbound delivery document. Out of this document we expect the ASN message.

The first activity before the outbound delivery document creation is still within the sales order document. As you may recall from earlier, we had a warning message on the sales order stating that there is an unconfirmed quantity. We will go to the sales order and then navigate to the item data. Next to the **Order Quantity** field with value **5** we also have the **Confirmed Quantity** field with value **0**. We will enter "5" in the **Confirmed Quantity** field to confirm that we are able to completely fulfill the order from our trading partner (see Figure 10.43).

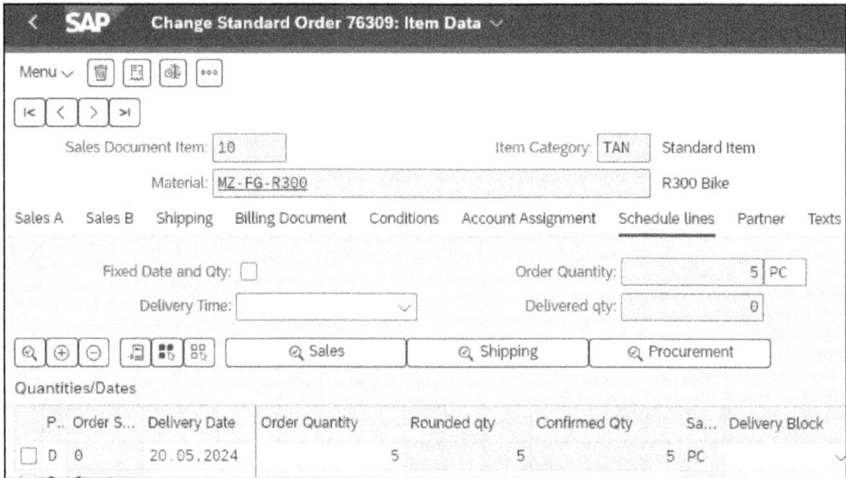

Figure 10.43 Changing Confirmed Quantity to Fulfill Sales Order

There is one important aspect worth noticing: in Chapter 7 when we did the output configuration, we also included and enabled output type ORDER_CONFIRMATION_CHANGE. This means that after each change in the existing sales order, the system retriggers the output and generates an additional EDI message with the most recent information. If we look now at the Message Monitoring Overview app, we can see there is indeed a second sales order response message successfully processed. That's the message triggered after the sales order update related to the confirmed quantity field (see Figure 10.44).

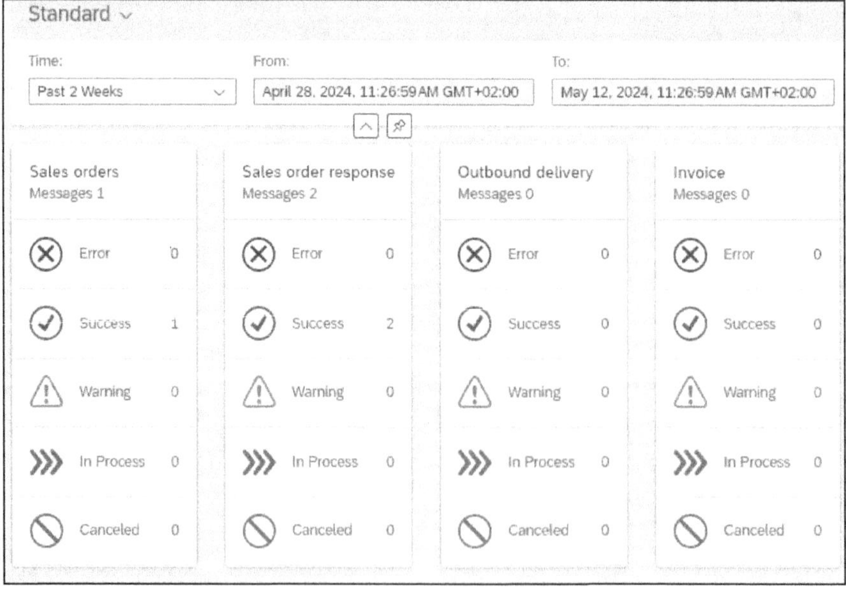

Figure 10.44 Message Monitoring Overview: Second Sales Order Response Triggered after Changes in Sales Order Document

10.2.2 Advanced Shipping Notification and Invoice EDI Transactions

Now we can make use of the Create Outbound Delivery with Order Reference app to perform the exact action that matches the name of this app. Depending on how the app is triggered, it can happen that the order reference is filled out automatically. This happens, for example, when you trigger the app from the sales order document itself. In other cases, you need to maintain at least the mandatory fields: **Shipping Point** and **Order** (see Figure 10.45).

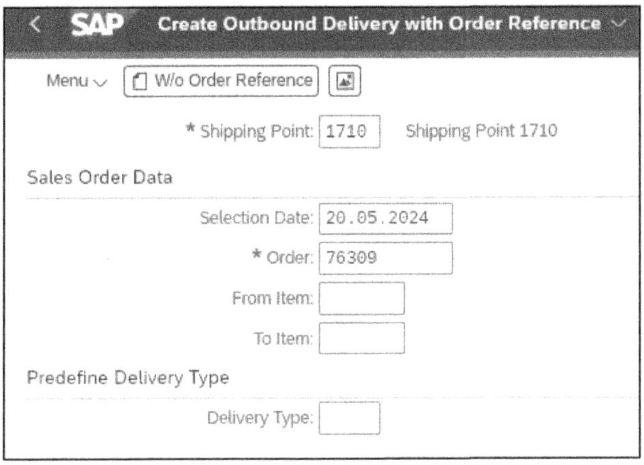

Figure 10.45 Creating Outbound Delivery with Order Reference

Figure 10.46 shows the newly created outbound delivery document.

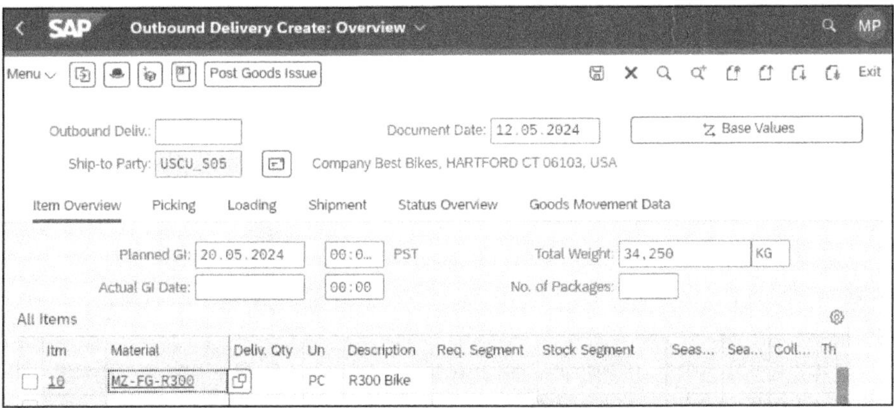

Figure 10.46 Overview of Newly Created Outbound Delivery Document

After creation of the outbound delivery, we will maintain the picked quantity; enter "5" in the **Picked Qty** field, as shown in Figure 10.47. After that, you can post the goods issue by clicking the **Post Goods Issue** button. Doing this will move the outbound delivery document to the shipped status. This will allow us to create a billing document and finalize the sales process.

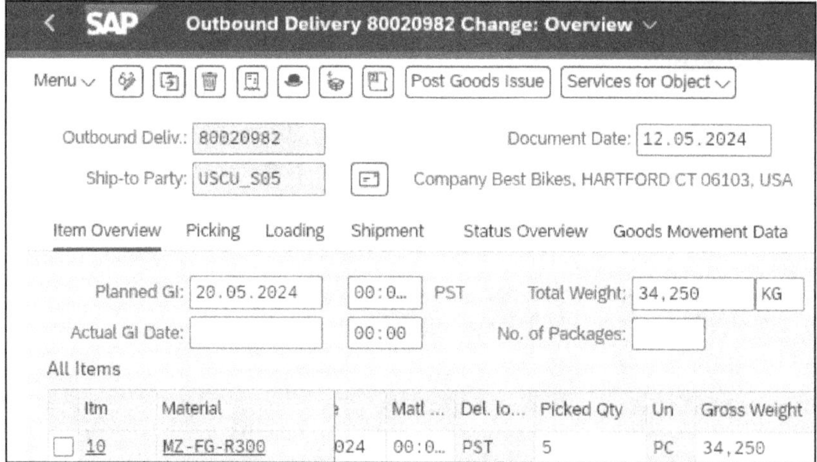

Figure 10.47 Outbound Delivery: Maintaining Picked Qty

Before checking the ASN EDI message, we will finalize the process and create a billing document. Start the Create Billing Document app; our outbound delivery **80020982** that we shipped a moment ago should be already in the list of documents to be processed (see Figure 10.48).

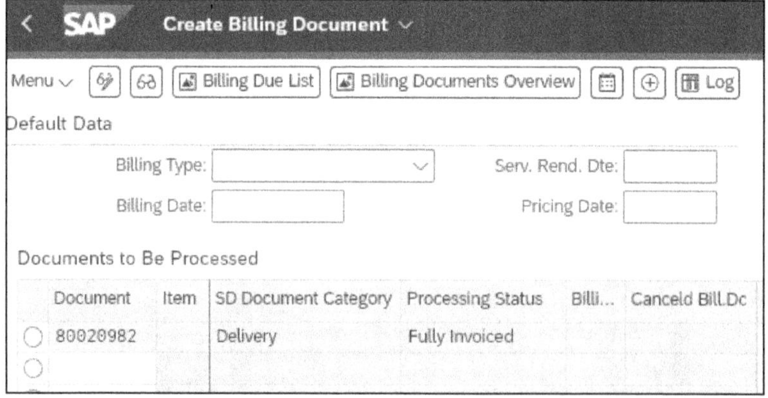

Figure 10.48 Billing Document Creation with Reference to Delivery 80020982

After creating the billing document, we have the whole process flow completed, as shown in Figure 10.49. (The accounting part is out of scope of this scenario.)

Now let's check the EDI processing related to outbound delivery 80020982 and invoice 90017453. We will start by looking at the outbound delivery output, which has been configured with the output control mechanism, not BRFplus. When we open the outbound delivery document and navigate to the header output, we can see that **ASNO Output** on **Partner USCU_S05** has been added and successfully processed (see Figure 10.50).

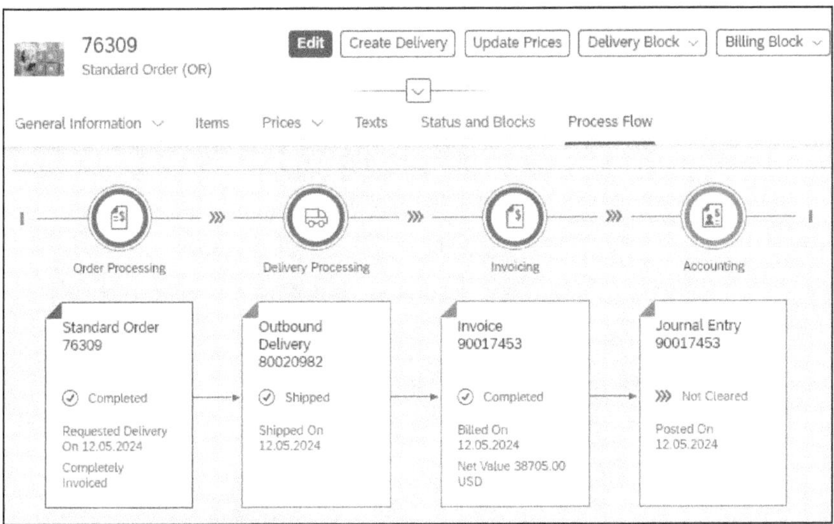

Figure 10.49 Complete Process Flow Related to First EDI Sales Order Message

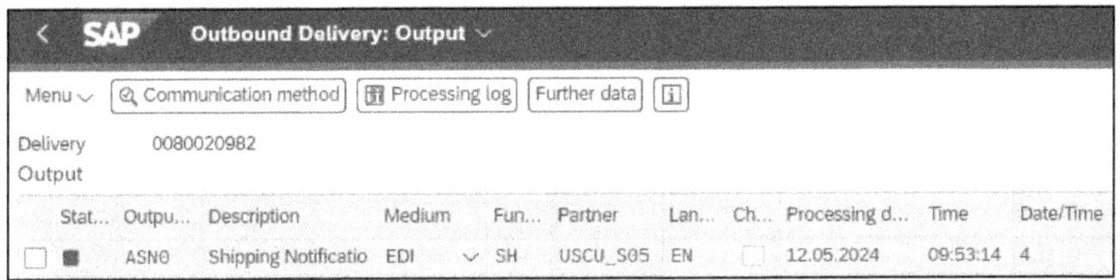

Figure 10.50 Outbound Delivery Document and Header Output ASN0

Next we can check the billing document output. We can look into the document itself and then the header outputs, but there is also an alternative option: use the Manage Output Items app. Start the app and search for an item with **Application Object** ID equal to **0090017453**, as shown in Figure 10.51, there successfully processed as output on the IDoc channel.

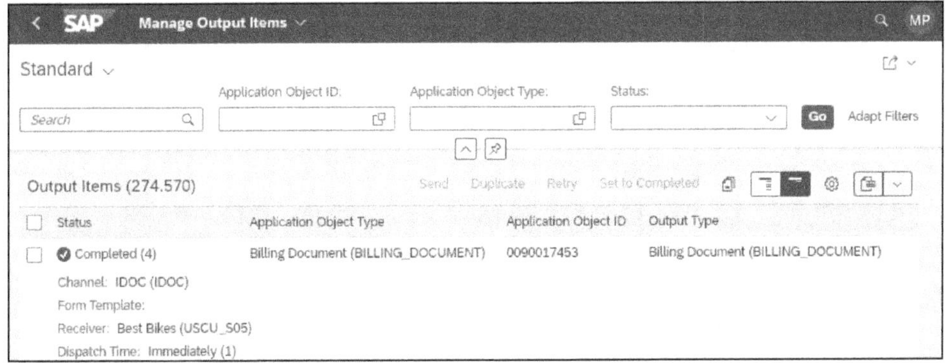

Figure 10.51 Manage Output Items

To confirm that the outbound delivery and invoice outputs have been successfully processed to IDocs and that the latter have been sent to Cloud Integration, we can have a look at the Message Monitoring overview app. Figure 10.52 shows that we have one successfully processed **Outbound Delivery** message and one successfully processed **Invoice** message. Both IDocs left the SAP S/4HANA system, and we can move to Cloud Integration and its B2B monitoring to check the EDI processing.

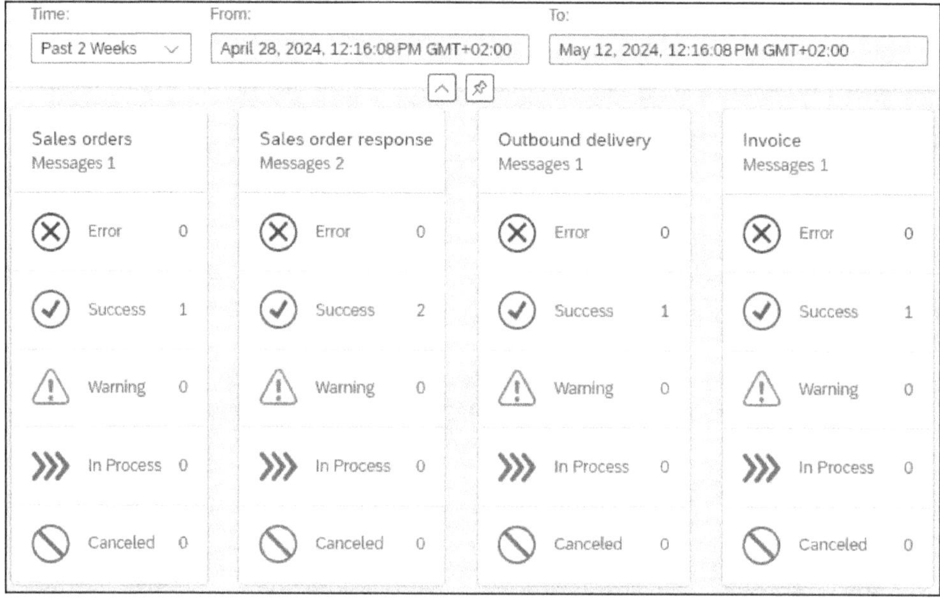

Figure 10.52 Message Monitoring Overview: All Transactions Related to Process Flow

After switching to Cloud Integration, navigate to **B2B Monitoring**. In the first step we will adapt filters to be able to search based on purchase order number, which is defined as a custom search attribute. Click the **Adapt Filters** button located on the right-hand side of the screen. On the selection screen like the one shown in Figure 10.53, choose **CSA Purchase Order Number** and confirm by clicking **OK**. This gives us an option to find all interchanges related to our initial EDI sales order; do that using the reference purchase order number, PO2400001.

After maintaining the appropriate time frame and the **CSA Purchase Order Number** field with value **PO2400001**, click **Go** to display all interchanges related to our EDI process flow (see Figure 10.54). Let's quickly review the key information we can collect from this overview:

- **Status**, **Sender Name**, and **Receiver Name** are self-explanatory.
- **Sender Message Type** is the inbound to Cloud Integration source message in the MAG.
- **Receiver Message Type** is the outbound from Cloud Integration target message in the MAG.

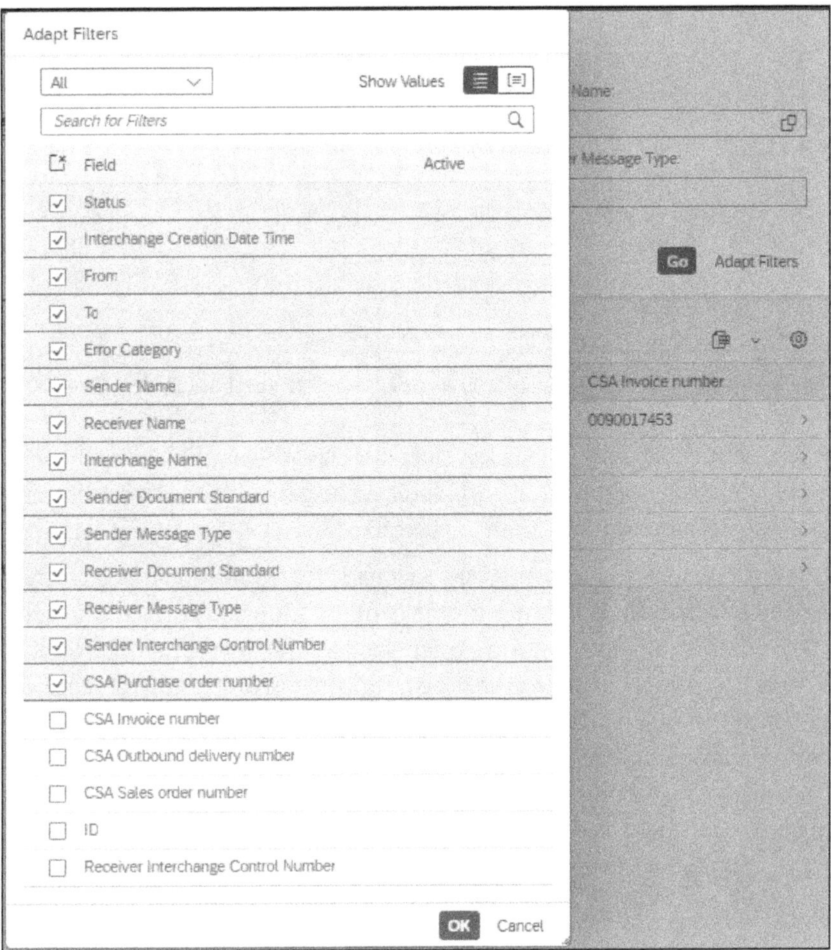

Figure 10.53 Adapting Filters: Adding CSA Purchase Order Number

What's powerful in this B2B monitoring is the possibility of adding custom search attributes, which can be built on any data from our interchange. In our case, we defined these attributes to contain document numbers; a simple query link in this example gives us very transparent view of the overall process flow and of each unique document. We can quickly compare this with the process flow in SAP S/4HANA (like the one shown earlier in Figure 10.49) to match each document with the relevant EDI transaction.

By making use of the settings, you can modify the view and include or exclude specific data as needed. Note that the sample view shown in Figure 10.54 is very transparent as it shows all interchanges related to a specific purchase order number. You can navigate to the details of each interchange by clicking on it.

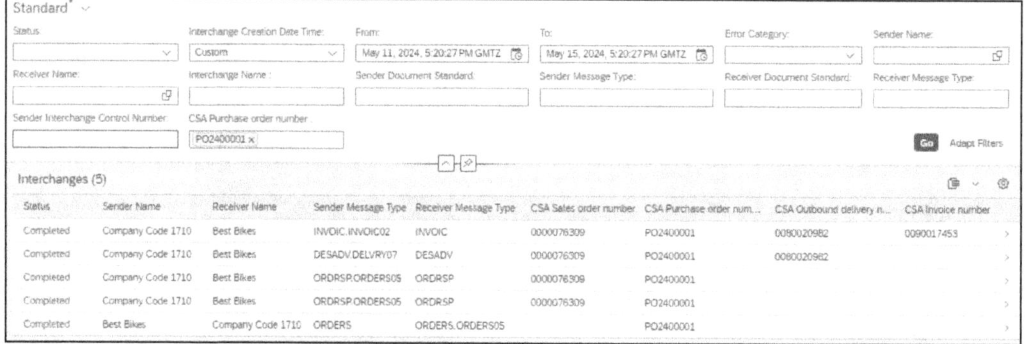

Figure 10.54 B2B Monitoring: View of All Interchanges Related to Purchase Order PO2400001

As you can see, all our EDI transactions have been processed successfully and ended with status **Completed**. Nevertheless, we will look at the details of the two outbound messages, first the ASN message and then the invoice.

Figure 10.55 shows details of the ASN message. First let's look at the sender and receiver details, where all information is in line with the configuration and correct. One addition in the fields that we did not mention earlier: **Sender Interchange Control Number** for **Adapter Type IDoc** simply contains the IDoc number from SAP S/4HANA. **Receiver Interchange Control Number** will be mapped to the UN/EDIFACT message; the value comes from number range we configured earlier in the agreement. Each trading partner and/or each EDI transaction can have its own number range. In our case, we maintain number range `ICN_DEFAULT` for each EDI transaction. The number range can be monitored and maintained in the **Manage Stores** section of the Cloud Integration monitoring.

Sender

Message Type:
DESADV.DELVRY07

Message Number:

Sender Identifier:
S42CLNT400

Group Control Number:

Sender Name:
Company Code 1710

Sender Identifier Qualifier:

Document Standard:
SAP_IDoc

System ID:

Receiver Identifier:
USCU_S05

Interchange Control Number:
0000000000417005

Adapter Type:
IDOC

Receiver Identifier Qualifier:

Receiver

Message Type:
DESADV

Message Number:
102

Sender Identifier:
EDI_COMPANY_1710

Group Control Number:

Receiver Name:
Best Bikes

Sender Identifier Qualifier:
Mutually defined

Document Standard:
UNEDIFACT

System ID:

Receiver Identifier:
EDI_BEST_BIKES

Interchange Control Number:
102

Adapter Type:
AS2

Receiver Identifier Qualifier:
Mutually defined

Figure 10.55 B2B Monitoring: Sender and Receiver Details of ASN Message

Figure 10.56 shows the ICN_DEFAULT number range.

Overview / Manage Number Ranges

Number Ranges (1) Add Edit Undeploy

Name	Minimum Value	Maximum Value	Next Value	Field Length	Rotate	Deployed By	Deployed On
ICN_DEFAULT	0	99999	104	0	☐	marek.piaseczny@int4.com	Mar 30, 2024, 13:11:41

Figure 10.56 Manage Number Ranges: ICN_DEFAULT Number Range Used in EDI Transactions

In the next step, we will look at the interchange payload, the UN/EDIFACT version that has been assembled and sent to the receiver. Figure 10.57 shows the DESADV D96A message as generated for our trading partner. As you can see, the UNB segment matches the information seen a while ago in the interchange details. The BGM segment contains the expected outbound delivery number 0080020982; the same for the RFF+ON segment with purchase order value PO2400001 and for RFF+VN segment with sales order value 0000076309. All identifiers in NAD segments have been properly converted to external identifiers.

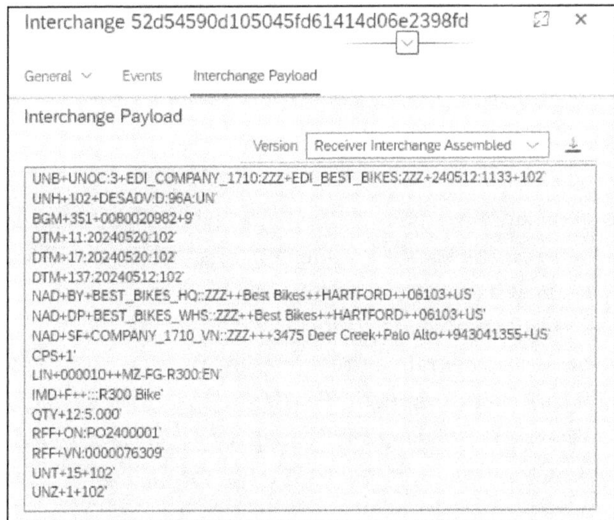

Figure 10.57 Interchange Payload for UN/EDIFACT DESADV D96A Message

The last part want to validate is the interchange payload related to the INVOIC message, but before we go there let's take a quick look at Figure 10.58, which shows the sender and receiver details related to the INVOIC message that we just sent to our trading partner. You may notice that there are no big differences compared to the sender/receiver details related to the ORDRSP and DESADV messages we have seen earlier. In our case, there are two attributes that are different across different message types; these are obvious and do not require any additional explanation:

- **Message Type** for both **Sender** and **Receiver**
- **Interchange Control Number** for both **Sender** and **Receiver**

You may also notice that the interchange control number for DESADV has value 102 and for INVOIC it is 103. That's because both messages use the same number range (see ICN_DEFAULT, mentioned earlier) and INVOIC has been sent directly after DESADV.

Sender

Message Type: INVOIC.INVOIC02	Message Number:	Sender Identifier: S42CLNT400
Group Control Number:	Sender Name: Company Code 1710	Sender Identifier Qualifier:
Document Standard: SAP_IDoc	System ID:	Receiver Identifier: USCU_S05
Interchange Control Number: 0000000000417007	Adapter Type: IDOC	Receiver Identifier Qualifier:

Receiver

Message Type: INVOIC	Message Number: 103	Sender Identifier: EDI_COMPANY_1710
Group Control Number:	Receiver Name: Best Bikes	Sender Identifier Qualifier: Mutually defined
Document Standard: UNEDIFACT	System ID:	Receiver Identifier: EDI_BEST_BIKES
Interchange Control Number: 103	Adapter Type: AS2	Receiver Identifier Qualifier: Mutually defined

Figure 10.58 B2B Monitoring: Sender and Receiver Details of Invoice Message

Now let's look at the interchange payload, UN/EDIFACT INVOIC, which has been assembled and sent to our trading partner. You can see the complete message in Figure 10.59. Like for other outbound messages, the UNB segment matches the information from the sender/receiver view. The BGM segment contains the correct invoice number, 90017453, and the following reference segments also contain correct references: segment RFF+ON with purchase order number PO2400001 and segment RFF+VN with sales order value 0000076309. External identifier COMPANY_1710_VN is properly mapped to the NAD+SU segment. The item and pricing details are also represented within the message as per the agreement with the trading partner.

At this point we are done checking all four EDI transactions related to the complete process flow. For checking the details on the SAP S/4HANA side, we can use the Message Monitoring app, which is part of SAP Application Interface Framework. For checking messages in Cloud Integration, we have made use of B2B monitoring, which coexists next to the standard message monitoring but offers additional features and views dedicated to B2B processing. When you are checking the processing on Cloud Integration, you can also look at the generic message monitoring; the B2B messages are also visible there, but the information is not complete and not so as transparent as in the specialized B2B monitoring views, especially if a single B2B message is processed by multiple different iFlows. Figure 10.60 shows the B2B message processing in the standard message monitoring.

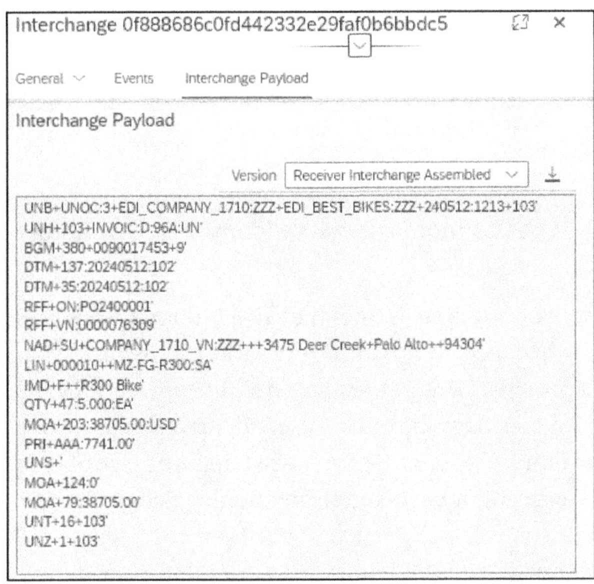

Figure 10.59 Interchange Payload for UN/EDIFACT INVOIC D96A Message

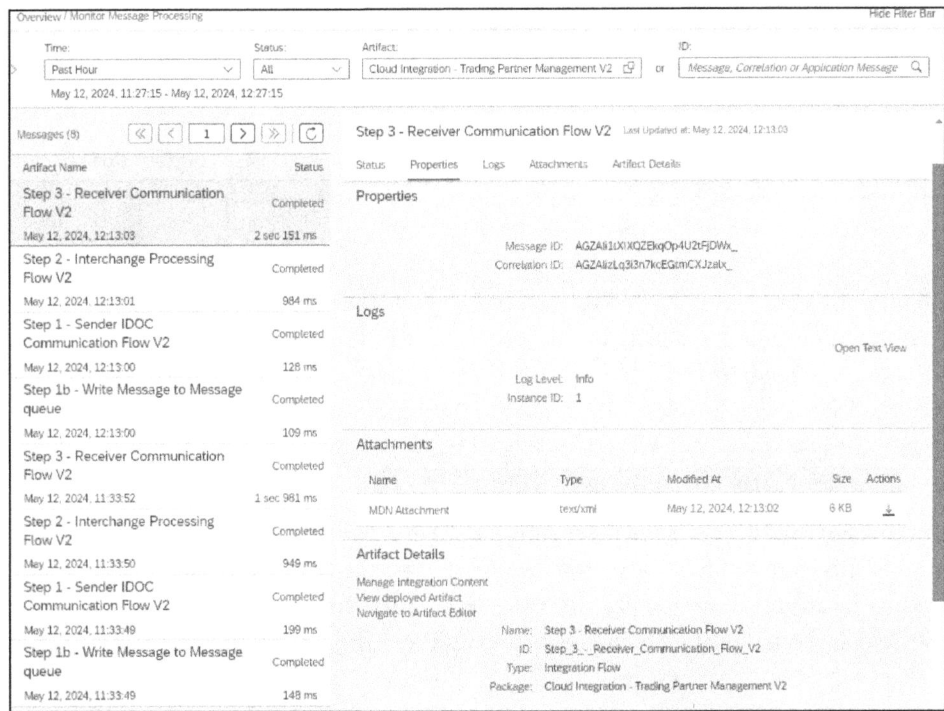

Figure 10.60 Checking B2B Message Processing in Standard Message Monitoring

10.3 Negative Testing, Alerting, and Error Handling

What we have gone through so far was only the happy flow scenario: each EDI message contained correct data, and processing of each EDI transaction went smoothly without any failures. In this section, we will look at two cases where we intentionally insert invalid data to see how the solution handles unexpected situations and the mechanisms to recover from it.

10.3.1 Receiving an EDI Message that Cannot Be Properly Identified by Trading Partner Management

Let's assume that we have a call from our trading partner that an EDI message has been sent to our system. They claim they had received an MDN for this AS2 transmission. To check this message, we start Cloud Integration and navigate to **B2B Monitoring**. On the main page we can see that there are messages in the **Unassigned Interchange** section. This is where all interchanges that cannot be identified are registered. The lack of identification means that the interchange cannot be linked to any trading partner and the EDI transaction cannot be configured in Trading Partner Management. Figure 10.61 shows a sample view with nine unassigned interchanges. Click the **Unassigned Interchanges** box on the right-hand side to see the specific interchanges.

Figure 10.61 B2B Monitoring: Nine Unassigned Interchanges

All nine interchanges are listed as shown in Figure 10.62. You will immediately notice that the information available in this view is very limited, especially compared to the view for (assigned) interchanges. This is because these messages do not have any link to a trading partner or to an agreement configuration. We will take a close look at the most recent message. Open its details by clicking the link in the **Message Processing Log ID** column at the top of the list.

Clicking any specific **Message Processing Log ID** link redirects to the related processing step. In this example, it is iFlow **Step 1b—Write Message to Message Queue**. When processing of an interchange fails, the reprocessing mechanism is triggered. The failed message is stored in the JMS queue, and from there the interchange processing iFlow tries to reprocess the message. The number of retries is set in a configuration parameter, which we looked at earlier. It is located in **Trading Partner Management • Configuration Manager • Maximum Retry Count • Interchagne Processing Flow**. Keep the default value **3** for **Maximum Retry**. Figure 10.63 shows the step where the message is stored in the JMS queue for reprocessing purposes.

Error Date Time: Past 24 Hours | From: May 17, 2024, 2:42:02 PM GMTZ | To: May 18, 2024, 2:42:02 PM GMTZ | Error Category: | Error Information: | Message Processing Log ID:

Go | Adapt Filters

Unassigned Interchanges (9)

Error Date Time	Error Category	Error Information	Message Processing Log ID
May 18, 2024, 10:59:19AM GMTZ	Agreement Configuration Access Failed	Incoming message could not be parsed.	AGZligYOF0IHYERhLF03SYnRC0v1
May 18, 2024, 10:58:18AM GMTZ	Agreement Configuration Access Failed	Incoming message could not be parsed.	AGZlicowRgfVK1VSSIFaXkb1hBmh
May 18, 2024, 10:57:02AM GMTZ	Agreement Configuration Access Failed	Incoming message could not be parsed.	AGZliX53_IYYswoQhndwl9scoT3Z
May 18, 2024, 10:09:16AM GMTZ	Interchange Transmission Failed	Error in receiving the document. Refer to the Message Processing Log for more details.	AGZlfkurbEk3OhJPO-imWEjwCyKD
May 18, 2024, 9:46:08AM GMTZ	Agreement Configuration Access Failed	Incoming message could not be parsed.	AGZleOA7qRr4M5N_GJWBXf79zc2S
May 18, 2024, 8:08:07AM GMTZ	Agreement Configuration Access Failed	Incoming message could not be parsed.	AGZlYedtsmcjzgEZyyfsX7EneKLP
May 18, 2024, 8:06:05AM GMTZ	Agreement Configuration Access Failed	Incoming message could not be parsed.	AGZlYW06ybJOSwe7E7n02hARWvR
May 18, 2024, 8:03:46AM GMTZ	Agreement Configuration Access Failed	Incoming message could not be parsed.	AGZlYOL3vSEQkZyLlYwwTgPWO21N
May 18, 2024, 8:03:00AM GMTZ	Agreement Configuration Access Failed	Incoming message could not be parsed.	AGZlYLQLEiA14IKVC4VosT0LmBDz

Figure 10.62 B2B Monitoring: List of Unassigned Interchanges

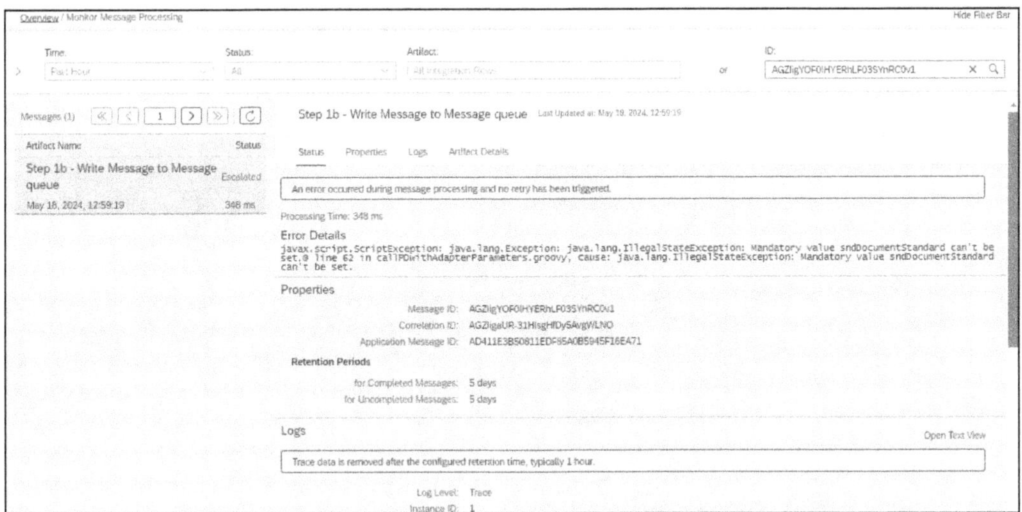

Figure 10.63 Failure on Unassigned Interchange as Seen in Standard Message Monitoring

We can look at the JMS queue and find the payload. The name of this queue is a configurable parameter in the iFlow itself; we keep the default value here. Figure 10.64 shows the content of the JMS queue and the single message which still shows a retry count of 3. From this view we can also download the message and analyze its content, what we found in our example is UN/EDIFACT ORDERS D96A message but with the UNB segment as shown in Figure 10.65. The receiver identifier, EDI_COMPANY_1810, is not correct and not configured in our system.

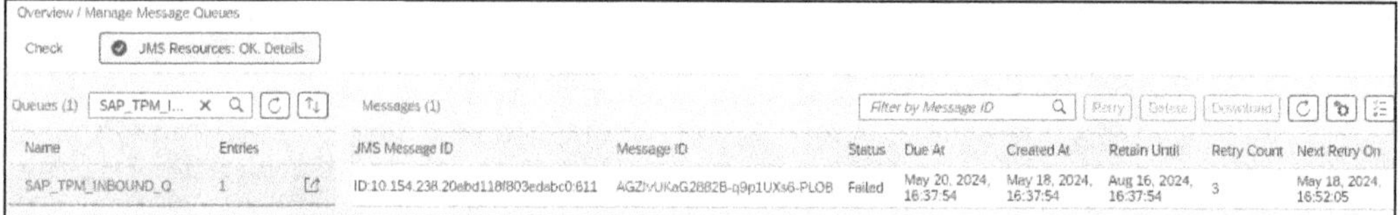

Figure 10.64 View of SAP_TPM_INBOUND_Q JMS Queue with Single Failed Message

```
UNB+UNOC:3+EDI BEST BIKES:ZZZ+EDI COMPANY 1810:ZZZ+240518:1630+25001'
```

Figure 10.65 UNB Segment from Failed UN/EDIFACT ORDERS D96A Message

We will not make any correction in Trading Partner Management; the message is incorrect and needs to be corrected on the sender side. After three retries, processing of this message will end with the Escalated status, as shown in Figure 10.66. This is in iFlow **Step 2—Interchange Processing Flow V2** and with additional details of the error itself, which points to a missing entry in the partner directory. At this stage, the JMS queue is empty.

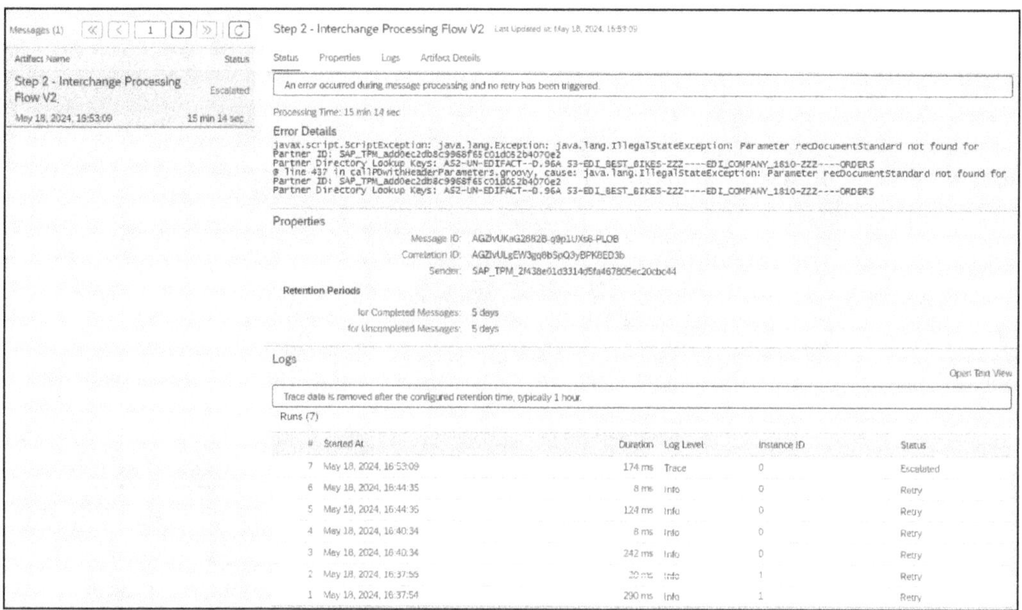

Figure 10.66 Message Monitoring: View of Message in Final Escalated Status

10.3.2 Receiving UN/EDIFACT ORDERS D96A that Contains Invalid Data (Material Number)

In the second negative testing scenario, an EDI message is successfully received and processed by the middleware but the processing fails in the ERP system. In our example, we intentionally send an EDI sales order message that contains an invalid material number. This material number does not exist in the SAP S/4HANA material master data and this is preventing the sales order creation. The IDoc is in Error status. Let's look at how such a case is visible in the system and then we will adjust the data and reprocess the IDoc. We will make use of the Message Monitoring app here.

The starting point is to find the failed message in Message Monitoring (see Figure 10.67). The log message is very clear about the root cause of the error.

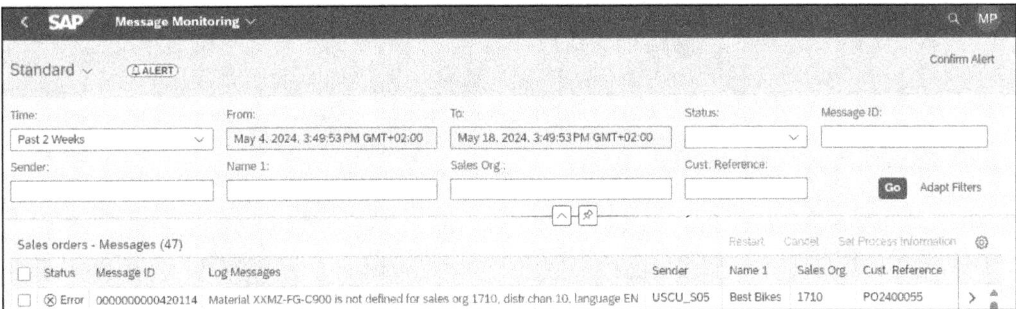

Figure 10.67 Message Monitoring: View of ORDERS IDoc Failing Due to Unknown Material Number

We can navigate to the details of this specific message, and now is the right time to review SAP Application Interface Framework alerting. At the very bottom of the log messages, we can see an alert-related notification: **Alert was created with ID 00215 (category /AIF/ALERT_CAT_DEF)**. When we configured SAP Application Interface Framework in Chapter 8, as shown in Figure 10.68, we also enabled alerting. Each alert is supposed to go to two recipient groups: IT_SUPPORT and a dynamically determined sales organization specific group. In our case, the latter would be BUSINESS_VKORG_1710. Our own SAP S/4HANA user is a member of the SUPPORT_IT recipient group.

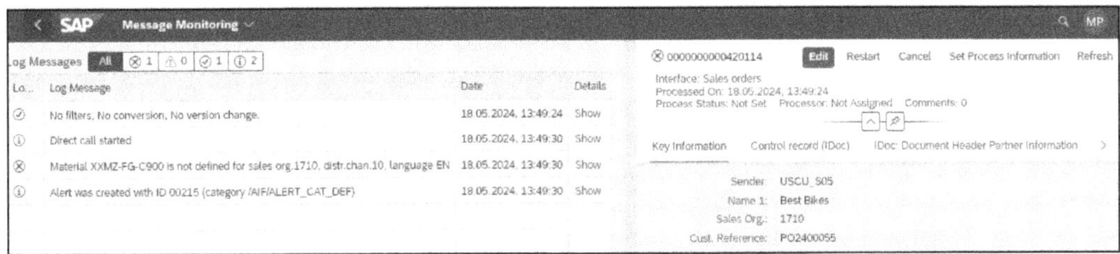

Figure 10.68 Message Monitoring: Detail View of ORDERS IDoc Failing Due to Unknown Material Number

The notification triggered from the alert is sent in email form. The destination email address is determined based on the user master data (SAP Application Interface Framework can also work with external recipients, but that's out of the scope of this setup). Figure 10.69 shows the notification generated for our failing message. It is the standard template from the alert category.

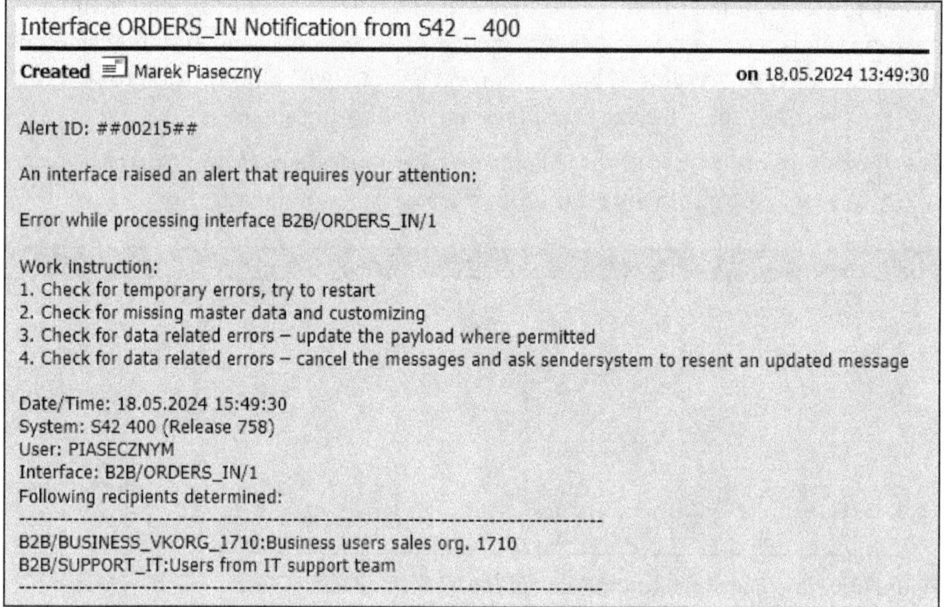

Interface ORDERS_IN Notification from S42 _ 400

Created Marek Piaseczny **on** 18.05.2024 13:49:30

Alert ID: ##00215##

An interface raised an alert that requires your attention:

Error while processing interface B2B/ORDERS_IN/1

Work instruction:
1. Check for temporary errors, try to restart
2. Check for missing master data and customizing
3. Check for data related errors – update the payload where permitted
4. Check for data related errors – cancel the messages and ask sendersystem to resent an updated message

Date/Time: 18.05.2024 15:49:30
System: S42 400 (Release 758)
User: PIASECZNYM
Interface: B2B/ORDERS_IN/1
Following recipients determined:

B2B/BUSINESS_VKORG_1710:Business users sales org. 1710
B2B/SUPPORT_IT:Users from IT support team

Figure 10.69 SAP Application Interface Framework Notification Triggered for Message Failing in ORDERS_IN Interface

At the bottom of the notification, there are two recipient groups mentioned. We can trace how these were determined. As a reminder, our dynamic recipient functionality for the sales order interface makes use of table ZAIF_ALRT_VKORG, which is shown in Figure 10.70.

Namespace	Interface	Version	Namespace	Recipient for Alert	Sales Org.
B2B	ORDERS_IN	1	B2B	BUSINESS_VKORG_1710	1710
B2B	ORDERS_IN	1	B2B	SUPPORT_IT	

Figure 10.70 Content of Table ZAIF_ALRT_VKORG for Dynamic Recipient Determination

In table /AIF/ALERT_IDX, as shown in Figure 10.71, we can see logs related to alert ID 215, which is the one related to our failing sales order message. Based on the content of this table, we can confirm that IDoc 420114 has been identified as the ORDERS_IN interface in the B2B namespace. There have been two recipient groups determined for alert number 215: BUSINESS_VKORG_1710 and SUPPORT_IT.

GUID 16	Number	Namespace	Interface	Version	Namespace	Recipient for Alert	Creation Date	Last Date	Status	Alert ID
0000000000420114	1	B2B	ORDERS_IN	1	B2B	BUSINESS_VKORG_1710	18.05.2024	18.05.2024	E	215
0000000000420114	2	B2B	ORDERS_IN	1	B2B	SUPPORT_IT	18.05.2024	18.05.2024	E	215

Figure 10.71 Content of Table /AIF/ALERT_IDX: History of All Triggered Alerts

After this brief look at the alerting and a sample notification, we can continue with fixing the problem by following these steps:

1. When you are still in the Message Monitoring application and displaying details of the failed message, navigate to the **IDOC: Document Item Object Identification** data section. This is the segment containing the material number. Behind this description there is a segment E1EDP19.
2. Click **Edit** located at the top, which allows you to edit the field containing the material number. Notice that you can edit only this field as it has been explicitly defined as changeable during SAP Application Interface Framework configuration (see Figure 10.72).
3. Change **Material** number XXMZ-FG-C900 to MZ-FG-C900.
4. Save the changes, which should be confirmed by a message: **Change was saved successfully.**
5. Click **Restart** located next to **Edit** to reprocess the IDoc. You will be asked to confirm the message reprocessing (see Figure 10.73).

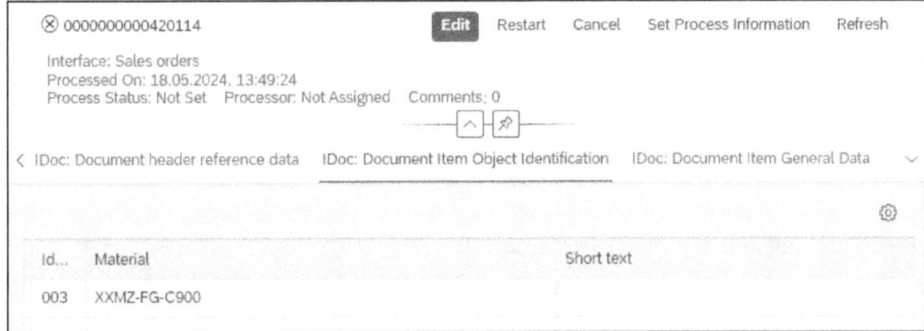

Figure 10.72 Message Monitoring: Editing Material Number in Failed Sales Order Message

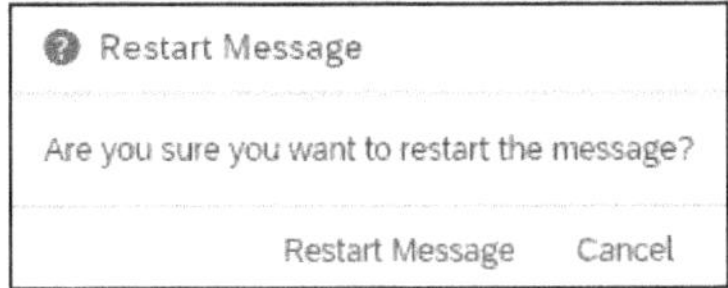

Figure 10.73 Message Monitoring: Prompt to Confirm Restarting Message

After all these steps, you should see the successfully processed IDoc and sales order number (see Figure 10.74).

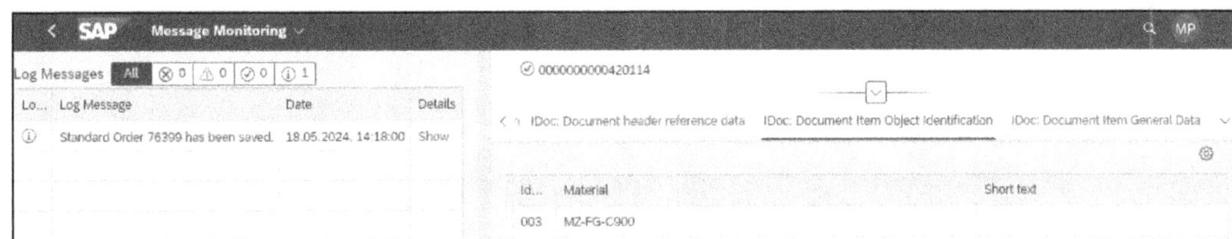

Figure 10.74 Message Monitoring: Successfully Reprocessed IDoc Resulting in Sales Order Creation

10.4 Summary

Finally, we can see results of the hard work performed in the earlier chapters. It has been even more than seeing: we have executed a full purchasing process supported by four EDI transactions running on Trading Partner Management. Because we established an AS2 connection with our dummy trading partner, we did not have to take any shortcuts and could trigger the whole process from the UN/EDIFACT ORDERS D96A message received via AS2 transmission. We even had a chance to see from the trading partner perspective how the successful transmission is confirmed by MDN and what the UN/EDIFACT CONTRL D96A functional acknowledgment looks like.

We went through the key features of B2B monitoring in Cloud Integration. As we experienced, the tool is specialized to analyze B2B processing. In a very flexible and transparent way, it delivers all the necessary details. With the out-of-the-box functionality called custom search attributes, we can easily extend the monitoring view with additional reference data coming from the sender or receiver message.

We were not limited to checking our process on Cloud Integration. We spent significant time looking at the processing within SAP S/4HANA. We have used the SAP Fiori–based Message Monitoring app, which involves SAP Application Interface Framework. This approach leaves behind the SAP GUI–based technical transactions used for decades (e.g., WE05). The Message Monitoring application can be very transparent and an efficient monitoring and error handling solution.

The overall process ended with successfully processed EDI transactions. To make things more challenging at the end of this chapter, we included two negative tests to demonstrate how unexpected situations are handled by the system.

At this stage, we have completed the B2B scenario. In the next chapter, we will discuss implementing automated testing for EDI.

Chapter 11
Test Automation

Building an EDI solution is one thing, but maintaining it through its lifecycle is another—especially with a landscape that has dozens of different EDI transactions and hundreds of trading partners. Implementing automated testing for the EDI area is definitely the way to go if we want to have a stable, efficient, and flexible IT solution.

Manual testing cannot be skipped and plays a vital role. However, it is hard to manually validate all the messages from different electronic data interchange (EDI) partners with the proper testing scope because of the time needed for such a validation. This is where automation comes into play. The goal of this chapter is to show you how we can automate the EDI testing in SAP Integration Suite and SAP S/4HANA and how to make sure that your automated tests can be reused whenever needed. We will start with some general concept introductions and then we will show you detailed component-based testing and end-to-end testing.

11.1 Overcoming the Challenges of EDI Testing

As we know, efficient EDI helps businesses send and receive information quickly and accurately. But before we reach this ultimate goal, testing EDI systems can be challenging.

There are many things that depend on each other, like different systems, data formats, and communication methods. We have to be very careful, because even a small mistake in the production environment can cause big problems. These mistakes can lead to expensive delays in goods delivery or even fines from our partners for sending wrong messages and generating additional effort on their side.

One big problem is that there are many different business partners, and each one might need something different. Even processes across one business partner might contain several variants. This means we have to test many different scenarios (practically speaking, hundreds or thousands).

Another issue is that there are no special internal test environments for EDI as the scope of EDI is beyond any single company. In other software systems, we can create test environments that look like the real ones, but with EDI, we often can't do that. This forces us to organize help from external partners to support our testing and usually limit the scope due to their low availability.

EDI testing takes a lot of time. As mentioned, we need external experts to perform full end-to-end EDI testing. There is another challenge: these experts are often very busy. They are usually focused on keeping their system running smoothly every day, so they don't have much time to help with testing new changes for their partners. This can delay our projects and increase the chance of mistakes.

In addition, EDI testing requires a specific skillset in an SAP team. EDI combines technical and functional skills. In most cases, EDI is implemented by technical teams with the support of business process experts. In many situations, neither a technical expert nor a functional expert can conduct the entire testing alone. It requires close cooperation between them, which also makes it difficult to plan these tests well.

But despite these challenges, we can still do effective EDI testing. This chapter will discuss ways to handle these problems and make sure our EDI systems work well. We will look at using an automated testing tool called Int4 Suite to address some of these concerns and to make testing easier and more reliable.

11.2 The Value of and Approaches to EDI Testing

There is a huge value to service virtualization and automation of EDI testing. This value will be described based on Int4 Suite, the solution available for testing EDI in SAP.

To present Int4 Suite, we must mention the differences between automated UI tests and API tests. In UI tests, we need to focus on testing the entire end-to-end scenario. We must record the sequence of screens at the partner's end to trigger a message (if that is even possible), and then in our system record the sequence of screens to validate the document created by EDI (or vice versa in case of an outbound scenario). If something stops during the test, we will not be aware of the fault's location (black box testing). Even if such automation is possible for EDI tests, it is very labor-intensive.

On the other hand, we have a set of well-known and widely available developer tools for API tests, such as Postman. These are of limited use due to the specific nature of EDI communication (AS2, flat files, etc.). In addition, even if we could simulate EDI partners in this way, we would not be able to perform validation. EDI is asynchronous, and API testing tools are based on validating responses generated by synchronous interfaces.

This is precisely why Int4 Suite was created. It is a dedicated tool for SAP that combines testing of the integration platform (SAP Process Orchestration, SAP Integration Suite, and others) with testing of the integration components in SAP S/4HANA, up to validation in the system of final transactional documents, such as sales orders, deliveries, or invoices.

The development of Int4 Suite was made possible by the decision to make it a dedicated tool only for SAP, where most business communication occurs synchronously. Thanks to close integration with integration platforms, Int4 Suite creates a database of test cases semi-automatically thanks to the persistence of historical transactions in the integration platform. The tests consist of a combination of original EDI documents, processed outgoing messages in an SAP format (e.g., IDoc), an a database extract of the final document in SAP S/4HANA.

When the test cases are created, they are available for fast, anytime execution. The tested and only needed components are SAP Integration Suite and, optionally, an SAP S/4HANA backend. In this case, EDI partners are simulated and the testing is performed without their involvement.

Thanks to this approach, EDI testing works great in a very limited time (a few weeks): it is possible to automate EDI scenarios for hundreds or thousands of partners and dozens of representative messages to cover all combinations of business scenarios.

The testing pyramid in Figure 11.1 shows how the EDI tests are positioned. It is a great simplification compared to the full end-to-end testing. With limited cost and effort, it allows wide coverage by thousands of test cases and a deep level of verification thanks to validation based on historical messages and documents posted already in the system.

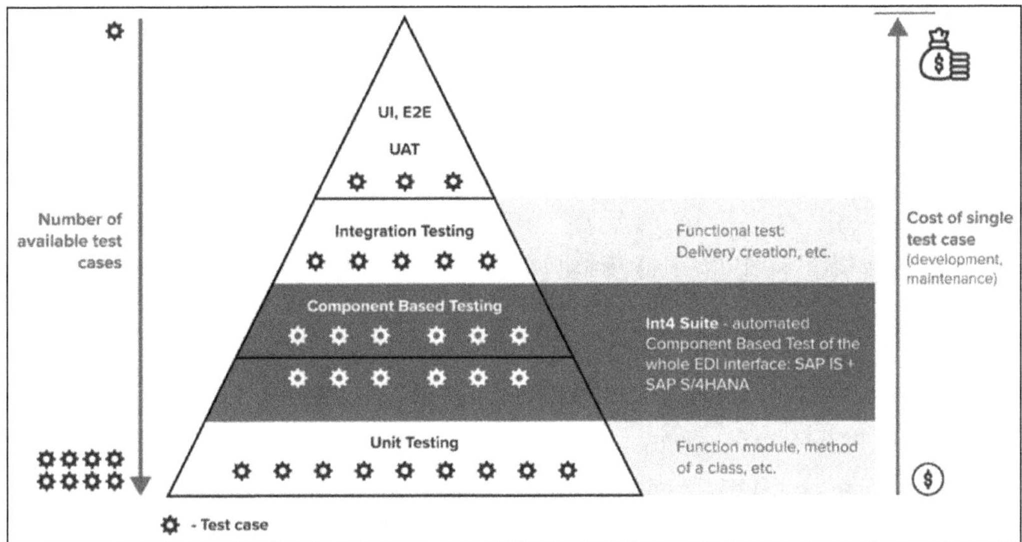

Figure 11.1 Testing Pyramid: Gradual Approach for End-to-End Testing

In conclusion, this way of EDI testing demonstrates significant value in service virtualization and automation for EDI testing. By eliminating external dependencies through the simulation of EDI partners, organizations can maintain control over project budgets and costs. The automation of test case creation and execution, based on historical transactions, reduces the time and effort required for testing, development, and maintenance, leading to lower overall costs. This efficient process enables a faster time to market, ensuring that business solutions are deployed swiftly. In addition, by integrating platform testing with the validation of final transactional documents in SAP S/4HANA, it provides high reliability, mitigating risks and ensuring that business operations remain unaffected. The comprehensive and practical approach of Int4 Suite offers businesses a cost-effective and efficient solution for thorough EDI testing, ensuring accuracy and reliability in electronic communications and supporting smooth business operations.

11.3 Automated Regression Testing for EDI with Int4 Suite

In this section, we will automate testing the whole EDI scenario presented in this book. The scope of testing will include the iFlows set up in SAP Integration Suite and the configuration done in the SAP S/4HANA system to validate the EDI process. We will automate the process that we validated manually in Chapter 10.

We will automate API testing as mentioned with the Int4 Suite solution. Int4 Suite helps eliminate dependencies from external systems and EDI partners during the build and testing phases. With Int4 Suite, developers and functional consultants can simulate the interfaces of any third-party, non-SAP systems and B2B partners.

Note

This book will focus only on the Int4 Suite capabilities to run the regression testing in SAP Integration Suite and SAP S/4HANA for the EDI setup. Int4 Suite can also be used in test-driven development approaches, build phases, and unit testing for all types of SAP projects, including SAP S/4HANA migrations and SAP Process Orchestration to SAP Integration Suite migration. To learn more about the various use cases where you can use Int4 Suite, visit the Int4 website at *https://int4.com/* or check out the book *Testing SAP APIs: Strategy and Execution* (SAP PRESS, *www.sap-press.com/5133*).

11.3.1 Int4 Suite Overview

Before jumping directly into automated testing, let's first understand how Int4 Suite works at a high level. Figure 11.2 shows how the Int4 Suite components are positioned within the SAP S/4HANA landscape.

In the first step, Int4 Suite removes the dependency on external systems and EDI partners via SAP-tailored service simulation. The simulation is done based on the historical messages available in the system. These messages can be SAP Integration Suite messages, SAP Process Orchestration messages, or backend interface documents like IDocs.

Int4 Suite can collect those messages and reuse them for testing at any given time. We will create test cases based on the already validated messages processed in Cloud Integration in Chapter 10. As we can reuse existing messages, there is no need to create manual scripts to execute such a simulation. Thus, the test case preparation is quick.

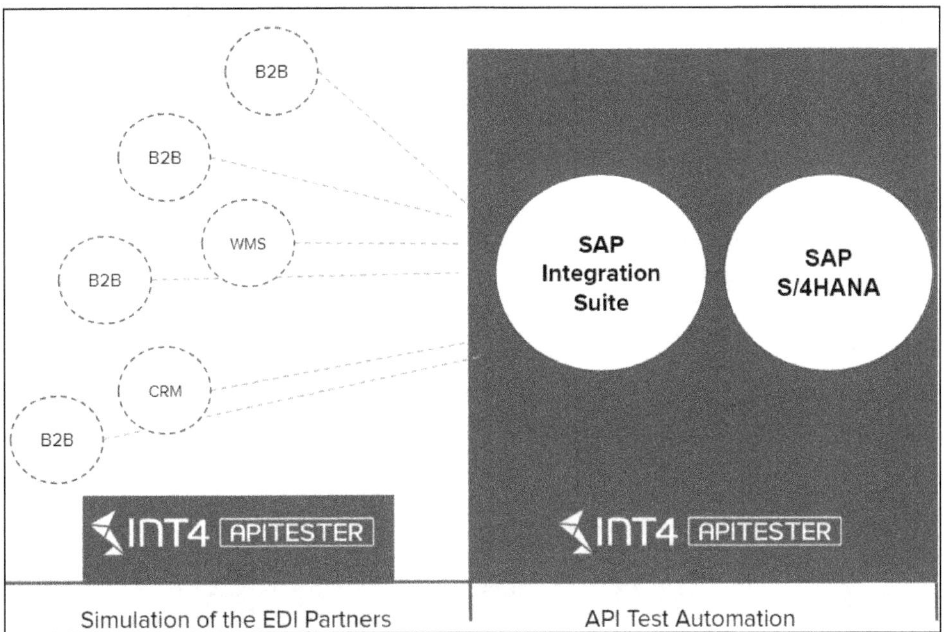

Figure 11.2 Simulation of EDI Partners and Scope of API Testing

In the second step, Int4 Suite can validate how the tested setup works in Cloud Integration and SAP S/4HANA. Int4 Suite will simulate a message from the EDI partner injecting the message to the specific iFlow and then it will validate if the simulated messages were processed properly by validation of the last step in the iFlow chain, ensuring that the IDoc is created correctly. In addition, it can validate whether the posting of the final business document was done correctly by comparing the reference business document with the tested one created with the simulated message.

This allows us not only to create test scenarios quickly but also to run them early in the build and testing phases, leveraging the shift-left approach.

In the following sections, we will focus on two approaches for EDI testing: component-based and end-to-end.

11.3.2 Automated Component-Based Testing of EDI

This section will focus on setting up the component-based testing for our EDI scenario. As mentioned in the previous sections, component-based testing allows for targeted testing of individual EDI components, such as inbound and outbound data flows in isolation. This focused approach enables thorough validation of specific functionalities and scenarios without the complexity of end-to-end processes.

This section will show you how component-based testing works for both inbound and outbound scenarios from the SAP S/4HANA perspective. This section will not focus on separate SAP S/4HANA testing or separate SAP Integration Suite testing, which can also be rewarding when one team is responsible for the middleware processing and one for the backend processing. In this section, we will show you how you can set up end-to-end testing for a specific interface, starting from the simulated message in Cloud Integration and ending with the posted sales order in the SAP S/4HANA system for the inbound flow. For the outbound flow, we'll start with the message trigger by the functional process in the SAP S/4HANA and end with the validation of the mapped IDoc to the EDI message sent to the EDI partner.

Inbound Scenario

For the inbound scenario, we want to eliminate the dependencies from the EDI partner while testing. To do this, we need to save the message sent by the EDI partner as a test case. Before we continue with the details, let's recall the inbound sales order scenario, as shown in Figure 11.3. We will use this as an example in this section.

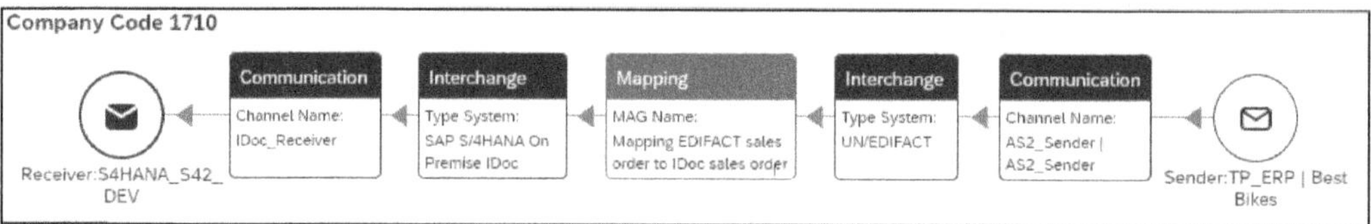

Figure 11.3 Inbound Sales Order Scenario Detail

Our example will look at the EDIFACT sales order from Best Bikes. In our example, we will save the reliable message from Chapter 10 because we are not yet live with the solution in a productive environment. If the solution is deployed in production, we could easily take the production messages as test cases considering all the GPDR and data masking–related concerns.

Our input test case data will be the EDIFACT sales order. The created test case will allow us to simulate messages whenever necessary. Because we don't want only to simulate the message but also to validate it, we will also save the output message to our test case, which, in our situation, is the ORDERS05 IDOC sent to the SAP S/4HANA system. Thanks to this, we can validate whether the tested input file we simulate always generates the same output IDoc ORDERS05 sent to the SAP S/4HANA system.

This will allow us to validate the whole configuration set in SAP Integration Suite with the Trading Partner Management configuration, with all the MIGs, MAGs, and iFlows.

In addition, our test case will validate the posting of the tested sales order in SAP S/4HANA created by a tested message. This will allow us to ensure that the whole configuration in SAP S/4HANA, like the partner profile and SAP Application Interface Framework, works fine and that the SAP customizing related to the sales order process works correctly.

We must first prepare the Int4 Suite configuration objects to start the test case creation. Open the Int4 Suite home screen and then open the **Automation Object** tile (see Figure 11.4).

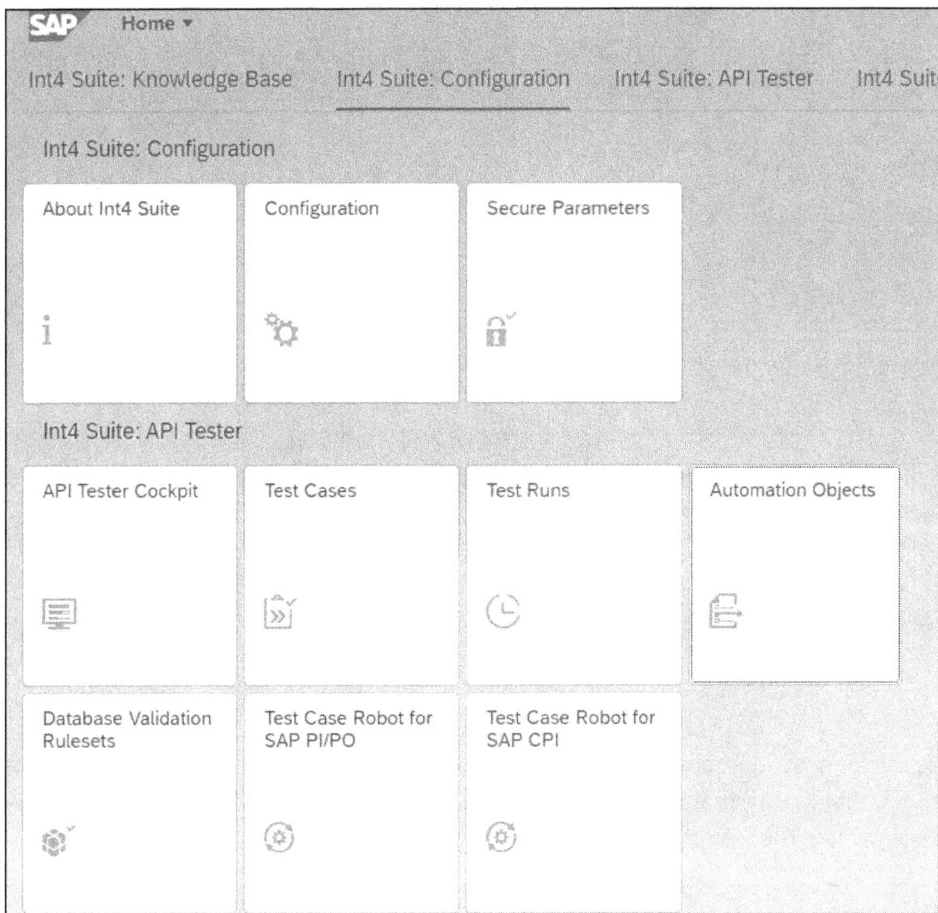

Figure 11.4 Int4 Suite Home Screen

The automation objects contain details about the tested iFlow and the configuration for each test execution. Automation objects define the interface under test, test conditions, and necessary data manipulation. They can also point to other objects, UI scripts, database validation definitions, and more.

In the Automation Objects app, click the **Create** button to create a new automation object. You will see a maintenance window as shown in Figure 11.5. Type the automation object name into the **Autom. Obj.** field, then choose **SAP CPI E2E Inbound** as the **Type**. Finally, fill in a description in the **Description** field and click the **Create** button.

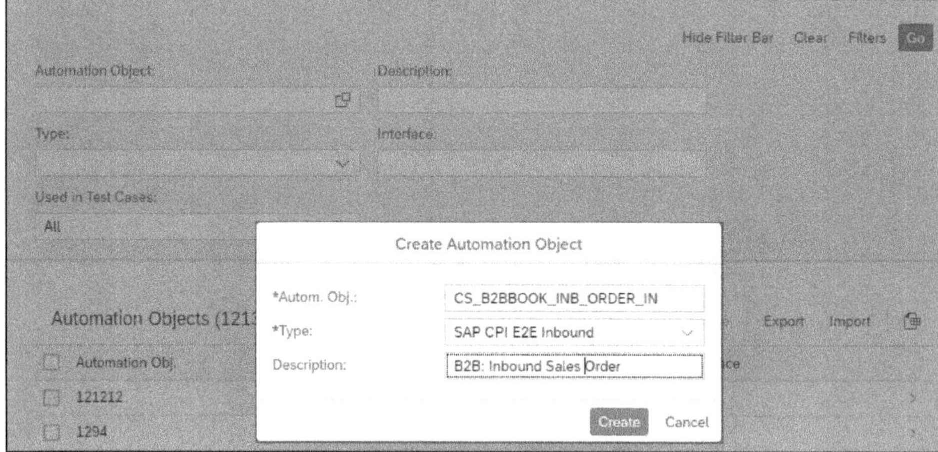

Figure 11.5 New Automation Object Creation in Automation Object Tile

> **Note**
>
> Note that the automation object and test case creation shown in this book will be relevant to all the Cloud Integration iFlows, including those used in the B2B/Trading Partner Management scenarios and the pipeline concept. The approach described in this book is independent of the scenario you may test. Remember that if SAP introduces specific APIs for getting the payloads from the B2B scenarios, then a new test type might become available in Int4 Suite to simplify the test case's configuration object creation.

In the automation object, click the **Import from CPI** button and choose the iFlow in which Int4 Suite will inject the message. Choose **Step_1b_-_Write_message_to_message_queue**. The selection of iFlows is shown in Figure 11.6.

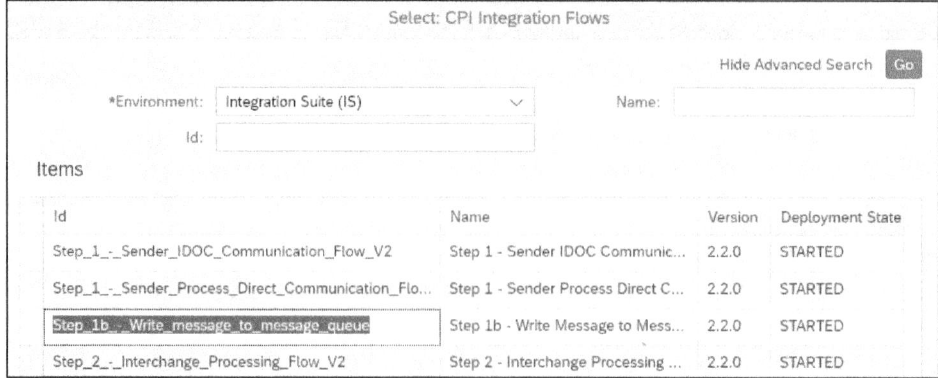

Figure 11.6 SAP Cloud Integration iFlows Search Help

Note

In our example, we will not use **Step 1—Sender AS2 Communication Flow V2**. Instead, we will use **Step 1b—Write a Message to a Message Queue**, which SAP designed for testing purposes. This also means that the AS2 certificates are not in the scope of the testing.

Next, we must choose the iFlow block to get the input payload. Click the **Search Help** button next to **CPI: Block in iFlow to Get Input Payload**, provide the SAP Integration Suite environment, and find the correct block. In our example, this is the **CallActivity_3411** block. Select it and confirm.

Next, we must choose **PROCESSDIRECT** as the **CPI: Adapter Type** in the **Test Case Execution** section. This is always the sender adapter of the iFlow in which we inject the message. In the same section, we can add optional HTTP header parameters. In our example, we want to set a proper EDI document standard, content type, and adapter type for the B2B/Trading Partner Management scenario we want to test. The complete configuration of the **PROCESSDIRECT Adapter Type** is shown in Figure 11.7.

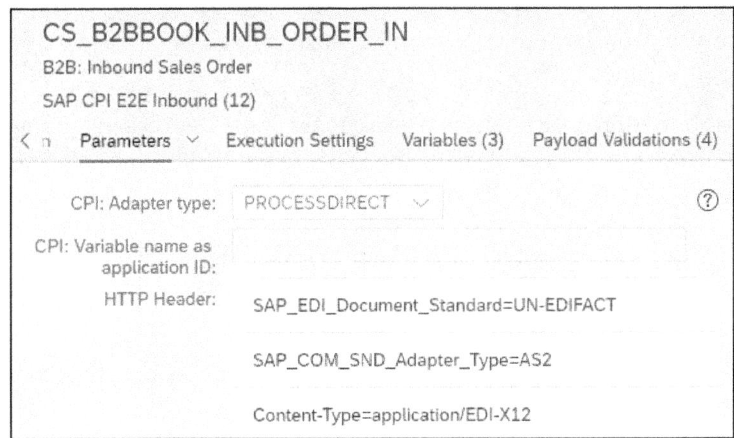

Figure 11.7 Adapter Type and HTTP Header Details in Automation Object

Remember that specific parameters must be filled out for each Cloud Integration adapter type. For the ProcessDirect adapter that we use, we have to go to the **ProcessDirect Adapter** section and fill in **CPI: ProcessDirect endpoint** as **/tpm/b2b/sap_tpm_internal_processdirect**, which is the endpoint to the iFlow in which messages will be injected (see Figure 11.8).

Note

You can find detailed descriptions of all the automation object parameters on the Int4 Suite wiki page at *http://s-prs.co/v573606*.

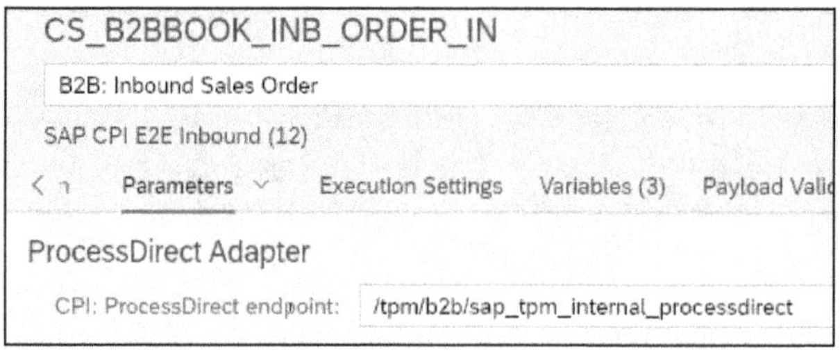

CS_B2BBOOK_INB_ORDER_IN
B2B: Inbound Sales Order
SAP CPI E2E Inbound (12)
Parameters | Execution Settings | Variables (3) | Payload Vali
ProcessDirect Adapter
CPI: ProcessDirect endpoint: /tpm/b2b/sap_tpm_internal_processdirect

Figure 11.8 ProcessDirect Endpoint

In the SAP Integration Suite scenarios where we test only one iFlow, we can add the block for the output payload in the same automation object. In our scenario, we inject the message into the **Step 1b** iFlow, but we want to retrieve the message from the **Step_3_-_Receiver_Communication_Flow_V2** iFlow where the mapped ORDERS05 XML IDoc is available. We must create an additional automation object for this iFlow and then use it in our main automation object for message injection.

To do this, return to the main Automation Objects app and select **Create a New Automation Object**. With this automation object, we will search for the message generated by the Cloud Integration processing after we inject the message with the first automation object. That is why we now have to choose **SAP CPI E2E Outbound** for **Type** while creating the automation object.

Similarly to the first automation object, for the iFlow, choose **Step_3_-_Receiver_Communication_Flow_V2**, then provide the block in the iFlow to get the output payload: **MessageFlow_197327** (see Figure 11.9).

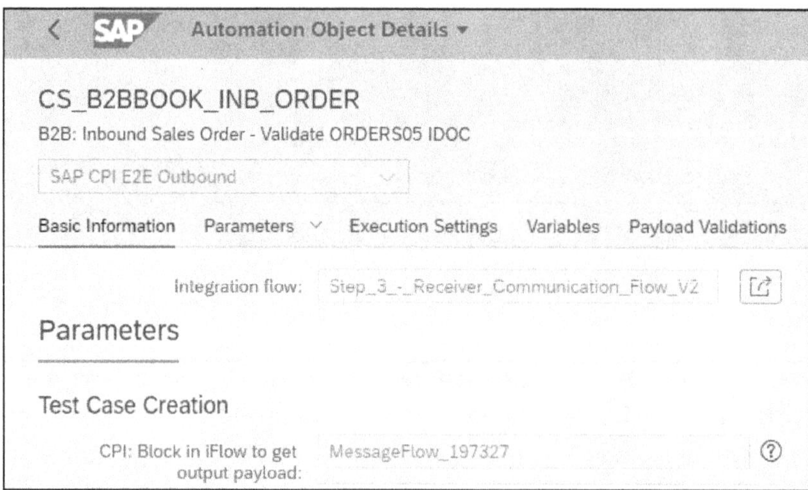

Figure 11.9 Automation Object Details Used as Output Automation Object

> **Note**
>
> To get the block in the iFlow for the output payload, you can use the search help directly in the Automation Objects app, or you can open the iFlow in Cloud Integration and take steps from there after you visualize and confirm the best iFlow block to make sure you get the results that you want.

Now let's go back to the initial automation object we created to inject the message and add the details of our second automation object in the **Output Automation Object** field (see Figure 11.10).

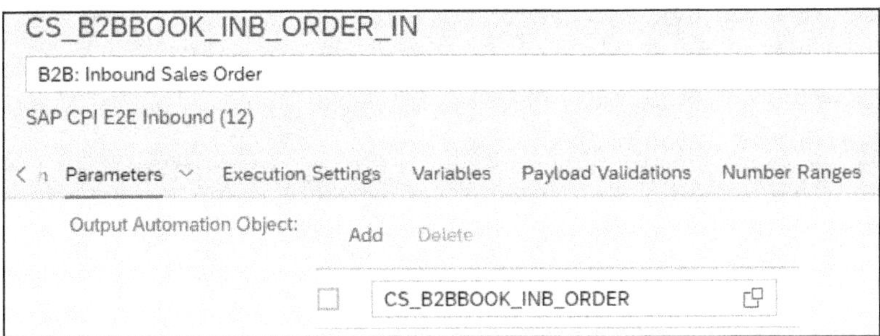

Figure 11.10 Output Automation Object to Get Details of Subsequent iFlow Output

We can create test cases based on the existing messages with the automation objects created and properly configured. To do this, we must go to the SAP Fiori home page for Int4 Suite and then click the Int4 Suite **API Tester Cockpit** tile.

We have to select the folder for our scenario, click the **Edit** button, and add a new entry. Provide the test case description, select **SAP CPI E2E Inbound** as the **Type**, select **CS_B2BBOOK_INB_ORDER_IN** as the **Automation Object** (the automation object used for message injection), and the set the **Origin** environment to your SAP Integration Suite (in our case, **IS**). This configuration is shown in Figure 11.11.

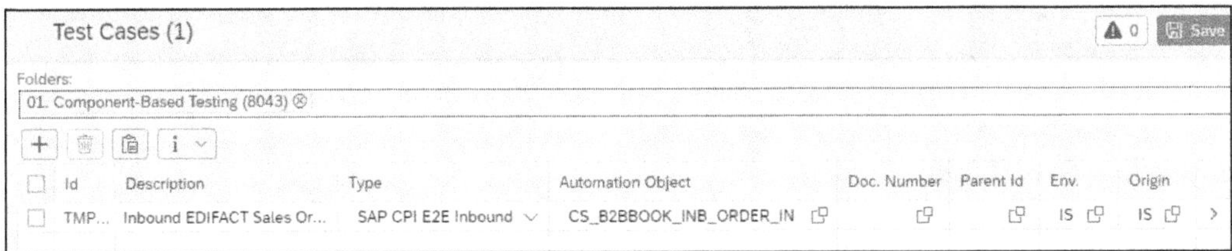

Figure 11.11 New Test Case Creation in Int4 Suite API Tester Cockpit

Next, click **Search Help** in the **Doc. Number** cell to select the relevant message from Cloud Integration. In our scenario, we are selecting an already processed message generated during the manual testing in Chapter 10. Choose the relevant message and then click the **Save** button (see Figure 11.12).

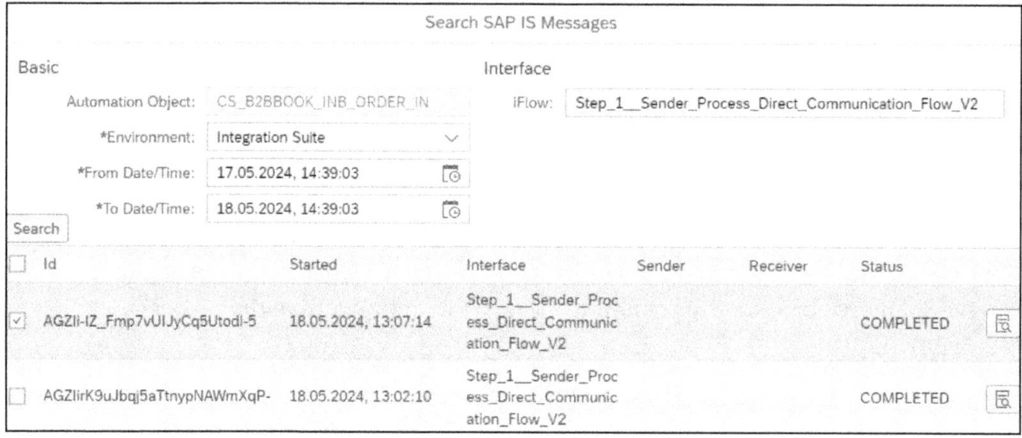

Figure 11.12 SAP Integration Suite Message Selector

> **Note**
>
> Currently, the prerequisite for saving the input and output iFlows payload is enabling the trace on the processed message; otherwise, the payload in Cloud Integration blocks cannot be read (the message processing logs [MPLs] will be empty). As mentioned earlier, for the B2B scenarios, this might be subject to change once SAP releases an API to get those details.

Once the test case is successfully saved, you can click the details to check whether the input and output payload were saved correctly. As a reminder, in our case, the input message was saved from **Step_1b_-_Write_message_to_message_queue**, whereas the output message was from the **Step_3_-_Receiver_Communication_Flow_V2** iFlow (see Figure 11.13).

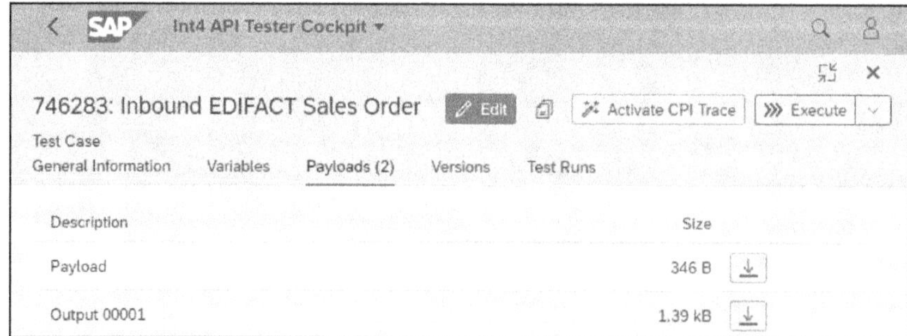

Figure 11.13 Details of Test Case with Listed Input and Output Payloads in Int4 Suite API Tester Cockpit

To ensure that the payloads are correct, click **Output 00001** and confirm if the output payload is the ORDERS05 XML IDoc sent to the SAP S/4HANA system. Sample XML is shown in Figure 11.14.

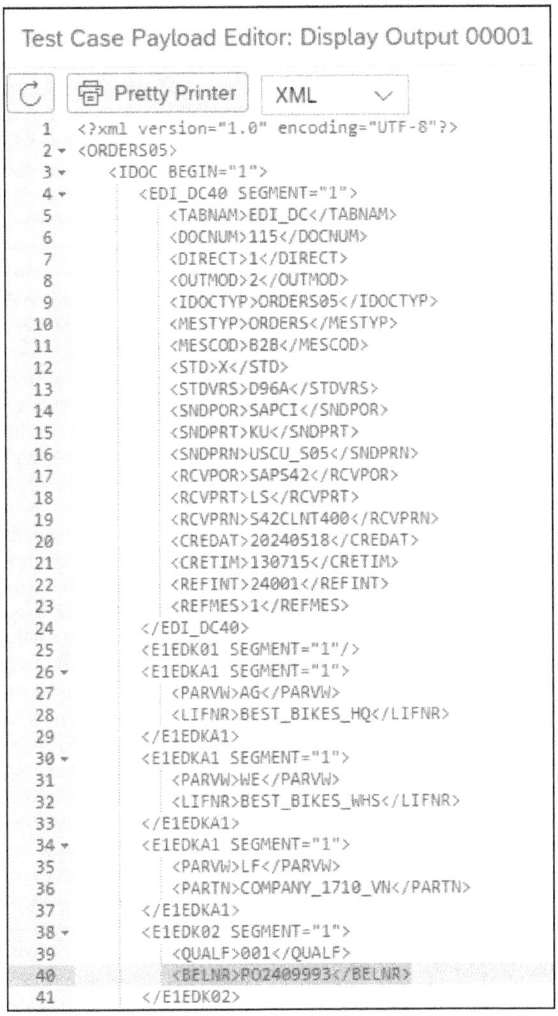

Test Case Payload Editor: Display Output 00001

Pretty Printer | XML


```xml
<?xml version="1.0" encoding="UTF-8"?>
<ORDERS05>
    <IDOC BEGIN="1">
        <EDI_DC40 SEGMENT="1">
            <TABNAM>EDI_DC</TABNAM>
            <DOCNUM>115</DOCNUM>
            <DIRECT>1</DIRECT>
            <OUTMOD>2</OUTMOD>
            <IDOCTYP>ORDERS05</IDOCTYP>
            <MESTYP>ORDERS</MESTYP>
            <MESCOD>B2B</MESCOD>
            <STD>X</STD>
            <STDVRS>D96A</STDVRS>
            <SNDPOR>SAPCI</SNDPOR>
            <SNDPRT>KU</SNDPRT>
            <SNDPRN>USCU_S05</SNDPRN>
            <RCVPOR>SAPS42</RCVPOR>
            <RCVPRT>LS</RCVPRT>
            <RCVPRN>S42CLNT400</RCVPRN>
            <CREDAT>20240518</CREDAT>
            <CRETIM>130715</CRETIM>
            <REFINT>24001</REFINT>
            <REFMES>1</REFMES>
        </EDI_DC40>
        <E1EDK01 SEGMENT="1"/>
        <E1EDKA1 SEGMENT="1">
            <PARVW>AG</PARVW>
            <LIFNR>BEST_BIKES_HQ</LIFNR>
        </E1EDKA1>
        <E1EDKA1 SEGMENT="1">
            <PARVW>WE</PARVW>
            <LIFNR>BEST_BIKES_WHS</LIFNR>
        </E1EDKA1>
        <E1EDKA1 SEGMENT="1">
            <PARVW>LF</PARVW>
            <PARTN>COMPANY_1710_VN</PARTN>
        </E1EDKA1>
        <E1EDK02 SEGMENT="1">
            <QUALF>001</QUALF>
            <BELNR>PO2409993</BELNR>
        </E1EDK02>
```


Figure 11.14 Output Payload: EDI File Mapped to ORDERS05 IDoc

Before beginning the simulation, we must consider ensuring that messages are processed and posted correctly in the SAP S/4HANA system. Sales orders typically have a duplicate check mechanism to validate whether an IDoc with the given purchase order number has been processed in the past. This means we need to modify the SAP purchase order each time we want to simulate a message to ensure that our test cases are reusable.

We can easily do this by introducing the variables in our automation object. We will create a PO variable, which will change the value of the purchase order number in the EDI file whenever we send a new message. We will also use the Int4 Suite number ranges in the automation objects to ensure the SAP purchase order number is incremented continuously without any manual effort with each test case execution.

To do this, open the first automation object we created for the message simulation, go to **Edit** mode, and click the **Number Ranges** Section. Click the **Create** button and provide details of the number range:

- **Number Range**
  B2B_NR_001 is the name of the number range we will use in the variable in the next step.
- **Prefix**
  POINT4 will be the prefix for all purchase order number values simulated by Int4 Suite.
- **Low Value**
  The number range start value is 1.
- **High Value**
  The number range end value is 999.
- **Add Zeros**
  If checked, this will add leading zeros to the value of the number range.

The configuration of the number range is shown in Figure 11.15.

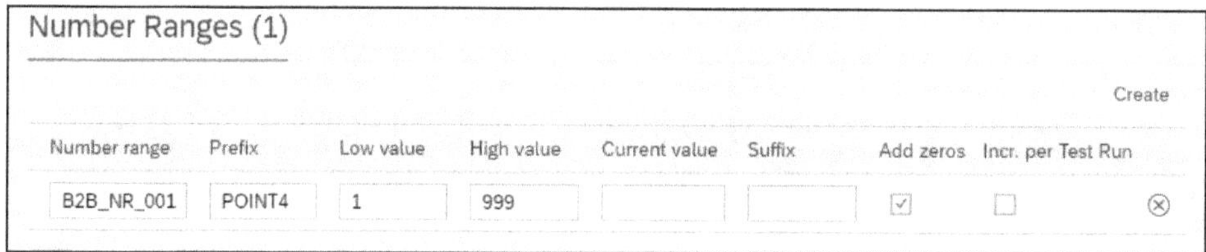

Figure 11.15 Number Range for Purchase Order in Automation Object

Now go to the **Variables** section. Create new variable with the following details:

- **Name: PO**
- **Description: Sap Purchase Order Number**
- **Type: Read & Replace**
- **Scope: Test Case**

The creation of the new variable is shown in Figure 11.16.

Variables (1)

Create

Name	Description	Type	Scope
PO	SAP Purchase Order Number	Read & Replace	Test Case

Payload Validations

Figure 11.16 Variable PO to Read and Replace Value of EDI Purchase Order Number

Now click the pencil icon (**Edit**) to jump into the variable's edit mode. In the new window, provide the following details:

- **Read action**
  Read Variable from Saved Message will assign the original value from the test case input message to the variable.
- **Expression**
  REGEX(BGM\+[0-9]{3}\+([A-Z0-9]+)) will read the purchase order number from the BGM segment in the EDIFACT file. From the BGM+220+P00000002+9' segment, value P00000002 will be extracted.
- **Replace action**
  Generate Value from Number Range will replace the original value fetched from the read action with the new value defined in the number range.
- **Number Range**
  Choose the **B2B_NR_001** range created in the previous step.

11

Note

To learn more about variable processing in Int4 Suite, read the **Variables & Variable** processing section in the Int4 Suite wiki at *http://s-prs.co/v573607.*

Once you save, you can provide the sample test case ID that saved before in the **Preview** section and click **Refresh**. The preview will confirm if the provided expression and actions work as expected. Figure 11.17 shows the complete **Edit Variable** window with all parameters mentioned previously.

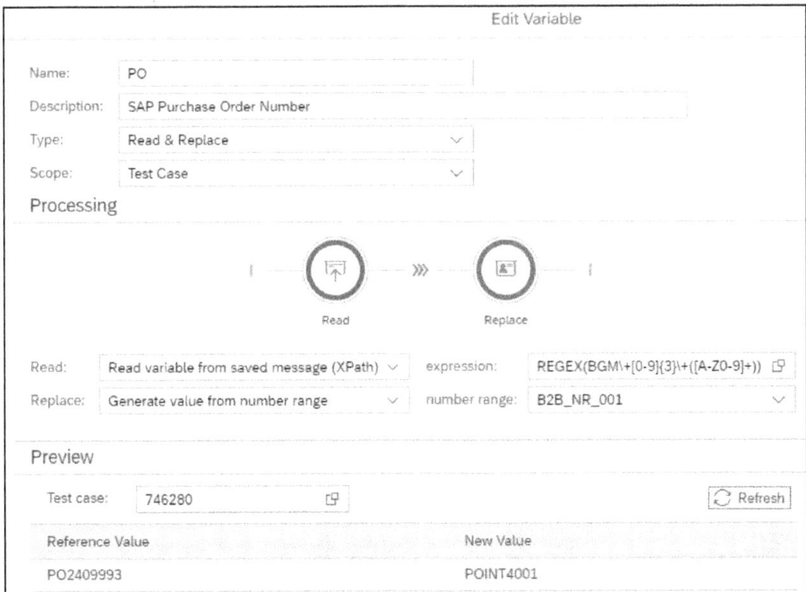

Figure 11.17 Variable Edit Screen with Details from Sample Test Case in Preview Section

Because we also want to read the details of the sales order by selecting data from the relevant entries in SAP tables, we have to define the details of the backend system and SAP tables we want to validate. In the automation object, jump to the **Backend Systems** and **Database Validations** section as shown in Figure 11.18. Fill in the following fields:

- **Backend system line for current doc**
  The **S4** system line will provide the details of the RFC connectivity to the system where the test cases will be executed (test system). In our scenario, this is the DEV SAP S/4HANA system.
- **Backend system line for reference doc**
  The **S4** system line will provide the details of the RFC connectivity to the system from which the test cases were stored (source system). In our scenario, this is a DEV SAP S/4HANA system.
- **Database validation ruleset**
  GENERIC_SO is the sample database validation ruleset delivered with the Int4 Suite add-on. You can create your database validation ruleset where you can define your tables, fields, and rules for selecting the data.

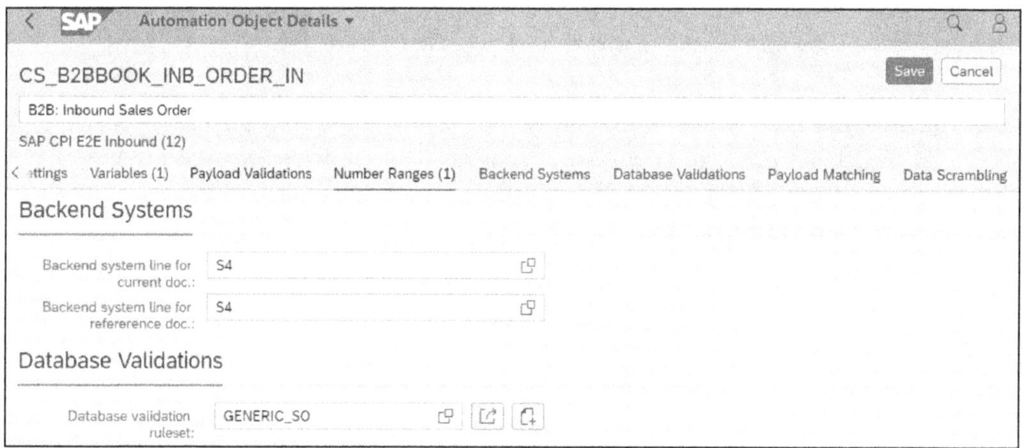

Figure 11.18 Backend Systems and Database Validation Section in Int4 Suite Automation Object

The database validation rulesets contain information about the tables that will be validated, including selection criteria, comparison details, joins, sorting criteria, matching criteria, and other rules. In our example, data from the SAP tables VBAK, VBAP, VBPA, and VBKD will be selected based on the BSTNK field in table VBAK, which should equal the value from the PO variable defined earlier.

With this setup, the reference sales order will be selected using the original purchase order value from the test case we saved, and the tested sales order will be chosen using the test value generated by the Int4 Suite variable based on the number range we created (see Figure 11.19).

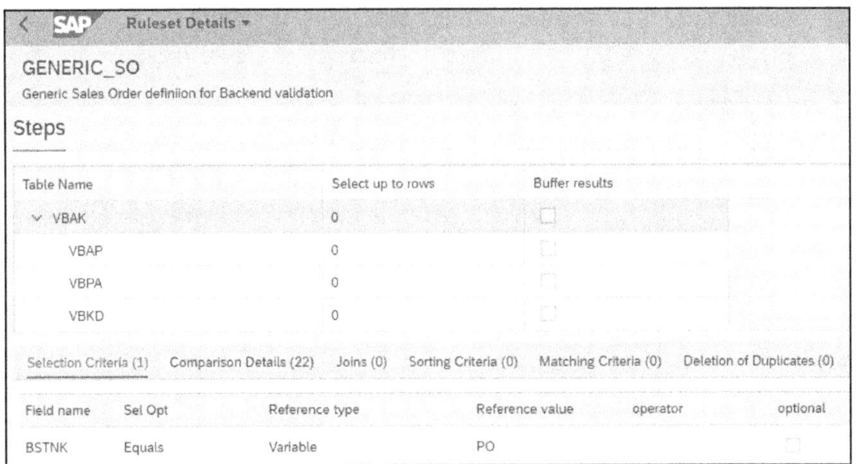

Figure 11.19 Sample Int4 Suite Ruleset Details Object for Sales Order

With the configuration done, we can go back to the Int4 Suite API Tester Cockpit, select the test case, and click the **Execute** button (see Figure 11.20).

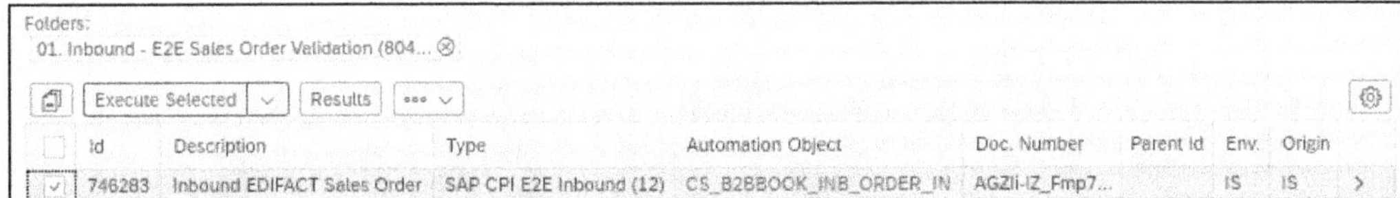

Figure 11.20 Int4 Suite API Tester Cockpit with First Test Case Creation

This will inject the message into the SAP Integration Suite into **Step_1b_-_Write_message_to_message_queue**, and subsequent processing steps will be executed. You can see in Cloud Integration that the injected message simulated the whole process (see Figure 11.21).

Figure 11.21 Tested Messages Simulated by Int4 Suite in Cloud Integration Monitoring

After you click to execute the test case, once the processing is completed, the test case report will be displayed, showing the status of the testing (see Figure 11.22).

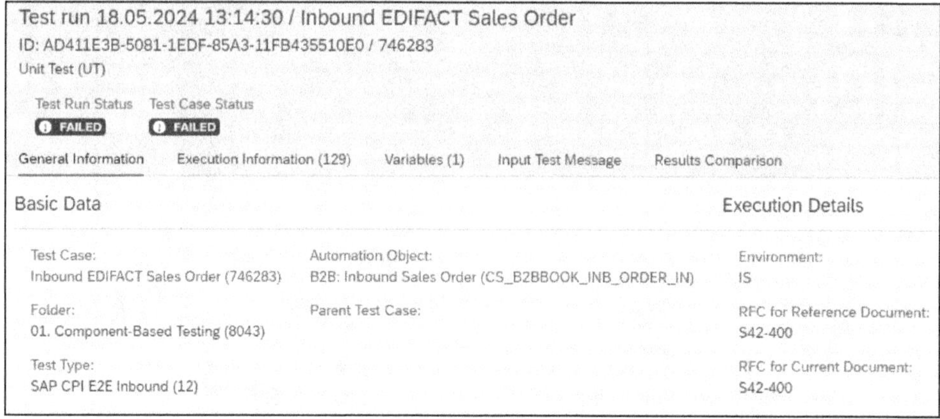

Figure 11.22 Test Run Status with General Information and Basic Data

You can scroll down to see why the test cases failed. First, the validation of the Cloud Integration output message is displayed. On the left-hand side, you can see the original ORDERS05 XML from the original message. On the right-hand side, you will see the tested ORDERS05 IDoc generated by the tested input message injected while executing the test case. Figure 11.23 shows the results of the XML comparison.

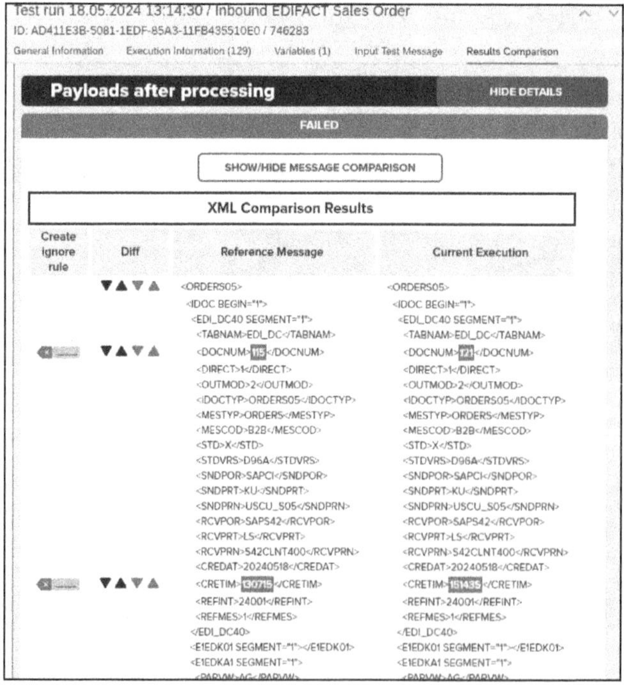

Figure 11.23 Original and Tested ORDERS05 Output Detailed Comparison

You can see that Int4 Suite noticed the differences in the following fields:

- `EDI_DC40-DOCNUM`
  IDoc number: The original IDoc number was *115* and the currently created number is *121*. This value will be naturally unique every time the test cases are executed.
- `EDI_DC-CRETIM`
  Creation time: This value will also be dynamically created based on the current time.
- `E1EDK02-BELNR`
  Purchase order number: Due to the rules we set in the automation object, we also expect that a different purchase order number will be generated with each test case execution.

The differences shown are false negative examples that we should mark as exceptions: we expect them to be different and do not want them to fail the test case. To do this, click the **Create Ignore Rule** icon next to each of the three errors. This will allow you to add warnings to these differences in the automation object, and these rules will be considered with all the other test case executions. Add the rules for all the above-mentioned differences, provide the descriptions, validate the XPath expressions proposed by Int4 Suite, and click the **Create** button (see Figure 11.24).

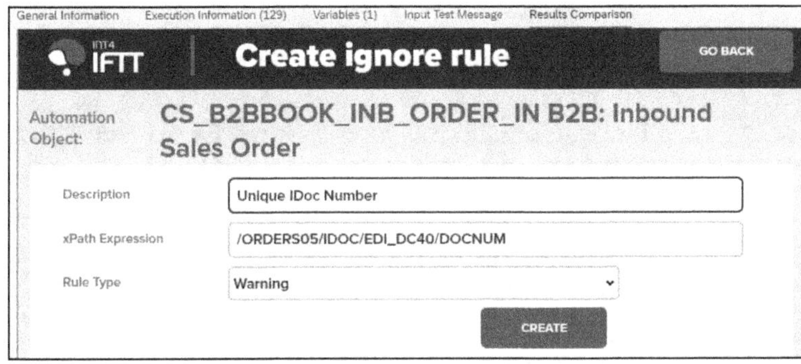

Figure 11.24 Exception Rule Creation Screen for IDoc Number in Test Case Result Report

After adding the exceptions, you can immediately see that all the rules have been added to the automation object (see Figure 11.25).

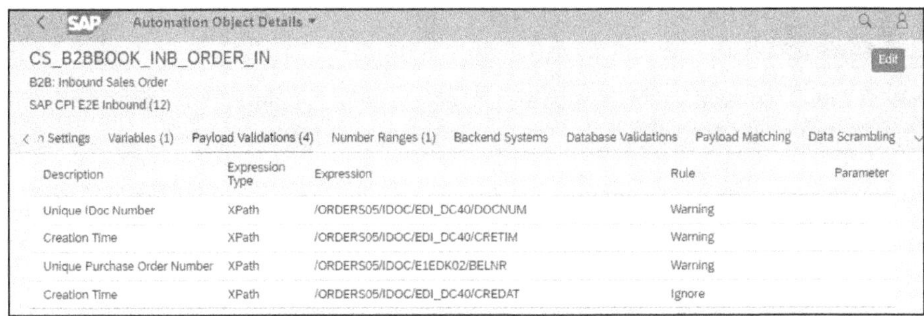

Figure 11.25 Updated Automation Object with Details of Payload Validation Rules

This is the end-to-end validation of the SAP Integration Suite processing, but we also want to see how the sales order was posted. For this purpose, we can scroll through the Int4 Suite test execution report to the **SAP Backend Document Validation** section. Here we will see all the SAP tables for which Int4 Suite found relevant entries for the source and test sales orders.

In the first row of each table line, you can see the original sales order, which Int4 Suite always compares to a newly tested sales order. In the second row, you see the sales order posted in the test message. You can see all the differences between the original and tested documents. Similar to the payload validation rulesets, you can add specific rules to each validated field. As you can see in Figure 11.26, fields **Sales Doc.** (VBELN; sales document number) and **Cust. Ref.** (BSTNK; customer reference) from the **Sales Document: Header Data** table VBAK have the ignore rule assigned so that differences in these fields don't fail the test case because we expect them to be different. All the other fields without such rules will be marked in red in case any discrepancies indicate a processing error.

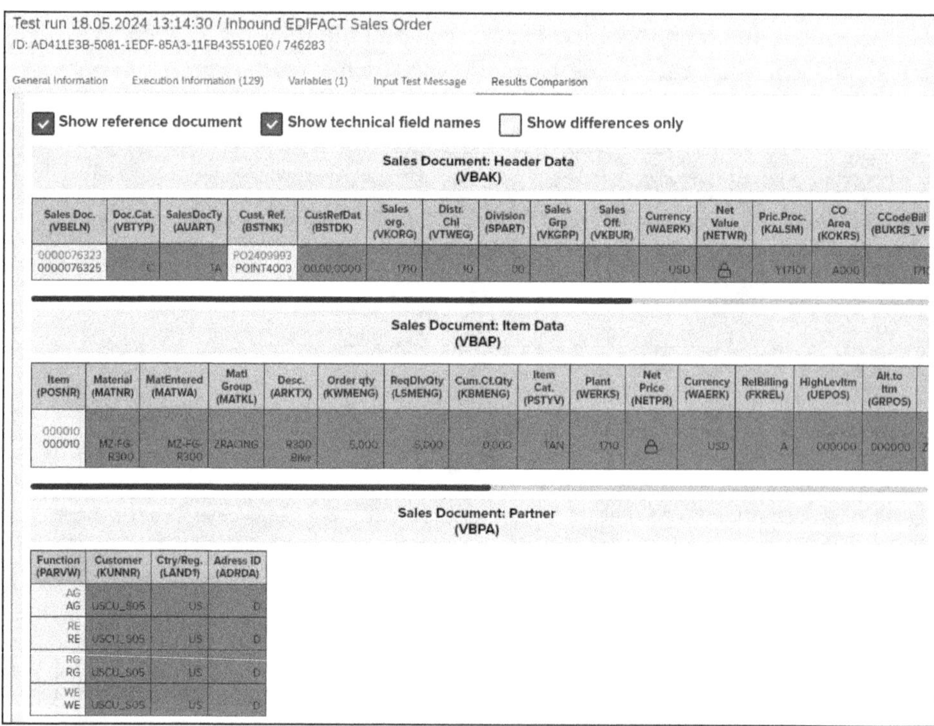

Figure 11.26 SAP Document Table Comparison in Int4 Suite Test Report Displayed Based on One-Time Automation Object Setup

We conduct this validation to ensure that all components of the SAP S/4HANA system are functioning correctly. We validate partner profiles, SAP customizing, SAP Application Interface Framework mappings, and rules, as these aspects impact the sales order creation. The advantage of this validation process is that we don't need to know the exact details of which parts of the code or customizing to validate. We assume that the final posting of the sales order and the detailed validation of the sales order will check all the components used in this process.

After we have added our rulesets, we can revalidate our execution to make sure that the payload validation exceptions we added are working fine. We can either execute the test case again or use the revalidation option (which does not inject the message but revalidates the old test case execution against newly added rules; this is beneficial at the beginning, when we're still working on test automation in Int4 Suite). You can see the two options in the top-right corner in Figure 11.27.

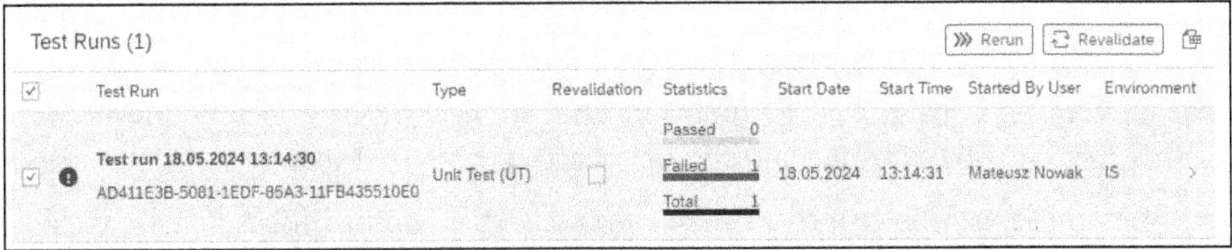

Figure 11.27 Past Test Run Details with Revalidation and Rerun Options

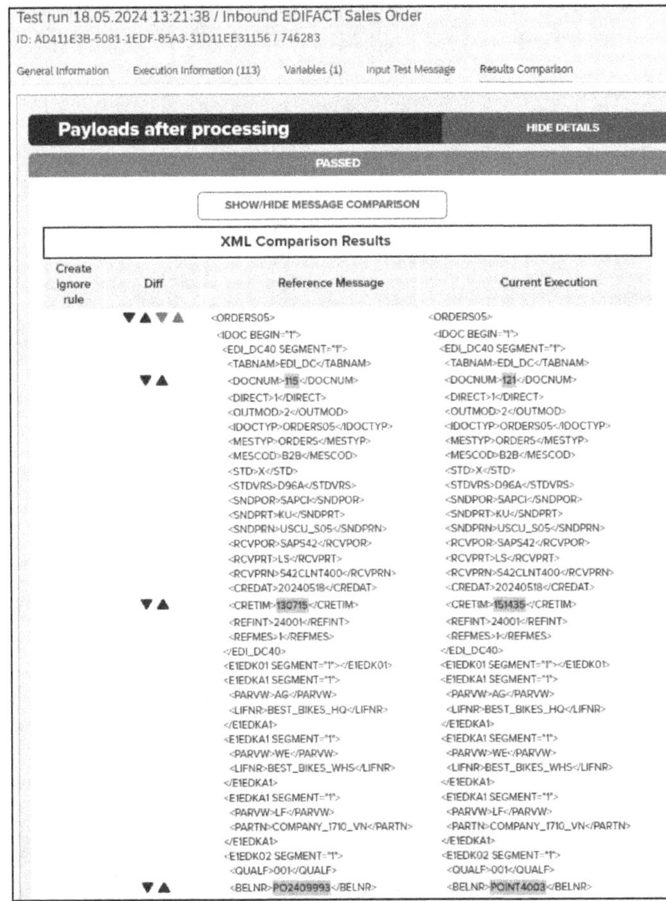

Figure 11.28 Details of Revalidated Test Case with Payload Validation Ignore Rules

As expected, the differences in the fields we added as exceptions are now shown as different, but they do not fail the test case and are marked in yellow, as shown in Figure 11.28.

You can reuse the created automation object for other messages in this process. In most cases, the automation object creation is a one-time process, and then we can reuse it to test the same flow in SAP Integration Suite and SAP S/4HANA with hundreds or even thousands of test cases. We can also retest the same messages over and over again to make sure that other changes done in the environment do not affect the working scenarios.

To show you how to leverage real regression testing, let's imagine a situation in which you have multiple EDI partners and you want to make sure that the changes you introduce in Trading Partner Management, in SAP Application Interface Framework, or directly in your iFlows for one partner do not affect other partners.

For this purpose, we will create some new test cases. Before doing this, let's create an additional variable in our automation object to identify the IDoc sender partner. You can create variable `SNDPRN` and assign the action to read the value from the `EDI_DC40-SNDPRN` field from the output message (after mapping). We will use this variable to describe the test cases automatically based on the content of the variable (see Figure 11.29).

Figure 11.29 Read Action for IDoc Sender Partner in Automation Object Variable

Now, in the Int4 Suite API Tester Cockpit, you can choose some other messages from SAP Integration Suite like last time. This time, instead of a single message, we will take multiple messages, as shown in Figure 11.30.

Variables

SAP Purchase Order Number (PO):

SAP EDI IDoc Partner (SNDPRN):

Search

	Id	Started	Interface	Sender	Receiver	Status	
☑	AGZIspabRd9lBdhY7Ae5YvKxxpql	18.05.2024, 15:52:22	Step_1__Sender_Process_Direct_Communication_Flow_V2			COMPLETED	
☑	AGZIspblEAkIGxVfOCGhU6LPZgT1	18.05.2024, 15:52:22	Step_1__Sender_Process_Direct_Communication_Flow_V2			COMPLETED	
☑	AGZIspYiERJmwodO1INtHPvyJrHc	18.05.2024, 15:52:22	Step_1__Sender_Process_Direct_Communication_Flow_V2			COMPLETED	
☑	AGZIspWkGFsaFcyl5ICm4tjEpuTP	18.05.2024, 15:52:21	Step_1__Sender_Process_Direct_Communication_Flow_V2			COMPLETED	
☑	AGZIspU5klP-ole6Iw9IRE8O_fiR	18.05.2024, 15:52:21	Step_1__Sender_Process_Direct_Communication_Flow_V2			COMPLETED	
☑	AGZIspQh1c13_WvQeDy93ndVu9lm	18.05.2024, 15:52:20	Step_1__Sender_Process_Direct_Communication_Flow_V2			COMPLETED	
☑	AGZIspRae0EytzwvHYri_BHt1I_R	18.05.2024, 15:52:20	Step_1__Sender_Process_Direct_Communication_Flow_V2			COMPLETED	
☑	AGZIspPTv1_h6CNBE7d2rySauHPO	18.05.2024, 15:52:19	Step_1__Sender_Process_Direct_Communication_Flow_V2			COMPLETED	
☑	AGZIspO3OSezObMzqLgm6G5xjKZr	18.05.2024, 15:52:19	Step_1__Sender_Process_Direct_Communication_Flow_V2			COMPLETED	

Figure 11.30 Multiple Messages Selection for Test Cases Collection

While creating the test cases in the description, we will add the following pattern: **Inbound EDIFACT Sales Order; Partner: [VAR_SNDPRN]; PO: [VAR_PO]**. With such a pattern, Int4 Suite will automatically replace **[VAR_SNDPRN]** and **[VAR_PO]** while creating test cases with the values directly coming from the source message, based on the variables we created. As shown in Figure 11.31, variable placeholders were replaced automatically with actual values. You will notice messages with different SAP purchase order numbers received by multiple partners.

Folders:
01. Inbound - E2E Sales Order Validation (804...

Execute All | Results

Id	Description	Type	Automation Object	Doc. Number	Parent Id	Env.	Origin
746283	Inbound EDIFACT Sales Order	SAP CPI E2E Inbound (12)	CS_B2BBOOK_INB_ORDER_IN	AGZli-IZ_Fmp7vUIJyCq5Utodi-5		IS	IS
746284	Inbound EDIFACT Sales Order; Partner: USCU_S10; PO: POINT4034	SAP CPI E2E Inbound (12)	CS_B2BBOOK_INB_ORDER_IN	AGZIshMyiE50IE6vGEQanWgXj8MX			IS
746285	Inbound EDIFACT Sales Order; Partner: USCU_S12; PO: POINT4043	SAP CPI E2E Inbound (12)	CS_B2BBOOK_INB_ORDER_IN	AGZIspabRd9lBdhY7Ae5YvKxxpql			IS
746286	Inbound EDIFACT Sales Order; Partner: USCU_S12; PO: POINT4042 #2	SAP CPI E2E Inbound (12)	CS_B2BBOOK_INB_ORDER_IN	AGZIspblEAkIGxVfOCGhU6LPZgT1			IS
746287	Inbound EDIFACT Sales Order; Partner: USCU_S05; PO: POINT4041 #3	SAP CPI E2E Inbound (12)	CS_B2BBOOK_INB_ORDER_IN	AGZIspYIERJmwodO1iNtHPvyJrHc			IS
746288	Inbound EDIFACT Sales Order; Partner: USCU_S05; PO: POINT4040 #4	SAP CPI E2E Inbound (12)	CS_B2BBOOK_INB_ORDER_IN	AGZIspWkGFsaFcyl5ICm4tjEpuTP			IS
746289	Inbound EDIFACT Sales Order; Partner: USCU_S92; PO: POINT4039 #5	SAP CPI E2E Inbound (12)	CS_B2BBOOK_INB_ORDER_IN	AGZIspU5kIP-ole6lw9IRE8O_fiR			IS
746290	Inbound EDIFACT Sales Order; Partner: USCU_S07; PO: POINT4038 #6	SAP CPI E2E Inbound (12)	CS_B2BBOOK_INB_ORDER_IN	AGZIspQh1c13_WvQeDy93ndVu9Im			IS
746291	Inbound EDIFACT Sales Order; Partner: USCU_S05; PO: POINT4037 #7	SAP CPI E2E Inbound (12)	CS_B2BBOOK_INB_ORDER_IN	AGZIspRae0EytzwvHYri_BHt1l_R			IS
746292	Inbound EDIFACT Sales Order; Partner: USCU_S05; PO: POINT4035 #8	SAP CPI E2E Inbound (12)	CS_B2BBOOK_INB_ORDER_IN	AGZIspPTv1_h6CNBE7d2rySauHPO			IS
746293	Inbound EDIFACT Sales Order; Partner: USCU_S05; PO: POINT4036 #9	SAP CPI E2E Inbound (12)	CS_B2BBOOK_INB_ORDER_IN	AGZIspO3OSezObMzqLgm6G5xjKZr			IS

Figure 11.31 Test Cases with Dynamically Created Test Case Descriptions Based on Content of Messages

Now, we could execute those test cases as before, or set up a daily regression run in which all 11 test cases for multiple partners could be executed daily. The latter approach will avoid manual test case execution and allow you to see the results of such testing every day, ensuring that your EDI scenarios are working seamlessly. Setting up such a scenario is not within the scope of this book, but you can see a sample report with such a daily execution in Figure 11.32, showing the overall status.

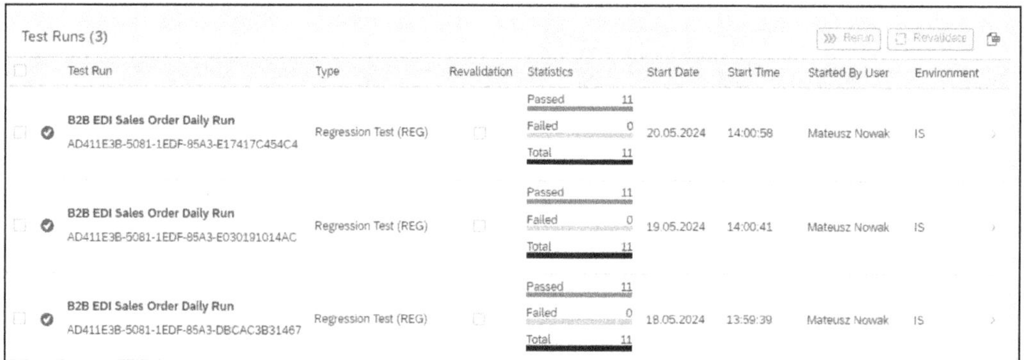

Test Runs (3)

Rerun | Revalidate

Test Run	Type	Revalidation	Statistics	Start Date	Start Time	Started By User	Environment
B2B EDI Sales Order Daily Run AD411E3B-5081-1EDF-85A3-E17417C454C4	Regression Test (REG)		Passed 11 Failed 0 Total 11	20.05.2024	14:00:58	Mateusz Nowak	IS
B2B EDI Sales Order Daily Run AD411E3B-5081-1EDF-85A3-E030191014AC	Regression Test (REG)		Passed 11 Failed 0 Total 11	19.05.2024	14:00:41	Mateusz Nowak	IS
B2B EDI Sales Order Daily Run AD411E3B-5081-1EDF-85A3-DBCAC3B31467	Regression Test (REG)		Passed 11 Failed 0 Total 11	18.05.2024	13:59:39	Mateusz Nowak	IS

Figure 11.32 Sample Report Showing Daily Test Run Statuses

In such scenarios, after selecting a specific run, you can see all the details about each test case status, as shown in Figure 11.33. You have the option to jump into the details of each test case execution run if needed—for instance, in the situation where out of 11 scheduled test cases, two fail.

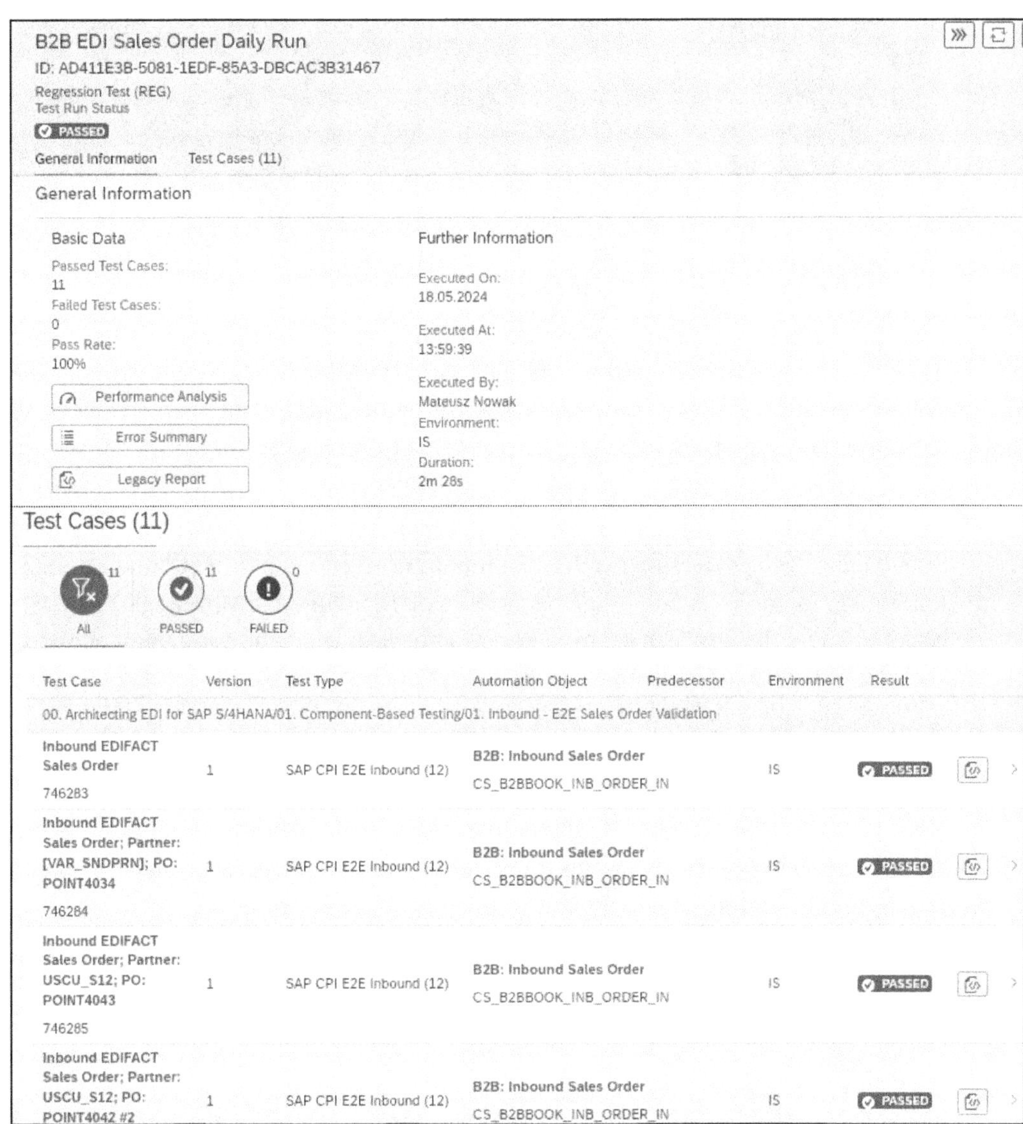

Figure 11.33 Regression Run with Details of All Executed Test Cases and Their Statuses

Outbound Scenario

In our EDI scenario, we have one inbound flow, for which we automated testing in the previous section, and three outbound flows: outbound orders response, outbound delivery, and outbound invoice.

This section will show you how to set up a sample test automation for the outbound EDI process based on one of our outbound flows: the outbound delivery message generated by the output control mechanism.

Note

Testing the EDI messages based on the output control mechanism is one of the options for how the Int4 Suite can automate testing. You can find more information how to test outbound interfaces on the Int4 home page at *https://int4.com*.

Before we proceed with the test automation process, let's explain the differences between the inbound and outbound scenarios that Int4 Suite can test.

For inbound messages, Int4 Suite's virtualization mechanism prepares and dispatches inbound messages to the tested system, simulating an EDI partner. Afterward, the message processing results are captured and cross-verified against predefined rules and reference data. This validation process allows us to assess the SAP Integration Suite processing and the creation of the business document in the SAP S/4HANA backend.

During outbound testing, we instruct the system under test to generate a message and then validate the processing results. It's important to clarify that Int4 Suite does not initiate any message for outbound messages; it does not inject any message. Instead, it either waits for the arrival of a message from a previous testing phase or explicitly triggers the system to send the message. In this type of test, our goal is to ensure that the message was generated and that it was done correctly.

In our example, Int4 Suite will use its capabilities to repeat existing output for the delivery created in Chapter 10. By doing so, the output repeat should generate exactly the same outbound DESADV.DELVRY07 IDoc, which will be our starting point. In the second step, Int4 Suite will validate whether the EDI ASN is adequately generated and sent to the EDI partner, validating the Cloud Integration setup.

First, we need to create the automation object that will trigger the output control message and then validate the relevant EDI message. To do this, we need the details of the output control that we can find either in table NAST or directly in the outbound delivery output, as shown in Figure 11.34.

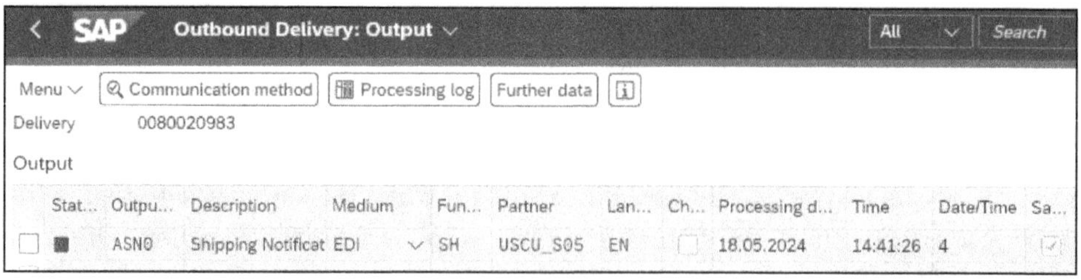

Outbound Delivery: Output

Menu | Communication method | Processing log | Further data

Delivery 0080020983

Output

Stat...	Outpu...	Description	Medium	Fun...	Partner	Lan...	Ch...	Processing d...	Time	Date/Time	Sa...
■	ASN0	Shipping Notificat	EDI	SH	USCU_S05	EN		18.05.2024	14:41:26	4	✓

Figure 11.34 Details of Outbound Delivery Output

We also need to define the iFlow in Cloud Integration and its block from which we will take the payload for the comparison. We will use **Step_2_-_Interchange_Processing_**

Flow_V2 and the **EndEvent_61790992** block that contains the EDI DESADV message (see Figure 11.35).

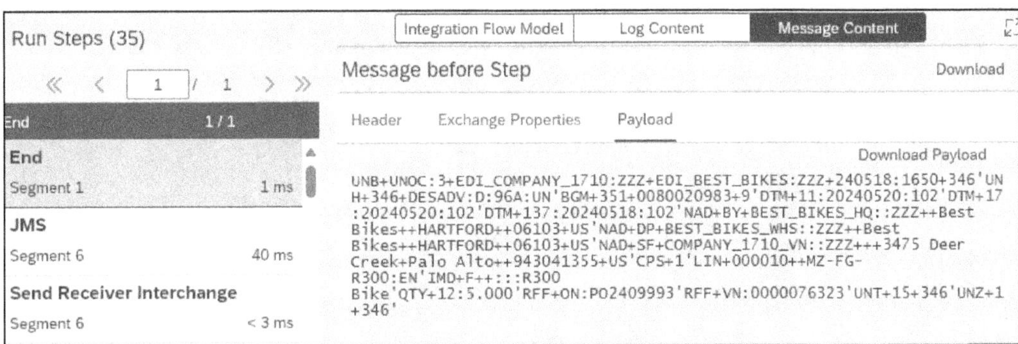

Figure 11.35 EndEvent_61790992 Block Containing EDI File Sent to EDI Partner after Mapping in Cloud Integration

This information allows us to create a new automation object in the Int4 Suite Automation Objects app with the SAP CPI E2E output type.

In our example, we will add **Step_2_-_Interchange_Processing_Flow_V2** to the iFlow and, in the **Test Cases Creation** section, add **EndEvent_61790992** as the block in the iFlow to get the output payload. Figure 11.36 shows the automation object tile with all these parameters entered.

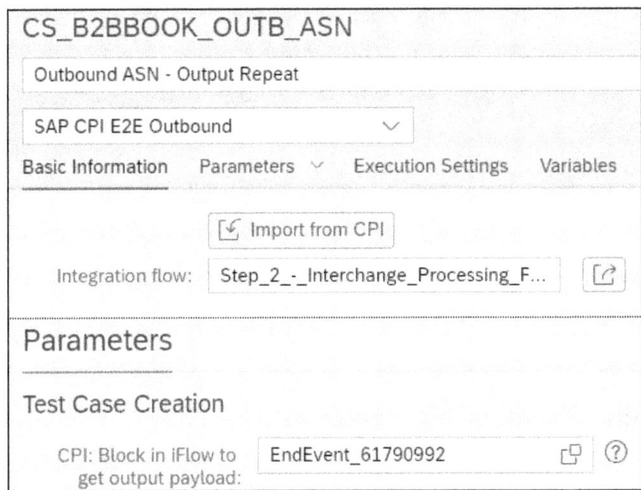

Figure 11.36 Basic Information and Test Case Creation Parameters in Automation Object

Next, in the **Test Case Execution** section, we have to choose the following options:

- **Trigger the output from message control (NAST)**
  When checked, this will indicate how Int4 Suite should trigger the message.

- **Flat Line Separator**
  Use the line separator used in our message—that is, the ' character.
- **Flat Files Compare with Payload Validation Applied**
  When checked, this indicates that the output payload we will compare is a flat file.

The configuration of the test case execution is shown in Figure 11.37.

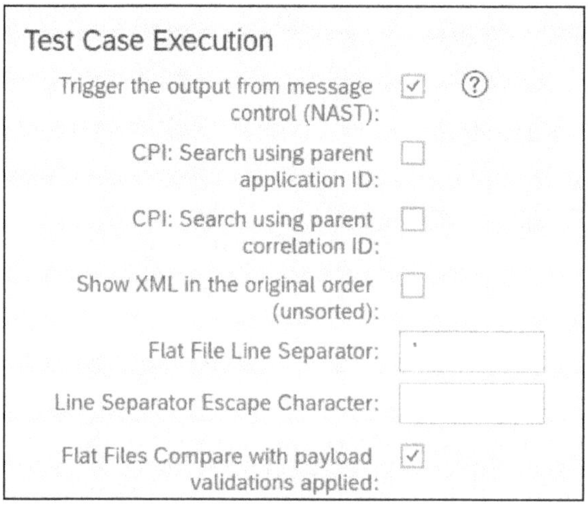

Test Case Execution
Trigger the output from message control (NAST): ☑
CPI: Search using parent application ID: ☐
CPI: Search using parent correlation ID: ☐
Show XML in the original order (unsorted): ☐
Flat File Line Separator: '
Line Separator Escape Character:
Flat Files Compare with payload validations applied: ☑

Figure 11.37 Test Case Execution Details in Automation Object

Because we will be triggering the output from the SAP S/4HANA system, we again have to add the details of the backend systems from which Int4 Suite will get the details of the RFC assigned to the tested and source environments during testing and test case creation, as shown in Figure 11.38.

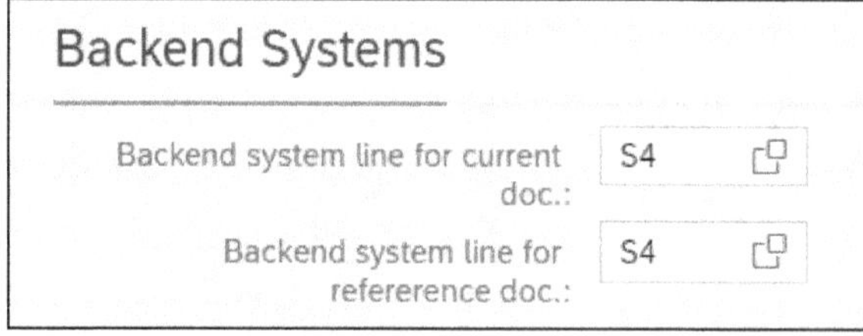

Figure 11.38 Backend System Information for Tested and Source Systems

As a next step, we must create the variables to let Int4 Suite repeat the proper output control. We do that in the **Variables** section of the automation object, where we have to create the following variables:

- **KAPPL**, **KSCHL**, and **NACHA** with **Custom Type** and **Test Case Scope**
- **OBJKY** with **Find Message Type** and **Test Case Scope**

The list of all created variables is shown in Figure 11.39.

Name	Description	Type	Scope	
KAPPL	Application	Custom	Test Case	>
KSCHL	Output Type	Custom	Test Case	>
NACHA	Tr. Medium	Custom	Test Case	>
OBJKY	Document Number	Find message	Test Case	>

Figure 11.39 Variables Created to Repeat Message Using Output Control

Variables KAPPL, KSCHL, and NACHA should have only one action: **Read: Constant**, as shown in Figure 11.40 for KAPPL; the details will be the same for KSCHL and NACHA.

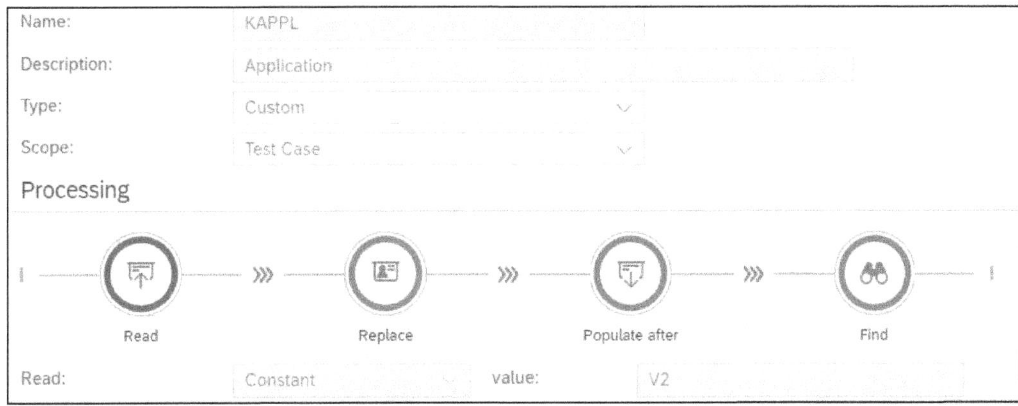

Figure 11.40 Details for KAPPL Variable

In the value for KAPPL, NACHA, and KSCHL, set the following constants:

- **KAPPL: V2**
- **NACHA: 6**
- **KSCHL: ASNO**

These values come directly from the previously mentioned output control details.

For OBJKY, set two actions:

- **Read variable from saved message** with **Expression**: **REGEX(BGM\+[0-9]{3}\+([0-9]+))**. This will read the delivery number from the EDI file and assign it to the OBJKY variable.
- **Find using XPath** with **Expression**: **REGEX(BGM\+[0-9]{3}\+([0-9]+))**. This action will check whether there is a message with the delivery number saved from the previous action in all the found messages from the tested iFlow. Thanks to this, we can make

sure that the proper messages are compared with themselves in situations where multiple outputs are generated at the same time for the same iFlow.

The processing details related to variable `OBJKY` are shown in Figure 11.41.

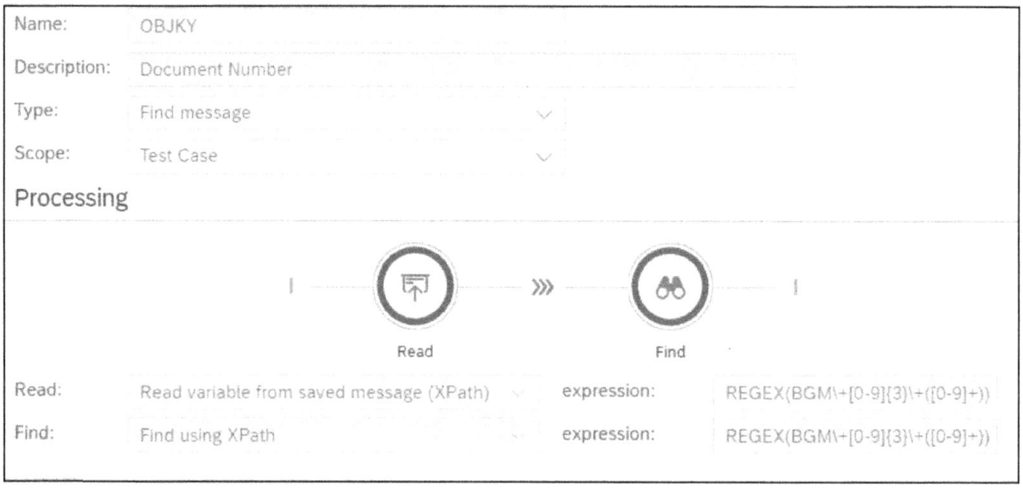

Name: OBJKY
Description: Document Number
Type: Find message
Scope: Test Case
Processing
Read
Find
Read: Read variable from saved message (XPath) expression: REGEX(BGM\+[0-9]{3}\+([0-9]+))
Find: Find using XPath expression: REGEX(BGM\+[0-9]{3}\+([0-9]+))

Figure 11.41 Variable Processing Details for Document Number Variable Used to Find Relevant Message in Cloud Integration Based on REGEX Expression

Note

A detailed explanation of all the options to repeat messages from output control is available in the Int4 Suite wiki at *http://s-prs.co/v573608*.

We also know that in the EDI file, some segments will either be different when we trigger a new message or depend on the creation date and time. This is why this time, we will jump immediately to the **Payload Validation** section to add warning rules:

- **UNB Date and time of preparation and interchange ID** with **Flat File Expression: REGEX(\d{6}:\d{4}\+\d+)**
- **UNH Message ref number** with **Flat File Expression: REGEX(UNH\+([0-9]{3})+)**
- **UNH Message ref number** with **Flat File Expression: REGEX(UNT\+[0-9]+\+([0-9]+))**
- **UNZ Interchange ID** with **Flat File Expression: REGEX(UNZ\+[0-9]+\+([0-9]+))**

The configuration of payload validations and related expressions is shown in Figure 11.42.

Payload Validations (4)

Create

Description	Expression Type	Expression	Rule	Parameter	
UNB Date and time of pr...	Flat...	REGEX(\d{6}:\d{4}\+\d+)	Warning		⊗
UNH Message ref number	Flat...	REGEX(UNH\+([0-9]{3})+)	Warning		⊗
UNT Message ref number	Flat...	REGEX(UNT\+[0-9]+\+([0-9]+))	Warning		⊗
UNZ interchange ID	Flat...	REGEX(UNZ\+[0-9]+\+([0-9]+))	Warning		⊗

Figure 11.42 Payload Validations Section in Int4 Suite Automation Object

We can enter the Int4 Suite API Tester Cockpit to create a new test case with the reference EDI DESADV triggered during manual validation in Chapter 10. Create a new entry and provide the test case details:

- **Description**
  Outbound Delivery: [VAR_OBJKY] will create a test case with the delivery number taken from the output.
- **Type**
  Set **SAP CPI E2E Outbound**.
- **Automation Object**
  CS_B2BBOOK_OUTB_ASN references teh automation object details we created for the outbound scenario.
- **Env. and Origin Env**
  Set **IS**.
- **Doc. Number**
  Use the message selector to find the relevant message in Cloud Integration.

The newly created test case is shown in Figure 11.43.

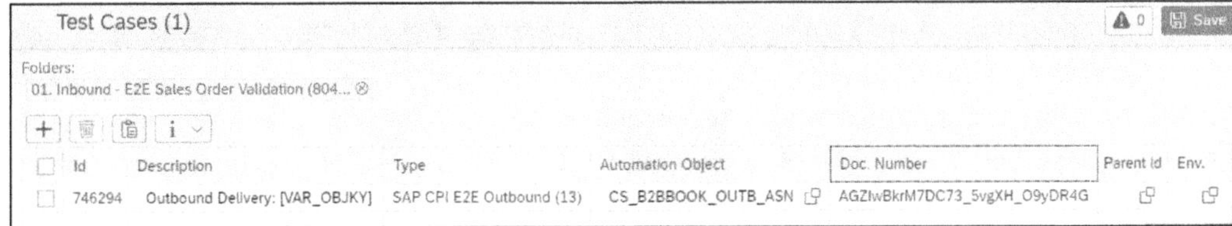

Figure 11.43 Test Case Creation in Int4 Suite API Tester Cockpit for Outbound Scenario

Execute the saved test case and check the overall status as shown in Figure 11.44.

Test run 18.05.2024 15:44:06
ID: AD411E3B-5081-1EDF-85A5-AEC0B333DA8B
Unit Test (UT)
Test Run Status
PASSED

General Information | Test Cases (1)

General Information

Basic Data
Passed Test Cases: 1
Failed Test Cases: 0
Pass Rate: 100%
Performance Analysis
Error Summary
Legacy Report

Further Information
Executed On: 18.05.2024
Executed At: 15:44:07
Executed By: Mateusz Nowak
Environment: IS
Duration: 4s

Test Cases (1)

All 1 | PASSED 1 | FAILED 0

Test Case	Version	Test Type	Automation Object	Predecessor	Environment	Result
00. Architecting EDI for SAP S/4HANA/01. Component-Based Testing/01. Inbound - E2E Sales Order Validation						
Outbound Delivery: 0080020983 746294	1	SAP CPI E2E Outbound (13)	**Outbound ASN - Output Repeat** CS_B2BBOOK_OUTB_ASN		IS	PASSED

Figure 11.44 Test Run Summary for Outbound EDI DESADV Message Validation

To see the comparison details, click the test case details. You can see the EDI DESADV message comparison, in which the flat files are compared differently, whereas each line is compared one by one. In our example, we can see that in the UNB segment, we marked some differences as exceptions because they show the creation date, creation time, and interchange ID. Similarly, the differences in the UNH, UNT, and UNZ segments are what we accepted in the automation object (see Figure 11.45).

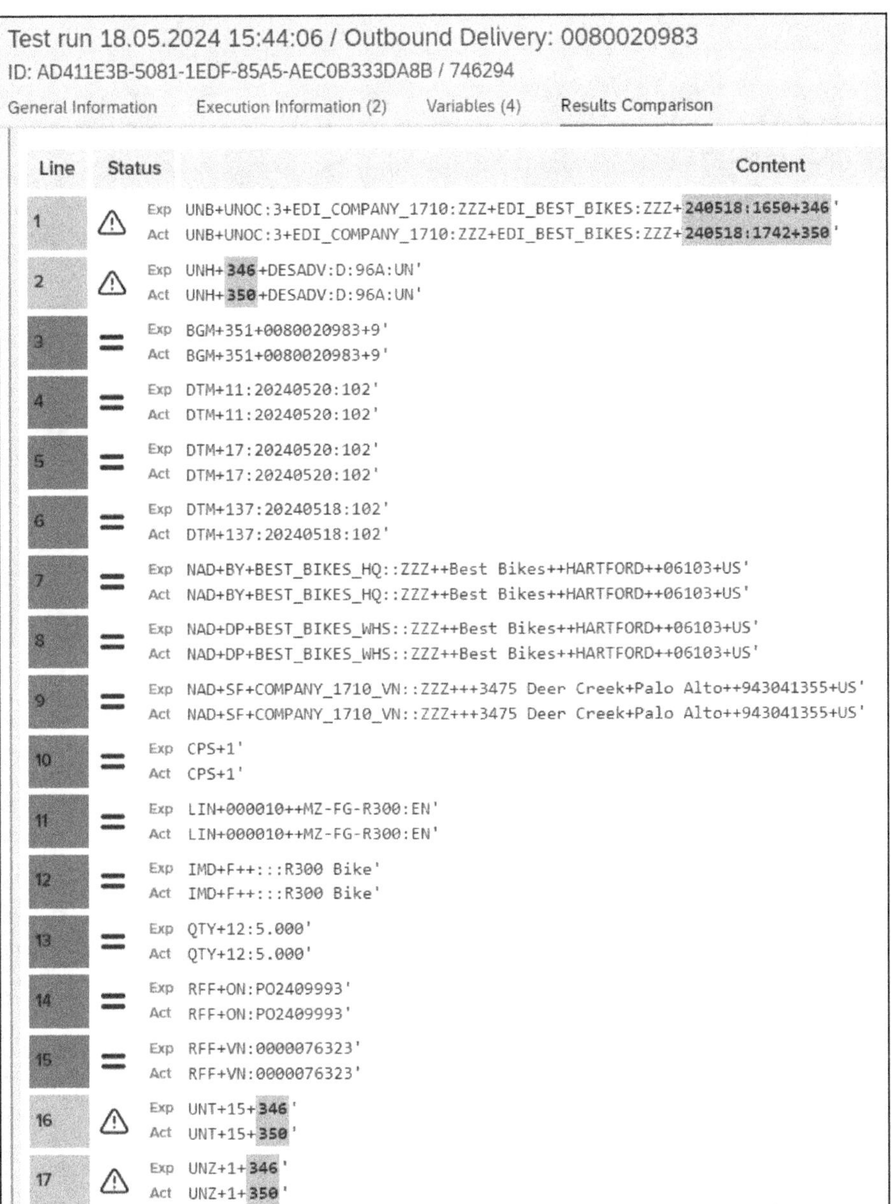

Figure 11.45 Detailed Comparison of Reference and Tested EDI Files

We can repeat the same test case multiple times, and each time we should see the new entry generated in the delivery output for the delivery number from our test case and the `OBJKY` variable. You can see this in Figure 11.46.

We should also see new messages in the B2B scenario monitor in SAP Integration Suite for the same delivery/purchase order number (see Figure 11.47).

Delivery 0080020983

Output

Stat...	Outpu...	Description	Medium	Fun...	Partner	Lan...	Ch...	Processing d...	Time	Date/Time	Sa...
■	ASN0	pping Notificat	EDI	SH	USCU_S05	EN		18.05.2024	15:44:10	4	✓
■	ASN0	Shipping Notificat	EDI	SH	USCU_S05	EN		18.05.2024	15:42:19	4	✓
■	ASN0	Shipping Notificat	EDI	SH	USCU_S05	EN		18.05.2024	15:40:17	4	✓
■	ASN0	Shipping Notificat	EDI	SH	USCU_S05	EN		18.05.2024	15:21:46	4	✓
■	ASN0	Shipping Notificat	EDI	SH	USCU_S05	EN		18.05.2024	15:20:00	4	✓
■	ASN0	Shipping Notificat	EDI	SH	USCU_S05	EN		18.05.2024	14:50:00	4	✓
■	ASN0	Shipping Notificat	EDI	SH	USCU_S05	EN		18.05.2024	14:41:26	4	✓

Figure 11.46 Repeated Outputs by Int4 Suite Visible in Delivery Header Output

Interchanges (10)

Status	Sender Name	Receiver Name	Sender Document Standard	Sender Message Type	CSA Purchase order ...
Completed	Company Code 1710	Best Bikes	SAP_IDoc	DESADV.DELVRY07	PO2409993
Completed	Company Code 1710	Best Bikes	SAP_IDoc	DESADV.DELVRY07	PO2409993
Completed	Company Code 1710	Best Bikes	SAP_IDoc	DESADV.DELVRY07	PO2409993
Completed	Company Code 1710	Best Bikes	SAP_IDoc	DESADV.DELVRY07	PO2409993
Completed	Company Code 1710	Best Bikes	SAP_IDoc	DESADV.DELVRY07	PO2409993
Completed	Company Code 1710	Best Bikes	SAP_IDoc	DESADV.DELVRY07	PO2409993
Completed	Company Code 1710	Best Bikes	SAP_IDoc	DESADV.DELVRY07	PO2409993
Completed	Company Code 1710	Best Bikes	SAP_IDoc	ORDRSP.ORDERS05	PO2409993
Completed	Company Code 1710	Best Bikes	SAP_IDoc	ORDRSP.ORDERS05	PO2409993
Completed	Best Bikes	Company Code 1710	UN-EDIFACT	ORDERS	PO2409993

Figure 11.47 List of DESADV Messages in Int4 Suite Cloud Integration Simulated and Tested by Int4 Suite

Similarly, as with the inbound sales order validated in the previous section, we could set up the regression run with different variations of the outbound deliveries from different EDI partners to ensure all of our scenarios are working fine.

With the option to execute the test cases continuously in the background, you can set up different regression runs depending on your needs. You could combine them by direction, interface, or all the scenarios belonging to a given EDI partner.

With the component-based testing, we can run testing in isolation and test each component of our EDI setup independently. This can make testing more reliable and easier to maintain.

Automated Test Cases Creation and Test Cases Analysis

In the previous sections, we focused on manual test case creation. In this section, we would like to show you at a high level an alternative: leveraging the Int4 Suite robotic crawler to create test cases and using test case analysis to analyze your test data coverage. Mass test case creation with the robotic crawler is helpful when you already have lots of reliable messages (e.g., from the production environment).

The robotic crawler can automatically create test cases based on those historical messages and categorize them according to the relevant scenario. It can continuously scan the Cloud Integration messages that are exchanged among your SAP system, EDI partners, and connected applications.

The robotic crawler can read messages sent by EDI partners and automatically save them in Int4 Suite as test cases, exactly like we did manually in previous sections. It can assign test cases to specific folders.

As you can see in Figure 11.48, in the robotic crawler, you can define the environment from which the robot will scan messages and create test cases. It will use the already created automation objects with details about from where the input and output messages should be taken. While executing the robot, you can define additional criteria like the connected application name, interface name, and all other details.

Figure 11.48 Test Case Creation for Cloud Integration Messages App: Robotic Crawler

You can also define how the test cases should be created and grouped into the folders and how they should be named. For example, you can put any value from the payload in the test case description. To identify it, you can set the sending partner's name, company code, and so on, as shown in Figure 11.49.

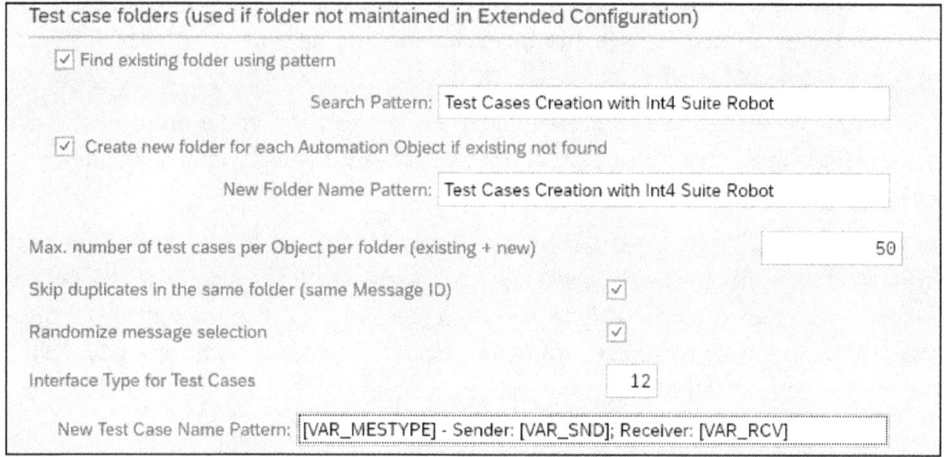

Figure 11.49 Definition of Test Cases Description Pattern and Folder Pattern in Robotic Crawler

Thanks to this, it will be easier to identify the test cases at a glance because they will contain essential information in the test case description about how to identify them from a business point of view, as shown in Figure 11.50.

Folders:
Test Cases Creation with Int4 Suite Robot (8060)

Execute All | Results

Id	Description	Type	Automation Object	Doc. Number
746368	ORDERS:D:93A - Sender: USCU_S08; Receiver: S42CLNT400	SAP CPI E2E Inbound (12)	CS_B2BBOOK_INB_ORDER_IN	AGZIi-IZ_Fmp7vUIJyCq5Utodi-5
746369	ORDERS:D:96A - Sender: USCU_S10; Receiver: S42CLNT400	SAP CPI E2E Inbound (12)	CS_B2BBOOK_INB_ORDER_IN	AGZIshMyiE50lE6vGEQanWgXj8MX
746370	ORDERS:D:96A - Sender: USCU_S05; Receiver: S42CLNT400	SAP CPI E2E Inbound (12)	CS_B2BBOOK_INB_ORDER_IN	AGZIspabRd9IBdhY7Ae5YvKxxpql
746371	ORDERS:D:96A - Sender: USCU_S05; Receiver: S42CLNT400	SAP CPI E2E Inbound (12)	CS_B2BBOOK_INB_ORDER_IN	AGZIspblEAkIGxVfOCGhU6LPZgT1
746372	ORDERS:D:96A - Sender: USCU_S10; Receiver: S42CLNT400	SAP CPI E2E Inbound (12)	CS_B2BBOOK_INB_ORDER_IN	AGZIspYIERJmwodO1iNtHPvyJrHc
746373	ORDERS:D:96A - Sender: USCU_S10; Receiver: S42CLNT400	SAP CPI E2E Inbound (12)	CS_B2BBOOK_INB_ORDER_IN	AGZIspWkGFsaFcyl5ICm4tjEpuTP
746374	ORDERS:D:93A - Sender: USCU_S08; Receiver: S42CLNT400	SAP CPI E2E Inbound (12)	CS_B2BBOOK_INB_ORDER_IN	AGZIspU5kIP-ole6lw9IRE8O_fiR
746375	ORDERS:D:96A - Sender: USCU_S05; Receiver: S42CLNT400	SAP CPI E2E Inbound (12)	CS_B2BBOOK_INB_ORDER_IN	AGZIspQh1c13_WvQeDy93ndVu9lm
746376	ORDERS:D:93A - Sender: USCU_S08; Receiver: S42CLNT400	SAP CPI E2E Inbound (12)	CS_B2BBOOK_INB_ORDER_IN	AGZIspRae0EytzwvHYri_BHt1l_R
746377	ORDERS:D:93A - Sender: USCU_S08; Receiver: S42CLNT400	SAP CPI E2E Inbound (12)	CS_B2BBOOK_INB_ORDER_IN	AGZIspPTv1_h6CNBE7d2rySauHPO
746378	ORDERS:D:93A - Sender: USCU_S08; Receiver: S42CLNT400	SAP CPI E2E Inbound (12)	CS_B2BBOOK_INB_ORDER_IN	AGZIspO3OSezObMzqLgm6G5xjKZr
746379	ORDERS:D:96A - Sender: USCU_S10; Receiver: S42CLNT400	SAP CPI E2E Inbound (12)	CS_B2BBOOK_INB_ORDER_IN	AGZNmhqupSgzko3HAt-AOiwdIbaP
746380	ORDERS:D:96A - Sender: USCU_S10; Receiver: S42CLNT400	SAP CPI E2E Inbound (12)	CS_B2BBOOK_INB_ORDER_IN	AGZNmhmhvd50iKJO61LMBmTX3MSX
746381	ORDERS:D:93A - Sender: USCU_S08; Receiver: S42CLNT400	SAP CPI E2E Inbound (12)	CS_B2BBOOK_INB_ORDER_IN	AGZIshMyiE50lE6vGEQanWgXj8MX
746382	ORDERS:D:93A - Sender: USCU_S08; Receiver: S42CLNT400	SAP CPI E2E Inbound (12)	CS_B2BBOOK_INB_ORDER_IN	AGZIspabRd9IBdhY7Ae5YvKxxpql
746383	ORDERS:D:93A - Sender: USCU_S08; Receiver: S42CLNT400	SAP CPI E2E Inbound (12)	CS_B2BBOOK_INB_ORDER_IN	AGZIspblEAkIGxVfOCGhU6LPZgT1
746384	ORDERS:D:96A - Sender: USCU_S10; Receiver: S42CLNT400	SAP CPI E2E Inbound (12)	CS_B2BBOOK_INB_ORDER_IN	AGZIspYIERJmwodO1iNtHPvyJrHc
746385	ORDERS:D:96A - Sender: USCU_S05; Receiver: S42CLNT400	SAP CPI E2E Inbound (12)	CS_B2BBOOK_INB_ORDER_IN	AGZIspWkGFsaFcyl5ICm4tjEpuTP
746386	ORDERS:D:96A - Sender: USCU_S10; Receiver: S42CLNT400	SAP CPI E2E Inbound (12)	CS_B2BBOOK_INB_ORDER_IN	AGZIspU5kIP-ole6lw9IRE8O_fiR
746387	ORDERS:D:96A - Sender: USCU_S10; Receiver: S42CLNT400	SAP CPI E2E Inbound (12)	CS_B2BBOOK_INB_ORDER_IN	AGZIspQh1c13_WvQeDy93ndVu9lm
746388	ORDERS:D:93A - Sender: USCU_S08; Receiver: S42CLNT400	SAP CPI E2E Inbound (12)	CS_B2BBOOK_INB_ORDER_IN	AGZIspRae0EytzwvHYri_BHt1l_R
746389	ORDERS:D:96A - Sender: USCU_S05; Receiver: S42CLNT400	SAP CPI E2E Inbound (12)	CS_B2BBOOK_INB_ORDER_IN	AGZIspPTv1_h6CNBE7d2rySauHPO

Figure 11.50 Test Cases Created by Robot with Detailed Descriptions Based on Content of Payloads

The robotic crawler can be scheduled to be executed in the background. Thanks to this, it can continuously scan new messages being exchanged with the connected applications and partners and save them as test cases, ensuring that your test data coverage is accurate. You can, for example, pick up the last month of production data and use actual data for testing.

Once the test cases are created automatically, it is crucial to be aware of the test case coverage. This is where Int4's test cases analysis comes into play.

In the test cases analysis, you can define the folder in which the Int4 robot created test cases. This will be the source for the analysis, and the test cases from the chosen folder will be analyzed based on the dimensions you can define. These dimensions are the data taken directly from the source input files from EDI partners or from mapped messages, such as IDocs. For example, you can create a dimension for the file type, message type, partner, material number (as shown in Figure 11.51), or any other value that is stored in the message and saved as a test case.

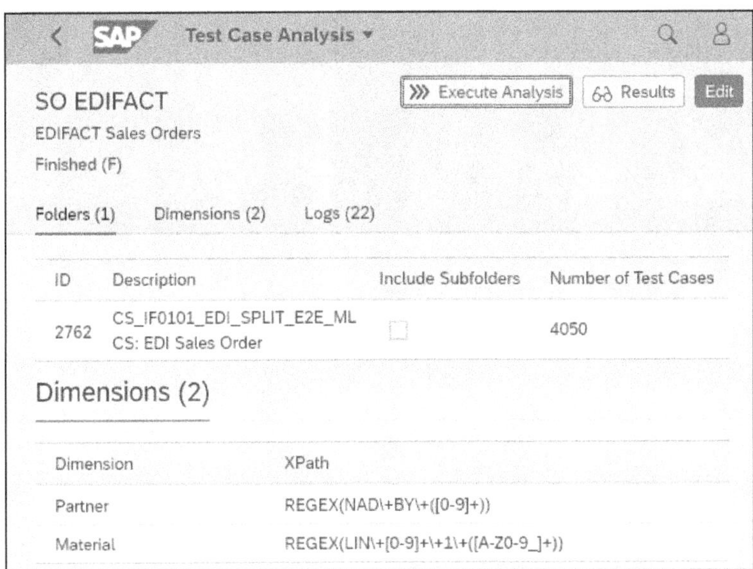

Figure 11.51 Test Cases Analysis Created for 4,050 EDI Messages Stored in Folder 2762 Based on Partner and Material Number Stored in EDI File

Once you define a specific source folder and dimensions for your analysis, click the **Execute Analysis** button to run the analysis in the background. Then, in the **Results** tab, you should see the breakdown of the test cases according to the defined dimensions. In our example, at a glance, we can identify the variations of the partners and materials sent by those partners, as shown in Figure 11.52.

Based on the analysis, you can dynamically create the folder structure for the exact breakdown in the test results.

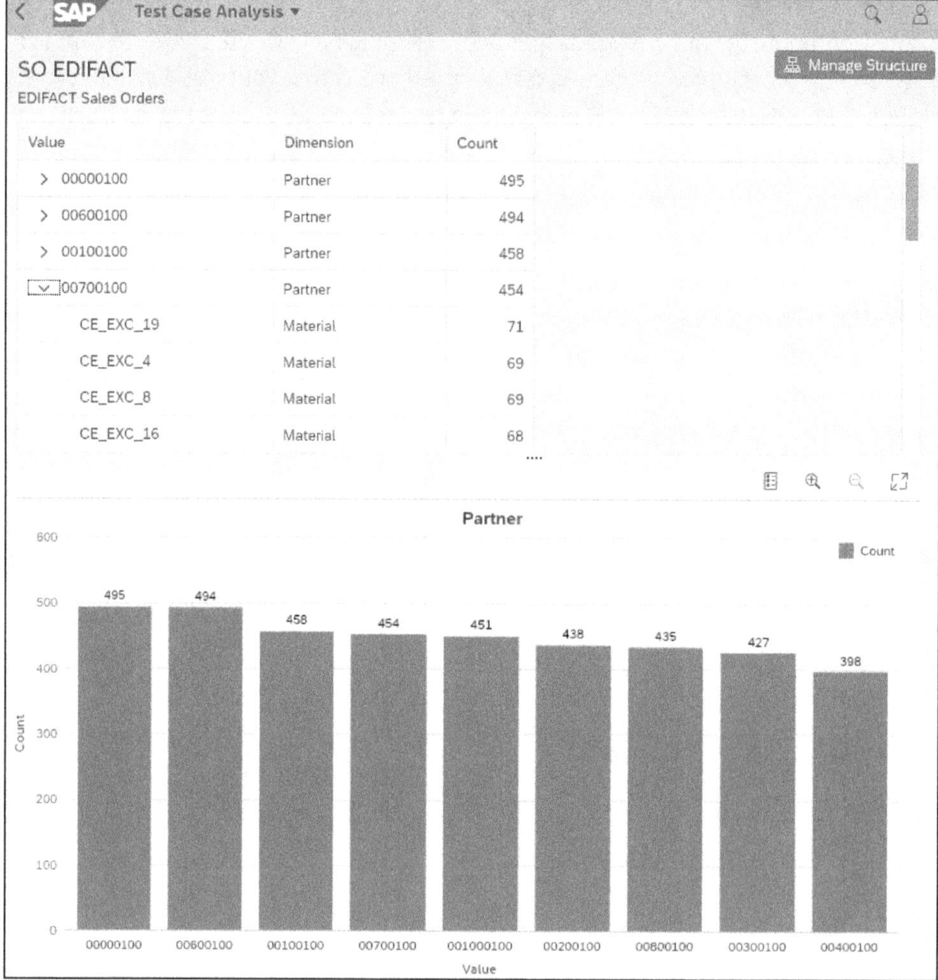

Figure 11.52 Test Cases Analysis with Breakdown for Partner and Material Numbers

The Int4 Suite test case robot allows you to automatically create test cases from the environment, minimizing the time needed to generate test data. The test cases analysis, however, can ensure that the collected data covers the scenarios you want to validate.

11.3.3 Automated End-to-End EDI Business Process Testing

This section will focus on understanding the end-to-end testing setup, where API testing is combined with UI testing. Int4 Suite can be integrated with any solution available in the market for UI testing that allows for such integration, like Worksoft, Tricentis Tosca, or UiPath. In such scenarios, values and test results for Int4 Suite and the UI testing tools are exchanged with each other.

In this section, you will learn what a sample flow can look like and how it can work as a whole. You will see an example based on Int4 Suite integration with the extended Computer-Aided Test Tool (eCATT).

Note

eCATT is used to automate the testing of SAP business processes. It allows for the creation and execution of automated test cases to validate various SAP transactions, ensuring that they function as intended.

In this section, we will automate the end-to-end business process testing that we validated manually in Chapter 10. For this purpose, in Int4 Suite, we created the following test cases:

- **Sales Order Simulation and Validation**
  This test case will simulate the message from the EDI partner, validate the Cloud Integration processing, and validate the sales order created in SAP S/4HANA. This test case's configuration and scope are precisely the same as the sales order in the component-based testing section.
- **Sales Order Confirmation—EDI Validation**
  This test case will validate whether the order response message was created automatically as the inbound sales order from the first test case posted. It will validate the outbound order response message.
- **Create Delivery UI Automation**
  This test case will execute the eCATT script that creates the delivery for the sales order made in the first test case. The sales order number will be passed from the first test case.
- **Validate Outbound ASN Sent to Partner**
  Once the third test case is executed and the delivery created, Int4 Suite will validate whether the output control triggered the message, and it will validate the EDI DESADV message sent to the partner.
- **PGI UI Automation**
  This test case will execute the eCATT script to post goods issues.
- **Invoice Creation UI Automation**
  This test case will execute the eCATT script to create the invoice sent to the customer.
- **Validate the Outbound Invoice Sent to the Partner**
  Once the invoice is created, Int4 Suite will validate whether the new INVOIC EDI message was processed correctly in Cloud Integration with the invoice number passed from Step 6.

Figure 11.53 shows all the test cases with their descriptions, types, automation objects names, document numbers, and parent IDs.

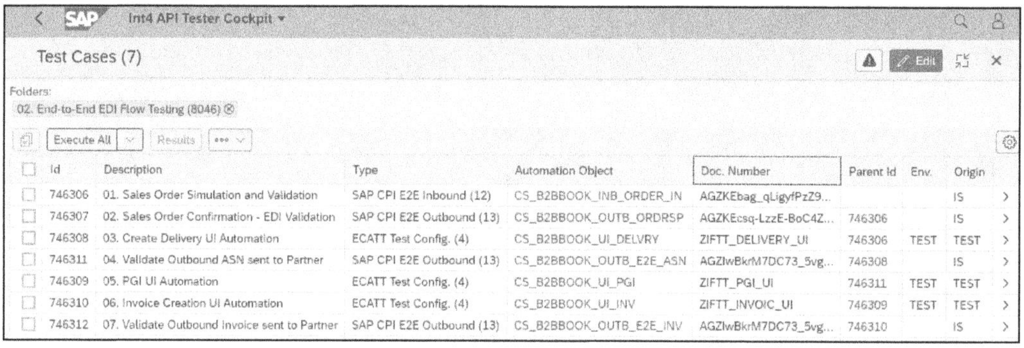

Int4 API Tester Cockpit

Test Cases (7)

Folders:
02. End-to-End EDI Flow Testing (8046)

Execute All | Results

Id	Description	Type	Automation Object	Doc. Number	Parent Id	Env.	Origin
746306	01. Sales Order Simulation and Validation	SAP CPI E2E Inbound (12)	CS_B2BBOOK_INB_ORDER_IN	AGZKEbag_qLigyfPzZ9...			IS
746307	02. Sales Order Confirmation - EDI Validation	SAP CPI E2E Outbound (13)	CS_B2BBOOK_OUTB_ORDRSP	AGZKEcsq-LzzE-BoC4Z...	746306		IS
746308	03. Create Delivery UI Automation	ECATT Test Config. (4)	CS_B2BBOOK_UI_DELVRY	ZIFTT_DELIVERY_UI	746306	TEST	TEST
746311	04. Validate Outbound ASN sent to Partner	SAP CPI E2E Outbound (13)	CS_B2BBOOK_OUTB_E2E_ASN	AGZIwBkrM7DC73_5vg...	746308		IS
746309	05. PGI UI Automation	ECATT Test Config. (4)	CS_B2BBOOK_UI_PGI	ZIFTT_PGI_UI	746311	TEST	TEST
746310	06. Invoice Creation UI Automation	ECATT Test Config. (4)	CS_B2BBOOK_UI_INV	ZIFTT_INVOIC_UI	746309	TEST	TEST
746312	07. Validate Outbound Invoice sent to Partner	SAP CPI E2E Outbound (13)	CS_B2BBOOK_OUTB_E2E_INV	AGZIwBkrM7DC73_5vg...	746310		IS

Figure 11.53 Int4 Suite Test Cases for End-to-End Test Automation with API Testing and UI Testing

As you might have noticed, you can see the parent IDs assigned to the test cases. This ensures that the test cases are executed in the correct order and that the values from the parent test cases will be transmitted to the child test cases. Thanks to this, we can make sure that the whole process is reusable whenever needed because all the documents will be created and validated against newly tested values, starting from the newly created sales order.

Now let's show you the exact setup based on the first three test cases. Once the first test case is executed, the new purchase order number is generated and the new sales order with the unique purchase order number is posted. Then we expect the order response message to be generated directly, and the second test case should validate whether the order response is appropriately generated. The first test case is a parent for the second test case with the validation of the sales order response message, and thanks to this, with variables, we can pass the new purchase order number.

The automation object for the sales order response validation has, in our case, two variables, as shown in Figure 11.54:

- **MSG_TYPE**
  EDI message type
- **PO**
  SAP purchase order number

Variable PO reads the value not from the message content but from the previous test case. Then this value is used to find the relevant message in the **Step_2_-_Interchange_Processing_Flow_V2** iFlow based on this SAP purchase order number. You can see this configuration in Figure 11.55.

Because we can also have other messages, not only order responses, in this iFlow, we added the variable for the message type. This variable is filled with the message type taken from the EDI content and is also used to find the right message. You can see this configuration in Figure 11.56.

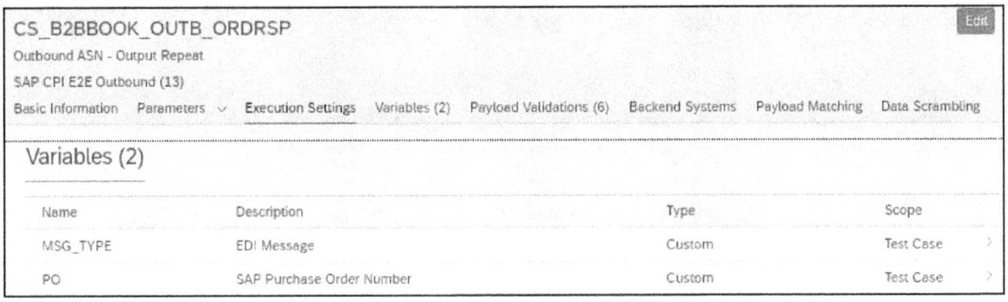

CS_B2BBOOK_OUTB_ORDRSP
Outbound ASN - Output Repeat
SAP CPI E2E Outbound (13)

Basic Information | Parameters | Execution Settings | Variables (2) | Payload Validations (6) | Backend Systems | Payload Matching | Data Scrambling

Variables (2)

Name	Description	Type	Scope
MSG_TYPE	EDI Message	Custom	Test Case
PO	SAP Purchase Order Number	Custom	Test Case

Figure 11.54 Variables in Automation Object for ORDRSP Message Validation

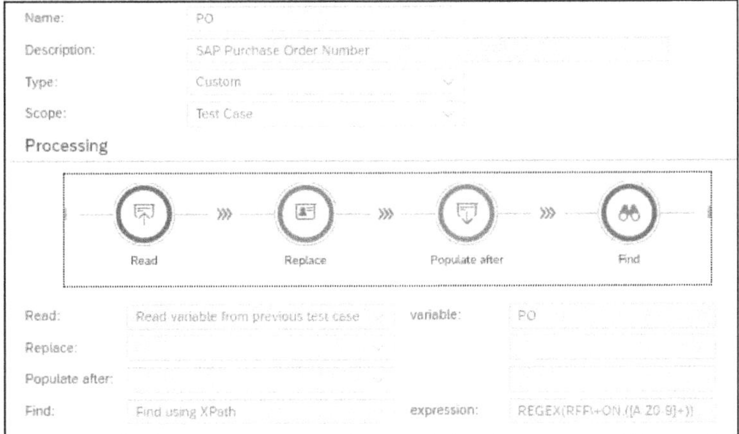

Figure 11.55 SAP Purchase Order Number Variable Read from Previous Test Case

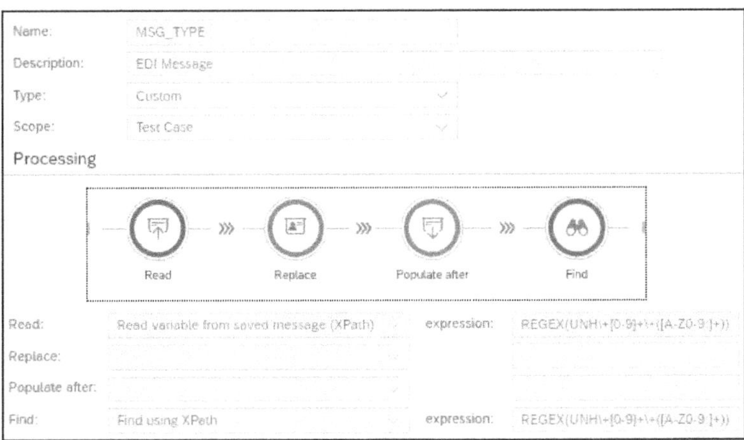

Figure 11.56 Message Type Variable that Reads Value from EDI Message and Uses it to Find Newly Created Message

The validation in the order response is done based on the EDI ORDRSP message, as shown in Figure 11.57.

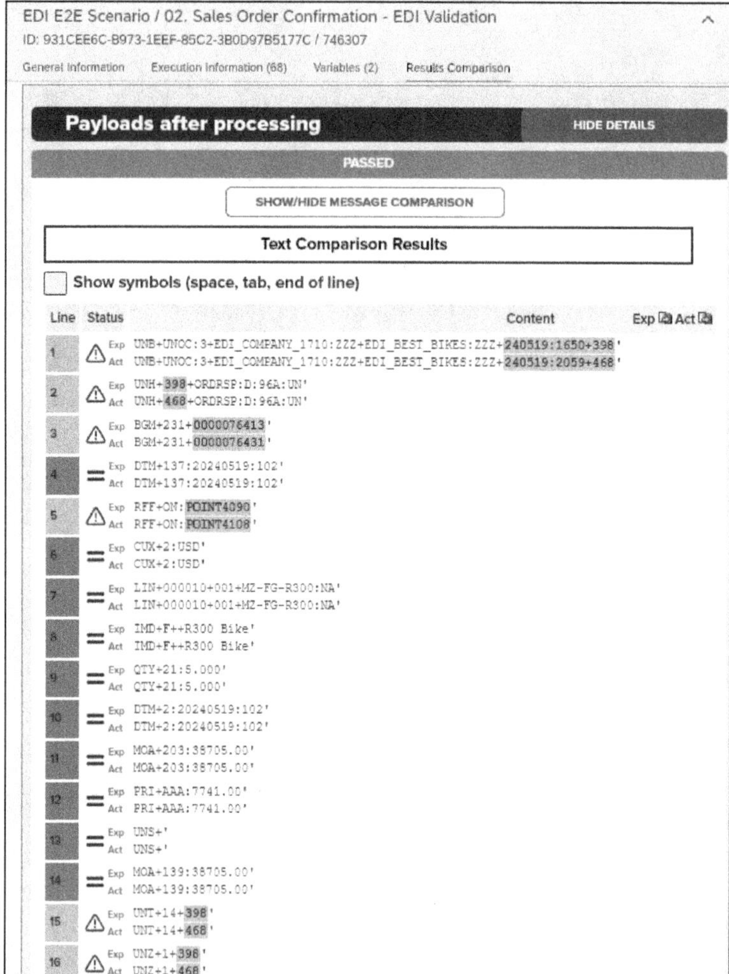

Figure 11.57 EDI Order Response Validation

Basic Data | Execution Details

Test Case:
03. Create Delivery UI Automation (746308)

Automation Object:
Delivery creation (CS_B2BBOOK_UI_DELVRY)

Environment:
TEST

Folder:
02. End-to-End EDI Flow Testing (8046)

Parent Test Case:
01. Sales Order Simulation and Validation (746306)

RFC for Reference Document:
ED1

Test Type:
ECATT Test Config. (4)

RFC for Current Document:
ED1

Execution Information (1)

Message

☑ eCATT Test Log Id: 0000004262

Variables (2)

Name	Description	Reference Value	Current Value
EXP_VBELN	Delivery (export)	0080020991	0080020991
IMP_VBELN	Sales Order (import)	0000076413	0000076431

Figure 11.58 eCATT Script Execution Details with Newly Created Sales Order and Delivery Order Numbers

The eCATT script in the third test case also reads values from the parent inbound sales order test case, and thanks to this, the script can always create the outbound delivery for a newly created sales order because this value is passed from the previous test case (see Figure 11.58).

The next test case validates whether the outbound DESADV message was created in Cloud Integration with the delivery number passed from the eCATT script once the delivery was created. This time again, the outbound DESADV is not triggered by Int4 Suite as in the component-based testing (where it was triggered based on the output control). Still, here it checks whether the message was created by some real action in the system (a delivery creation automated with eCATT script; see Figure 11.59).

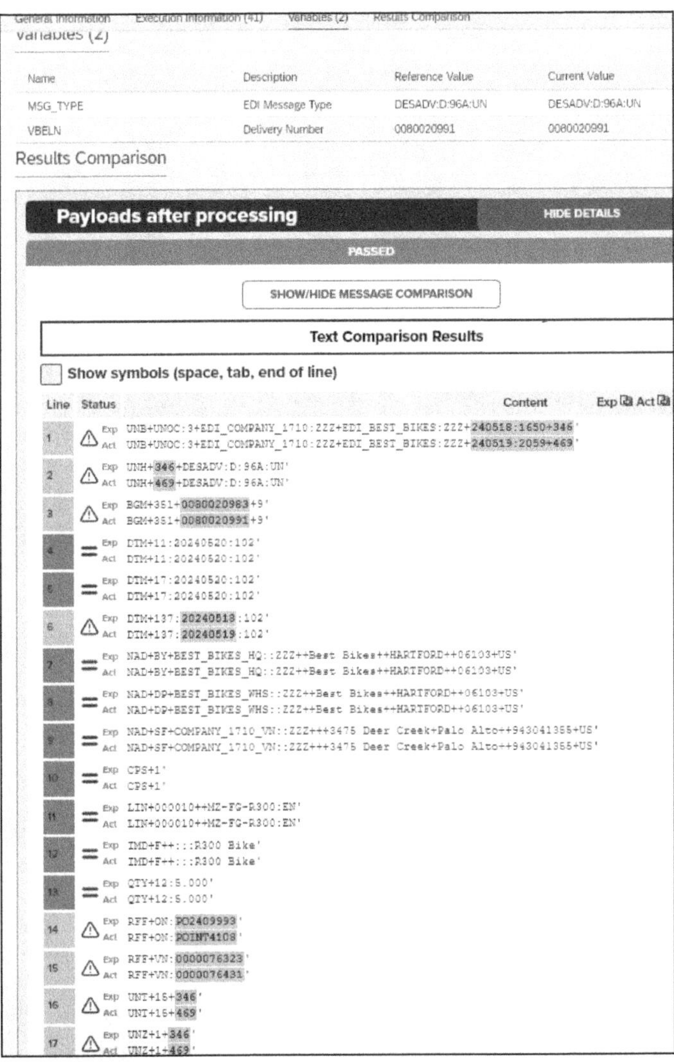

Figure 11.59 EDI DESADV Message Validation

All the other steps in this whole end-to-end testing process work in exactly the same way. The test case results show the exact details of all the test case reports from the component-based testing, as shown in Figure 11.60.

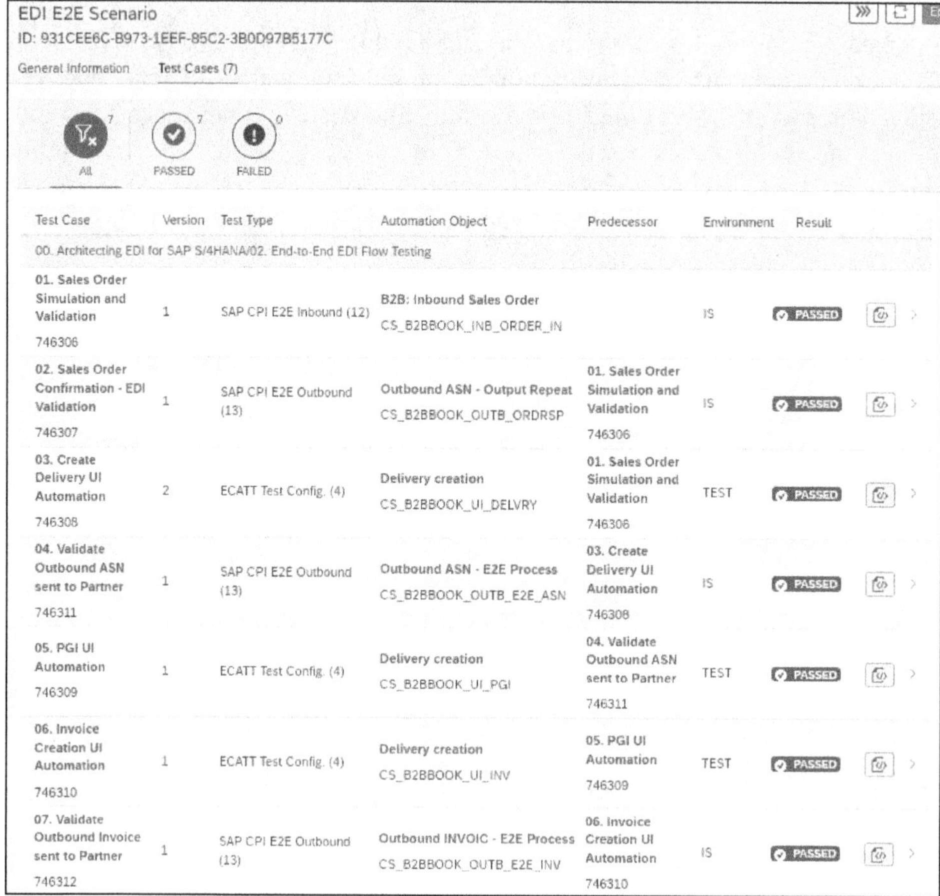

Figure 11.60 Test Case Results Report with All End-to-End Test Cases

11.4 Summary

In this chapter, you reviewed SAP API automation with Int4 Suite for the EDI scenarios in Cloud Integration and SAP S/4HANA. You also learned the differences between component-based testing and end-to-end business flow testing.

You configured the Int4 Suite automation objects from scratch for the component-based testing. You also learned about the robotic crawler, which creates test cases automatically, and about using test case analysis to validate the test case coverage. In addition, you saw how the configuration for the end-to-end testing with the UI automation tool can be set.

In the next chapter, we'll take an in-depth look at APIs.

Chapter 12
APIs

In this chapter we will look at APIs and teach you how to use OData APIs for B2B integration in certain situations. We will look at API concepts and characteristics, examine APIs in SAP S/4HANA, and extend our EDI scenario.

The purpose of this chapter is to look at an alternative approach to classic, message-based EDI. As mentioned in Chapter 4 when presenting the technical solution design, the involved components and presented approach is not the only way to integrate EDI with trading partners. Among several other options that could either slightly or completely alter the existing solution, we will focus on the characteristics of application programming interfaces (APIs). The use case from previous chapters is that we would like to have an integration that can allow sharing product-related information with our trading partner. This product information could be then used within the EDI sales order that the trading partner generates and sends to us. In the first part of this chapter, we will give a short introduction to the API concept, and then we will look how APIs are supported within the SAP ecosystem. In the second part of this chapter, we will extend our existing, classic EDI scenario by adding an API to it. We will provide step-by-step instructions for all the components and activities necessary to enable reliable and secure API service. EDI and using APIs are both methods for exchanging data between systems, but they differ significantly in their approaches and use cases. EDI is a more rigid, standardized approach suitable for high-volume, predictable data exchanges, while APIs offer greater flexibility and real-time capabilities for diverse and interactive system integrations.

12.1 API Concept and Characteristics

An API comprises protocols and definitions that enable diverse applications or programs to communicate with each other. This communication facilitates tasks such as sending and receiving data and enabling software components to interact securely and efficiently, ensuring safe and controlled access to functionality and data.

An API consists of routines, protocols, and tools utilized by programmers during application development to facilitate communication between applications. APIs specify how one application can access data or functions provided by another software program. By establishing standardized methods and tools, APIs ensure reliable interaction between different software systems, regardless of their underlying architectures or technologies. For example, imagine that a travel agent based in Sydney is searching for flights to Singapore for a customer. Using their system, they can query multiple airlines to retrieve various flight routes, times, and prices for a specific date. The agent's internal system communicates with APIs provided by different airlines, enabling seamless integration and access to flight information directly within their own interface.

APIs work by displaying a specific number of actions and data points that external software can interact with. When a software system wants to access a resource provided by another system (such as data or functionality), it sends a request detailing the action it needs to perform through an API. If the API request is authorized, the system processes it and returns a response.

In application and website development, APIs are needed to fetch data without having to build and manage a separate database. APIs use a request-response model of communication, where a client sends a request or API call to an API, which processes the request and provides a response. APIs typically use standard data formats like JavaScript Object Notation (JSON) or XML to exchange data.

In this section, we'll take a closer look at the key components of APIs and the most common types of APIs, as well as the different architectural and protocol styles. We'll close this section by reviewing the benefits and limitations of APIs.

12.1.1 Key Components and Types

The key components and structure of an API include endpoints, methods, requests, and responses. *Endpoints* are specific addresses (i.e., URLs in the case of web APIs) where APIs can be accessed. *Methods* define allowable actions (e.g., GET, POST, PUT, or DELETE for HTTP APIs) that can be executed at these endpoints. *Requests* include the necessary data and parameters to execute actions, while *responses* consist of data returned by the API.

Modern APIs are primarily designed to adhere to a schema that defines the interaction rules, formatting, validation, and documentation in order to ensure secure, reliable, and efficient communication between systems. APIs include specifications for routines, data structures, object classes, and variables that are essential for API functionality.

Figure 12.1 shows a generic API flow for accessing data. The API consumer searches for information from a database (endpoint), which is obtained through the cloud (public internet) using an API and a web/application server.

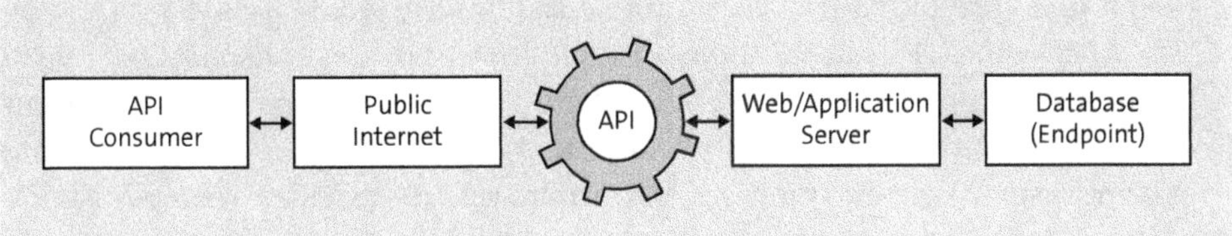

Figure 12.1 Generic API Flow

There are three categories of APIs that can be classified by *audience*:

- **Public APIs**
  Public APIs, also referred to as *developer-facing* or *external* APIs, are accessible to any third-party developers. A well-executed public API program can enhance brand awareness and create new revenue streams. Public APIs are easy to access as they are intended for public use and are designed to boost new use cases and integrations. There is minimal to zero user restriction with APIs, although many require registration and authentication, often via an easy-to-grab API key.

  There are two types of public APIs:

  - *Open* APIs are freely accessible without restrictive terms or conditions, as defined by the OpenAPI specification. Developers can build applications using these APIs without explicit approval or mandatory licensing fees. Open APIs must provide openly available descriptions and documentation, and developers can freely use them for application creation and testing. An example is a Department of Transportation vehicle registration API provided by the Australian government. All information is public and ready to be used as the government wants to ensure that it is widely accessible.
  - A *commercial* API is one that's provided by a company or service with the intention of generating revenue. These APIs typically offer premium features or enhanced capabilities compared to free or open-source alternatives.

- **Private APIs**
  Private APIs, also known as *internal* APIs, are application interfaces designed to enhance organizational solutions and services. These APIs are utilized by in-house developers or contractors to integrate a company's IT systems, develop new systems, or create customer-facing applications using existing infrastructure and are intended for use internally by each of the organization's developers and within a closed group of API consumers, such as an institution or private company.

  Accessing data from a private API typically requires developers to obtain explicit permission due to the proprietary nature of the data and functionalities it offers, so they are restricted to individuals directly affiliated with the API publisher. This ensures complete control over the use of the API. Private APIs are configured with robust logging and load-balancing capabilities to ensure high fault tolerance and security as per requirements.

Private APIs are typically used in large organizations. These APIs facilitate communication between various applications and systems within these organizations, creating an abstraction layer that allows them to function as independent units in terms of modifications and maintenance. Integrating new technologies and services into existing systems becomes more straightforward with the assistance of internal APIs.

- **Partner APIs**

  Partner APIs are openly promoted but accessible only to business partners who have signed agreements with the publisher. Typically used for software integration between two parties, partner APIs allow companies to share data or capabilities with partners. Publishers can monitor how their digital assets are utilized by third-party solutions, ensure a positive user experience, and maintain a corporate identity within partner applications.

  Partner APIs occupy a middle ground between public and private APIs, focusing on securely sharing data between two companies or organizations for specific business purposes while maintaining strict privacy safeguards. These APIs encompass diverse services such as identity verification, payment processing, and data syndication integration.

As shown in Figure 12.2, there are also four API types that can be defined by their *use cases*:

- **Database APIs**

  Database APIs facilitate communication between applications and database management systems. Developers interact with databases by writing queries to retrieve data, modify tables, and perform other operations. For example, the MS SQL Database API allows users to write unified queries that work across various databases, including Oracle, MongoDB, PostgreSQL, MySQL, CouchDB, and others.

- **Operating system APIs**

  Operating system APIs define how applications utilize the resources and services of operating systems. Each OS provides a set of APIs, such as the Windows API or the Linux API (including the kernel user-space API and kernel internal API). Apple offers API references for macOS and iOS in its developer documentation. Developers building applications for macOS use the Cocoa set of developer tools, while those creating apps for iOS use Cocoa Touch, a variant of Cocoa tailored for mobile platforms.

- **Remote APIs**

  Remote APIs establish interaction standards for applications running on separate machines, allowing one software product to access resources located on another device via a network connection, typically the internet. Remote APIs are often based on web standards due to their reliance on communications networks.

- **Web APIs**

  Web APIs are the most prevalent type, facilitating machine-readable data and functionality exchange between web-based systems in a client-server architecture. These APIs handle requests from web applications and provide responses from servers via HTTP.

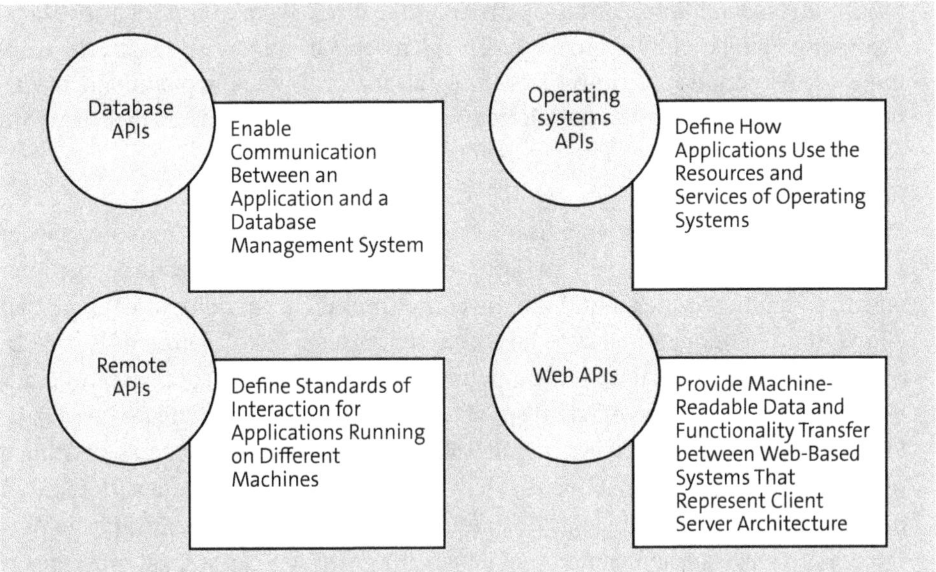

Figure 12.2 Brief Descriptions of APIs by Purpose

API Example

One example of an API is the Google Maps API. This is a mapping services API that is widely integrated into websites for location-based functionalities. Google Maps stores its mapping data in a private cloud. The Google Maps API lets you interact with Google's knowledge of the real world via a public RESTful API interface.

12.1.2 Architectural and Protocol Styles

An API can be built with different architectural styles as per your needs. There are four API different types of API architecture:

- **Monolithic APIs**
  Most public APIs are structured as monolithic APIs, which means they are designed as a single, cohesive codebase that grants access to a complex data repository. These APIs are well-known among web developers and typically mirror the architecture found in relational databases or model-view-controller (MVC) applications. They offer consistent functionality across various resources and tend to maintain stability over time due to their widespread adoption across diverse user needs. Monolithic APIs can pose challenges when it comes to scalability or refactoring, primarily because of the interconnected nature of their data; even minor adjustments can lead to unexpected repercussions.
- **Microservices APIs**
  The primary alternative to a monolithic architecture is a microservices API architecture, where each API serves a specific and focused function. This approach is com-

monly utilized for internal and partner APIs, although organizations may also incorporate public APIs within a broader microservices framework. Many development teams adopting a continuous integration/continuous deployment (CI/CD) methodology employ multiple microservices throughout their code lifecycle, with each serving a distinct and independent role.

- **Composite APIs**
  Microservices introduce a significant challenge by generating a large number of individual API calls. To address this issue, two additional API architecture types offer effective solutions. One solution is the composite API, a specialized API type that allows multiple API endpoints to be accessed with a single call. Often APIs contain overlapping data across different endpoints, and a composite API optimizes API calls by determining the most efficient set of calls needed to retrieve the required data. Moreover, composite APIs can bundle calls for common tasks, such as creating a new user account, thus enabling the delivery of comprehensive data with reduced data volume through a single payload sent to multiple endpoints. Composite APIs typically manage authentication and data formats through an API gateway, ensuring coordinated handling of requests and responses across various endpoints.
- **Unified APIs**
  A unified API operates similarly to a composite API, but instead of consolidating calls to multiple endpoints within a single API, it consolidates related calls across multiple distinct APIs. Unified APIs are commonly used in partner API scenarios, particularly in applications like personal finance tools.

  A unified API acts as a central hub, allowing the financial planning application to send a single HTTP request to a unified endpoint. Secure servers then handle the task of mapping the data from this request into the necessary formats for each financial institution. Subsequently, the data is returned to the client-side application in a controlled and predictable manner, ensuring efficient data retrieval without compromising security or integrity.

APIs can also be categorized by the nature of the protocol style, as follows:

- **REST APIs**
  REST APIs (also called RESTful APIs or RESTful web APIs), based on the Representational State Transfer architecture, excels in scalability and secure data transfer by managing requests over HTTP and maintaining statelessness, where no client content is stored on servers between requests. The REST protocol, widely recognized in API development, defines routes using URLs, which simplifies development compared to protocols requiring XML wrapping.

  However, REST APIs are limited to transmitting information solely through the HTTP protocol, restricting their capability to transfer text and similar data types. To accommodate more complex data like images or audio files, developers can utilize formatting parameters, such as specifying content types in headers. Despite encoding these files as text, typically in JSON or XML formats, REST APIs remain versatile and are extensively used across a diverse range of functionalities due to their simplicity and scalability. Adapting to REST and HTTP constraints requires innovative approaches from API developers to effectively leverage their capabilities. Figure 12.3 shows the flow of RESTful APIs. This flow uses standard HTTP methods (GET, POST, PUT, DELETE) for communication and supports multiple data formats like JSON, XML, and HTML.

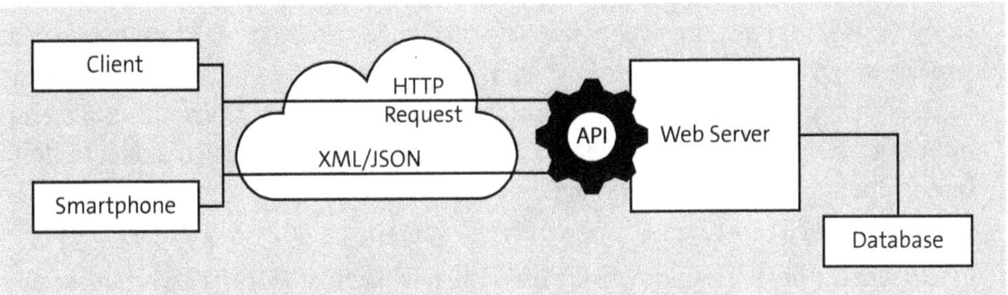

Figure 12.3 RESTful API Flow Diagram

- **SOAP APIs**

  Simple Object Access Protocol (SOAP) facilitates data sharing among applications in different environments or written in different languages. SOAP APIs govern how data is transmitted over networks, the structure of messages, and their content. Unlike REST, SOAP supports communication over various internet protocols such as HTTP, TCP, and SMTP, enhancing its flexibility.

  However, SOAP APIs are more restrictive compared to REST APIs. They exclusively handle XML data and impose stringent requirements on request formats. SOAP requests typically consume more bandwidth than REST requests due to their verbose XML format. Developing and maintaining SOAP code is also more complex. It is often used in enterprise systems, where reliability and security are critical.

  One significant advantage of SOAP is its use of metadata files that describe requests, ensuring predictability in data exchanges. SOAP also supports stateful requests, unlike stateless REST, which enhances its capability for managing session information. SOAP's standardized protocol enables reliable communication of complex data across diverse channels beyond HTTP. Its service-oriented architecture fosters greater discoverability for knowledgeable users, moving beyond a simple URL-based organization. Figure 12.4 shows the process of a SOAP API, where SOAP is a protocol itself with a strict specification. It relies on XML for its message format and usually operates over HTTP or SMTP.

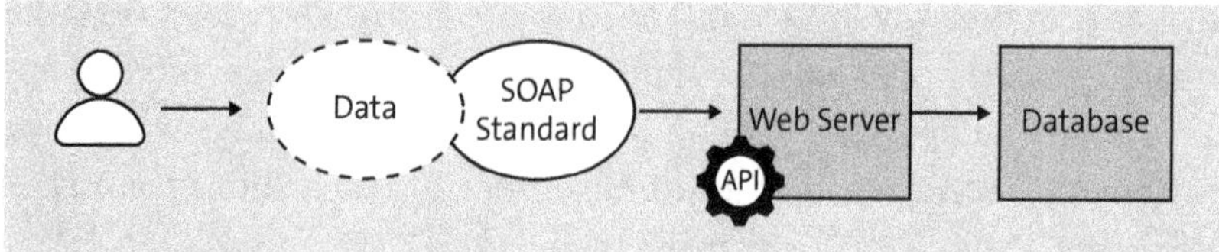

Figure 12.4 SOAP API Flow Diagram

- **GraphQL APIs**
  GraphQL is an open-source query language designed to reduce the number of round trips between clients and servers, which is advantageous for applications operating on slow or unreliable connections. Unlike REST APIs, which utilize multiple endpoints, each representing different data schemas, GraphQL APIs consolidate these into a single endpoint. This unified endpoint offers virtually unlimited data schemas, allowing clients to specify precisely the data they require in a single query.

  Queries in GraphQL are submitted within the payload of an HTTP POST request, and the data returned conforms to the structure defined by the query's schema. This flexibility allows users to combine fields in any order, providing significant flexibility compared to the rigid routing requirements of REST APIs.

  However, GraphQL introduces challenges such as caching data effectively and requiring API consumers to maintain a consistent query syntax for consistent data retrieval. Users must also possess knowledge of the available fields to construct queries, necessitating comprehensive custom documentation compared to REST APIs. Moreover, automation tools for GraphQL are less abundant than those available for REST APIs.

  Figure 12.5 shows GraphQL as a query language for APIs and a runtime for executing those queries by using a type system you define for your data.

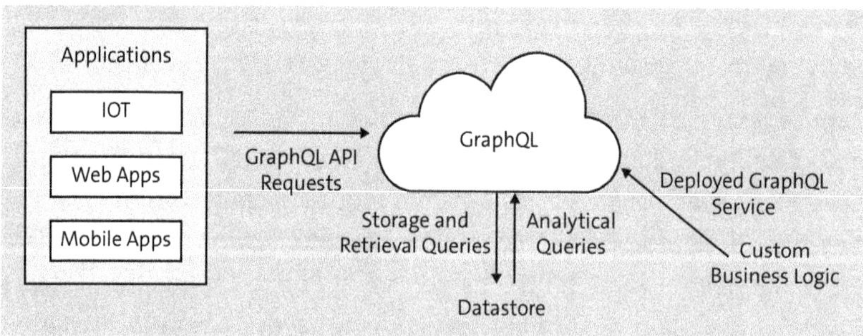

Figure 12.5 GraphQL API Flow Diagram

- **Remote procedure call APIs**
  Remote procedure call (RPC) APIs utilize RPCs to execute code on remote networks. An RPC enables a computer program to invoke a subroutine in a different address space, often on another computer within a shared network. This interaction is coded

similarly to a local procedure call, without requiring explicit coding of the remote communication details by the programmer.

Unlike the SOAP and REST APIs, RPC APIs focus on calling methods rather than accessing data resources. While a RESTful API typically returns a document, an RPC API confirms whether a function was successfully triggered or returns an error indicating why it failed to execute. In essence, REST APIs manage resources, whereas RPC APIs handle actions.

RPC protocols can deliver responses in XML or JSON formats. Unlike REST APIs that display both the server and query parameters in their routes, RPC APIs only identify the server in their uniform resource identifiers (URIs). Consequently, RPC APIs are rarely made public due to concerns about allowing method invocations on remote servers for the general public. This method invocation changes the server's state, distinguishing RPC from the stateless/stateful distinction found in REST and SOAP APIs. Therefore, RPC APIs necessitate high levels of security and trust between producers and consumers, making them predominantly private APIs.

Discoverability and predictability are less critical for RPC APIs compared to REST or SOAP APIs, whereas reliability and performance take precedence. RPC APIs are commonly used in distributed client-server applications where payloads are lightweight, limited to method parameters. Frontend developers can invoke server methods without managing connection details or input parsing intricacies. RPC also facilitates simplified task threading by enabling multithreaded processes to run on remote servers independently from client applications. Figure 12.6 shows the flow of an RPI API with a protocol that allows a program to execute code on a remote server as if it were a local procedure call. RPC APIs offer a method-centric approach to remote communication in tightly coupled systems or scenarios requiring efficient data serialization.

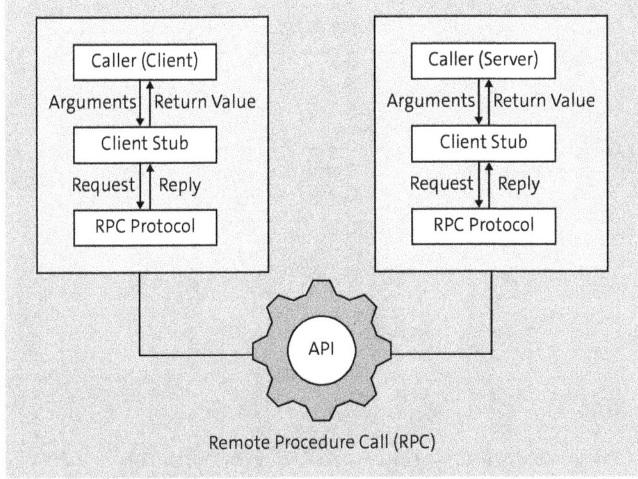

Figure 12.6 RPC API Flow Diagram

- **Webhooks and WebSocket APIs**
  Webhooks, also known as *event-driven APIs* or *asynchronous APIs*, are integral to event-driven architectures where servers automatically send requests in response to specific events. They enable clients to subscribe to events and receive data updates as they occur. For instance, a webhook might notify a client when a new order is placed on an e-commerce site.

  In contrast, WebSocket APIs facilitate bidirectional communication between client applications and servers, enabling servers to send call back messages to connected clients efficiently. This approach contrasts with REST APIs, as WebSocket APIs allow for event-driven responses without requiring clients to continuously poll the server for updates.

  Event-driven APIs, such as webhooks and WebSocket APIs, excel in transmitting information in near real time, making them ideal for applications requiring constantly updated data, such as stock market trackers and IoT devices monitoring real-time events.

  Webhooks automate the delivery of real-time data to applications, enabling them to remain current without relying on periodic polling. This contrasts with traditional methods where clients query applications to check for changes. By leveraging webhooks, applications can enhance functionality such as notifications, automation, and real-time synchronization across systems, without the complexity of deep API integrations. Figure 12.7 shows the difference between WebSocket and webhook APIs. WebSockets are used for real-time, bidirectional communication, while webhooks are used for notifying systems about specific events asynchronously.

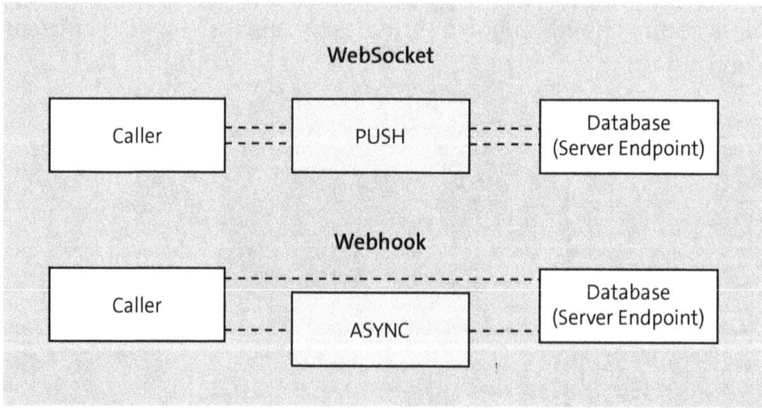

Figure 12.7 WebSocket versus Webhook

12.1.3 Benefits and Limitations

APIs are necessary as they enable seamless integration between different software applications, which helps companies build complex systems that incorporate multiple applications. APIs also make it easier to access data and resources from other applications without the need for a thorough understanding of how they work, which simplifies the development of new applications that use data from other sources. APIs enable the possibility of creating reusable code that can be shared across multiple applications, and this can save organizations significant time and effort.

Key reasons to use APIs include the following:

- **Automation**
  APIs enable software to manage tasks more efficiently. By taking advantage of APIs, organizations can streamline workflows to enhance speed and productivity.
- **Applications**
  APIs provide access to app components, enhancing the flexibility of service delivery and sharing of information.
- **Efficiency**
  With API access, generated content can be automatically published and made available across various channels, facilitating easier sharing and distribution.
- **Integration**
  APIs allow seamless embedding of content from any site or application, ensuring smooth information delivery and a cohesive user experience.
- **Adaptation**
  APIs help anticipate evolving needs by supporting better data migration and closer scrutiny of information, thus enhancing flexibility in service provisioning.
- **Security**
  APIs necessitate robust security measures, including authentication, authorization, and data encryption, in order to safeguard sensitive information and prevent unauthorized access. Regular security audits and compliance with data protection regulations are essential.
- **Customer experiences**
  APIs enable businesses to create seamless interactions with customers by integrating services and data, resulting in quicker response times, personalized experiences, and innovative features that boost customer satisfaction.
- **Modernizing legacy systems**
  APIs play a pivotal role in digital transformation by facilitating the integration of legacy systems with modern applications. This allows organizations to leverage existing investments while adopting new technologies and methodologies.

As with any technology, there can be challenges with APIs. While APIs provide numerous advantages to businesses, they also present challenges that organizations need to manage effectively. Addressing these challenges is crucial for maximizing the value of API integration. Here are some common issues associated with APIs:

- **Data privacy concerns**
  APIs can expose sensitive data and functionalities. Ensuring security and data privacy for APIs involves a multifaceted approach that includes robust authentication and authorization mechanisms to verify and control access, along with encrypting data both in transit and at rest to safeguard against unauthorized data interception and breaches.
- **Reliability and performance**
  APIs need to perform reliably under varying loads and conditions. Downtime or slow response times can disrupt operations and impact user experience, requiring thorough testing and monitoring.
- **Scalability**
  As usage grows, APIs must scale to handle increased traffic and data volumes efficiently. Scalability issues can lead to performance degradation or system failures without proper planning.
- **Compatibility and versioning**
  Changes to APIs can affect compatibility with existing applications and integrations. Managing versioning effectively is crucial to ensure backward compatibility when introducing new features.
- **Documentation and support**
  Comprehensive and up-to-date documentation is vital for developers to understand API functionalities and integration requirements. In addition, responsive support channels help resolve issues promptly.
- **Governance and compliance**
  APIs must adhere to regulatory requirements and industry standards to protect user data and ensure legal compliance. Organizations need to implement governance frameworks to manage API usage and access securely.
- **Integration complexity**
  Integrating diverse systems and applications through APIs can be complex and time-consuming. It requires careful planning, clear communication, and sometimes customization to align with specific business needs.
- **Monitoring and analytics**
  Continuous monitoring of API performance and usage analytics is essential for optimizing performance, identifying bottlenecks, and detecting potential security threats or anomalies.
- **Cost management**
  API usage can incur costs related to data transfer, usage limits, or premium features. Effective cost-management strategies are necessary to avoid unexpected expenses and optimize resource allocation.
- **Vendor lock-in**
  Dependence on external APIs can lead to vendor lock-in, limiting flexibility and scalability options. Organizations should consider diversifying API sources or implementing fallback mechanisms.

- **Maintenance and upgrades**
  Maintaining and upgrading APIs involves ongoing efforts to fix bugs, address security vulnerabilities, and implement new features. Handling multiple API versions and ensuring seamless transitions between them can become intricate, particularly as a user base expands.

API Use Cases

The following are a few examples of the use cases for APIs and how APIs are being used today:

- **Finance**
  APIs have revolutionized the way banks and fintech companies operate, enabling services like online payments, real-time analytics, and automated trading.
- **Healthcare**
  APIs are used to connect patient records, hospital management systems, and telehealth services, improving care coordination and patient outcomes.
- **Travel**
  APIs are used to connect agencies and online booking platforms to airline databases, hotel reservation systems, and transportation services.
- **E-commerce**
  APIs enable seamless integration of payment gateways, inventory management, and customer service tools, creating a smooth shopping experience.
- **Social media integration**
  APIs allow developers to integrate social features into their applications, including features like logging in with social media credentials, sharing content, and retrieving user data.

12.2 APIs in SAP S/4HANA

In this section, we look at the APIs in the SAP ecosystem. As this suggests, our review will not be limited to SAP S/4HANA itself. We'll begin with an overview of the API concept, then take a look at different examples of how APIs are used within SAP S/4HANA, such as in API Management and Graph in SAP Integration Suite, as well as in other SAP solutions. We'll discuss the SAP Gateway component of SAP S/4HANA, then close the section with a sample API.

12.2.1 Overview

In SAP S/4HANA, APIs play a crucial role in enabling connectivity, interoperability, and extensibility of the system. API-based integration involves connecting various software applications through APIs to enable them to work together smoothly. By using

APIs, different systems can exchange data and commands, facilitating interoperability and enhancing functionality across platforms. This integration approach allows businesses to streamline workflows, automate processes, and leverage capabilities from multiple sources without needing to build everything from the start.

API-based integration in SAP S/4HANA is essential for enabling smooth connectivity with both SAP and external systems. Let's review the key components of API-based integration in SAP S/4HANA:

- **Standard OData APIs**
  SAP S/4HANA provides standard OData APIs for various business objects and processes. These APIs follow OData protocol standards, offering a RESTful interface for accessing data.
- **Custom OData APIs**
  Organizations can also develop custom OData services to meet specific business needs using tools like SAP Gateway and SAP Fiori elements.
- **System-to-system integration**
  APIs facilitate integration between SAP S/4HANA and other SAP systems (e.g., SAP SuccessFactors, SAP Ariba) or external systems (e.g., customer relationship management [CRM] platforms, e-commerce systems).
- **Cloud integrations**
  APIs enable connections with cloud-based applications and services, utilizing SAP Business Technology Platform (SAP BTP) integration services or third-party middleware solutions.
- **Extension support**
  APIs allow for the extension and enhancement of SAP S/4HANA's functionality. This includes adding custom fields, business logic, and workflows through APIs that interact with SAP S/4HANA's core data and processes.
- **Event-driven architecture**
  APIs support event-driven architectures, where SAP S/4HANA can publish events (using APIs or other methods like Cloud Integration) to trigger actions in connected systems, ensuring real-time data synchronization and process automation.
- **Security standards**
  SAP S/4HANA APIs comply with strict security standards, often using OAuth 2.0 for authentication to secure API endpoints and restrict access to authorized users and systems.
- **Governance**
  Governance frameworks oversee API usage to ensure it is monitored, is managed, and complies with organizational policies and regulations.
- **SAP Business Accelerator Hub**
  SAP Business Accelerator Hub is a central repository for discovering and using SAP APIs, offering documentation, sample code, and integration patterns.

- **Prebuilt integration content**
  Cloud Integration provides prebuilt integration content, connectivity options, and monitoring capabilities for linking SAP S/4HANA with other systems.

Figure 12.8 shows how APIs work in SAP S/4HANA. The SAP HANA platform and the SAP S/4HANA ABAP application are connected via core data services (CDS). The frontend system is connected to the ABAP platform via OData. SAP provides various APIs, including OData services, RESTful APIs, and SOAP-based web services, to connect to other private and on-premise systems.

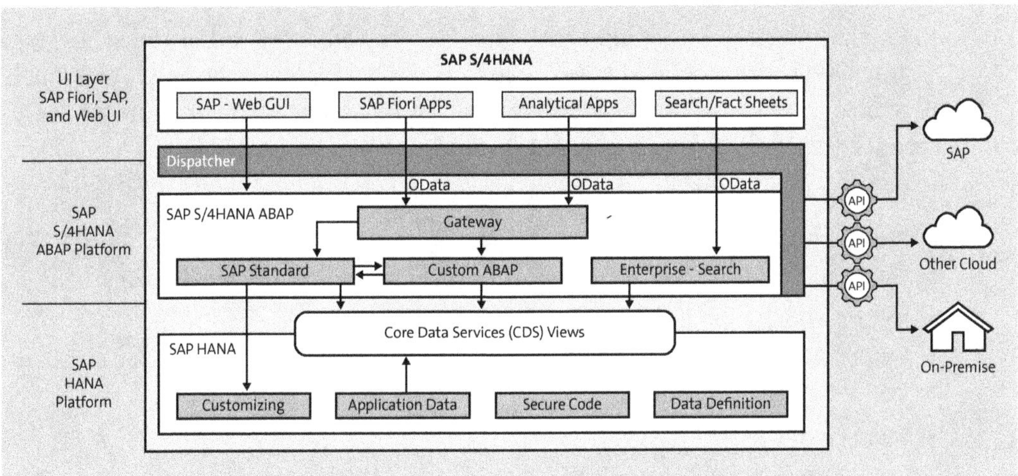

Figure 12.8 Examples of APIs in SAP S/4HANA

12.2.2 API Management

API Management involves the creation, publication, and oversight of API connections between various applications and systems within an organization. It provides a scalable, unified platform that enables organizations to share and manage their API configurations, control access, gather and analyze usage data, and enforce security policies.

SAP emphasizes that an effective API management strategy allows organizations to deliver seamless digital experiences for consumers, partners, and employees. By exposing processes and data as APIs through their entire lifecycle, organizations can unlock new opportunities and maximize their business potential in the API economy. API Management facilitates the publication, promotion, and management of APIs in a secure and scalable environment, enhancing the digital experience for all stakeholders.

API Management can be broadly categorized into two main areas:

- **API portal**
  In the API portal, the API builder can create API proxies or facade patterns for existing APIs or SOAP web services. These proxies act as the sole access point for the

actual API implementation, concealing internal details and incorporating functionalities such as security, traffic management, and data transformation. This separation allows API developers to concentrate on business features rather than technical aspects.

- **Developer portal**
  The self-service developer portal allows developers to discover APIs, review their documentation, test them, and subscribe as needed. Integrated closely with the API portal, the developer portal streamlines the API publishing process and enhances the developer experience by simplifying API usage.

API Management does not alter the actual API implementation but rather involves defining API proxies, or facades, for existing APIs. This approach allows API developers to remain focused on building effective business APIs while API Management handles security, traffic management, and API consumption centrally.

With API Management, you can do the following:

- **Proxy APIs**
  Establish a unified and harmonized API presence using your own domain.
- **Secure APIs**
  Protect APIs from unauthorized access and potential threats by defining a standardized set of security policies.
- **Manage traffic**
  Configure caching, control traffic quotas, and handle traffic spikes through traffic management policies.
- **Govern APIs**
  Discover, document, and manage the lifecycles of APIs while applying various policies.
- **Gain business insights**
  Utilize usage analytics, logs, events, and triggers to monitor performance and leverage business insights for monetizing APIs.
- **Transform APIs**
  Apply advanced modifications to headers and payloads as needed.
- **Engage developers**
  Use SAP Business Accelerator Hub, a feature-rich, customizable portal tailored for application developers. It offers comprehensive API documentation, code snippets, and more, allowing developers to easily discover, subscribe to, and use APIs directly.

Figure 12.9 provides a detailed overview of how API Management ensures that APIs are safeguarded against unauthorized access and potential threats. You can implement API Management to define standardized policies that protect both your APIs and their underlying systems, configure caching, and manage traffic quotas and spikes through traffic management policies.

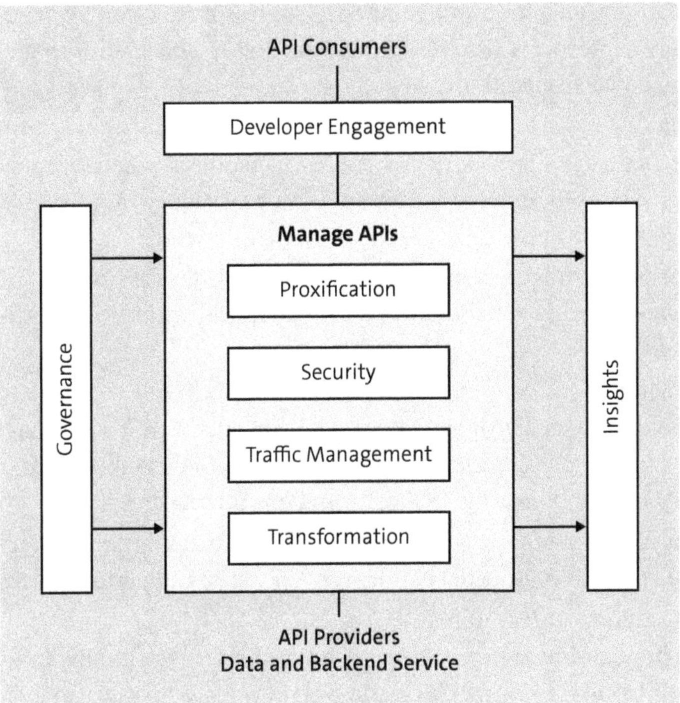

Figure 12.9 API Management Overview

The key capabilities of API Management include the following:

- **API lifecycle management**
  API Management enables API providers to create APIs using a visual designer or to import existing API definitions. It also facilitates automatic documentation generation, ensuring clear understanding and effective usage.

  APIs can also be deployed across multiple environments (such as development, testing, and production) with versioning support to manage updates and maintain backward compatibility.
- **Security and access control**
  The platform offers robust security features, including OAuth 2.0, API keys, and JSON Web Tokens (JWT), to secure API endpoints and regulate access based on roles and permissions.

  API Management also ensures data privacy by supporting encryption for both data in transit and data at rest, helping organizations comply with regulatory standards.
- **API Gateway**
  API Gateway efficiently handles incoming API requests by implementing policies for rate limiting, throttling, and caching, optimizing performance and safeguarding backend systems from overload.

It offers protocol mediation capabilities, allowing for seamless translation between different communication protocols (e.g., from SOAP to REST) and data formats, ensuring compatibility between clients and APIs.

- **Developer engagement**
  API Management features a developer portal where API consumers, whether internal or external, can discover APIs, explore documentation, and request API keys for authentication.

  Developers can independently register on the portal, request API access, and manage their applications, promoting collaboration and innovation within the developer community.
- **Analytics and monitoring**
  The platform provides comprehensive insights into analytics such as API usage, performance metrics, error rates, and trends through built-in analytics dashboards and reports. This visibility aids in monitoring API health and optimizing usage.

  API Management also supports proactive monitoring with alerts and notifications based on predefined thresholds or unusual patterns in API traffic and performance.
- **Integration with SAP and non-SAP systems**
  The solution includes prebuilt integration adapters and connectors to simplify integration with SAP systems (like SAP S/4HANA and SAP SuccessFactors) as well as third-party applications and services.

  It also supports standards such as OData, REST, SOAP, and GraphQL, enabling integration across a wide array of systems and ensuring interoperability among diverse technologies.
- **Governance and compliance**
  Administrators can centrally define and enforce policies (including security, traffic management, and data transformation) across all APIs, ensuring consistency and compliance with organizational standards and regulations.

  API Management also maintains logs of API transactions, policy enforcement actions, and user activities for auditing purposes, supporting compliance audits and governance needs.

12.2.3 Graph

Graph is a groundbreaking feature of API Management within SAP Integration Suite. Building upon traditional API management, Graph allows users to present all business data via a semantically connected data graph, accessible through a single, powerful API.

As a unified API for SAP BTP, Graph streamlines data access across various SAP solutions. It offers a single entry point for interacting with multiple SAP systems and services through a consistent and standardized interface.

Graph can be found in SAP Integration Suite under **Discover (APIs) • SAP Graph**, as shown in Figure 12.10.

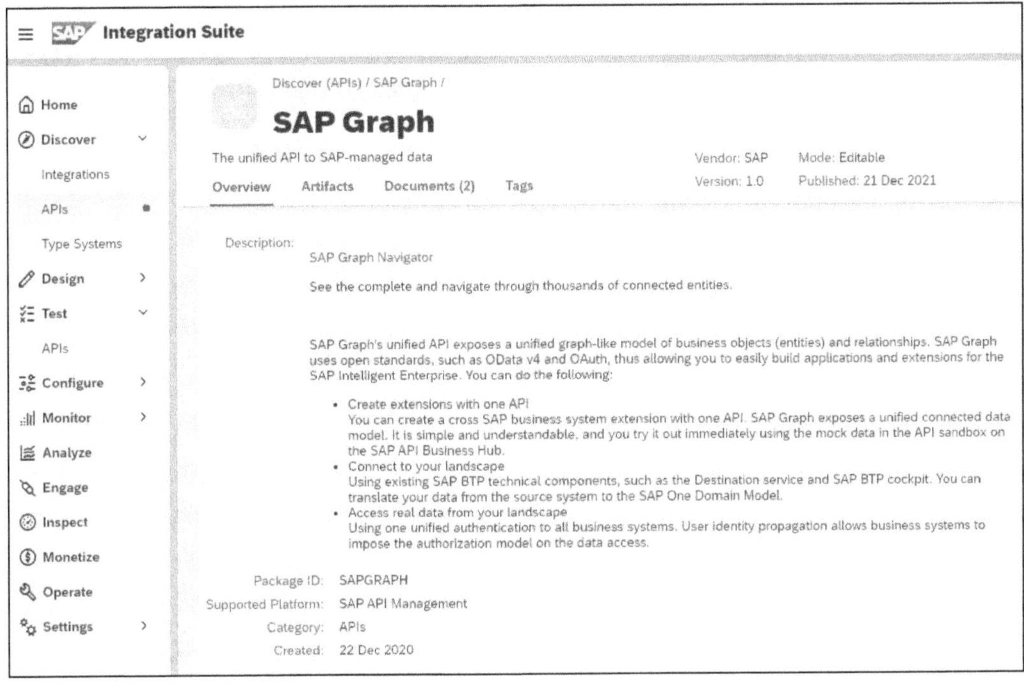

Figure 12.10 Graph in SAP Integration Suite

From a customer perspective, accessing SAP-managed data has become more challenging. Data may be distributed across hybrid networks of on-premise and cloud systems, each with different security protocols, replication processes, and multiple copies of master data. Developers must not only focus on the specific data they need but also understand which data copy to use, its precise location, and how to connect to these diverse customer systems. Each product comes with its own connection protocols, security requirements, and APIs, exposing data through inconsistent models. Consequently, creating SAP extension applications requires mastering a wide range of skills, even for the simplest data queries, and these applications are sensitive to even minor changes in product and landscape configurations. This is where Graph comes into play.

Graph tackles the challenges of API complexity and integration by providing developers with a single, connected, and unified view of all their business data. It consolidates the data models of various sources, such as SAP S/4HANA, SAP Sales Cloud, and SAP SuccessFactors, into one cohesive data model known as the business data graph.

A *data graph* is composed of interconnected nodes representing data entities and edges that denote relationships (referred to as *associations*) between these entities. In the Business Data Graph, the nodes correspond to common enterprise business objects like the customer, supplier, product, sales order, purchase order, and general ledger.

As a developer, knowing whether the concepts of a product or customer are prevalent across multiple business systems within the landscape may be a challenge. Which definition should you use, and where is this data managed? Graph addresses this issue by introducing unified entities that encapsulate the most common attributes of a business model for easy consumption while also linking to the corresponding system-specific business objects for a comprehensive 360-degree view. This means an organization can simplify its use cases with unified entities while having easy access to over 1,500 additional system-specific entities for more advanced needs.

Key aspects of Graph include the following:

- **Unified API layer**
  Graph centralizes access to data and services from various SAP systems, including SAP S/4HANA, SAP SuccessFactors, and SAP Ariba, through a single API layer. This simplifies development by removing the need to interact with multiple APIs individually.

  It provides a consistent API interface based on OData, enabling uniform data access and manipulation across different SAP solutions.
- **Comprehensive data access**
  Graph facilitates scenarios requiring access to and integration of data from multiple SAP applications. For instance, it allows a single query to retrieve information from both SAP S/4HANA and SAP SuccessFactors.

  Developers can create complex, customizable queries to obtain specific datasets or execute operations that span various SAP solutions, enhancing flexibility and efficiency.
- **Developer experience**
  Integrated with SAP Business Accelerator Hub, Graph offers a developer portal where APIs, documentation, and resources are centrally managed and readily accessible.

  SAP also provides software development kits (SDKs) and tools to support integration and development with Graph, catering to a variety of programming languages and environments.
- **Security and governance**
  Graph utilizes SAP BTP's authentication mechanisms, such as OAuth 2.0, to secure API access, ensuring that only authorized users and applications can interact with the exposed data and services.

  The platform complies with data privacy regulations through encryption, data masking, and other security measures designed to protect sensitive information.
- **Integration with SAP and non-SAP systems**
  Graph enhances interoperability by enabling a seamless integration between SAP solutions and third-party applications or services, thus supporting hybrid IT environments where various systems coexist.

As shown in Figure 12.11, Graph is a unified API for SAP, using modern open standards like OData v4. Graph is a connection to your business data. It is a new, unified API to access all business data as a single and semantically sound connection.

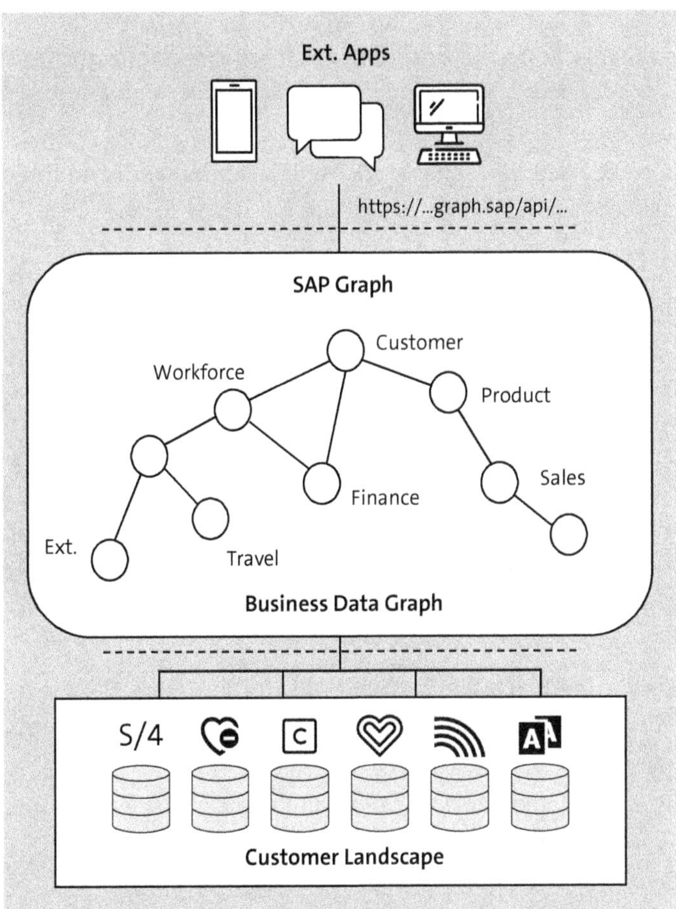

Figure 12.11 Overview of Graph

12.2.4 APIs in Other SAP Solutions

As a leading provider of enterprise software, SAP has made significant investments in API development to enhance the integration capabilities of its products. These APIs serve as a vital connection between SAP systems and external applications, facilitating seamless data exchange and process automation.

SAP offers a range of API types to address different integration needs:

- **OData APIs**
  The preferred choice for modern integrations, providing a RESTful interface for create, read, update, and delete (CRUD) operations.
- **SOAP APIs**
  Although older, these APIs are still widely used for existing integrations, particularly in complex scenarios.

- **BAPIs**
  These are legacy APIs primarily designed for specific business processes within SAP systems.
- **IDocs**
  IDoc is a file-based format used for asynchronous data exchange, commonly employed for batch processes.

APIs are essential across various SAP products, enabling integration, extensibility, and interaction with SAP systems. APIs are utilized in different SAP offerings like the following:

- **SAP S/4HANA**
  SAP S/4HANA predominantly exposes its functionalities through OData APIs, which provide RESTful endpoints for accessing and manipulating business objects such as sales orders, purchase orders, customers, and materials.

  The APIs within SAP S/4HANA facilitate seamless integration with other SAP systems (such as SAP SuccessFactors and SAP Ariba) and external applications, supporting real-time data exchange and process automation throughout the enterprise.
- **SAP SuccessFactors**
  SAP SuccessFactors provides both SOAP and OData APIs for accessing HR-related data and processes, allowing integration with other HR systems, payroll systems, and external applications.

  Its APIs enable customizations and extensions tailored to specific business requirements, such as adding custom fields or integrating with third-party services.
- **SAP Ariba**
  SAP Ariba offers RESTful APIs to support procurement processes, including supplier management, sourcing, procurement, and invoicing. These APIs facilitate integration with ERP systems and third-party procurement platforms.

  SAP Ariba also supports event-driven integration patterns through its APIs, ensuring real-time synchronization of procurement activities and data across systems.
- **SAP Integration Suite**
  SAP Integration Suite includes API Management for overseeing APIs across both SAP and non-SAP systems. It provides tools for designing, publishing, securing, and monitoring APIs.

  SAP Integration Suite offers prebuilt connectors and integration adapters for connecting SAP applications (like SAP S/4HANA and SAP SuccessFactors) with non-SAP applications through APIs.
- **SAP HANA**
  SAP HANA allows developers to create and expose XS JavaScript (XSJS) and REST APIs for accessing data stored in SAP HANA databases. These APIs support custom application development and integration with external systems.

- **SAP Fiori**
  SAP Fiori applications are built on OData services, which are consumed by SAP Fiori UI components. These services expose backend data and business processes through a standardized API interface, enhancing the user experience and system responsiveness.
- **SAP Analytics Cloud**
  SAP Analytics Cloud provides RESTful APIs for embedding analytics, managing users and permissions, importing data, and automating administrative tasks. These APIs facilitate integration with other applications and services, enhancing data visualization and analysis.
- **SAP Customer Experience (formerly SAP C/4HANA)**
  SAP Customer Experience includes APIs for integrating e-commerce capabilities, customer data management, marketing automation, and sales processes. These APIs empower businesses to deliver personalized customer experiences across various channels.
- **SAP Business Technology Platform**
  SAP BTP provides a suite of integration and development tools, including API management and integration capabilities.

APIs are integral to SAP's strategy for fostering connectivity, flexibility, and innovation across its product portfolio. They support digital transformation initiatives and enhance business agility. Figure 12.12 shows that APIs are an integral part of SAP S/4HANA and in SAP BTP, ABAP, SAP Analytics Cloud, SAP HANA, on-premise systems, and other systems.

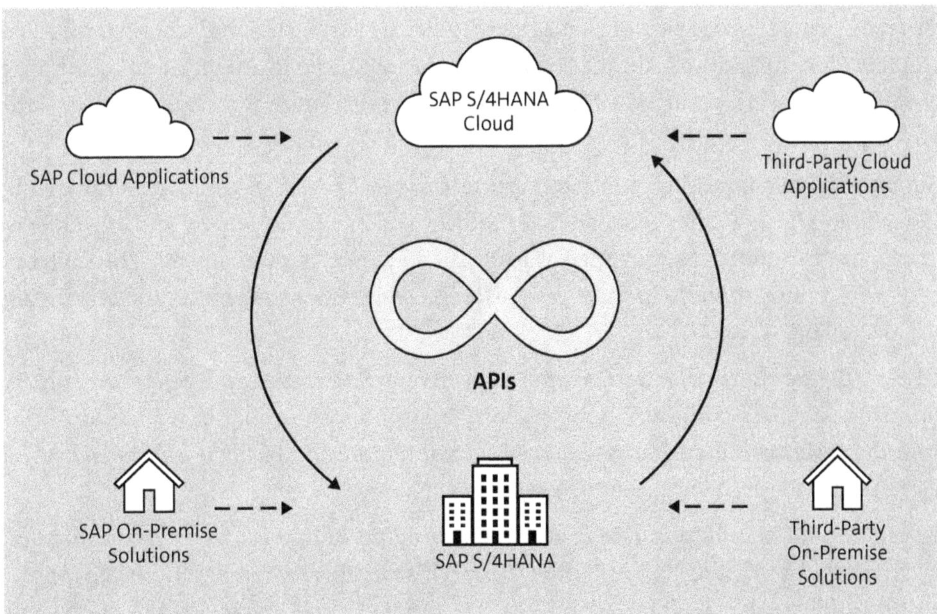

Figure 12.12 SAP Business Accelerator HUB

SAP Business Accelerator Hub is a web application that allows users to test and share different APIs of different products—for example, SAP S/4HANA, SAP SuccessFactors, or SAP BTP. Users can use these sandbox APIs to create extensions, applications, or integrations and test these with testing data provided through the sandbox API. For authentication, users need an API key that can be generated inside SAP Business Accelerator Hub. Users can get their API keys by logging in with an SAP ID account on SAP Business Accelerator Hub and choosing **Show API Key**, which enables users to discover, test, and use different APIs.

SAP Business Accelerator Hub is a key component of SAP's digital ecosystem, designed to facilitate the integration and innovation of SAP solutions.

SAP Business Accelerator Hub is a pivotal element of SAP's digital ecosystem, designed to enhance integration and drive innovation across SAP solutions. One of its primary goals is to simplify integration by providing a centralized platform where users can easily discover, explore, and access a diverse array of SAP APIs and integration scenarios. This centralized access streamlines the process of connecting SAP systems with other applications and services, both within the organization and externally.

Another key goal of SAP Business Accelerator Hub is to promote innovation. It encourages developers and businesses to create new applications and extend existing ones by offering comprehensive access to SAP's APIs, data, and services. By providing the necessary tools and resources, the hub supports the development of new functionalities and integrations, fostering a more dynamic and innovative environment.

The hub also focuses on enabling easy access to its resources. It features a user-friendly interface where developers and business users can easily find and understand APIs. Detailed documentation, API specifications, and sample code are included to simplify the process of working with SAP technologies, making it more accessible for users with varying levels of expertise.

Supporting collaboration is another important aspect of SAP Business Accelerator Hub. It facilitates interaction between SAP and its partners, customers, and developers through community features such as forums and feedback mechanisms. This collaborative environment helps in sharing knowledge, addressing challenges, and enhancing the overall ecosystem.

Driving digital transformation is a core objective of SAP Business Accelerator Hub. By enabling seamless connectivity between SAP systems and other cloud-based or on-premise systems, it accelerates the creation of integrated and agile business processes, essential for modern digital enterprises.

Furthermore, the hub promotes best practices by providing access to recommended approaches for utilizing SAP APIs effectively. This ensures that integrations are robust, secure, and scalable, aligning with industry standards and enhancing overall performance.

SAP Business Accelerator Hub is designed to make it easier for businesses to leverage SAP's technology ecosystem, integrate diverse systems, and drive innovation through API-based solutions. SAP Business Accelerator Hub can be accessed at *https://api.sap.com/*.

SAP Business Accelerator Hub offers several key features designed to support businesses in accelerating their digital transformation and optimizing their SAP solutions, such as the following:

- **Search and discovery tools**
  You can search for and discover SAP solutions and services tailored to your specific business needs using search and discovery tools.
- **Solution comparison**
  Compare different SAP solutions to identify the best fit for your organization.
- **Implementation guides**
  Review detailed implementation guides and best practices for various SAP solutions.
- **Industry standards**
  SAP Business Accelerator Hub offers insights into industry-specific practices and standards to help align SAP solutions with industry requirements.
- **Templates and accelerator**
  You have access to preconfigured templates and accelerators that simplify and expedite the deployment of SAP solutions.
- **Rapid deployment**
  SAP Business Accelerator Hub supports faster and more efficient implementation by using predefined configurations and scenarios.
- **Training resources**
  Access training materials, courses, and certification programs to help users gain the necessary skills and knowledge.
- **Learning paths**
  Review curated learning paths for different SAP solutions and roles within the organization.
- **Expert network**
  Connect with SAP experts, consultants, and partners for advice and support.
- **Forums and discussion groups**
  Access forums and discussion groups where users can share experiences and solutions.
- **Latest updates**
  SAP Business Accelerator Hub keeps users informed about the latest innovations, updates, and trends in SAP technology.

- **Future trends**
  SAP Business Accelerator Hub offers insights into emerging technologies and future trends that may impact your SAP strategy.
- **Consulting services**
  Access a network of consulting services and partners who can assist with SAP implementation and optimization.
- **Support services**
  Various support options help resolve issues and ensure smooth operation.
- **ROI calculation tools**
  Use these tools to estimate the return on investment (ROI) and business value of different SAP solutions.
- **Case studies**
  Review success stories and case studies from other businesses that have implemented SAP solutions.

12.2.5 API Consumption

API consumption refers to the process of using APIs to interact with and utilize the services or data provided by an API. In the context of SAP, API consumption involves integrating SAP's APIs into applications, systems, or workflows to leverage SAP's functionalities, data, and services. Let's look at the steps of API consumption:

1. **Understand API requirements**
   Start by reviewing the API documentation available in SAP Business Accelerator Hub. This documentation includes details about endpoints, methods (GET, POST, PUT, DELETE), request/response formats, authentication, and usage limits. Familiarize yourself with the technical specifications and requirements, such as authentication methods (OAuth, basic authentication), data formats (JSON, XML), and any constraints or rate limits.
2. **Obtain access**
   For most APIs, you need an API key or token to authenticate requests. This usually involves registering the application with SAP or a specific API provider to obtain these credentials. Also, ensure the necessary subscription or permissions to access the API are set, particularly for premium or restricted APIs.
3. **Set up the integration API client**
   Use an API client or library that matches the API's specifications and protocols. Many programming languages and environments offer libraries or tools for interacting with APIs. Then you can configure the API client with the necessary credentials, endpoints, and request parameters.
4. **Make API requests**
   Build the API requests according to the API documentation. This involves specifying the correct HTTP method (GET, POST, PUT, DELETE), setting headers (e.g., for authen-

tication), and including any required parameters or request bodies. Process the API responses, which may include handling data in JSON or XML format, error messages, or status codes.

5. **Test and debug**
   Use tools like Postman or Swagger UI to test API requests and responses before integrating them into an application. This helps ensure that requests are correctly formatted and that you understand the responses. Implement error handling to manage issues such as invalid requests, rate limits, or authentication errors.
6. **Integrate into the application and monitor usage**
   Incorporate the API calls into an application's codebase, handling responses appropriately and integrating data or functionalities as needed. Keep track of API usage, monitor performance, and ensure compliance with any usage limits or terms of service.
7. **Optimize API requests and perform maintenance as needed**
   Optimize API requests for performance, such as by reducing the number of calls, handling large data efficiently, and caching responses where appropriate. Stay updated with any changes or deprecations in the API, and make necessary adjustments to the integration as the API evolves.

12.2.6 SAP Gateway

SAP Gateway is a key component in the SAP S/4HANA landscape, playing a crucial role in enabling integration and communication between SAP systems and external applications or devices. The SAP Gateway hub provides a way to expose SAP data and functionality to external systems and applications via APIs. It acts as a central hub for managing these APIs and ensuring smooth integration.

Its key features and functions are as follows:

- **OData services**
  SAP Gateway Hub primarily uses OData to expose SAP data and functionality. OData is a standard protocol for creating and consuming query-able and interoperable RESTful APIs. This allows external systems to interact with SAP data in a standardized way.
- **Centralized service management**
  The hub provides a centralized point for managing and monitoring OData services. This includes service creation, deployment, and monitoring.
- **Integration**
  It facilitates integration between SAP S/4HANA and external applications, mobile apps, or other SAP systems. This integration is crucial for real-time data access and transactional operations.

- **Performance optimization**
  The hub includes mechanisms to optimize performance and ensure efficient communication between SAP systems and external applications.
- **Security**
  SAP Gateway Hub ensures secure communication by implementing authentication and authorization mechanisms. It supports various security protocols to protect data and transactions.
- **User-friendly development**
  For developers, the hub provides tools and frameworks to simplify the development of OData services. This includes SAP Gateway Service Builder (Transaction SEGW) for creating and managing services.
- **Customizability**
  Organizations can customize and extend standard OData services to meet specific business requirements.

In an SAP S/4HANA environment, the SAP Gateway hub enables seamless interaction between the SAP core system and other systems or applications. For example, for a custom application or a third-party system that needs to access SAP S/4HANA data, the hub provides the APIs and manages the communication between these systems.

Its common use cases include the following:

- **Mobile applications**
  Enabling mobile apps to interact with SAP S/4HANA data
- **Third-party integrations**
  Integrating with external systems or services
- **Custom web applications**
  Building web applications that need to access SAP data

The SAP Gateway hub is an essential component for enabling modern, flexible integration scenarios in SAP S/4HANA. It provides the tools and infrastructure necessary to expose and manage APIs, ensuring that SAP systems can interact effectively with external applications and services.

12.2.7 SAP Gateway Deployment Options

There are four deployment options for SAP Gateway:

- **Central hub deployment**
  This is where customers can only access SAP Gateway functionalities on a dedicated hub system. The SAP Business Suite backend system is the platform for developing services before exposing them on the hub server. This deployment alternative supports the composition and routing of multiple systems while allowing access only via SAP Gateway Hub, thus boosting security.

Central hub deployment should be used in situations where either access to the backend system is restricted or development on the backend isn't possible. It should also be used on releases prior to SAP NetWeaver 7.40, when an installation of the IW_BEP component on the backend isn't allowed. In this form of hub deployment, server functionalities are only used on the SAP Gateway Hub system. The benefit of this hub deployment is that the backend systems don't have to be touched; that is, there's no need to install or upgrade SAP Gateway add-ons in the backend. In addition, services developed by third-party developers don't need a deployment on the backend system.

In the scenario shown in Figure 12.13, multiple backend systems want to use the central hub as a central entry point through which to expose the SAP Fiori UI as the frontend.

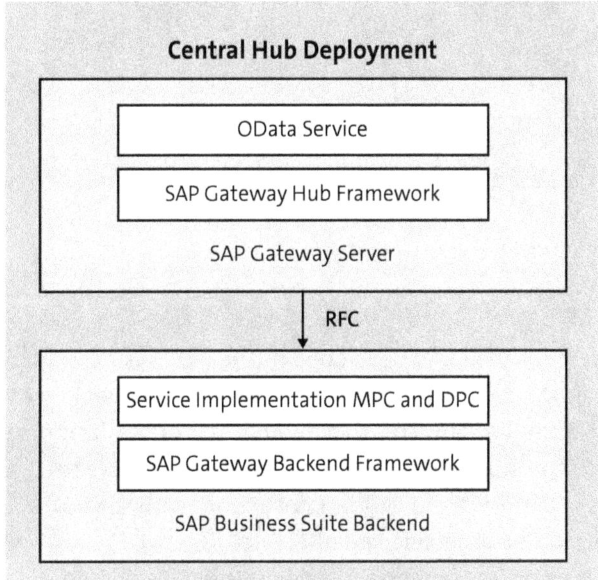

Figure 12.13 Central Hub Deployment Option

- **Central hub with development in hub**
  Here the routing will be easy in case of multiple systems: the development of the project and the registration service will be handled in the frontend where the SAP Gateway component is installed, while data retrieval (physical data from database tables) will be happening in the backend via RFC.

  This method can be chosen when there is no chance of installing SAP Gateway components in the backend system as they came before the release of SAP 7.0 but data resides only in this backend system. In this situation, the class created in the backend system cannot be addressed directly; it has to be exposed only via interfaces such as RFC or BAPI.

As shown in Figure 12.14, the registration service will be handled in the frontend where the SAP Gateway component is installed. Data retrieval will occur in the backend via RFC.

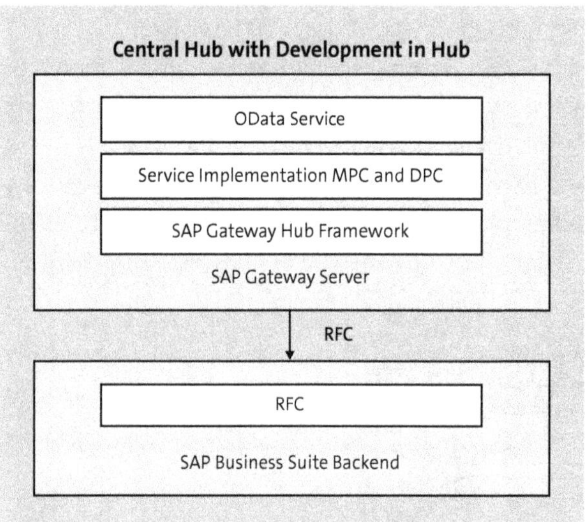

Figure 12.14 Central Hub with Development in Hub Option

- **Embedded deployment**

  In this option, the services are published and developed in the SAP Business Suite backend system. The organization's SAP Gateway components will also be directly deployed to the SAP Business Suite system. This deployment option has minimal runtime overhead and a reduced number of remote calls. However, you have to configure the SAP Gateway system, and an organization may end up with several SAP Fiori launchpads if it uses SAP Fiori on embedded architecture.

  As shown in Figure 12.15, the SAP Gateway project creation, registration, and consumption of service is done in one SAP backend system.

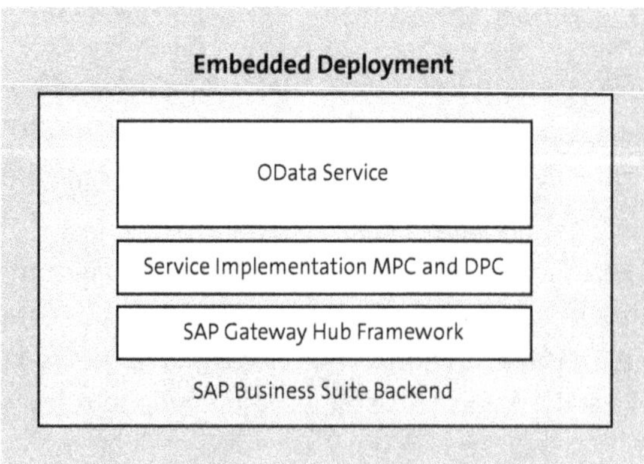

Figure 12.15 Embedded Development Option

- **SAP Gateway as a service**

  OData provisioning is a technology that exposes the business logic and data in the SAP cloud via OData services. The biggest advantage of this deployment option is that customers need not run their own SAP Fiori frontend server or SAP Gateway Hub as the OData services and SAP Fiori apps are deployed on SAP BTP.

  OData provisioning is offered as part of SAP BTP deployment, as shown in Figure 12.16. This allows for access of the data from within the cloud. It's basically like the development on the hub option, with the difference that it's managed by SAP BTP.

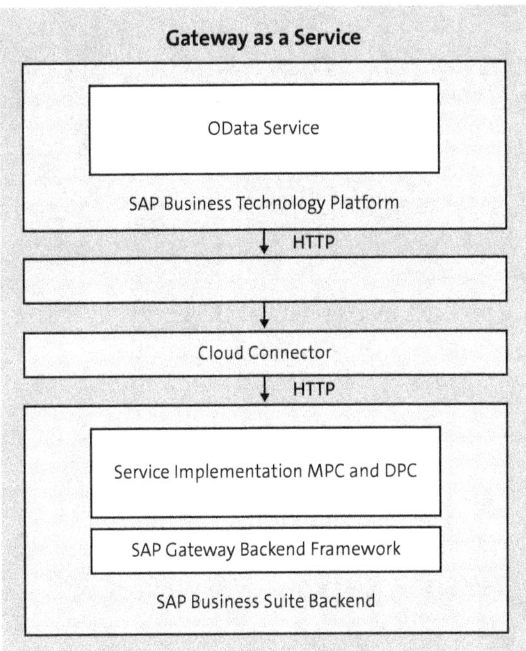

Figure 12.16 Gateway as a Service Option

12.2.8 Sample API

Before we begin, let's look at a sample API: Product Master (A2X), technical name API_PRODUCT_SRV. This API can be viewed in SAP Business Accelerator Hub. This is the API used later on in Section 12.3.

This synchronous inbound service enables the user to create, read, update, and delete the master data for products by exposing the remote API views through OData.

> **Note**
>
> This service is published on SAP Business Accelerator Hub. For more information about APIs, see *https://api.sap.com/*.

Table 12.1 describes the available operations and sample URLs.

Operation	Sample URLs
GET (to read product master data)	/sap/opu/odata/SAP/API_PRODUCT_SRV/A_Product (Product='TG-17')
POST (to create product master data)	/sap/opu/odata/SAP/API_PRODUCT_SRV/A_Product
PATCH, PUT (to update product master data)	/sap/opu/odata/SAP/API_PRODUCT_SRV/A_Product (Product='TG-17')
DELETE (to delete product master data)	/sap/opu/odata/SAP/API_PRODUCT_SRV/A_Product Description(Product='TG-17', Language='EN')
BATCH (to update multiple product master data records)	/sap/opu/odata/sap/API_PRODUCT_SRV/$batch

Table 12.1 Sample Operations and URLs

You can create, read, update, and delete product master data at the entity level. Note that you can delete master data for the product only in the following entities:

- Product description
- Sales tax
- Units of measure
- Units of measure EAN
- Basic text
- Inspection text
- Purchase text
- Plant text
- Sales text

You can extend the following entities for product master data using the Custom Fields and Logic app (available via the SAP Fiori launchpad) to add custom fields as per requirements:

- `A_Product`
- `A_ProductPlant`
- `A_ProductSalesDelivery`
- `A_ProductStorageLocation`
- `A_ProductUnitsOfMeasure`

Service Structure

All information from this OData service is utilized when processing data, including appropriate checks for restricted values and authorizations. If any issues arise during an operation, the system displays error messages in the response. This service is built using the CDS views based on product master data tables via the gateway service builder and Service Adaptation Description Language (SADL).

12.3 Extending the EDI Scenario with APIs

Although we do not recommend mixing different integration patterns, technologies, or standards within a specific solution, we will make an exception here. The EDI scenario configured in Chapters 6 through 9 will be now extended by an API-based interface. The approach is that we keep all four message-based transactions as is, with no modifications. Next to these classic EDI interfaces, we will enable an API via which our trading partner will have the possibility for a real-time query and check of products that we are selling. From a business perspective, both sides will benefit from such functionality as this would have a significant, positive impact on data quality and correctness in EDI sales orders.

If you have studied in detail the earlier part of this chapter, then the fact that we will not be developing an API should not surprise you. The development is not necessary as there is a whole set of predefined, standard APIs available within SAP S/4HANA. We will make use of one of such existing API, but there will be still configuration activities that we need to perform within SAP S/4HANA to enable SAP Gateway and this specific API. In our example, because our trading partner cannot access our SAP S/4HANA system directly and we have some additional security requirements, we will also make use of the cloud connector and API Management. Let's go through the necessary configuration step by step in the following sections.

12.3.1 SAP Gateway Activation

In our scenario, we will make use of SAP Gateway in an embedded deployment scenario. In this section, we will go through the main configuration steps necessary for activation of the embedded SAP Gateway:

1. In the first step, we will activate the SAP Gateway service. To perform this operation, start Transaction SPRO and navigate to the appropriate option in the SAP Gateway customizing area, as shown in Figure 12.17. The exact path is as follows: **ABAP Platform • SAP Gateway • OData Channel • Configuration • Activate or Deactivate SAP Gateway**.

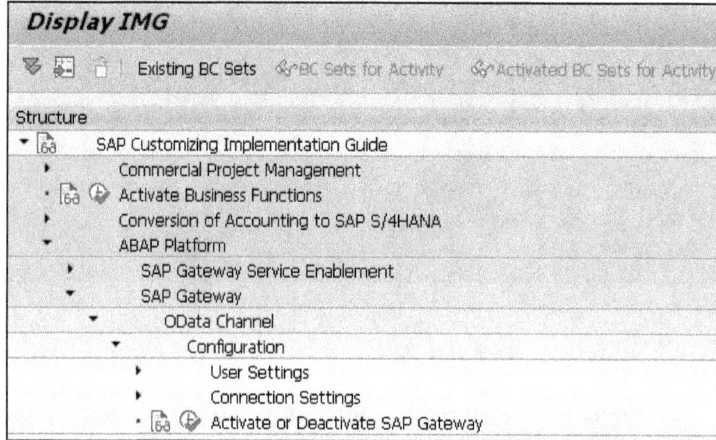

Figure 12.17 Transaction SPRO Path for Activation of SAP Gateway

It may be that SAP Gateway is already activated in your SAP S/4HANA system, but in most default cases it is deactivated, in which case you should see a window as shown in Figure 12.18.

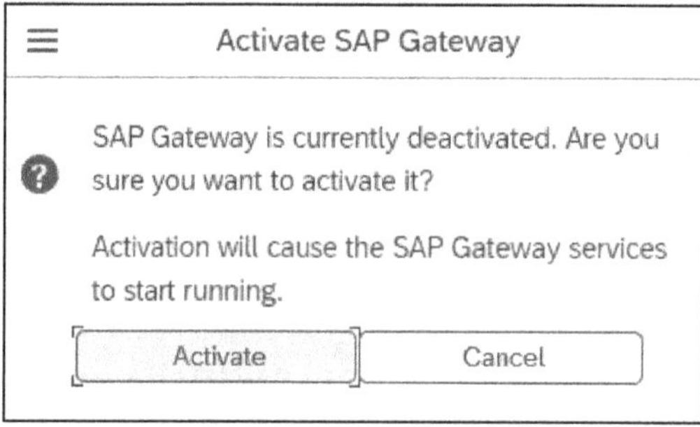

Figure 12.18 Confirmation Prompt for SAP Gateway Activation

Click **Activate**. This should result in the SAP Gateway activation and a short confirmation, as shown in Figure 12.19.

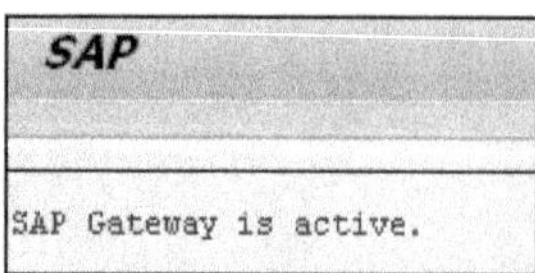

Figure 12.19 Confirmation of SAP Gateway Status

2. In the next step, we need to create a system alias, the purpose of which is to make a connection to the SAP backend system. In our embedded deployment scenario, it is enough to create an alias pointing to our local system. Using Transaction SPRO, navigate to the appropriate function: **ABAP Platform • SAP Gateway • OData Channel •**

Configuration • Connection Settings • SAP Gateway to SAP System • Manage SAP System Alias. Maintain the parameters as follows:

- **System Alias: LOCAL**
- **RFC Destination: NONE**
- **Local Gateway: set**
- **Software Version: DEFAULT**
- **Description: Local System Alias**

All the other values can be left empty. The final configuration should look as shown in Figure 12.20.

Update System Alias	
System Alias	LOCAL
RFC Destination	NONE
Local Gateway	☑
For Local App	☐
OData on Backend	☐
Target System ID	
Target Client	
Software Version	DEFAULT
Web Service Provider System	
Description	Local System Alias

Figure 12.20 Adding System Alias LOCAL

3. The next step is to configure at least one destination system in the SAP Gateway settings. This will again be pointing to our SAP S/4HANA system. Navigate through Transaction SPRO: **ABAP Platform • Gateway Service Enablement • Backend OData Channel • Connection Settings to SAP Gateway • SAP Gateway Settings.** Maintain the settings as follows:
   - **Destination system: S42**
   - **Client: 400**
   - **System Alias: S42**
   - **RFC Destination: S42**

   The settings may vary depending on your system, but the configuration is usually straightforward. The final configuration is shown in Figure 12.21.

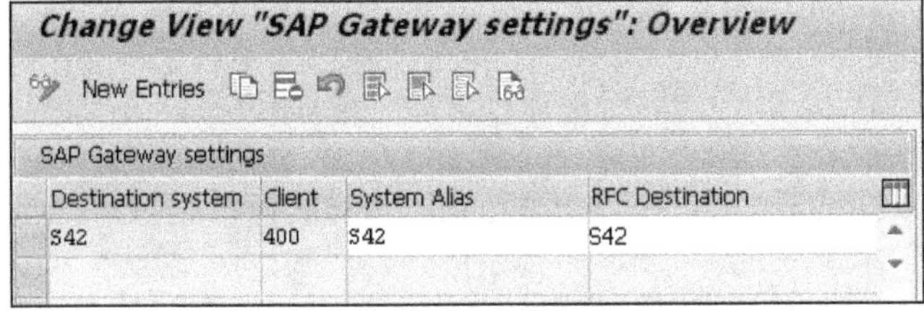

Change View "SAP Gateway settings": Overview

New Entries

SAP Gateway settings

Destination system	Client	System Alias	RFC Destination
S42	400	S42	S42

Figure 12.21 Adding Connection Settings to SAP Gateway

4. In the last step, we must activate services in the OPU node. Start Transaction SICF and navigate to the appropriate node: **default_host • sap • opu**. Choose the last node, right-click it, and choose **Activate Service**. The mentioned OPU node is shown in Figure 12.22.

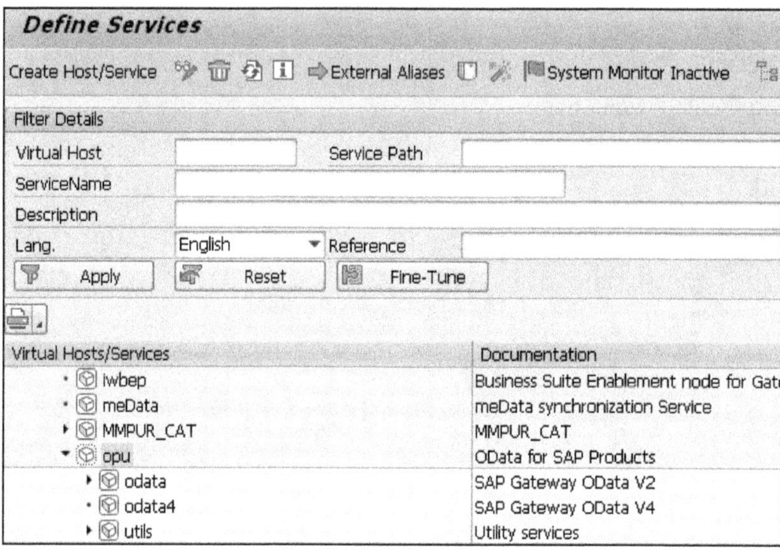

Figure 12.22 Transaction SICF Services and OPU Node

These steps are related only to a basic activation of the SAP Gateway service, but this should allow you to enable and publish services on SAP Gateway. For production usage, there may be additional configuration activities depending on your specific requirements, landscape setup, security, or logging/tracing requirements. It is recommended to consider that and go through all the possible configuration parameters and options.

SAP Gateway is now active and configured for managing OData services. We can go to the next section, where we will publish specific a OData service on SAP Gateway.

12.3.2 Enabling the Product API

We will start with a reminder that we are using embedded deployment, and our system configured in the previous section is LOCAL. The intention of this section is to activate a standard SAP OData API related to a business object called *product*. The technical name of this API is API_PRODUCT_SRV, and its title is Product API.

To activate the OData service, follow these steps:

1. Start Transaction /IWFND/MAINT_SERVICE (Activate and Maintain Services). You should see a screen like that shown in Figure 12.23. In this example, you can see a few services that are already active on SAP Gateway, such as /BDTS/COMMAND_SRV and /BDTS/INSTALLED_SAP_NOTE_SRV at the top of the list.

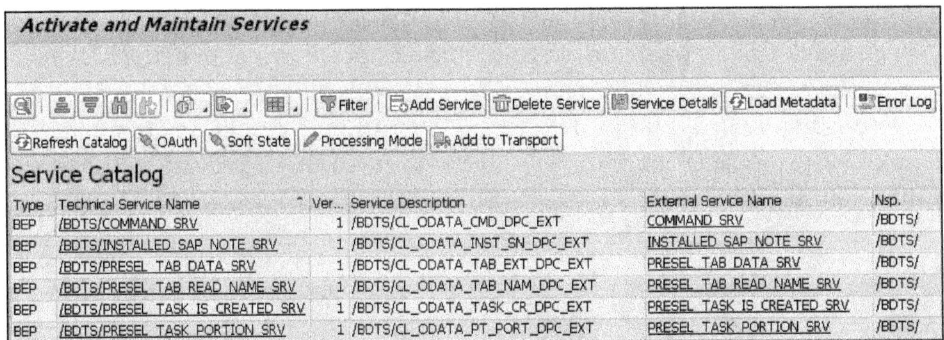

Figure 12.23 Activate and Maintain Services Transaction

2. Click the **Add Service** button located in the top panel. This will open a new window like that shown in Figure 12.24. Maintain two fields in the **Filter** section of the **Add Selected Services** transaction screen:
   - **System Alias: LOCAL**
   - **Technical Service Name: API_PRODUCT_SRV**

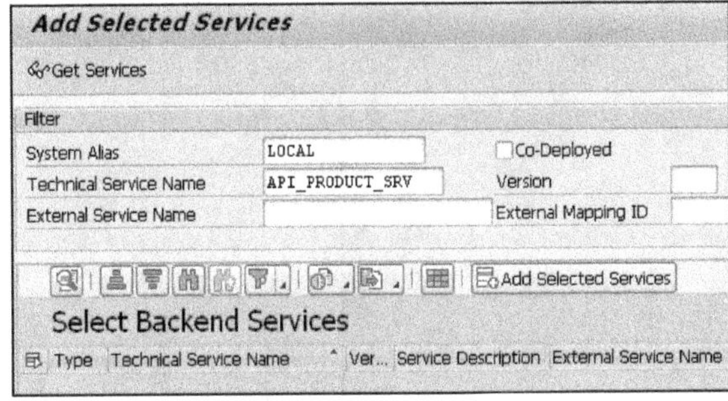

Figure 12.24 Add Selected Service Selection Screen with Filter Parameters

3. With the **Filter** parameters maintained, click the **Get Services** button. This will display services that match the selection criteria. In our case, it is only one specific service, as shown in Figure 12.25.

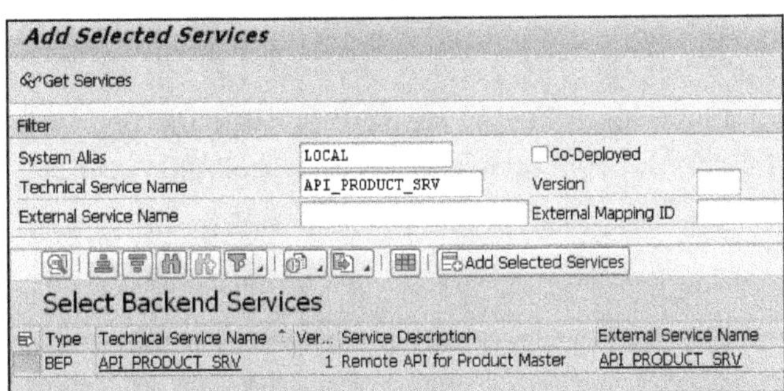

Figure 12.25 Add Selected Services: API_PRODUCT_SRV API

4. Select the appropriate service from the list and click **Add Selected Services.** This will open a new window where we can adjust the technical parameters of the added service. Figure 12.26 shows the default values proposed by SAP S/4HANA. All these values can be accepted; for this demo, there is no need to change any of them. One value that is missing and must be added to allow successful activation is **Package Assignment**. Maintain a value appropriate for your system and confirm by clicking the green icon located in the right corner at the bottom of this screen.

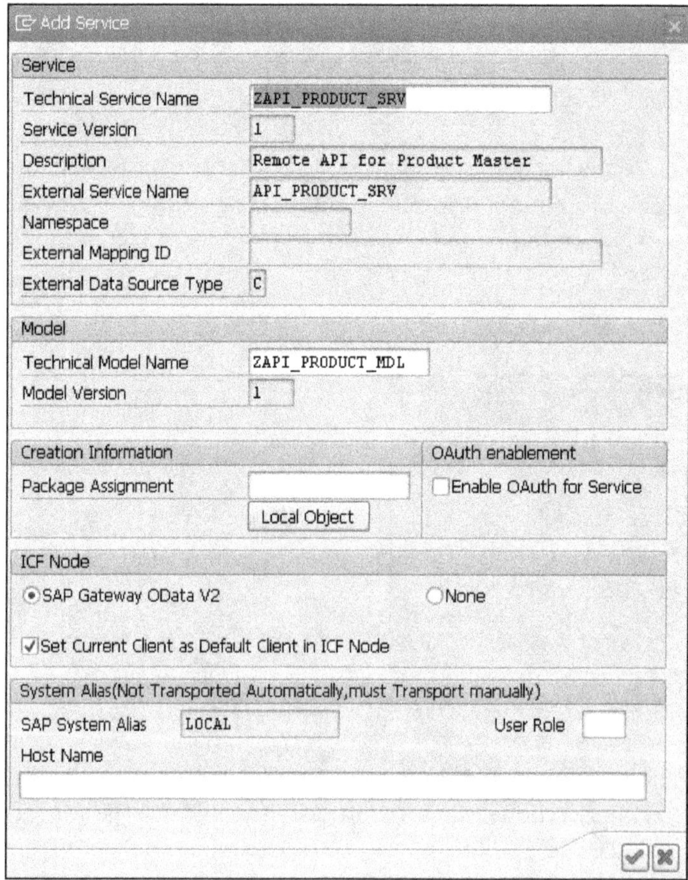

Figure 12.26 Add Service: Technical Parameters Related to Service Activation

The newly added service should be now visible in Transaction /IWFND/MAINT_SERVICE (Activate and Maintain Services), under **Technical Service Name: ZAPI_PRODUCT_SRV**, as shown in Figure 12.27.

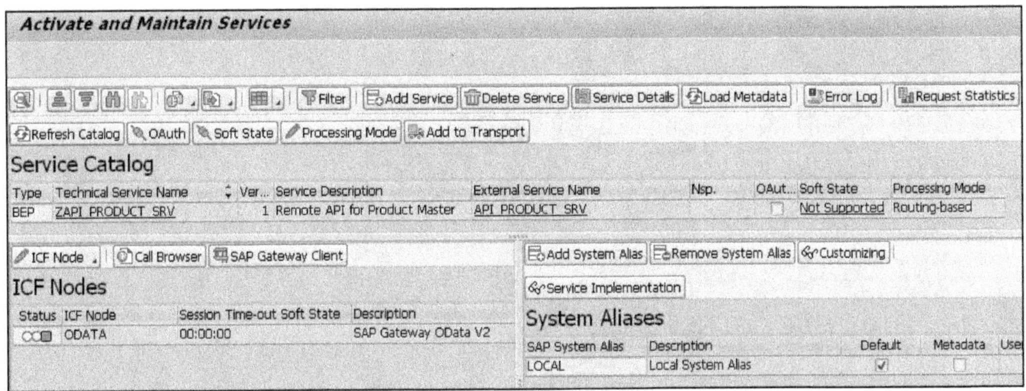

Figure 12.27 ZAPI_PRODUCT_SRV Activated on SAP Gateway as OData V2 Service

In the lower part of the screen shown in Figure 12.27, you can see two sections: the one on the left shows information related to ICF nodes. By clicking **ICF Node**, you can navigate directly to Transaction SICF and this exact service, as shown in Figure 12.28. Another powerful tool is hiding behind SAP Gateway Client, and we will look at it in Section 12.3.3.

Virtual Hosts/Services	Documentation	Reference Service
default_host	VIRTUAL DEFAULT HOST	
sap	SAP NAMESPACE; SAP IS OBLIGED NOT T...	
opu	OData for SAP Products	
odata	SAP Gateway OData V2	
sap	Namespace	
api_product_srv	API_PRODUCT_SRV	

Figure 12.28 View of OData API Service in Transaction SICF

The left section shows information related to system aliases. You can see that this OData service is pointing to the local system via **SAP System Alias: LOCAL**, which we configured in previous section. At any time, you can see details of the selected service by choosing **Service Details** located at the top of the screen. Details related to the API_PRODUCT_SRV service are shown in Figure 12.29.

At this stage, we have the OData API service active on SAP Gateway embedded in our SAP S/4HANA system. It is already possible to call this OData API, and this is something we will show in the next section.

Display Service

Cleanup Cache | Configuration

Service Information	
Technical Service Name	API_PRODUCT_SRV
Service Version	1
Description	Remote API for Product Master
SAP Object Type	Product
External Service Name	API_PRODUCT_SRV
Namespace	
Data Provider Class	CL_API_PRODUCT_DPC_EXT
Created By	SAP
Changed By	SAP
Package	VDM_MD_PRODUCT_API_V1
Service Type	WebAPI

Extension for Service	
Technical Service Name	
Service Version	0

Model Information	
Technical Model Name	API_PRODUCT_MDL
Model Version	1
Description	Remote API for Product Master
Model Provider Class	CL_API_PRODUCT_MPC_EXT
Created By	RAYAVARAPUAR
Changed By	
Package	VDM_MD_PRODUCT_API_V1

Create Model | Unassign Model | Assign Model | Annotation Model

Figure 12.29 API_PRODUCT_SRV Service Details

12.3.3 SAP Gateway Client

Although the key idea behind and benefits of OData API services are related to communication among remote components, we will stay on the SAP S/4HANA system and run a couple of local test. This will be possible due to the fact that the SAP S/4HANA and SAP Gateway components come with SAP Gateway Client, as shown in Figure 12.30. This tool can be started via Transaction /IWFND/GW_CLIENT, and it gives the possibility to call OData services running on a local gateway. The tool supports all key HTTP methods and has great flexibility when it comes to preparation and configuration of requests.

Another option for starting the SAP Gateway client is to trigger it from the Activate and Maintain Services transaction mentioned in the previous section. If you look back at Figure 12.30, you can see the **SAP Gateway Client** button in the middle of the screen. Clicking this button will start the SAP Gateway Client with **Requested URI** already prefilled with the selected OData service.

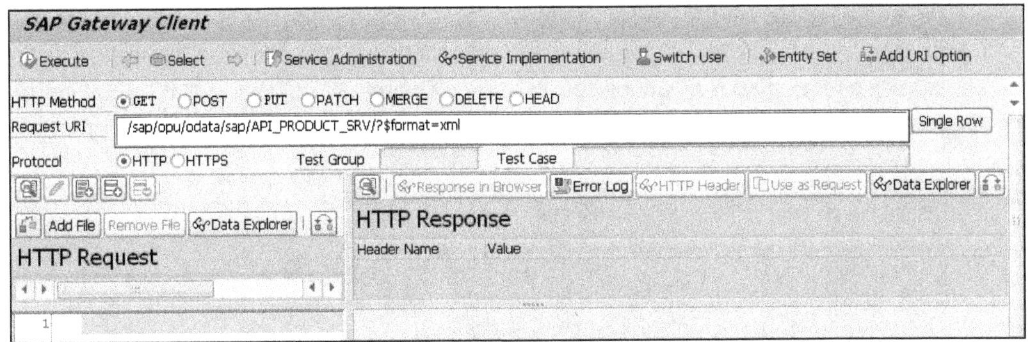

Figure 12.30 Sample View of SAP Gateway Client Transaction

The aim of our quick test will be to check if the OData service actually works and if it returns the expected result (product data). We will limit our test calls to the GET method, as we and our trading partners are only interested in fetching the product data out of the SAP S/4HANA system.

The first request will be for fetching service *metadata*, a static resource that describes the data model and type system. The metadata tells us how to navigate between entities in the system. The request and (part of the) response are shown in Figure 12.31. We will not go into details of the metadata itself, but in the response header you can see HTTP status code 200, which confirms the successful execution of the service.

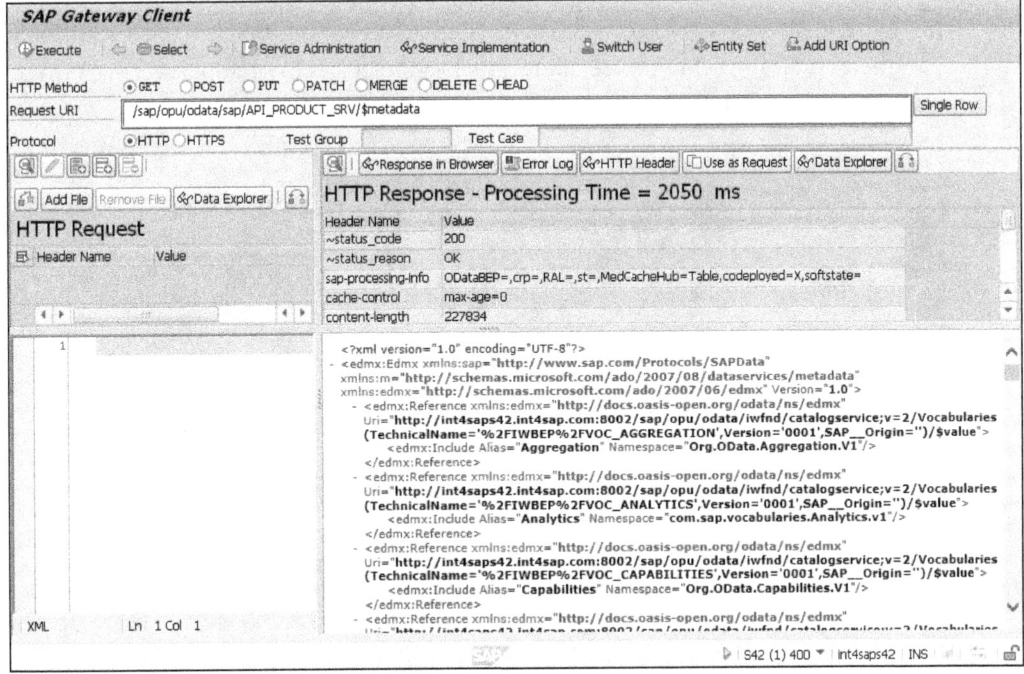

Figure 12.31 SAP Gateway Client: Request for API_PRODUCT_ SRV Service Metadata

We will make one additional call to check if the service returns top-level data for a specific product, which will be the product used earlier in the EDI scenario. In Figure 12.32, you can see a request with a GET method on a specific product number, MZ-FG-R300. In the response, you can see the details of this product: note the response is in XML format.

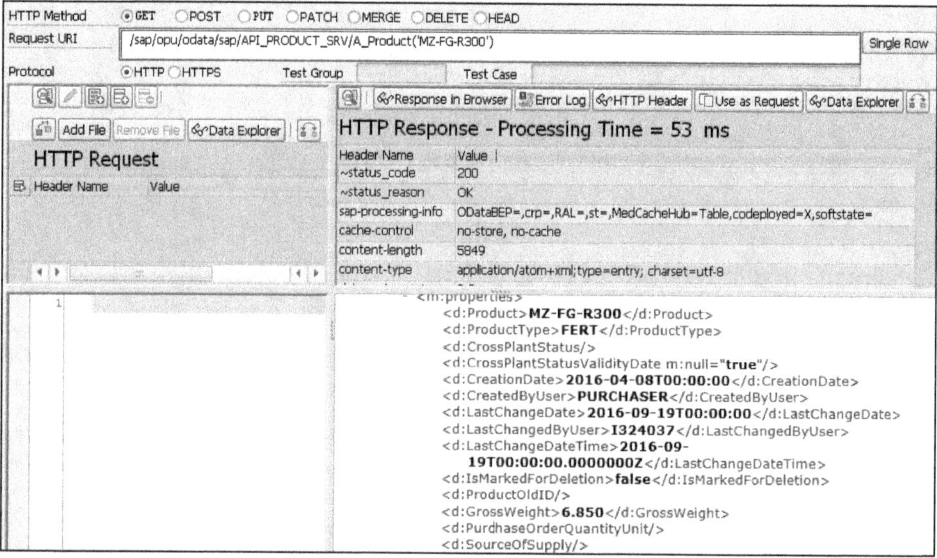

Figure 12.32 Request on API_PRODUCT_SRV Service for Specific Product

In Figure 12.33, we have exactly the same request as earlier, but with one additional parameter indicating that we are requesting JSON format. OData supports two formats for representing resources: XML and JSON.

HTTP Method GET POST PUT PATCH MERGE DELETE HEAD
Request URI /sap/opu/odata/sap/API_PRODUCT_SRV/A_Product('MZ-FG-R300')?$format=json
HTTP Response - Processing Time = 51 ms

Header Name	Value
~status_code	200
~status_reason	OK
sap-processing-info	ODataBEP=,crp=,RAL=,st=,MedCacheHub=Table,codeployed=X,softstate=
cache-control	no-store, no-cache
content-length	4153
content-type	application/json; charset=utf-8


```
},
"Product" : "MZ-FG-R300",
"ProductType" : "FERT",
"CrossPlantStatus" : "",
"CrossPlantStatusValidityDate" : null,
"CreationDate" : "\/Date(1460073600000)\/",
"CreatedByUser" : "PURCHASER",
"LastChangeDate" : "\/Date(1474243200000)\/",
"LastChangedByUser" : "I324037",
"LastChangeDateTime" : "\/Date(1474243200000+0000)\/",
"IsMarkedForDeletion" : false,
"ProductOldID" : "",
"GrossWeight" : "6.850",
```


Figure 12.33 Request on API_PRODUCT_SRV Service for Specific Product in JSON Format

There is no need for additional development or configuration on the SAP S/4HANA side for this, and the selection of an appropriate format can be defined in the request itself.

By performing a few simple calls, we proved that the product-related OData API is properly activated and returns the expected data. But this only happens within SAP S/4HANA itself, whereas our requirement is to make this OData API available for our trading partner, which has no access to our SAP S/4HANA on-premise system. We will explain the situation and look at a possible solution in the next section.

12.3.4 Exposing the API: Security Aspects

Let's look at the simplified diagram of the landscape, which is shown in Figure 12.34. At the bottom part of the diagram, we have an SAP S/4HANA on-premise system with an OData API exposed on embedded SAP Gateway. This part is within an internal network that is protected by a firewall and other security measures; direct access from the public network to SAP S/4HANA is not possible. 12

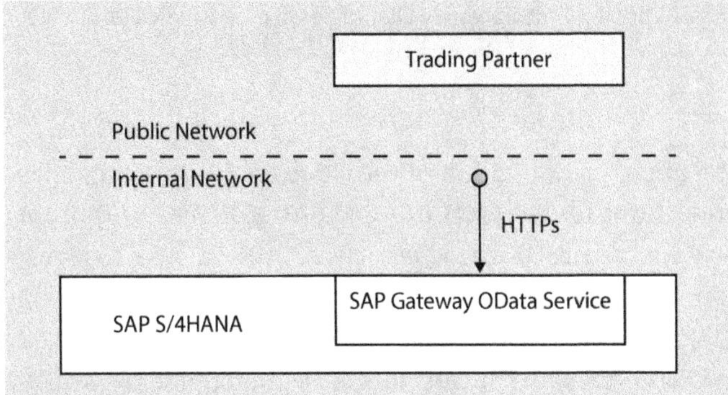

Figure 12.34 Starting Point for Exposing SAP Gateway OData Service on Public Network

We will make use of API Management on SAP BTP to expose the OData service for our trading partner. This component of SAP Integration Suite has many additional security features that let us protect our service after its exposure on a public network. But first we have to make sure that SAP BTP itself has secure access to the SAP Gateway instance located in an internal network. This will be managed by the cloud connector. The overall solution is presented in Figure 12.35.

The setup presented here guarantees security for both the SAP S/4HANA system in our internal network and the API endpoint on the public network. Within API Management, we do not change any logic or the functionality of the original OData service; we only create a facade to control access to it. In the next section, we will look at configuration within the cloud connector.

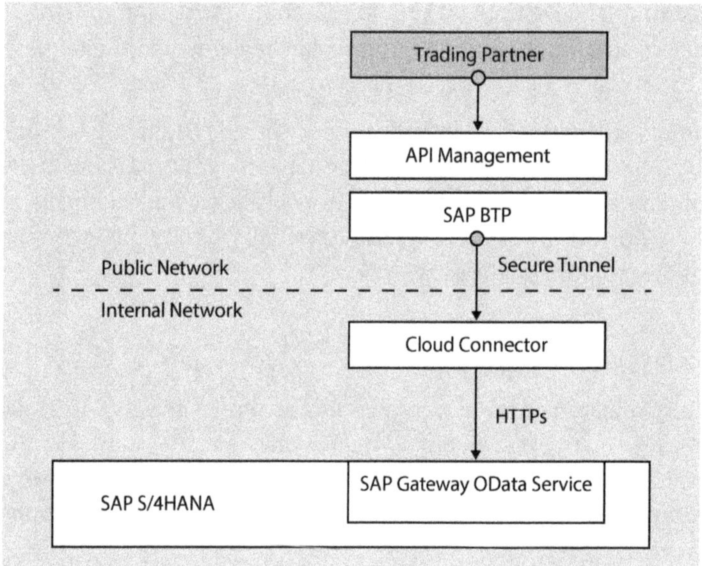

Figure 12.35 Overall Solution for Exposing SAP Gateway OData Service on Public Network

12.3.5 Cloud Connector

In Chapter 6, we covered the initial configuration of the cloud connector component and we also configured connections to SAP S/4HANA and SAP BTP. As you probably recall, within our EDI scenario we used a SOAP IDoc service on SAP S/4HANA to receive IDocs from Cloud Integration. This SOAP IDoc service is exposed on HTTPS, and for this protocol we also configured a connection from SAP BTP to SAP S/4HANA. SAP Gateway also exposes OData services via HTTPS, which means that both the standard SOAP IDoc service and the product OData API service are reachable on the same internal host and the same HTTPs port, 44301. This is something what we already maintained in the cloud connector in Chapter 6. As a reminder, let's look at the existing cloud connector configuration in Figure 12.36.

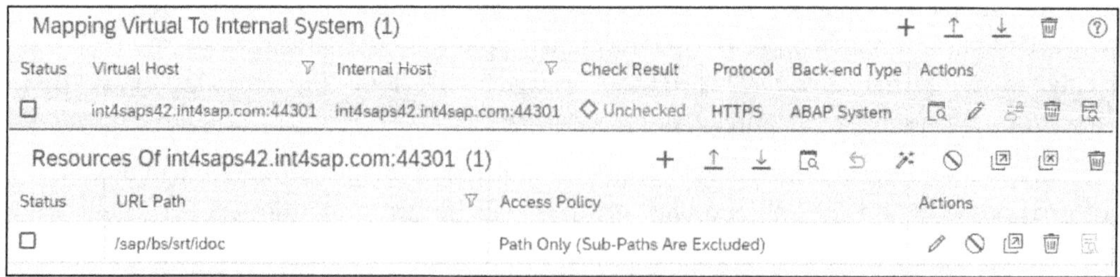

Figure 12.36 Internal SAP S/4HANA System Connected via Cloud Connector

What indicates the exact service being called is the content of the **URL Path** field. The SOAP service related to processing inbound IDocs is in path **/sap/bc/srt/idoc**. This is the only URL path that we enabled during the initial configuration. In the cloud connector configuration related to our SAP S/4HANA internal host, we need to add a URL path pointing to the OData service. We will maintain the generic URL path */sap/opu/odata/sap/* while also allowing all subpaths. Click + located in the middle of the screen (**Resources...** level) and select **Add Resource**. In the newly appearing window, maintain the following fields:

- **URL Path**: **/sap/opu/odata/sap/**
- **Active**: Select this checkbox
- **Access Policy**: **Path and All Sub-Paths**

The complete configuration should look as shown in Figure 12.37. Confirm by clicking **Save**.

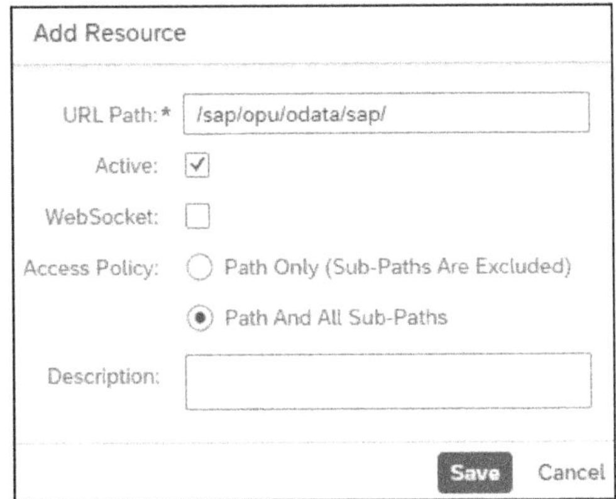

Figure 12.37 Adding URL Path Related to OData Services

The cloud connector configuration for our internal SAP S/4HANA host should contain two URL paths, as shown in Figure 12.38.

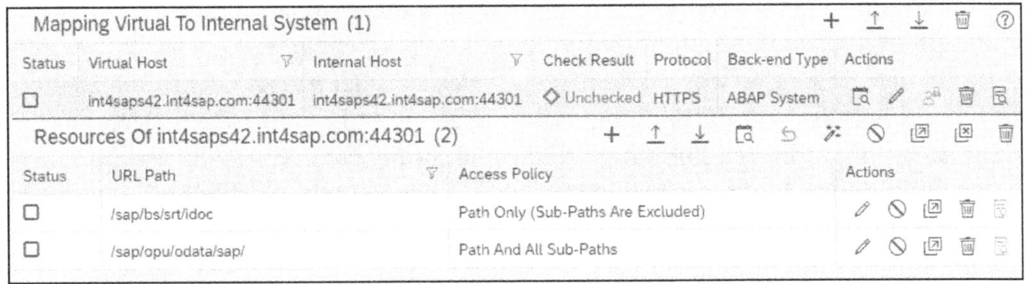

Figure 12.38 Final Configuration of URL Paths in Cloud Connector

SAP BTP and its components should have now the access to OData Services exposed on SAP S/4HANA. We can leave the cloud connector now and move to API Management to configure the API endpoint for our trading partner.

12.3.6 API Management

Trying to demonstrate all the capabilities and functions of API Management could be a topic for at least one additional book. We will limit the activities to basic functionality, where we expose the internal service publicly with some basic security measures. This setup will allow our trading partner to fetch the product-related information. But we need to keep in mind that we operate on a public network, so in the actual project, where you expose actual business data, the whole setup and especially the security measures should be carefully planned and, in most cases, also more strict.

In this section, we'll walk through configuration of our demo scenario in API Management and then discuss API Management's API analytics capabilities.

Configuration

Let's start with configuration of our demo scenario. Navigate to the API configuration area in SAP Integration Suite via **SAP Integration Suite Home Page • Configure • APIs**, as shown in Figure 12.39.

Figure 12.39 Initial Screen of API Configuration Area within SAP Integration Suite

In the first step, we will create an *API provider*, a configuration entity that defines the connectivity details for the target backend system on which the actual services are running. In our case, the SAP S/4HANA system is the API provider. We will need to configure the details necessary to establish the connection to the SAP backend. Navigate to the **API Providers** tab, as shown in Figure 12.40.

Click the **Create** button located in the middle of the screen. This will open a new window where you can maintain the necessary details. In the first tab, called **Overview**, enter "S4HANA_BackendSystem" as the name for the API provider in the **Name** field. Optionally, you can add a description (see Figure 12.41).

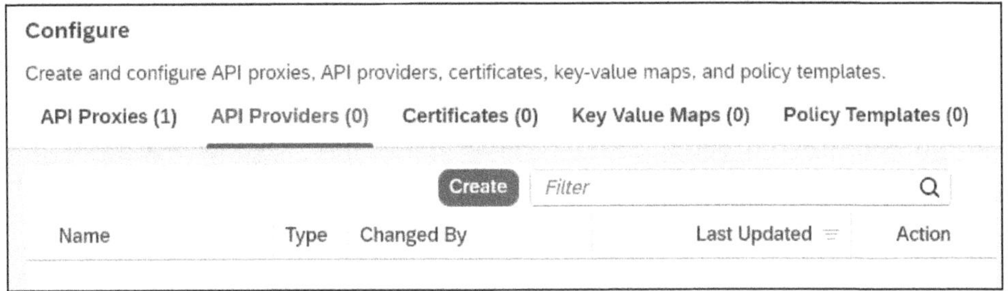

Figure 12.40 API Provider: Test Connection with Successful Response Code

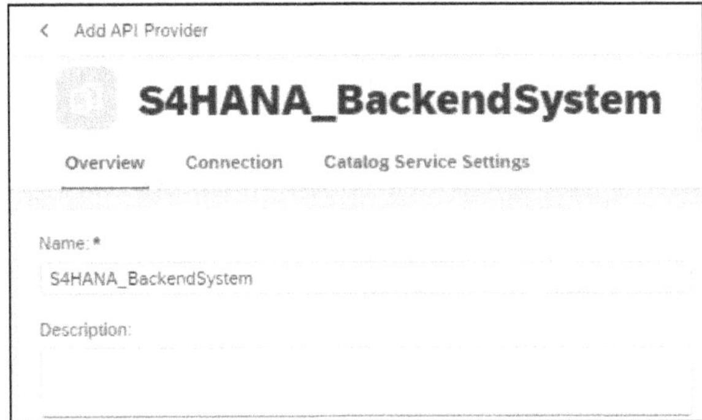

Figure 12.41 API Provider: Overview

In the next tab, **Connection**, you maintain connectivity details for the SAP S/4HANA system. **Type** indicates the location of the system; **On Premise** is the correct option for our case, where the system is located in an internal network and the access is arranged via the cloud connector. **Host** and **Port** should be maintained as per your system properties. **Location ID** is also related to the cloud connector properties. Leave **Authentication** as **None.** At the end, in the **Additional Properties**, add property "sap-client" with value **400**, which is the client number in our SAP S/4HANA system. The sample configuration is shown in Figure 12.42.

The last tab, **Catalog Service Settings**, is optional and lets us connect to the catalog service running on SAP Gateway. With this configuration in place, we will be able to browse through the catalog of all services available on SAP Gateway. The values maintained in **Path Prefix** and **Service Collection URL** are used to build the **Catalog URL**. The default value for **Path Prefix** is **/sap/opu/odata**, and for the **Service Collection URL** we should enter "/iwfnd/catalogservice/ServiceCollection". At the end, set **Authentication Type** as **Basic** and maintain the appropriate credentials. The complete configuration is shown in Figure 12.43.

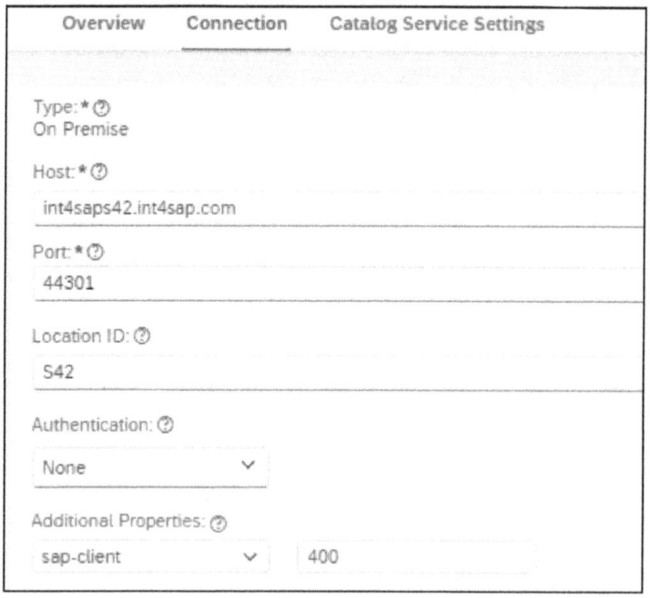

Figure 12.42 API Provider: Connection

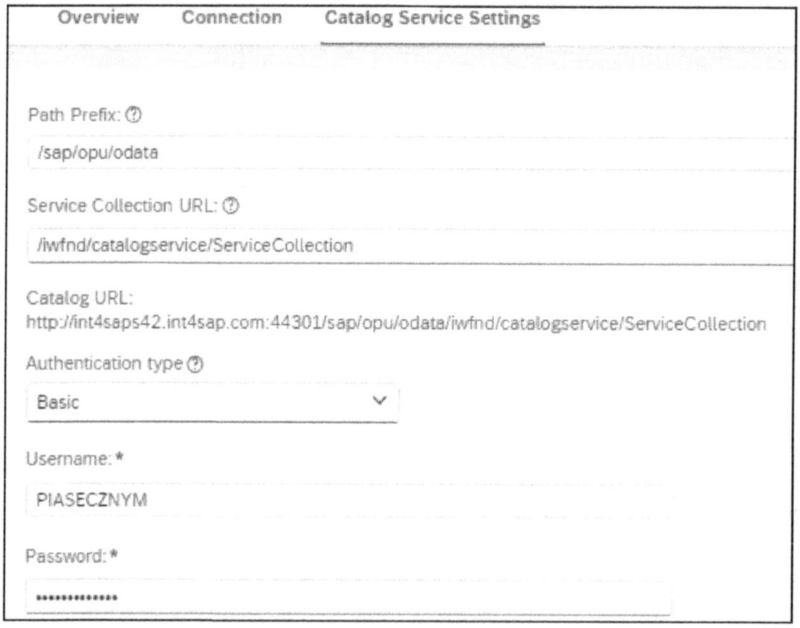

Figure 12.43 API Provider: Catalog Service Settings

After saving the API provider configuration, in display mode, you can find a **Test Connection** option. Figure 12.44 shows a sample execution with a successful response code.

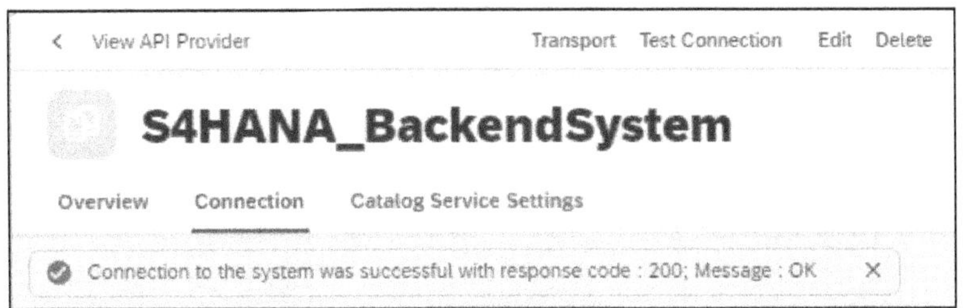

Figure 12.44 API Provider: Test Connection with Successful Response Code

After confirming that the connection to the SAP S/4HANA system works, you can navigate back to the **API Proxies** tab as shown previously in Figure 12.39. Once again, we make use of the **Create** button, which starts the process of API proxy creation. In the newly opened window, fill in the configuration parameters as follows:

- **Select**
  Choose **API Provider** as the API type
- **API Provider**
  Select **S4HANA_BackendSystem**, which we configured a while ago
- **URL**
  /sap/opu/odata/sap/API_PRODUCT_SRV
- **Name**
  API_PRODUCT_SRV
- **Title**
  Product API
- **Host Alias**
  This does not need to be maintained as the value comes from API Management itself; in our example it is **int4-apim.test.apimanagement.eu10.hana.ondemand.com**
- **API Base Path**
  /API_PRODUCT_SRV
- **Service Type**
  OData

The other parameters can stay with the default values or no value. At the end, the configuration screen should look like Figure 12.45. Confirm the creation of the API proxy by clicking **Create** at the bottom of this window.

The API proxy is already created, but the configuration process is still ongoing, giving us the option to go through all the parameters and make additional adjustments. There are four tabs related to the newly created API proxy, which we will review in the next steps. Figure 12.46 shows the parameters we maintained in the initial creation window. We don't apply any additional adjustments.

Figure 12.45 Creation of API under API Proxy

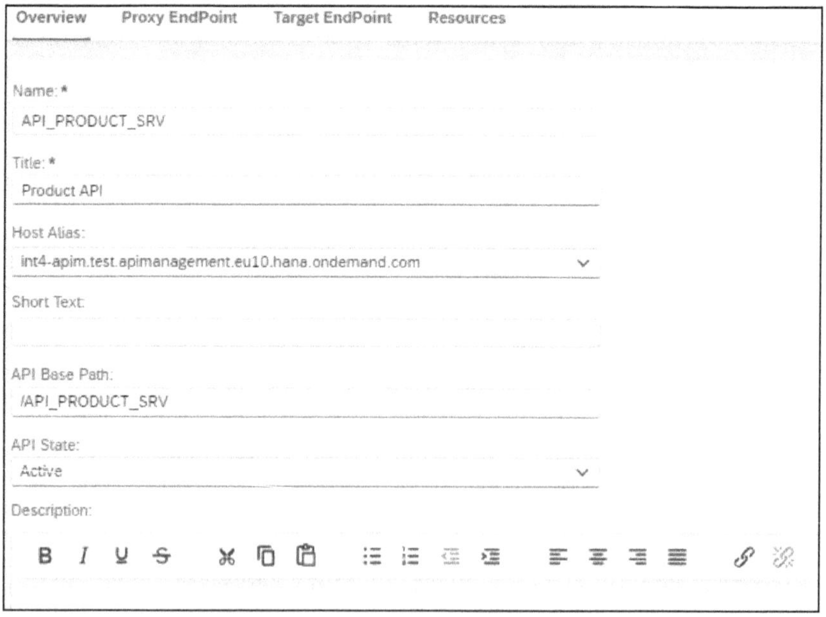

Figure 12.46 API Proxy: Overview

The next tab, **Proxy EndPoint**, allows us to maintain additional **Proxy Endpoint Properties** and **Route Rules** related to the published API facade—that is, the endpoint called by our trading partner. There is no need to maintain any of these properties for our Product API; you can leave the screen empty as shown in Figure 12.47.

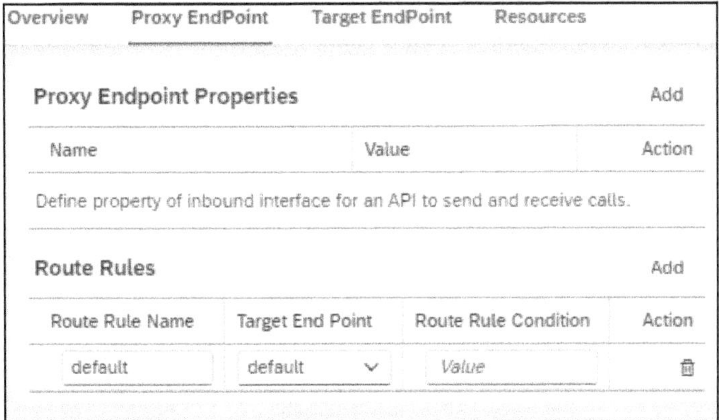

Figure 12.47 API Proxy: Proxy EndPoint Tab

The **Target EndPoint** tab is related to the system that runs the actual service and the exact service itself. On this tab you can see details related to the API provider selected earlier. There is no need to maintain any additional properties. Leave the default values as shown in Figure 12.48.

12

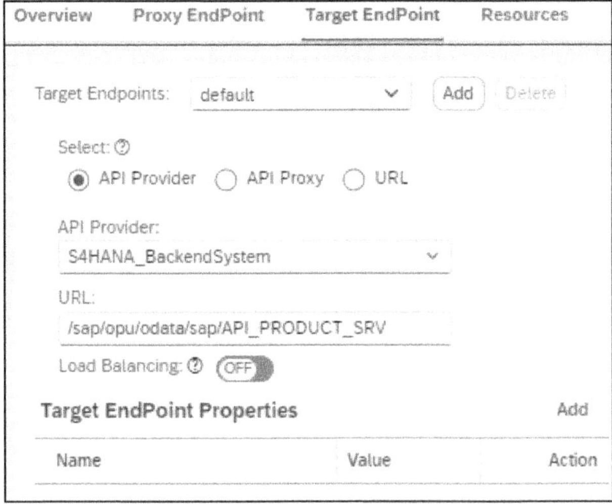

Figure 12.48 API Proxy: Target EndPoint Tab

The last tab, **Resources**, gives an overview of methods that can be invoked on the API. It is also possible to try out the listed methods from here as behind this option there is a basic API client tool. Figure 12.49 shows the **Resources** tab and a list of operations available for the product API.

At this stage, our first API proxy is configured and ready for deployment, which can be triggered via the menu in the top-right-hand corner (see Figure 12.50). Only when the API is deployed can it be called.

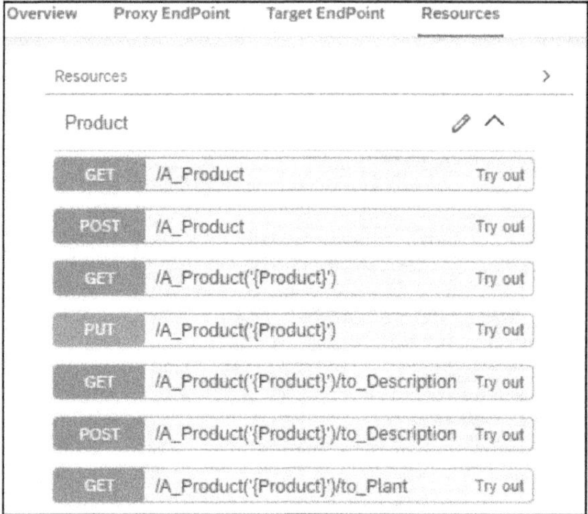

Figure 12.49 API Proxy: Resources

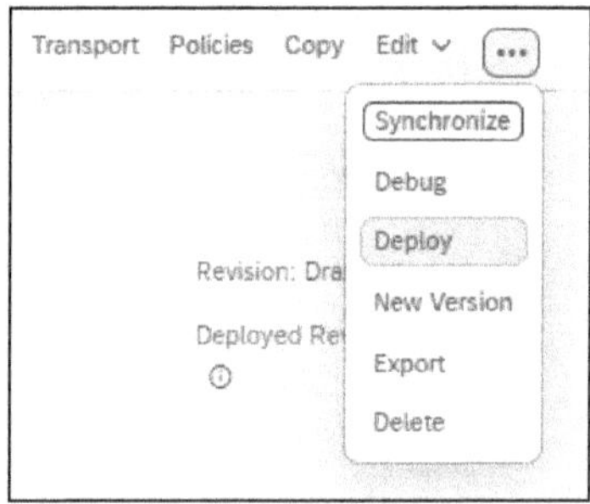

Figure 12.50 Operations Available on API Proxy

Figure 12.51 shows the API proxies overview screen and the product API just created. It offers some key information, including that **Deployed** is set as the **Status** value.

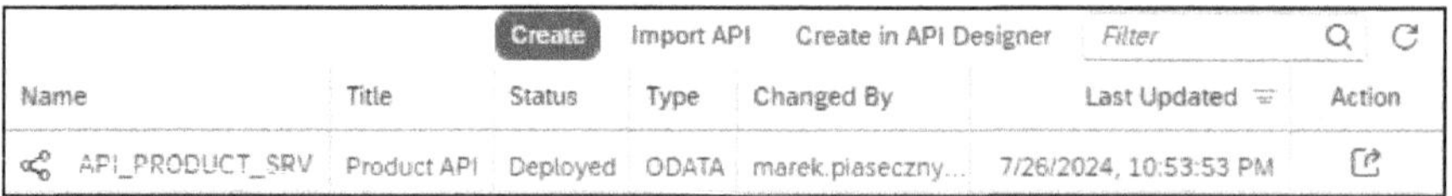

Figure 12.51 API Proxies Overview Screen: Deployed API_PRODUCT_SRV

The product API is deployed and ready to be tested by us or even to be called by the trading partner. But before the final verification, let's take a quick look at policies, which are the most powerful capability of API Management. Policies are sets of rules and functions that can be applied to APIs and to change their behavior at runtime. API Management offers a set of predefined policies from four categories:

- Security policies
- Traffic management policies
- Mediation policies
- Extension policies

To better explain the concept and to improve the security of our product API, we will introduce two policies to our API. The first is the AccessControl policy from the security polices, which introduces IP whitelisting functionality, allowing us to block or allow specific IP addresses or IP ranges. The second one will be Quota from the traffic management policies, which defines the number of requests that can be sent to an API over a specific time. To maintain the policies, we use the policy editor, which you can start by opening Transaction API_PRODUCT_SRV from the **API Proxies** tab and clicking **Policies** at the top-right corner, as shown in Figure 12.52.

Figure 12.52 View of API_PRODUCT_SRV Header from API Proxies Tab

The initial policy editor screen is shown in Figure 12.53.

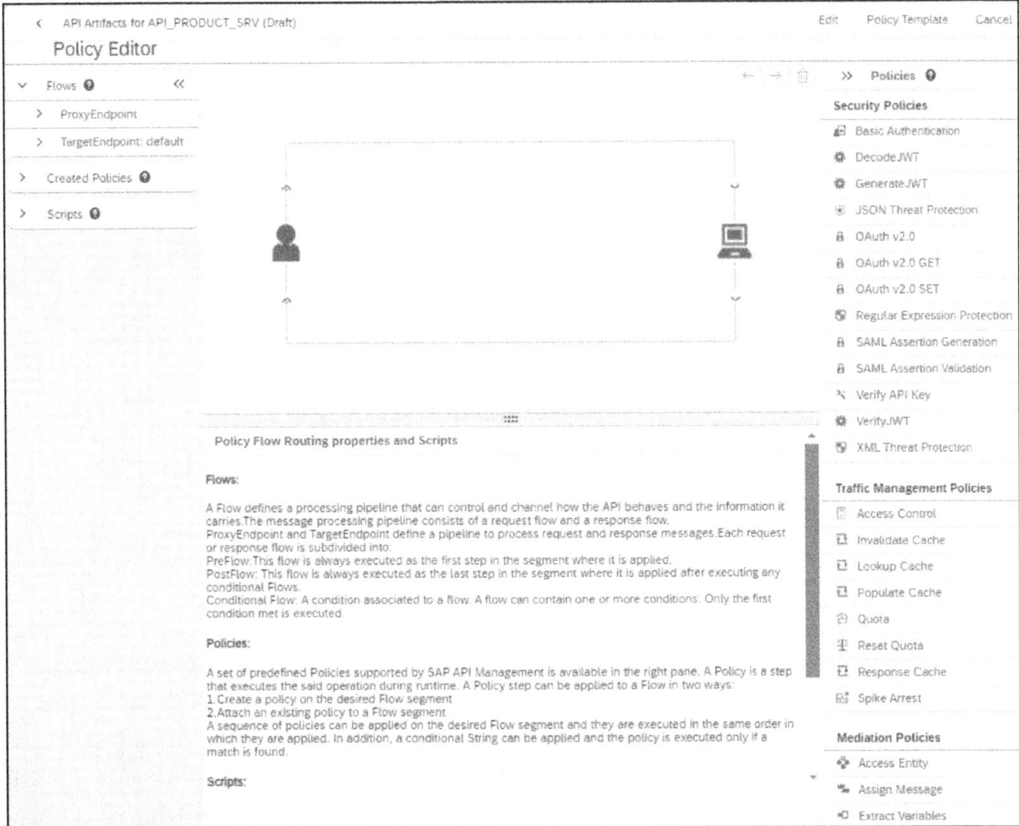

Figure 12.53 API Management: Initial Policy Editor Screen

The concept is as follows:

- On the left side of the screen, we choose the flow and the appropriate part of the flow. The policies can be added to the **ProxyEndPoint** or **TargetEndpoint** area. In each of these, we can also distinguish preflows and postflows. This is where we can also choose other elements already added to the API, like created policies or scripts.
- In the middle of the screen, we have a graphical representation of the flow and the implemented policies. In the bottom part of the screen, we can maintain details of the selected elements.
- On the right side of the screen, we have four sections with the predefined policies. From here we choose the policies to be added to the flow.

  In our case, we will be adding policies to the preflow in the **ProxyEndPoint**; this is the part of the flow triggered immediately after the API is called. Click **Flows • ProxyEndpoint • PreFlow** on the right side of the screen to indicate the working area. After that, navigate to the left side of the screen, find **Access Control Policy** in the **Traffic Management Policies** section, and click the **Add** icon + next to it. You should see the **Create Policy** window, as shown in Figure 12.54.

Enter "AccessControl_Whitelisting" as the value for **Policy Name** and then confirm by clicking **Add.**

Figure 12.54 API Management: Creating Access Control Policy

The access control policy should be already visible as a part of the flow in the middle of the screen. At the bottom of the screen, you can see additional configuration related to this specific policy. In the example in Figure 12.55, you can already see the adjusted configuration of the access control policy, which by default blocks connections from all IP addresses and allows only one specific IP address. You can imagine that this is a scenario where we only allow this API to be called from a single IP address declared and registered by our trading partner.

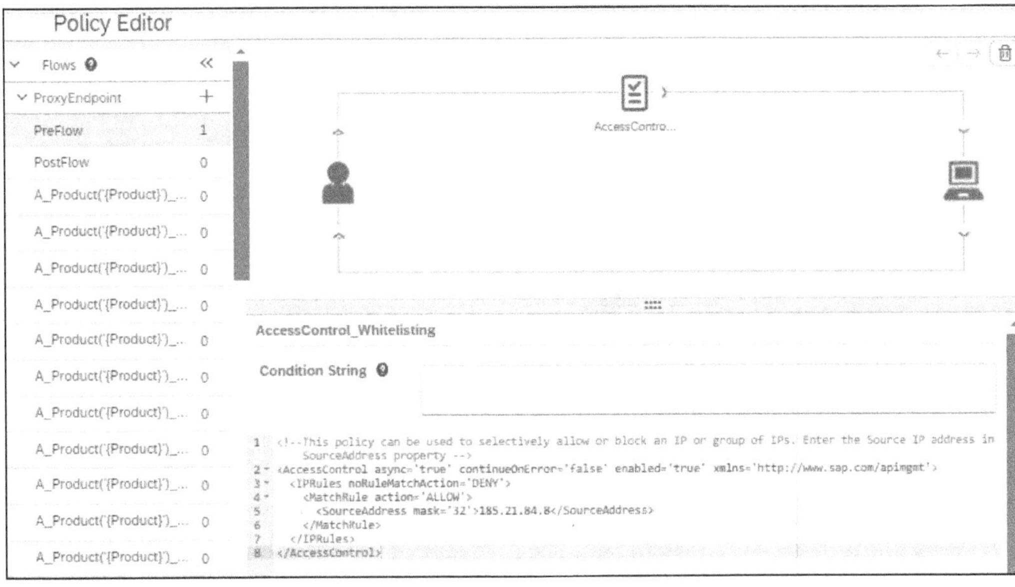

Figure 12.55 SAP API Management: Policy Editor

Listing 12.1 shows the complete configuration code for the access control policy.

```
<!--This policy can be used to selectively allow or block an IP or group of IPs.
Enter the Source IP address in SourceAddress property -->
<AccessControl async='true' continueOnError='false' enabled='true' xmlns='http:/
/www.sap.com/apimgmt'>
  <IPRules noRuleMatchAction='DENY'>
    <MatchRule action='ALLOW'>
      <SourceAddress mask='32'>185.21.84.8</SourceAddress>
    </MatchRule>
  </IPRules>
</AccessControl>
```

Listing 12.1 Access Control Policy: Sample Configuration that Allows Only One Specific IP Address

Now we will add the second policy, called **Quota**. As we are already working in ProxyEndpoint and its preflow, we can go directly to the selection of the policy on the righ side of the screen. Click + next to **Quota Policy** in the **Traffic Management Policies** section. In the newly opened window, shown in Figure 12.56, enter "Quota_5calls" as the value for **Policy Name** and then confirm by clicking **Add.**

Create Policy

Policy Type:
Quota

Policy Name: *
Quota_5calls

Endpoint Type:
ProxyEndpoint

Flow Type:
Preflow

Stream: *
Incoming Request

Add Cancel

Figure 12.56 SAP API Management: Creating Quota Policy

The policy should appear in the flow diagram with the name **Quota_5calls**, just behind the **AccessControl_Whitelisting** policy added earlier. This is shown in Figure 12.57.

To adjust the quota policy to our needs, we slightly modified its configuration. Listing 12.2 shows the complete configuration code for a functionality in which we only allow five requests per minute. Any requests to the API above this number will be blocked.

```
<!-- can be used to configure the number of request messages that an app is
allowed to submit to an API over a course of unit time -->
<Quota async="false" continueOnError="false" enabled="true" type="calendar"
xmlns="http://www.sap.com/apimgmt">
    <!-- specifies the number of requests allowed for the API Proxy -->
        <Allow count="5"/>
    <!-- the interval of time for which the quota should be applied -->
    <Interval>1</Interval>
    <!-- used to specify if a central counter should be maintained and
continuously synchronized across all message processors -->
    <Distributed>true</Distributed>
    <!-- Use to specify the date and time when the quota counter will begin
counting,
        regardless of whether any requests have been received from any apps -->
    <StartTime>2015-2-11 12:00:00</StartTime>
    <!-- if set to true, the distributed quota counter is updated synchronously.
This means that
        the update to the counter will be made at the same time the API call is
quota-checked -->
    <Synchronous>true</Synchronous>
    <!-- Use to specify the unit of time applicable to the quota. Can be second,
```

```
minute, hour, day, or month -->
    <TimeUnit>minute</TimeUnit>
</Quota>
```

Listing 12.2 Quota Policy: Sample Configuration that Introduces Limit of Five Calls per Minute

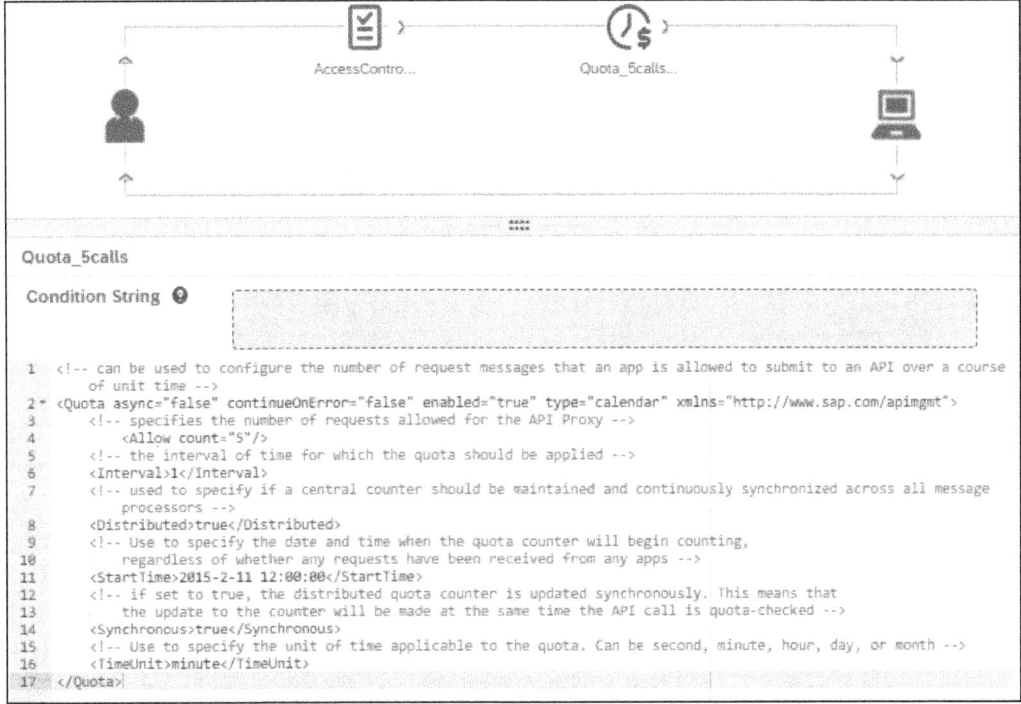

Figure 12.57 SAP API Management: Policy Editor

Implementing these two policies should give you a good understanding of the concept. An additional benefit is that our product API is now much more secure as only one specific IP address can send requests to it. The quota policy we implemented will also prevent uncontrolled load on our systems. Figure 12.58 shows the overall solution. Compared to the earlier design from Figure 12.35, you can see the additional security and traffic management layers enabled in API Management through policies.

It is finally time for a quick verification of the API published via API Management. We will make use of the Postman tool here, from which we send a few GET requests to the Product API exposed on the public network. The test calls will be made from the IP address that is allowed in the AccessControl policy. We will also test the access control policy by sending a request from an unknown IP address, which should fall into the DENY category.

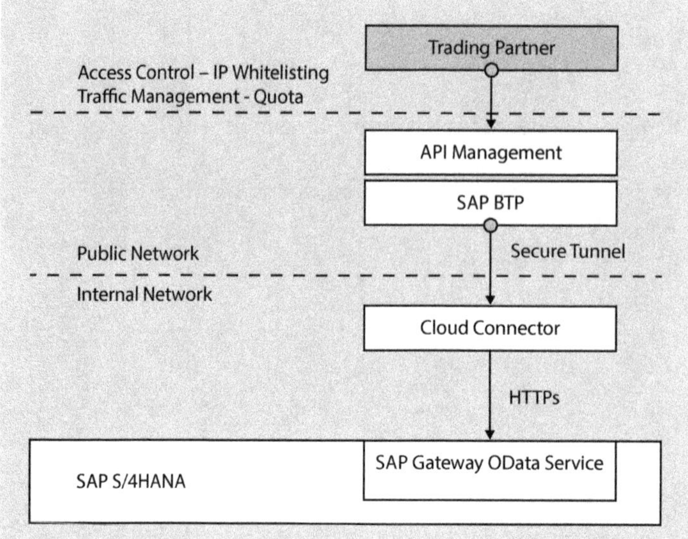

Figure 12.58 Overall Solution for Exposing SAP Gateway OData Service with API Policies

Postman, HTTP Requests and Responses, XML and JSON Formats

On the following pages, we will test the API using Postman. To fully understand the testing process, we assume that the reader should have at least foundational knowledge of the tool and also of HTTP, especially the syntax and content of the HTTP requests and responses. Understanding XML and JSON standards would be also useful. Knowledge related to these topics goes far beyond the scope of this book, but we encourage you to look at the publicly available resources if you feel uncomfortable in this area.

We will start with a GET request to fetch the metadata of the API OData service. This is something we have already done in Section 12.3.3, but with the use of the local SAP Gateway Client. Figure 12.59 shows the request and response as seen in the Postman tool. At the top of the screen, you can see the GET method and the URL pointing to API_PRODUCT_SRV published via API Management. As you can see, the endpoint is exposed on a public network through one of the SAP BTP regions and cloud infrastructure.

The main conclusion from this call is that the whole concept works. We send a request to an API proxy, which acts only as a façade, and finally the request is forwarded to the main service, which is located on the SAP backend system. The additional facade does not influence the main logic and functionality of the OData service.

In another request, we try to fetch data related to a specific product. This is exactly the type of request that can add value for our trading partner and for our company. Figure 12.60 shows a request where we fetch data for a specific product, MZ-FG-R300. As you can see, the request has been executed successfully, which is confirmed by HTTP code 200.

Details of this specific product are displayed in JSON format as this was the format requested via the URL parameter.

GET https://int4-apim.test.apimanagement.eu10.hana.ondemand.com/API_PRODUCT_SRV/?$metadata Send

Params • Authorization • Headers (8) Body Scripts Settings Cookies

Body Cookies (2) Headers (16) Test Results Status: 200 OK Time: 377 ms Size: 8.23 KB Save as example

Pretty Raw Preview Visualize XML


```
<?xml version="1.0" encoding="utf-8"?>
<app:service xml:lang="en" xml:base="https://int4saps42.int4sap.com:44301/sap/opu/odata/sap/API_PRODUCT_SRV/"
    xmlns:app="http://www.w3.org/2007/app" xmlns:atom="http://www.w3.org/2005/Atom" xmlns:m="http://schemas.
    microsoft.com/ado/2007/08/dataservices/metadata" xmlns:sap="http://www.sap.com/Protocols/SAPData">
    <app:workspace>
        <atom:title type="text">Data</atom:title>
        <app:collection sap:deletable="false" sap:content-version="1" href="A_Product">
            <atom:title type="text">A_Product</atom:title>
            <sap:member-title>A_ProductType</sap:member-title>
        </app:collection>
        <app:collection sap:content-version="1" href="A_ProductBasicText">
            <atom:title type="text">A_ProductBasicText</atom:title>
            <sap:member-title>A_ProductBasicTextType</sap:member-title>
        </app:collection>
```


Figure 12.59 Postman: Response with Service Metadata

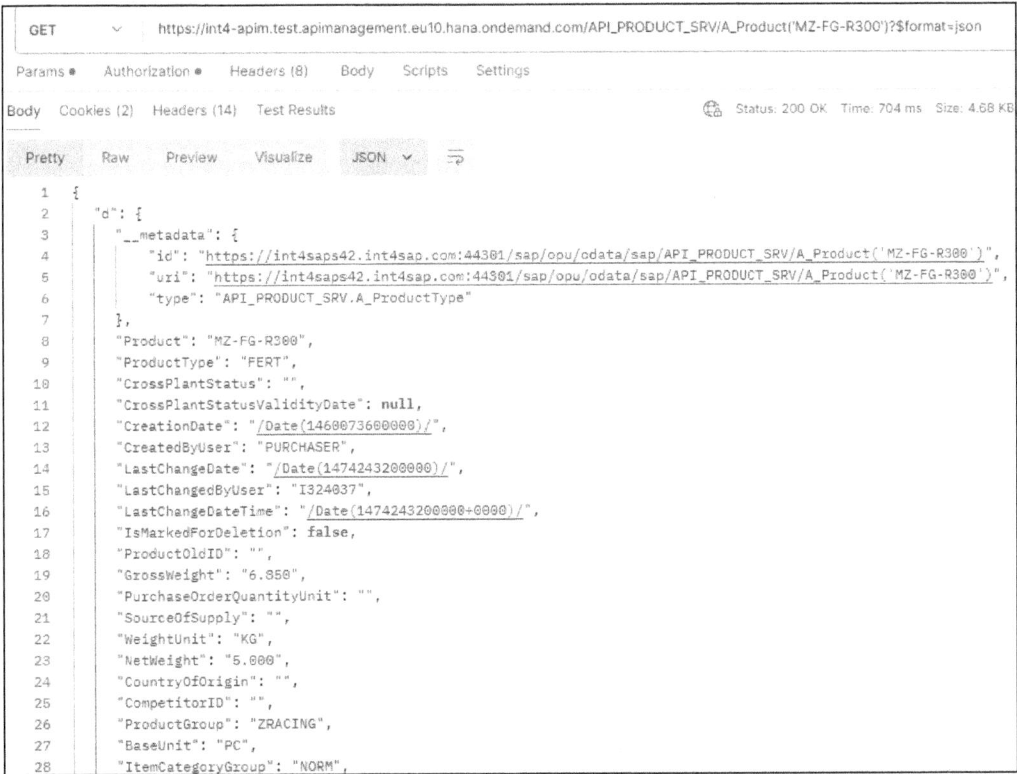

Figure 12.60 Postman: Response with Product Data

The product master data that can be created, read, updated, or deleted through this OData service is not limited to the master records: data related to the plant, procurement, sales organization, valuation area, unit of measure, and much more also can be handled. Figure 12.61 shows one more request where we fetch the English description of the known product.

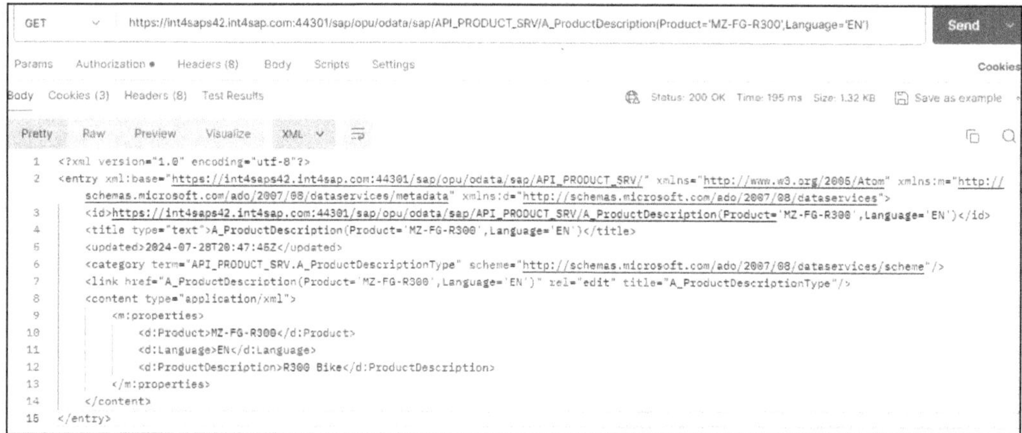

Figure 12.61 Postman: Response with Product Description

At the end, we will verify the two policies that we added to the API proxy. What you can see in Figure 12.62 is the result of sending more than five requests within a minute. This matches the configuration of the quota policy, where we configured a limit of five calls per minute. This part of our API proxy works as expected.

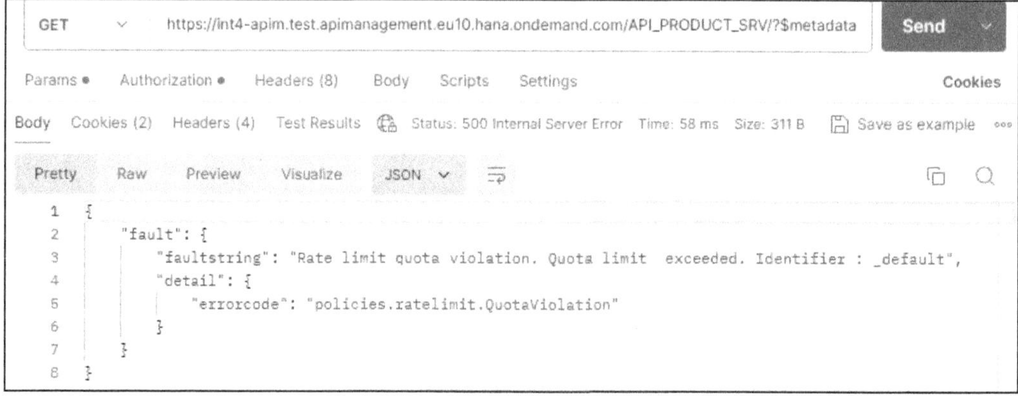

Figure 12.62 Postman: Response from Quota Policy

Figure 12.63 shows a response for a request coming from an unknown IP address. This is also correct behavior from our API proxy as in the AccessControl policy we have maintained one specific IP address as allowed; any other IP addresses are denied by default.

GET https://int4-apim.test.apimanagement.eu10.hana.ondemand.com/API_PRODUCT_SRV/?$metadata Send

Params • Authorization • Headers (8) Body Scripts Settings Cookies

Body Cookies (2) Headers (4) Test Results Status: 403 Forbidden Time: 983 ms Size: 263 B Save as example

Pretty Raw Preview Visualize JSON


```
{
    "fault": {
        "faultstring": "Access Denied for client ip : 37.30.4.166",
        "detail": {
            "errorcode": "accesscontrol.IPDeniedAccess"
        }
    }
}
```


Figure 12.63 Postman: Response from Access Control Policy

Although the section related to API Management has been simplified to a minimum, it should hopefully give you an overall idea of the concept of an API facade and the capabilities that come with policies. In the next section, we will take a quick look at the analytics features within API Management and SAP Gateway.

12

API Analytics

API Management comes with predelivered features related to analytics in the area of APIs. The analytics-related information, highlights, and KPIs can be found in the **Analyze** section. You can navigate there via **SAP Integration Suite Home Page • Analyze**.

After opening the analyze area, you will be welcomed by an **Overview** tab that gives the high-level information related to API processing and consumption in the last seven days. Figure 12.64 shows a sample view of the **Overview** tab.

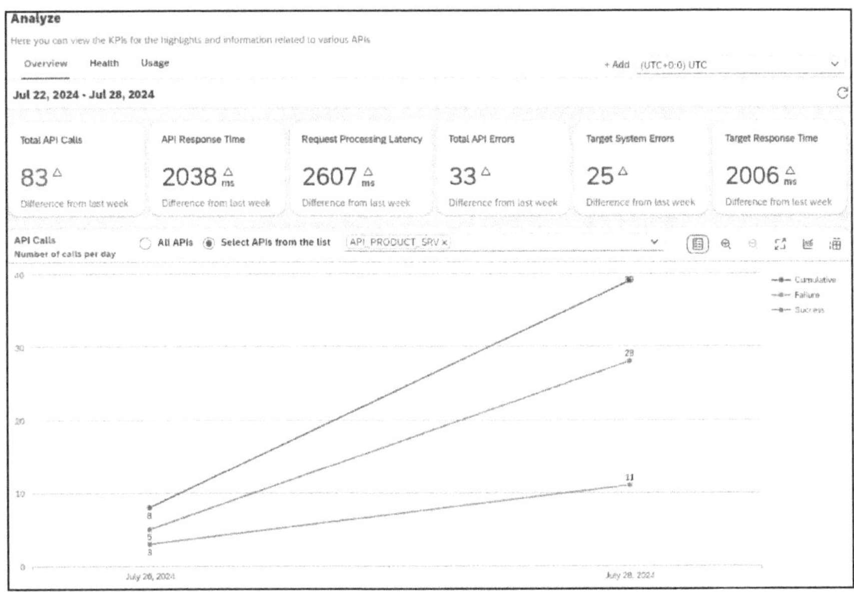

Figure 12.64 API Management Analytics: Overview Tab

In total, the **Overview** tab contains 18 predefined charts and diagrams, which are displayed automatically after opening the tab. You can scroll down the page to see them. The names and default sequence of the charts here are as follows:

- **Total API Calls**
  Number of calls per day
- **Top APIs of the Week**
  Number of calls per API (shows maximum five APIs)
- **Top Products of the Week**
  Number of calls per product (shows maximum five APIs)
- **Top Applications of the Week**
  Number of calls per application (shows maximum five APIs)
- **Top API Providers of the Week**
  Number of calls per API Provider (shows maximum five APIs)
- **Top Developers of the Week**
  Number of calls per developer (shows maximum five APIs)
- **Top APIs**
  Number of calls per API per day
- **Top Products**
  Number of calls per product per day
- **Top Applications**
  Number of calls per application per day
- **Top API Providers**
  Number of calls per API provider per day
- **Top Developers**
  Number of calls per developer per day
- **Top Response Codes**
  Number of calls per response code
- **API Errors**
  Number of errors per API
- **API Response Time**
  Average response time (in ms) per API per day
- **Developer Engagement**
  Number of calls per application per developer
- **Top Backend Errors**
  Number of backend call errors per URL
- **Top Proxy Errors**
  Number of error calls per proxy per day
- **Slowest APIs**
  Average response time (in ms) per API per day

The next tab, **Health**, gives detail information mainly related to the performance and errors of the APIs. Figure 12.65 shows a sample view of the **Health** tab.

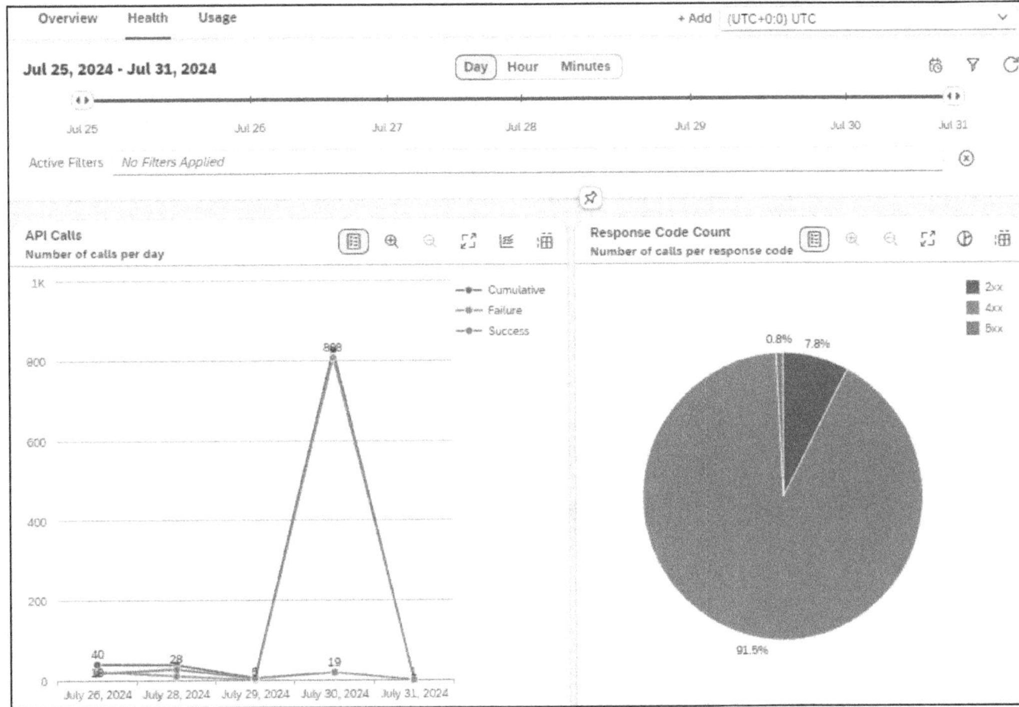

Figure 12.65 API Management Analytics: Health Tab

In total, the **Health** tab has nine predefined charts and diagrams that are displayed automatically here, as follows:

- **API Calls**
  Number of calls per day
- **Response Code Count**
  Number of calls per response code
- **Cache Response**
  Number of calls hitting or missing cache
- **Backend Error Call Count**
  Number of backend call errors per URL
- **Backend Response Time**
  Average backend response time (in ms)
- **Proxy Error Call Count**
  Number of error calls per proxy
- **Proxy Response Time**
  Average proxy response time (in ms)

- **Error Count Per Response Code**
  Number of call errors per response code
- **Average Response Time**
  Average response time against cache (in ms)

The last tab is called **Usage**, and the detailed information that can be found here is mainly related to the consumption of the APIs. Figure 12.66 shows a sample view of the **Usage** tab.

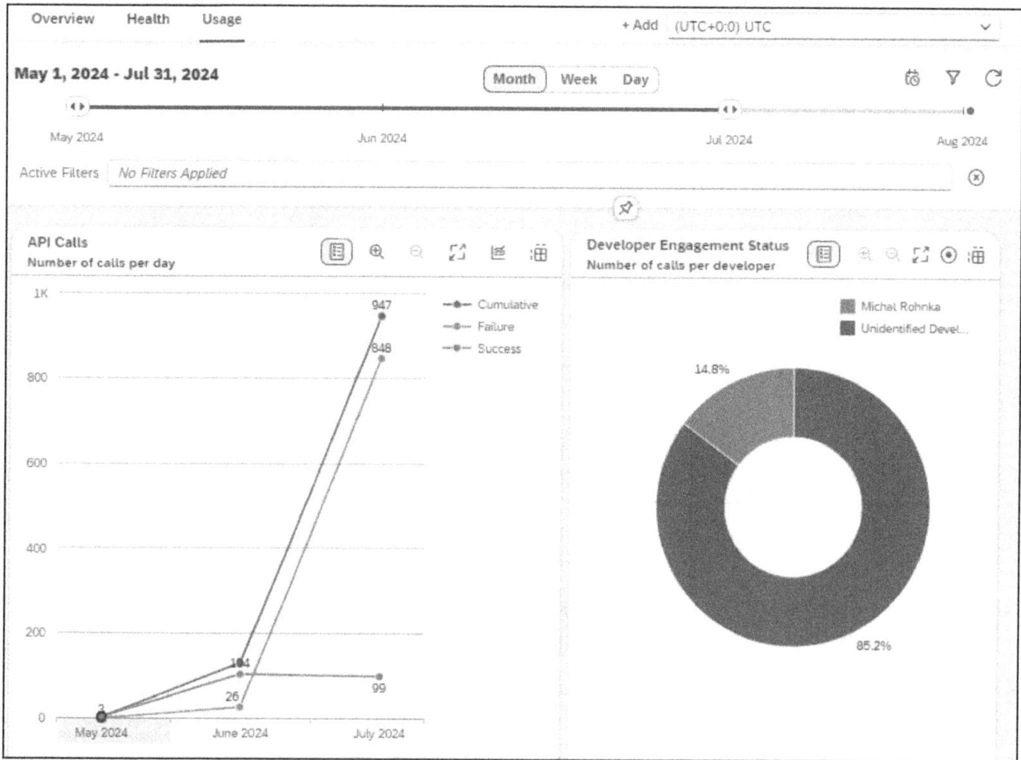

Figure 12.66 API Management Analytics: Usage Tab

In total, the **Usage** tab has 13 predefined charts and diagrams displayed automatically, as follows:

- **API Calls**
  Number of calls per day
- **Developer Engagement Status**
  Number of calls per developer
- **New Developers**
  Recently onboarded developers
- **New Applications**
  Recently created applications

- **Top Browsers**
  Most used browsers
- **Top Agents**
  Most used agents
- **Top Operating Systems**
  Most used operating systems
- **Top Device Types**
  Most used devices
- **Browser Call Count**
  Number of calls per browser
- **Agent Call Count**
  Number of calls per agent
- **Operating Systems Call Count**
  Number of calls per operating system
- **Device Types Call Count**
  Most used devices
- **Request Verb Call Count**
  Number of calls per request verb

Apart from statistics collected in API Management, SAP Gateway running on SAP S/4HANA also has capabilities for collecting and sharing statistics. You can trigger SAP Gateway statistics from Transaction /IWFND/MAINT_SERVICE (Activate and Maintain Services), in which we worked earlier. After opening this transaction, select the appropriate service from the list and click the **Request Statistics** button located in the top panel. Figure 12.67 shows the transactions and the **Request Statistics** button at the end on the right side of the window.

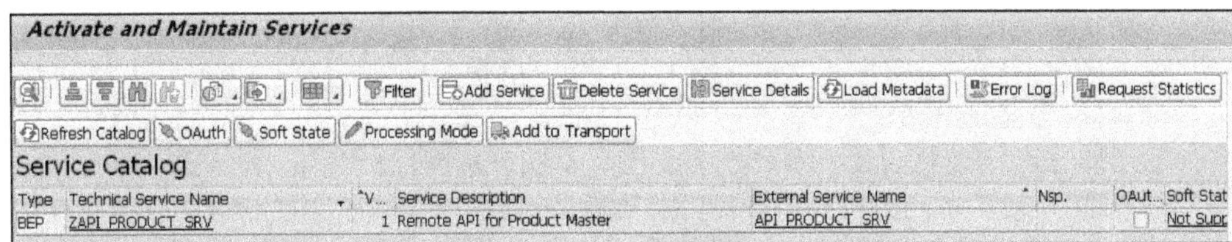

Figure 12.67 Activate and Maintain Services: Request Statistics

Figure 12.68 shows the SAP Gateway statistics transaction with a view of the statistics related to the preselected service. You can find details related to each request, which is especially useful for performance-related investigations.

SAP Gateway Statistics

Re-Select | Aggregated Statistics

Request URI | Summarize | Performance Trace | Hub System Statistics | Backend System Statistics

Request Statistics

Line	Service Name	V...	Operation	Entity Set or Fun...	Processin...	Backen...	Applicat...	Non-G...	Fra...	Parallel...	Sum of ...	Request ...	Response S...	Format	No. ...	User ID	Date
1	API_PRODUCT_SRV	1	document		18	14	4	0	78 ...	0	4	0	7.682	xml	1	PIASECZNYM	28.07.20...
2	API_PRODUCT_SRV	1	document		20	16	4	0	80 ...	0	4	0	7.682	xml	1	PIASECZNYM	28.07.20...
3	API_PRODUCT_SRV	1	document		20	16	4	0	80 ...	0	4	0	7.682	xml	1	PIASECZNYM	28.07.20...
4	API_PRODUCT_SRV	1	document		21	17	4	0	81 ...	0	4	0	7.682	xml	1	PIASECZNYM	28.07.20...
5	API_PRODUCT_SRV	1	document		21	17	4	0	81 ...	0	4	0	7.682	xml	1	PIASECZNYM	28.07.20...
6	API_PRODUCT_SRV	1	document		19	15	4	0	79 ...	0	4	0	7.682	xml	1	PIASECZNYM	28.07.20...
7	API_PRODUCT_SRV	1	document		28	21	7	0	75 ...	0	7	0	7.682	xml	1	PIASECZNYM	28.07.20...
8	API_PRODUCT_SRV	1	document		22	18	4	0	82 ...	0	4	0	7.682	xml	1	PIASECZNYM	28.07.20...
9	API_PRODUCT_SRV	1	document		23	19	4	0	83 ...	0	4	0	7.682	xml	1	PIASECZNYM	28.07.20...
10	API_PRODUCT_SRV	1	read feed	A_Product	80	49	31	0	61 ...	0	31	0	651.879	xml	100	PIASECZNYM	28.07.20...
11	API_PRODUCT_SRV	1	read feed	A_Product	105	50	55	0	48 ...	0	55	0	651.903	xml	100	PIASECZNYM	28.07.20...
12	API_PRODUCT_SRV	1	read entry	A_ProductDescri...	41	28	13	0	68 ...	0	13	0	996	xml	1	PIASECZNYM	28.07.20...
13	API_PRODUCT_SRV	1	read entry	A_ProductDescri...	40	22	18	0	55 ...	0	18	0	996	xml	1	PIASECZNYM	28.07.20...
14	API_PRODUCT_SRV	1	read entry	A_ProductPlant	148	27	121	0	18 ...	0	121	0	5.632	xml	1	PIASECZNYM	28.07.20...
15	API_PRODUCT_SRV	1	read feed	A_ProductPlant	213	39	174	0	18 ...	0	174	0	6.474	xml	1	PIASECZNYM	28.07.20...
16	API_PRODUCT_SRV	1	read entry	A_Product	154	42	112	0	27 ...	0	112	0	4.183	json	1	PIASECZNYM	28.07.20...
17	API_PRODUCT_SRV	1	read feed	A_ProductDescri...	39	23	16	0	59 ...	0	16	0	16.837	json	38	PIASECZNYM	28.07.20...

Figure 12.68 SAP Gateway Statistics: Overview

From the overview screen shown in Figure 12.68, it is also possible to navigate to the details. Figure 12.69 shows details related to **Request URI**.

Request URI | Summarize | Performance Trace | Hub System Statistics | Backend System Statistics

Request Statistics

Line	Service Name	V...	Operation	Entity Set or Fun...	Processin...	Backen...	Applicat...	Non-G...	Fra...	Parallel...	Sum of ...	Request ...	Response S...	Format	No. ...	User ID	Date
1	API_PRODUCT_SRV	1	document		18	14	4	0	78 ...	0	4	0	7.682	xml	1	PIASECZNYM	28.07.20
2	API_PRODUCT_SRV	1	document		20	16	4	0	80 ...	0	4	0	7.682	xml	1	PIASECZNYM	28.07.20
3	API_PRODUCT_SRV	1	document														
4	API_PRODUC																20
5	API_PRODUC																20
6	API_PRODUC																20
7	API_PRODUC																20
8	API_PRODUC																20
9	API_PRODUC																20
10	API_PRODUCT_SRV	1	read feed	A_Product	80	49	31	0	61 ...	0	31	0	651.879	xml	100	PIASECZNYM	28.07.20
11	API_PRODUCT_SRV	1	read feed	A_Product	105	50	55	0	48 ...	0	55	0	651.903	xml	100	PIASECZNYM	28.07.20
12	API_PRODUCT_SRV	1	read entry	A_ProductDescri...	41	28	13	0	68 ...	0	13	0	996	xml	1	PIASECZNYM	28.07.20

Request URI

Request URI `/sap/opu/odata/sap/API_PRODUCT_SRV/A_ProductDescription(Product='MZ-FG-R300',Language='EN')`

Figure 12.69 SAP Gateway Statistics: Details of Request URI

We end this section with the knowledge that with every API being processed via API Management, there is also a set of data collected for analytics purposes, and this data can be analyzed to a great extent by predelivered charts, diagrams, and KPIs. To a certain extent, these can be slightly adjusted, and this goes even beyond the basic time frame selection and formatting of the charts and diagrams. In addition to this, we can also find statistics for the SAP Gateway component located on our SAP S/4HANA system.

12.4 Summary

In some ways, we've proven that there is a place for both the classic EDI and the new API approaches. They do not have to compete with each other; they can complement each other.

In this chapter, we looked at the API concept and characteristics and at APIs in SAP S/4HANA, which included the API catalog, reviewing a sample API, and API consumption. We looked at extending our EDI scenario, which included enabling the Product API, exposing the API, delving into security aspects, and setting the configuration in the cloud connector and in API Management. Finally, we discussed API monitoring, including analytics data in API Management.

In the next chapter, we will conclude the book, discussing items that have been intentionally or necessarily overlooked. We will also provide a roadmap and final summary.

Chapter 13
Conclusion and Outlook

So far, this book has provided a complete B2B EDI solution for an SAP S/4HANA landscape, as well as a way to implement it. In this final chapter, we'll share some additional information and advice to help you maintain and extend this solution.

This book has provided you with a complete business-to-business (B2B) electronic data interchange (EDI) solution and a way to implement it. However, it's nearly impossible to elaborate in detail on each possible variant and case that you may encounter in your SAP S/4HANA or B2B EDI project. In this chapter, we'll take a quick look at a few additional aspects that are important to EDI implementation, such as component versions, as well as review topics that were intentionally or necessarily overlooked in previous chapters.

13.1 Component Versions

In this section, we'll look at the versions of SAP S/4HANA, SAP Integration Suite, and the cloud connector used in our demo implementation.

> **Note**
> SAP is investing a lot of effort into delivering additional functionality and improvements to both SAP S/4HANA and SAP Integration Suite. On top of that, these updates also contain fixes for existing functionality as well as stability- and security-related improvements. We recommend following up on the latest updates and releases related to each component. Make sure to plan and execute their updates as soon as possible. Even if the immediate update of a specific component is impossible due to specifics of an internal change process or maintenance procedure, you should be aware of the backlog and what limitations and risk it brings to your landscape.

13.1.1 SAP S/4HANA

The EDI scenario has been defined in a landscape with an SAP S/4HANA on-premise deployment as the core ERP system. Also, all the necessary configuration steps are specific to SAP S/4HANA on-premise. This choice was made based on the fact that this is the most popular deployment option. The solution design and the required configuration steps would be slightly different in a landscape with SAP S/4HANA Cloud Public Edition as the core ERP system. Figure 13.1 shows the installed product versions in the SAP S/4HANA on-premise system used for the configuration presented in this book.

Installed Software Component Versions | Installed Product Versions

Product	Release	SP / FP Stack	Vendor	Short Description of Product Version
SF EMPLOYEECENTRAL INTEGRATION	1210	36 (11/2023)	sap.com	SFSF EC INTEGRATION 1210
SCREEN PERSONAS	3.0	SP18	sap.com	SAP SCREEN PERSONAS 3.0
APPLIC INTERFACE FW EXT	01	Initial Shipment Stack	sap.com	APPLIC INTERFACE FW EXT 01
S4HANA ON PREMISE	2023	01 (02/2024) FPS	sap.com	SAP S/4HANA 2023
SAP S/4HANA FOUNDATION	2023	01 (02/2024) FPS	sap.com	SAP S/4HANA FOUNDATION 2023
SAP FIORI FES FOR S/4HANA	2023	01 (02/2024)	sap.com	SAP FIORI FES 2023 FOR S/4HANA
SAP_S4HANA_ECTR	1.0	01 (09/2016)	sap.com	SAP ECTR 1.0 FOR SAP S/4HANA
SAP MRS FOR S4HANA	1.0	10 (11/2023)	sap.com	SAP MRS FOR S4HANA 1.0
SAP MII	15.5	01	sap.com	SAP MII 15.5
SAP FIORI FOR S4HANA	2023	01 (02/2024) FPS	sap.com	SAP FIORI FOR SAP S/4HANA 2023
ABAP PLATFORM	2023	01 (02/2024)	sap.com	ABAP PLATFORM 2023

Figure 13.1 Installed Product Versions on Demonstration SAP S/4HANA On-Premise System

13.1.2 SAP Integration Suite

There are two SAP Integration Suite version aspects to consider:

- **The version of each specific technical component from SAP Integration Suite itself**
  This is the design and runtime environment of our integration-platform-as-a-service (iPaaS) platform. For the SAP Integration Suite instance used for setting up the demonstration EDI scenario, we relied on the standard version delivered by SAP, including all monthly updates and increments from SAP. At the final stage, when the solution was fully functional, the version details matched what's shown in Figure 13.2.

Figure 13.2 Version of SAP Integration Suite Instance Used for Demonstration EDI Scenario

You can check the content of the latest features and updates related to SAP Business Technology Platform (SAP BTP) on SAP Help Portal at the following URL: *http://s-prs.co/v573609*. On the main page, navigate to the **Stay Current** section and click the

What's New for SAP BTP link. Using the filter options at the top of the new screen, you can define the exact scope for your search. For SAP Integration Suite, you can set a filter for each of its main components: Integration Advisor, Cloud Integration, Trading Partner Management, and so on. Sample results are shown in Figure 13.3.

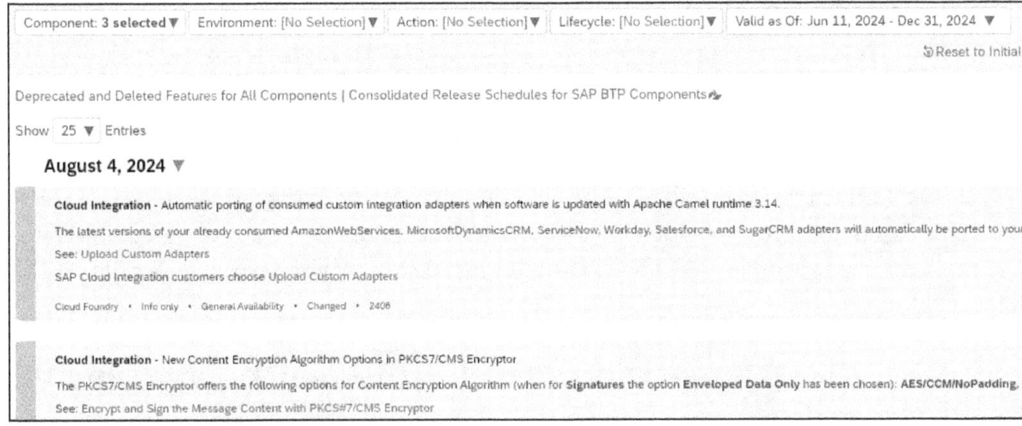

Figure 13.3 What's New for SAP Business Technology Platform: Filtered Results

- **The version of the integration solution**
  In our case, this is the predelivered Cloud Integration—Trading Partner Management V2 package. As you may have noticed in Chapter 6 and Figure 6.2, the version initially deployed was 2.2.1. This version did not change and was used throughout the whole configuration and testing as presented in the book.

  Since the initial deployment, there were a few updates to the Cloud Integration—Trading Partner Management V2 package. You can see this under **Design • Integration and APIs** within SAP Integration Suite, where all packages relevant to your specific Cloud Integration instance are stored. Figure 13.4 shows the **Update Available** indicator.

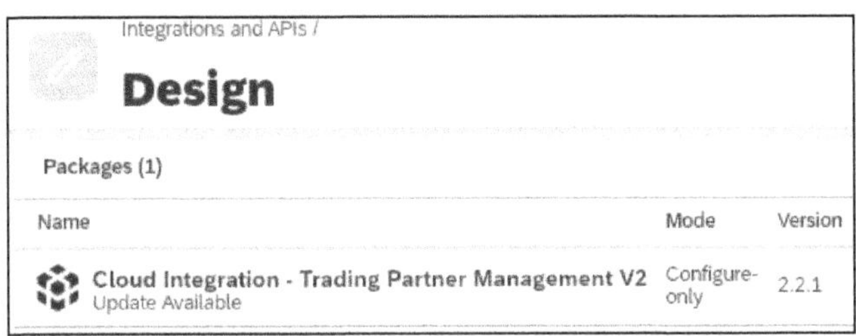

Figure 13.4 Package with Update Available Indicator

By clicking the package, you can not only see its details and content but also get the **Update Package** option in the top-right corner of the screen. Clicking **Update Package** triggers the update, preceded by a confirmation prompt. Figure 13.5 shows the option available within the Cloud Integration—Trading Partner Management V2 package.

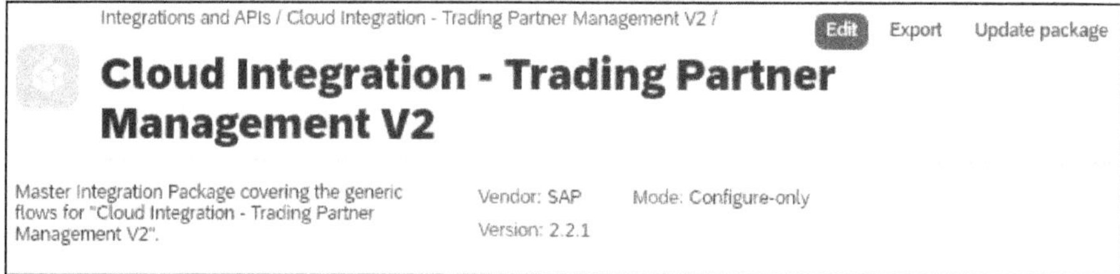

Figure 13.5 Details of Package with Update Package Option

Before performing the update itself, it is always recommended to review the changes that the update introduces. The common practice for packages delivered by SAP is that in the **Documents** section there will be a text file called the *change log* that traces the changes. Figure 13.6 shows the mentioned section and the change log file.

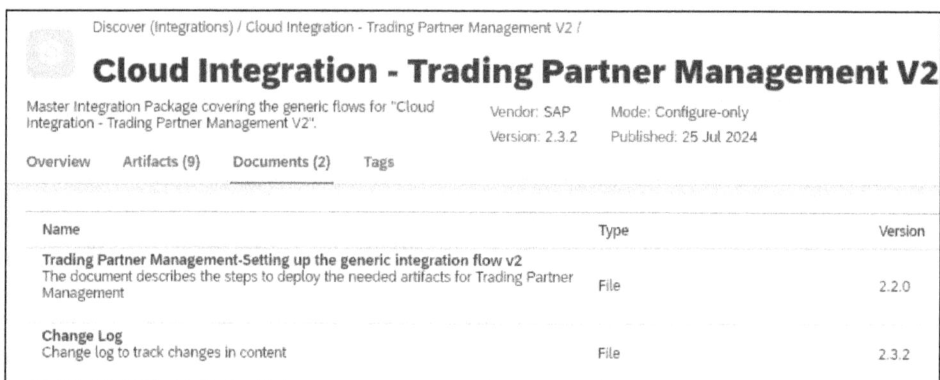

Figure 13.6 Change Log for Cloud Integration—Trading Partner Management V2 Package

Figure 13.7 shows the content of the change log file related to the Cloud Integration—Trading Partner Management V2 package.

```
23.07.2024
2.3.2 - Support TPM Dead-Letter Queue
        Support hard stop for payload validation error
        Support Target Decimal Character for UN/EDIFACT
18.06.2024
2.3.1 - Support Communication Partner (VAN)
        Update EDI Splitter to V2.4, EDI to XML Converter to V2.3, XML to EDI Converter version to V2.3
        Enable the configuration of syntax validation
        Enhance to support sending sender functional acknowledgement via Process Direct Receiver communication channel
        Fix message retry causing B2B Monitor interchange status shows "Retry" or "Processing" for completed interchange bug
        Fix message retry causing B2B Monitor displays wrong error category bug
        Fix Functional Acknowledgement payload cannot be correctly stored bug
20.05.2024
2.3.0 - Support Identifier Groups
        Support B2B Archiving
        Support displaying Business Transaction Activity Name and Business Transaction Activity Direction in B2B Monitor
        Support Code Pages for UN/EDIFACT
        Update EDI Extractor to V2.1
        Update EDI Splitter, EDI to XML Converter, XML to EDI Converter version to V2.2
19.01.2024
2.2.1 - To follow the SAP_TPM naming convention, ACTUAL_PARTNER_ID is now renamed to SAP_TPM_ACTIVITYPARTNER_ID. It is recommended
```

Figure 13.7 Change Log for Cloud Integration—Trading Partner Management V2 Package

13.1.3 Cloud Connector

The cloud connector is the third key component in our landscape. If you refer back to Chapter 6 and look at Figure 6.32, you'll see that version 2.15.1 was deployed during the initial configuration. Similar to other components, we also recommend updating the cloud connector to the latest available version as soon as possible. The main motivation in this case is the security and stability of the connectivity between SAP BTP services and local systems. With the high-availability setup consisting of master and shadow instances of Cloud Integration, the update can be performed with zero downtime.

The exact scope and content of updates related to SAP Cloud Connector can be traced on the **What's New for SAP BTP** portal. To display changes related to the cloud connector, you need to apply a filter for **Component** with the value **Connectivity.** A sample change log is shown in Figure 13.8.

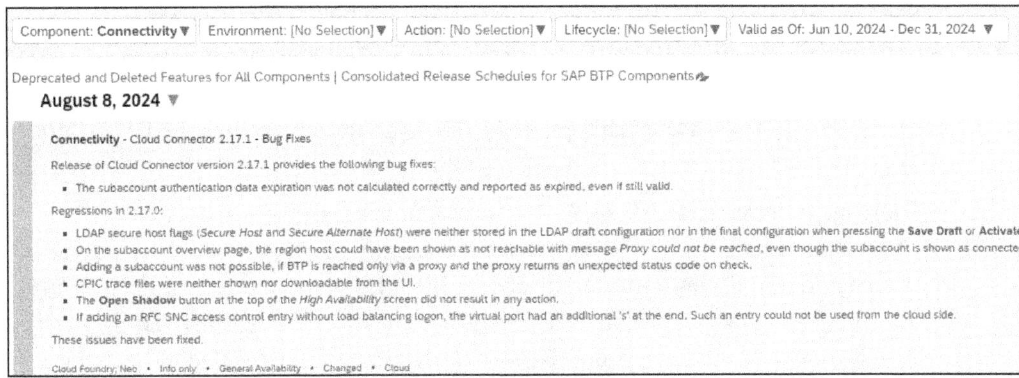

Figure 13.8 What's New for SAP Business Technology Platform Portal: Cloud Connector

13.2 Topics Intentionally or Necessarily Overlooked

In this section, we'll review topics that have been intentionally or necessarily overlooked in previous chapters, such as custom requirements and enhancements as well as the operational aspects of monitoring data and archiving.

13.2.1 Custom Requirements and Enhancements

It's rare to work on an SAP project without a discussion of nonstandard requirements and the need for a custom enhancements. EDI implementation is no different, and at some point there may be a strong motivation and solid arguments to apply custom enhancements within the solution—that is, to add something on top of the out-of-the-box functionality coming from SAP components, replace part of the standard functionality with a custom element, or any other mix or variant of these two approaches. Especially in a B2B EDI solution that supports a very complex environment, there may exist custom enhancements, and that does not mean that the solution is poorly designed or that we have technical debt. The most critical factor is how these custom enhancements are introduced to the standard functionality and how they are managed. The same rule applies as for any other component and concept within B2B EDI integration: that is, keep everything as standardized and harmonized as possible. In this case, we may deviate from the standard, but that does not mean that the nonstandard pieces cannot be harmonized. The custom enhancements should still follow the same policies as any other development or configuration activity; at a minimum, that means they should be transparent and reusable.

In the whole EDI scenario presented in this book, we only used standard functionality available within SAP Integration Suite. On the SAP S/4HANA on-premise side, there were some ABAP function modules created for the purposes of SAP Application Interface Framework monitoring and some key fields, but we would not classify these as actual custom enhancements. These were objects that fall under SAP Application Interface Framework and were developed within a given template. More important is that only standard IDocs were used. And when we think about custom enhancements in the area of B2B EDI, these are usually related to the two areas we mentioned. Let's briefly look at the enhancement options available for our solution:

- **Message processing and transformation in middleware**
  For the EDI solutions running on Trading Partner Management in Cloud Integration, the custom enhancements are quite specific as the framework already harmonizes the processing on the middleware platform. If you look at the Trading Partner Management V2 package, you will notice that it is not possible to edit any of the artifacts from the package as it is set in configure-only mode. The first impression can be that it is impossible to modify anything in the given processing pipeline. This is luckily misleading as the generic standard iFlows have multiple enhancement points where it is possible to trigger custom iFlows via a ProcessDirect mechanism. The mentioned enhancement points are available in every main step of the transaction, as represented in **Agreement • B2B Scenarios**. There are in total five steps where custom processing can be linked:
  - Custom sender communication
  - Custom preprocessing before mapping
  - Custom main mapping
  - Custom postprocessing after mapping
  - Customer receiver communication

  The triggers for custom preprocessing, custom mapping, and custom postprocessing are located in the agreement. For each of the mentioned steps, we have an option to link a custom iFlow with the necessary custom logic. The trigger and the link to the iFlow is available via a ProcessDirect adapter.

Figure 13.9 shows the **Mapping** step for a sales order transaction. In the bottom-right corner, you can see that there is a customized mapping process enabled. In this case, the mapping guideline (MAG) transformation will not be performed; instead, the custom iFlow with ProcessDirect address */tpm/CustomMappingBestBikes* will be triggered. In the latter, we may have any kind of operations as per our needs, such as Groovy or graphical mapping.

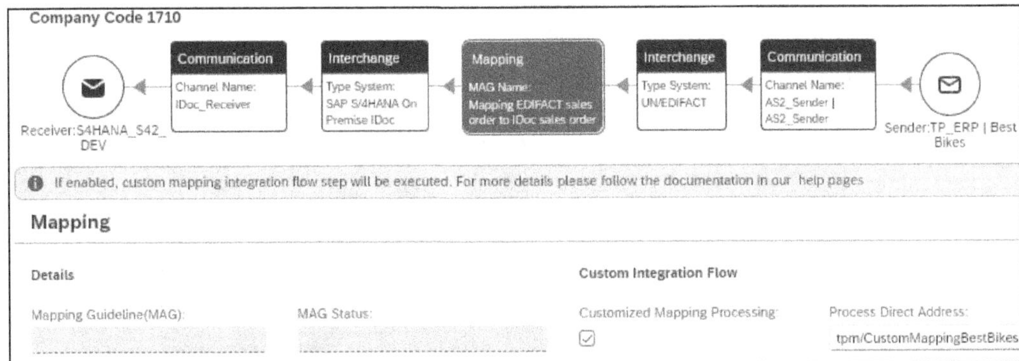

Figure 13.9 Custom iFlow Option in Mapping Step (Custom Main Mapping)

Figure 13.10 shows the **Customized Postprocessing** option. This is related to the sales order transaction and gives the option to add any additional custom logic after the MAG-based mapping.

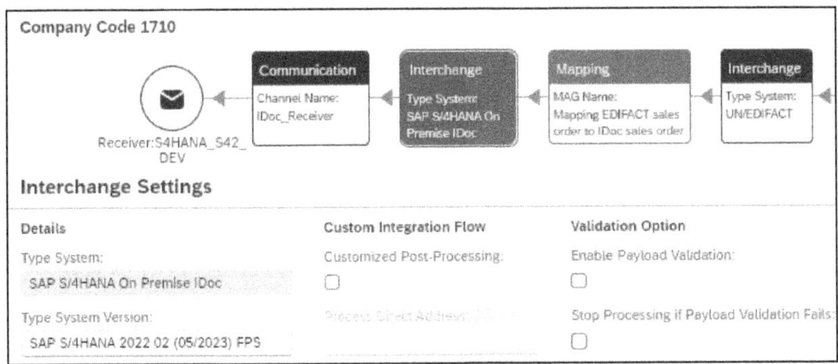

Figure 13.10 Custom iFlow Option in Interchange Step (Custom Postprocessing after Mapping)

- **IDoc structure, content, and processing logic within SAP ERP**
  It is often the case that the standard IDoc structure and the standard processing logic within the SAP ERP system cannot cover all business requirements. Extending a standard IDoc is the usual and preferred approach for solving limitations related to IDoc structures. While the activity of extending an IDoc may seem pretty easy, it is important to define a reusable and harmonized approach for extending IDocs in our system. There is nothing wrong with making extensions; there is even nothing wrong with making dozens of extensions—but only when all the changes are per-

formed within a set of predefined rules and conditions. Harmonization is the key to keeping things under control. There are design approaches that, when properly followed, can make the work related to IDoc enhancements very efficient and transparent. The same applies to the custom enhancements related to the processing logic, and the recommended solution for keeping custom logic under control is using SAP Application Interface Framework with its capability to add preprocessing or postprocessing to the standard processing logic.

When you already have a custom IDoc defined within SAP S/4HANA, you can look at how it can be handled by Trading Partner Management in SAP Integration Suite. After navigating from the SAP Integration Suite home page to **Design • Custom Type Systems**, you can find the **Library of Custom Type Systems** screen, with an option for uploading **Custom IDocs**. This option is shown in Figure 13.11.

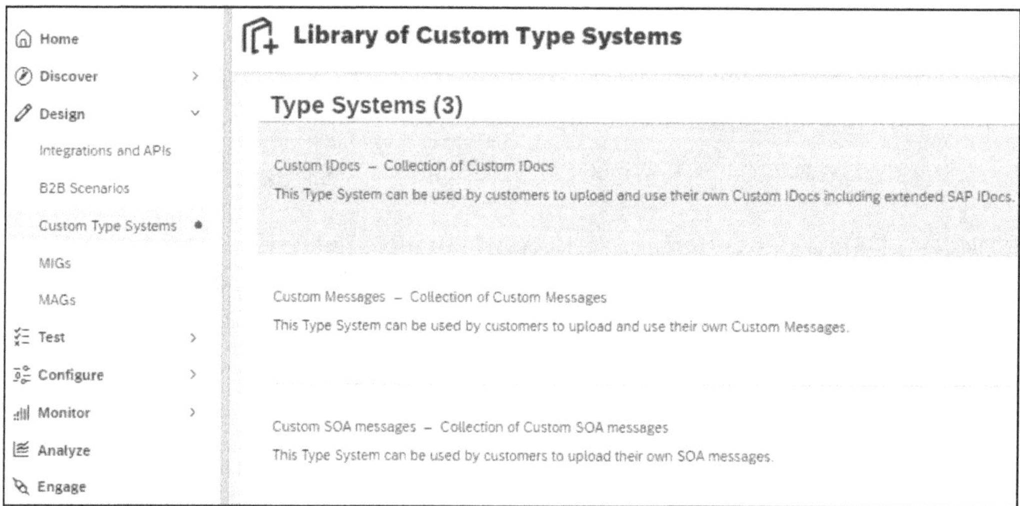

Figure 13.11 Library of Custom Type Systems

This feature allows you to upload XSD files with an IDoc structure definition. There are three cases in which you would use this option:

- You have standard SAP IDocs with custom extensions.
- Your IDoc structures are not based on standard SAP IDocs but are customer-proprietary structures.
- You use standard SAP IDocs that are not available in Integration Advisor. There can be some industry- or solution-specific standard IDocs that are missing in the catalog; with this option, you can choose to add them.

After successful load of an XSD file, the custom IDoc is visible as a message implementation guideline (MIG). It then can be used for the definition of a MAG, and the B2B scenario and activities within these steps do not different from the ones related to a standard IDoc.

13.2.2 Monitoring Data and Archiving

One of the operational aspects that you should be aware of in regard to the present solution is that during the processing of the EDI message, some of the data is temporarily or permanently stored. This is pertinent for a couple of components and types of data:

- **SAP S/4HANA IDocs**
  All the intermediate messages in IDoc form are stored in the database. There is no default process in the system via which to delete or archive them. There is a set of standard transactions that can perform these tasks, but the definition and implementation of this housekeeping activity must be done by an appropriate operation team on your side.
- **SAP Application Interface Framework messages and logs**
  Each message processed via SAP Application Interface Framework also creates entries in multiple index and log tables. These should be deleted or archived on a regular basis. There is also a set of transactions that allows for easy and efficient reorganization of the data.
- **Cloud Integration**
  This includes the following:
  - **Standard message monitoring**
    During message processing, the monitoring data is written to the database and stored for 30 days. This is the data with which we can find the messages and their statuses in the Message Monitoring app. There is no option to change the default 30 day period; there is also no need to perform any additional housekeeping activities.
  - **B2B scenario monitoring**
    Data related to the processing of B2B messages through Trading Partner Management is stored separately and for a longer period than the data in standard message monitoring. As you have already noticed, there are much richer details available for B2B messages than for standard messages. That also includes payload details. The monitoring data related to B2B messages is available for 90 days. There is no option to change the default value; there is also no need to perform any additional housekeeping activities.

To summarize:

- You need to implement a housekeeping process on the intermediate processing data within SAP S/4HANA.
- There is nothing permanently stored in Cloud Integration in relation to the processed B2B messages.

The other conclusion from this is that there is a possible gap as the original EDI messages are also not stored permanently, and after a certain period of time they will be unavailable. Having access to the original EDI messages is a key requirement from the perspective of the IT support team, but it can be also be a requirement of the finance, legal, or compliance team. Luckily, an archiving feature has recently become available on SAP Integration Suite. It allows you to store message processing logs in an archiving systems such as a content management system (CMS).

The archiving feature allows you to store message processing logs and data like persisting step content, adapter payloads, and attachments in an archiving system that supports the Content Management Interoperability Services (CMIS) API standard, version 1.1. The CMS isn't part of SAP Integration Suite; it needs to be arranged separately. But the archiving option is available and would greatly improve the transparency and efficiency of any support-related activities, not to mention that in some cases it may be obligatory to store specific data for years or decades.

13.3 Summary

This was a short chapter, but hopefully it has given you some additional useful information. Section 13.1 clarified on which component versions the demo EDI implementation has been built and deployed. We also showed you where you can find information on latest releases and their content. That should help you with planning updates to fix or improve existing functionality or to introduce new features. Section 13.2 started with a short elaboration on options and recommendations related to custom enhancements within the EDI solution. The main outcome is that these customizations are possible and supported by all the components involved in the presented EDI solution. Nevertheless, the exact design and implementation of custom enhancements should be well planned and all pros and cons should be checked. After that, we discussed the need for housekeeping activities on the SAP S/4HANA side in relation to IDocs and SAP Application Interface Framework logs. Last but not least, we discussed the archiving feature in SAP Integration Suite, which allows for a connection to a CMS and permanent storage of data processed through Cloud Integration and its adapters.

This is the last chapter of the book. Thank you for taking the time to explore this topic with us. We wish you continued success in your journey of learning and discovery!

The Authors

Marek Piaseczny has more than 25 years of experience in IT, with more than 20 years working in various roles within the SAP and non-SAP integration areas. He has played leading roles in the integration stream of dozens of global-scale SAP projects, including implementations, migrations, and rollouts. Marek demonstrates deep theoretical knowledge and practical experience in most aspects of end-to-end system integration, from business processes to technology.

Agasthuri Doss is a seasoned IT professional with more than 20 years of progressive experience in the areas of architecture, solutioning, development, planning, digitization, and support in SAP and cloud-based implementation. He worked with several multinational corporations in various countries as an expert in more than 17 ERP implementations. He is recognized as one of the leading SAP experts with more than 30 articles and blogs on SAP Community. He was a speaker at SAP TechED and an SAP Mentor from 2014 to 2018, and he also won the top contributor/topic leader award from SAP Developer Network Portal.

Contributor

Mateusz Nowak is an experienced SAP Integration Architect and the coauthor of the SAP PRESS E-Bite *Interface Monitoring and Error Handling with SAP AIF*. He has contributed to numerous SAP implementation and migration projects, recently focusing on complex SAP S/4HANA implementations and transitions from SAP Process Orchestration to SAP Integration Suite for large corporations. Mateusz has a strong focus on integration automation in SAP, leveraging tools and methodologies that streamline and simplify integration processes.

Mateusz is also the coauthor of courses on SAP Learning and openSAP, where he supports the professional development of SAP experts worldwide. Mateusz contributed to Chapter 11 of this book.

Index

B

C

D

E

F

G

H

I

J

K

L

M

T

U

V

W

X

Interested in reading more?

Please visit our website for all new book and e-book releases from SAP PRESS.

www.sap-press.com